W9-CSF-108

P. Wasserscheid and T. Welton (Eds.)
Ionic Liquids in Synthesis

Related Titles from WILEY-VCH

P.G. Jessop and W. Leitner (eds.)

Chemical Synthesis Using Supercritical Fluids

1999. 500 pages.
Hardcover.
ISBN 3-527-29605-0

F. Zaragoza Dörwald (ed.)

Organic Synthesis on Solid Phase
Supports, Linkers, Reactions

2nd Edition,
2002. ca. 580 pages.
Hardcover.
ISBN 3-527-30603-X

A. Loupy (ed.)

Microwaves in Organic Synthesis

2002. ca. 350 pages.
Hardcover.
ISBN 3-527-30514-9

K. Tanaka (ed.)

Solvent-free Organic Synthesis

2003. ca, 300 pages.
Hardcover.
ISBN 3-527-30612-9

P. Wasserscheid and T. Welton (Eds.)

Ionic Liquids in Synthesis

Volume Editors

Dr. Peter Wasserscheid
RWTH Aachen
Institute for Technical and
Macromolecular Chemistry
Worringer Weg 1
52074 Aachen
Germany

Dr. Thomas Welton
Imperial College of Science,
Technology and Medicine
Department of Chemistry
South Kensington
London SW7 2AY
UK

Library of Congress Card No.: applied for
A catalogue record for this book is available from the British Library.

Bibliographic information published by Die Deutsche Bibliothek
Die Deutsche Bibliothek lists this publication in the Deutsche Nationalbibliografie; detailed bibliographic data is available in the Internet at http://dnb.ddb.de

Printed in the Federal Republic of Germany.
Printed on acid-free paper.

Typsetting Hagedorn Kommunikation, Viernheim
Printing Strauss Offsetdruck GmbH, Mörlenbach
Bookbinding J. Schäffer GmbH & Co. KG, Grünstadt

ISBN 3-527-30515-7

Contents

Preface

"We prided ourselves that the science we were doing could not, in any conceivable circumstances, have any practical use. The more firmly one could make that claim, the more superior one felt."

The Two Cultures, C.P. Snow (1959)

A book about ionic liquids? Over three hundred pages? Why? Who needs it? Why bother? These aren't simply rhetorical questions, but important ones of a nature that must be addressed whenever considering the publication of any new book. In the case of this one, as two other books about ionic liquids will appear in 2002, the additional question of differentiation arises – how is this distinctive from the other two? All are multi-author works, and some of the authors have contributed to all three books.

Taking the last question first, the answer is straightforward but important. The other two volumes are conference proceedings (one of a NATO Advanced Research Workshop, the other of an ACS Symposium) presenting cutting-edge snapshots of the state-of-the-art for experts; this book is *structured*. Peter Wasserscheid and Tom Welton have planned an integrated approach to ionic liquids; it is detailed and comprehensive. This is a book designed to take the reader from little or no knowledge of ionic liquids to an understanding reflecting our best current knowledge. It is a teaching volume, admirable for use in undergraduate and postgraduate courses, or for private learning. But it is not a dry didactic text - it is a user's manual! Having established a historical context (with an excellent chapter by one of the fathers of ionic liquids), the volume describes the synthesis and purification of ionic liquids (the latter being crucially important), and the nature of ionic liquids and their physical properties. Central to this tome (both literally and metaphorically) is the use of ionic liquids for organic synthesis, and especially green organic synthesis, and this chapter is (appropriately) the largest, and the *raison d'être* for the work. The book concludes with much shorter chapters on the synthesis of inorganic materials and polymers, the study of enzyme reactions, and an overview and prospect for the area. This plan logically and completely covers the whole of our current knowledge of ionic liquids, in a manner designed to enable the tyro reader to feel confident in using them, and the expert to add to their understanding. This is the first book to attempt this task, and it is remarkably successful for two reasons. Firstly, the volume has been strongly and wisely directed, and is unified despite being a multi-author work. Secondly, the choice of authors was inspired; each one writes with authority and clarity within a strong framework. So, yes, this book is more than justified, it is a crucial and timely addition to the literature. Moreover, it is written and edited by the key people in the field.

Are ionic liquids really green? A weakly argued letter from Albrecht Salzer in *Chemical and Engineering News* (**2002**, 80 [April 29], 4-6) has raised this nevertheless valid question. Robin Rogers gave a tactful, and lucid, response, and I quote directly from this: "Salzer has not fully realized the magnitude of the number of potential ionic liquid solvents. I am sure, for example, that we can design a very toxic ionic liquid solvent. However, by letting the principles of green chemistry drive this research field, we can ensure that the ionic liquids and ionic liquid processes developed are in fact green. [. . .] The expectation that real benefits in technology will arise from ionic liquid research and the development of new processes is high, but there is a need for further work to demonstrate the credibility of ionic liquid-based processes as viable green technology. In particular, comprehensive toxicity studies, physical and chemical property collation and dissemination, and realistic comparisons to traditional systems are needed. It is clear that while the new chemistry being developed in ionic liquids is exciting, many are losing sight of the green goals and falling back on old habits in synthetic chemistry. Whereas it is true that incremental improvement is good, it is hoped that by focusing on a green agenda, new technologies can be developed that truly are not only better technologically, but are cleaner, cheaper, and safer as well."

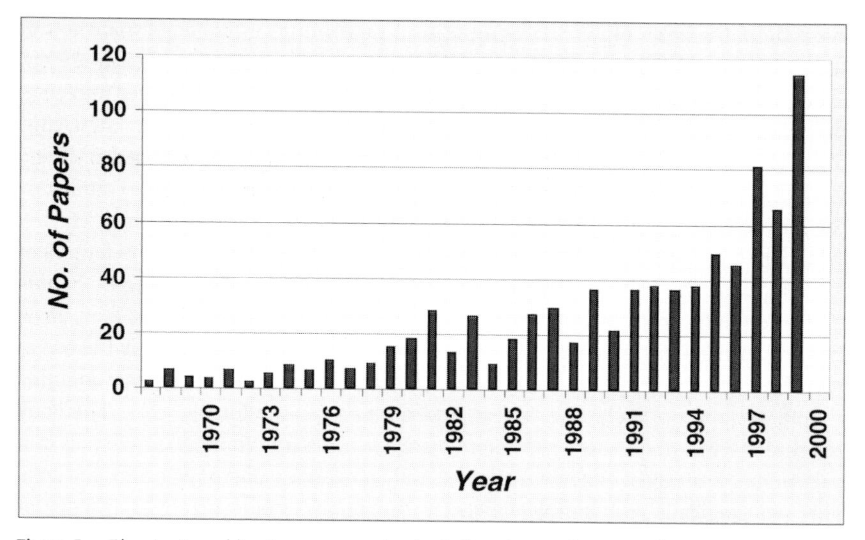

Figure 1: The rise in publications concerning ionic liquids as a function of time, as determined using SciFinder.

Robin's response is insightful. It reflects, in part, the burgeoning growth of papers in this area (see Figure 1) combined with the inevitable (and welcomed) rise in new researchers entering the area. However, with increasing activity comes the inevitable increasing "garbage" factor. In recent years we have (unfortunately) seen papers reporting physical data on ionic liquids that were demonstrably impure, liquids reported as solids and solids reported as liquids because of the impurity level, communications "rediscovering" and publishing work (without citation) already published in the patent literature, the synthesis of water-sensitive ionic liquids under conditions that inevitably result in hydrolysis, and academically weak publications appearing in commercial journals with lax refereeing standards. I truly believe that this book will help combat this; it should, and will, be referred to by all workers in the field. Indeed, if the authors citing it actually read it too, then the garbage factor should become insignificantly small!

In conclusion, this volume reflects well the excitement and rapid progress in the field of ionic liquids, whilst effectively providing an invaluable hands-on instruction manual. The lacunae are emphasised, and the directions for potential future research are clearly signposted. Unlike Snow in his renowned *Two Cultures* essay, many of us (Mamantov, Osteryoung, Wilkes, and Hussey, to name but a few of the founding fathers) who entered this area in its early (but not earliest!!) days prided ourselves that the science we were doing could not fail to have a practical use. Whether that use was battery applications, fuel cells, electroplating, nuclear reprocessing, or green industrial synthesis, we all believed that ionic liquids (or room-temperature molten salts, as they were then commonly known) offered a unique chemical environment that would (*must*) have significant industrial application. Because of this, we suffered then (and to some extent now) from the disdain of the "pure" scientists, who failed (and still fail) to appreciate that, if selecting an example to study to illustrate a fundamental scientific principle, there is actually some merit in selecting a product manufactured at the one million ton per annum level, rather than an esoteric molecule of no use and even less interest. Unfortunately, the pride and superiority Snow refers to is still alive and well, and living in the hearts of some of the academic establishment. I believe that this book will help tackle this prejudice, and illustrate that useful practical applications and groundbreaking fundamental science are not different, opposing areas, but synergistic sides of the same coin.

K.R. Seddon
May, 2002

A note from the editors

This book has been arranged in several chapters that have been prepared by different authors, and the reader can expect to find changes in style and emphasis as they go through it. We hope that, in choosing authors who are at the forefront of their particular specialism, this variety is a strength of the book. The book is intended to be didactic, with examples from the literature used to illustrate and explain. Therefore, not all chapters will give a comprehensive coverage of the literature in the area. Indeed, with the explosion of interest in some applications of ionic liquids comprehensive coverage of the literature would not be possible in a book of this length. Finally, there is a point when one has to stop and for us that was the end of 2001. We hope that no offence is caused to anyone whose work has not been included. None is intended.

Acknowledgements

We would like to sincerely thank everyone who has been involved in the publication of this book. All of our authors have done a great job in preparing their chapters and it has been a pleasure to read their contributions as they have come in to us. When embarking on this project we were both regaled with stories of books that never saw the light of day because of missed deadlines and the general tardiness of contributors. All of our colleagues have met their commitments in the most timely and enthusiastic manner. We are truly grateful for them making our task so painless. We would also like to thank the production team at VCH-Wiley, particularly Dr. Karen Kriese.

Finally, in a project like this, someone must take responsibility for any errors that have crept in. Ultimately we are the editors and this responsibility is ours. So we apologise unreservedly for any mistakes that have found their way into the book.

P. Wasserscheid, T. Welton
August, 2002

Contributors

Prof. Dr. Joan F. Brennecke
Department of Chemical Engineering
University of Notre Dame
Notre Dame, IN 46556

Prof. James H. Davis, jr.
Department of Chemistry
University of South Alabama
Mobile, Alabama 36688
USA

Dr. Martyn J. Earle
School of Chemistry
The Queen's University of Belfast
Stranmills Road
Belfast BT9 5AG, Northern Ireland

Dr. Frank Endres
Institut für Physikalische Chemie
Universität Karlsruhe
D-76128 Karlsruhe

PD Dr. Andreas Dölle
Institut für Physikalische Chemie
RWTH Aachen
D-52056 Aachen
Germany

Dr. Charles M. Gordon
University of Strathclyde
Department of Pure and
Applied Chemistry
Thomas Graham Building
295 Cathedral Street
Glasgow G1 1XL
Scotland, UK

Prof. Dr. David M. Haddleton
University of Warwick
Department of Chemistry
Coventry CV4 7AC
U.K.

Dr. Chris Hardacre
Physical Chemistry
School of Chemistry
The Queen's University of Belfast
Stranmillis Road
BELFAST BT9 5AG
Northern Ireland

Dr. Claus Hilgers
Solvent Innovation GmbH
Alarichstraße 14-16
D-0679 Köln
Germany

Dr. John Holbrey
Department of Chemistry
The University of Alabama
Tuscaloosa, AL 35487
jholbrey@bama.ua.edu

Prof. Dr. Udo Kragl
Technische Chemie,
Universität Rostock
Fachbereich Chemie
Buchbinderstr. 9
D-18051 Rostock

Dr. Hugh C. de Long
Department of Chemistry
US Naval Academy
572 Holloway RD
Annapolis, MD 21402-5026
USA

Dr. Hélène Olivier-Bourbigou
Division cinetique a Catalyse
Institut Francais du Petrole
129 Av. DeBois-Preau
92852 Rueil-Malmaison, France

Prof. Dr. Robin D. Rogers
Professor of Chemistry
Director, Center for
Green Manufacturing
Department of Chemistry
The University of Alabama
Tuscaloosa, AL 35487

Prof. K. R. Seddon
Chair of Inorganic Chemistry
School of Chemistry
The Queen's University of Belfast
Stranmillis Road
BELFAST BT9 5AG
Northern Ireland

Dr. Paul C. Trulove
AFOSR/NL
801 North Randolph Street
Arlington, VA 22203-1977
USA

Dr. Peter Wasserscheid
Institut für Technische Chemie und
Makromolekulare Chemie der
RWTH Aachen
Worringer Weg 1
D-52074 Aachen
Germany

Dr. Tom Welton
Department of Chemistry
Imperial College
London SW7 2AY UK

Prof. John S. Wilkes
Department of Chemistry
2355 Fairchild Drive, Suite 2N255
USAF, Colorado 80840-6230
USA

1
Introduction

John S. Wilkes

Ionic liquids may be viewed as a new and remarkable class of solvents, or as a type of materials that have a long and useful history. In fact, ionic liquids are both, depending on your point of view. It is absolutely clear though, that whatever "ionic liquids" are, there has been an explosion of interest in them. Entries in Chemical Abstracts for the term "ionic liquids" were steady at about twenty per year through 1995, but had grown to over 300 in 2001. The increased interest is clearly due to the realization that these materials, formerly used for specialized electrochemical applications, may have greater utility as reaction solvents.

For purposes of discussion in this volume we will define ionic liquids as salts with a melting temperature below the boiling point of water. That is an arbitrary definition based on temperature, and says little about the composition of the materials themselves, except that they are completely ionic. In reality, most ionic liquids in the literature that meet our present definition are also liquids at room temperature. The melting temperature of many ionic liquids can be problematic, since they are notorious glass-forming materials. It is a common experience to work with a new ionic liquid for weeks or months to find one day that it has crystallized unexpectedly. The essential feature that ionic liquids possess is one shared with traditional molten salts: a very wide liquidus range. The liquidus range is the span of temperatures between the melting point and boiling point. No molecular solvent, except perhaps some liquid polymers, can match the liquidus range of ionic liquids or molten salts. Ionic liquids differ from molten salts in just where the liquidus range is in the scale of temperature.

There are many synonyms used for ionic liquids, which can complicate a literature search. "Molten salts" is the most common and most broadly applied term for ionic compounds in the liquid state. Unfortunately, the term "ionic liquid" was also used to mean "molten salt" long before there was much literature on low-melting salts. It may seem that the difference between ionic liquids and molten salts is just a matter of degree (literally); however the practical differences are sufficient to justify a separately identified niche for the salts that are liquid around room temperature. That is, in practice the ionic liquids may usually be handled like ordinary solvents. There are also some fundamental features of ionic liquids, such as strong ion–ion interactions that are not often seen in higher-temperature molten salts. Synonyms in the literature for materials that meet the working definition of ionic liquid are: “room temperature molten salt”, “low-temperature molten salt”, “ambient-temperature molten salt”, and “liquid organic salt.”

Our definition of an ionic liquid does not answer the general question, “What is an ionic liquid?” This question has both a chemical and a historical answer. The details of the chemical answer are the subject of several subsequent chapters in this book. The general chemical composition of ionic liquids is surprisingly consistent, even though the specific composition and the chemical and physical properties vary tremendously. Most ionic liquids have an organic cation and an inorganic, polyatomic anion. Since there are many known and potential cations and anions, the potential number of ionic liquids is huge. To discover a new ionic liquid is relatively easy, but to determine its usefulness as a solvent requires a much more substantial investment in determination of physical and chemical properties. The best trick would be a method for predicting an ionic liquid composition with a specified set of properties. That is an important goal that awaits a better fundamental understanding of structure–property relationships and the development of better computational tools. I believe it can be done.

The historical answer to the nature of present ionic liquids is somewhat in the eye of the beholder. The very brief history presented here is just one of many possible ones, and is necessarily biased by the point of view of just one participant in the development of ionic liquids. The earliest material that would meet our current definition of an ionic liquid was observed in Friedel–Crafts reactions in the mid-19th century as a separate liquid phase called the “red oil.” The fact that the red oil was a salt was determined more recently, when NMR spectroscopy became a commonly available tool. Early in the 20th century, some alkylammonium nitrate salts were found to be liquids [1], and more recently liquid gun propellants based on binary nitrate ionic liquids have been developed [2]. In the 1960s, John Yoke at Oregon State University reported that mixtures of copper(I) chloride and alkylammonium chlorides were often liquids [3]. These were not as simple as they might appear, since several chlorocuprous anions formed, depending on the stoichiometry of the components. In the 1970s, Jerry Atwood at the University of Alabama discovered an unusual class of liquid salts he termed “liquid clathrates” [4]. These were composed of a salt combined with an aluminium alkyl, which then formed an inclusion compound with one or more aromatic molecules. A formula for the ionic portion is $M[Al_2(CH_3)_6X]$, where M is an inorganic or organic cation and X is a halide.

None of the interesting materials just described are the direct ancestors of the present generation of ionic liquids. Most of the ionic liquids responsible for the burst of papers in the last several years evolved directly from high-temperature molten salts, and the quest to gain the advantages of molten salts without the disadvantages. It all started with a battery that was too hot to handle.

In 1963, Major (Dr.) Lowell A. King (Figure 1.1) at the U.S. Air Force Academy initiated a research project aimed at finding a replacement for the LiCl/KCl molten salt electrolyte used in thermal batteries.

Figure 1.1: Major (Dr.) Lowell A. King at the U.S. Air Force Academy in 1961. He was an early researcher in the development of low-temperature molten salts as battery electrolytes. At that time "low temperature" meant close to 100 °C, compared to many hundreds of degrees for conventional molten salts. His work led directly to the chloroaluminate ionic liquids.

Since then there has been a continuous molten salts/ionic liquids research program at the Air Force Academy, with only three principal investigators: King, John Wilkes (Figure 1.2), and Richard Carlin. Even though the LiCl/KCl eutectic mixture has a low melting temperature (355 °C) for an inorganic salt, the temperature causes materials problems inside the battery, and incompatibilities with nearby devices. The class of molten salts known as chloroaluminates, which are mixtures of alkali halides and aluminium chloride, have melting temperatures much lower than nearly all other inorganic eutectic salts. In fact NaCl/$AlCl_3$ has a eutectic composition with a melting point of 107 °C, very nearly an ionic liquid by our definition [5]. Chloroaluminates are another class of salts that are not simple binary mixtures, because the Lewis acid-base chemistry of the system results in the presence of the series of the anions Cl^-, $[AlCl_4]^-$, $[Al_2Cl_7]^-$, and $[Al_3Cl_{10}]^-$ (although fortunately not all of these in the same mixture). Dr. King taught me a lesson that we should take heed of with the newer ionic liquids: if a new material is to be accepted as a technically useful material, the chemists must present reliable data on the chemical and physical properties needed by engineers to design processes and devices. Hence, the group at the Air Force Academy, in collaboration with several other groups, determined the densities, conductivities, viscosities, vapor pressures, phase equilibria, and electrochemical behavior of the salts. The research resulted in a patent for a thermal battery that made use of the NaCl/$AlCl_3$ electrolyte, and a small number of the batteries were manufactured.

Early in their work on molten salt electrolytes for thermal batteries, the Air Force Academy researchers surveyed the aluminium electroplating literature for electrolyte baths that might be suitable for a battery with an aluminium metal anode and chlorine cathode. They found a 1948 patent describing ionically conductive mixtures of $AlCl_3$ and 1-ethylpyridinium halides, mainly bromides [6]. Subsequently, the salt 1-butylpyridinium chloride/$AlCl_3$ (another complicated pseudo-binary)

Figure 1.2: Captain (Dr.) John S. Wilkes at the U.S. Air Force Academy in 1979. This official photo was taken about when he started his research on ionic liquids, then called "room-temperature molten salts."

Figure 1.3: Prof. Charles Hussey of the University of Mississippi. The photo was taken in 1990 at the U.S. Air Force Academy while he was serving on an Air Force Reserve active duty assignment. Hussey and Wilkes collaborated in much of the early work on chloroaluminate ionic liquids.

was found to be better behaved than the earlier mixed halide system, so its chemical and physical properties were measured and published [7]. I would mark this as the start of the modern era for ionic liquids, because for the first time a wider audience of chemists started to take interest in these totally ionic, completely nonaqueous new solvents.

The alkylpyridinium cations suffer from being relatively easy to reduce, both chemically and electrochemically. Charles Hussey (Figure 1.3) and I set out a program to predict cations more resistant to reduction, to synthesize ionic liquids on the basis of those predictions, and to characterize them electrochemically for use as battery electrolytes.

Figure 1.4: Dr. Michael Zaworotko from Saint Mary's University in Halifax, Nova Scotia. He was a visiting professor at the U.S. Air Force Academy in 1991, where he first prepared many of the water-stable ionic liquids popular today.

We had no good way to predict if they would be liquid, but we were lucky that many were. The class of cations that were the most attractive candidates was that of the dialkylimidazolium salts, and our particular favorite was 1-ethyl-3-methylimidazolium [EMIM]. [EMIM]Cl mixed with $AlCl_3$ made ionic liquids with melting temperatures below room temperature over a wide range of compositions [8]. We determined chemical and physical properties once again, and demonstrated some new battery concepts based on this well behaved new electrolyte. We and others also tried some organic reactions, such as Friedel–Crafts chemistry, and found the ionic liquids to be excellent both as solvents and as catalysts [9]. It appeared to act like acetonitrile, except that is was totally ionic and nonvolatile.

The pyridinium- and the imidazolium-based chloroaluminate ionic liquids share the disadvantage of being reactive with water. In 1990, Mike Zaworotko (Figure 1.4) took a sabbatical leave at the Air Force Academy, where he introduced a new dimension to the growing field of ionic liquid solvents and electrolytes.

His goal for that year was to prepare and characterize salts with dialkylimidazolium cations, but with water-stable anions. This was such an obviously useful idea that we marveled that neither we nor others had tried to do it already. The preparation chemistry was about as easy as the formation of the chloroaluminate salts, and could be done outside of the glove-box [10]. The new tetrafluoroborate, hexafluorophosphate, nitrate, sulfate, and acetate salts were stable (at least at room temperature) towards hydrolysis. We thought of these salts as candidates for battery electrolytes, but they (and other similar salts) have proven more useful for other applications. Just as Zaworotko left, Joan Fuller came to the Air Force Academy, and spent several years extending the catalog of water-stable ionic liquids, discovering better ways to prepare them, and testing the solids for some optical properties. She made a large number of ionic liquids from the traditional dialkylimidazolium cations, plus a series of mono- and trialkylimidazoliums. She combined those cations with the water-stable anions mentioned above, *plus* the additional series of bromide, cyanide, bisulfate, iodate, trifluoromethanesulfonate, tosylate, phenyl-phosphonate, and tartrate. This resulted in a huge array of new ionic liquids with anion sizes ranging from relatively small to very large.

It seems obvious to me and to most other chemists that the table of cations and anions that form ionic liquids can and will be extended to a nearly limitless number. The applications will be limited only by our imagination.

References

1 Walden, P., *Bull. Acad. Imper. Sci. (St Petersburg)* **1914**, 1800.
2 CAS Registry Number 78041-07-3.
3 Yoke, J. T., Weiss, J. F., Tollin, G., *Inorg. Chem.* **1963**, *2*, 1210–1212.
4 Atwood, J. L., Atwood, J. D., *Inorganic Compounds with Unusual Properties*, Advances in Chemistry Series No. 150, American Chemical Society: Washington, DC, **1976**, pp 112–127.
5 For a review of salts formerly thought of as low-temperature ionic liquids, see Mamantov, G., *Molten salt electrolytes in secondary batteries*, in *Materials for Advanced Batteries* (Murphy, D. W., Broadhead, J., and Steele, B.C. H. eds.), Plenum Press, New York, **1980**, pp. 111–122.
6 Hurley, F. H., U.S. Patent 4,446,331, **1948**. Wier, T. P. Jr., Hurley, F. H., U.S. Patent 4,446,349, **1948**. Wier, T. P. Jr., US Patent 4,446,350, **1948**. Wier, T. P. Jr., US Patent 4,446,350, **1948**.
7 Gale, R. J., Gilbert, B., Osteryoung, R. A., *Inorg. Chem.*, **1978**, *17*, 2728–2729. Nardi, J. C., Hussey, C. L., King, L. A., U.S. Patent 4,122,245, **1978**.
8 Wilkes, J. S., Levisky, J. A., Wilson R. A., Hussey, C. L. *Inorg. Chem.* **1982**, *21*, 1263.
9 Boon, J., Levisky, J. A., Pflug, J. L., Wilkes, J. S., *J. Org. Chem.* **1986**, *51*, 480–483.
10 Wilkes, J. S., Zaworotko, M. J., *J. Chem. Soc., Chem.* Commun. **1992**, 965–967.

2 Synthesis and Purification of Ionic Liquids

James H. Davis, Jr., Charles M. Gordon, Claus Hilgers, and Peter Wasserscheid

2.1 Synthesis of Ionic Liquids

Charles M. Gordon

2.1.1 Introduction

Despite the ever-growing number of papers describing the applications of ionic liquids, their preparation and purification has in recent years taken on an air of "need to know". Although most researchers employ similar basic types of chemistry, it appears that everyone has their own tricks to enhance yields and product purity. This chapter is an attempt to summarize the different methods reported to date, and to highlight the advantages and disadvantages of each. The purity of ionic liquids is also an area of increasing interest as the nature of their interactions with different solutes comes under study, so the methods used for the purification of ionic liquids are also reviewed. The aim is to provide a summary for new researchers in the area, pointing to the best preparative methods, and the potential pitfalls, as well as helping established researchers to refine the methods used in their laboratories.

The story of ionic liquids is generally regarded as beginning with the first report of the preparation of ethylammonium nitrate in 1914 [1]. This species was formed by the addition of concentrated nitric acid to ethylamine, after which water was removed by distillation to give the pure salt, which was liquid at room temperature. The protonation of suitable starting materials (generally amines and phosphines) still represents the simplest method for the formation of such materials, but unfortunately it can only be used for a small range of useful salts. The possibility of decomposition through deprotonation has severely limited the use of such salts, and so more complex methods are generally required. Probably the most widely used salt of this type is pyridinium hydrochloride, the applications of which may be found in a thorough review by Pagni [2].

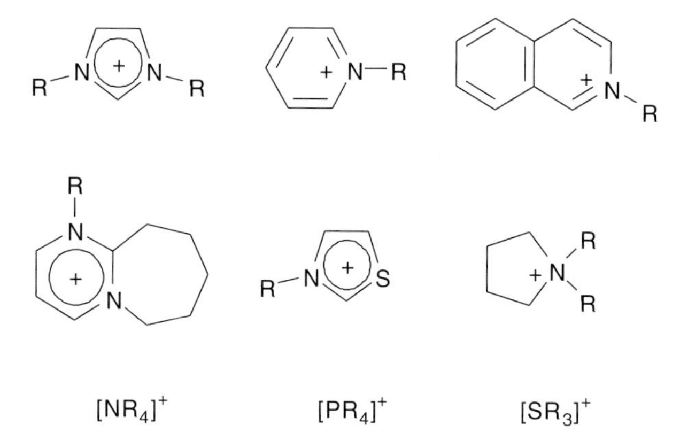

Figure 2.1-1: Examples of cations commonly used for the formation of ionic liquids.

Thus, most ionic liquids are formed from cations that do not contain acidic protons. A summary of the applications and properties of ionic liquids may be found in a number of recent review articles [3]. The most common classes of cations are illustrated in Figure 2.1-1, although low melting point salts based on other cations, such as complex polycationic amines [4] and heterocycle-containing drugs [5], have also been prepared.

The synthesis of ionic liquids can generally be split into two sections: the formation of the desired cation, and anion exchange where necessary to form the desired product (demonstrated for ammonium salts in Scheme 2.1-1).

In some cases only the first step is required, as with the formation of ethylammonium nitrate. In many cases the desired cation is commercially available at reasonable cost, most commonly as a halide salt, thus requiring only the anion exchange reaction. Examples of these are the symmetrical tetraalkylammonium salts and trialkylsulfonium iodide.

This chapter will concentrate on the preparation of ionic liquids based on 1,3-dialkylimidazolium cations, as these have dominated the area over the last twenty

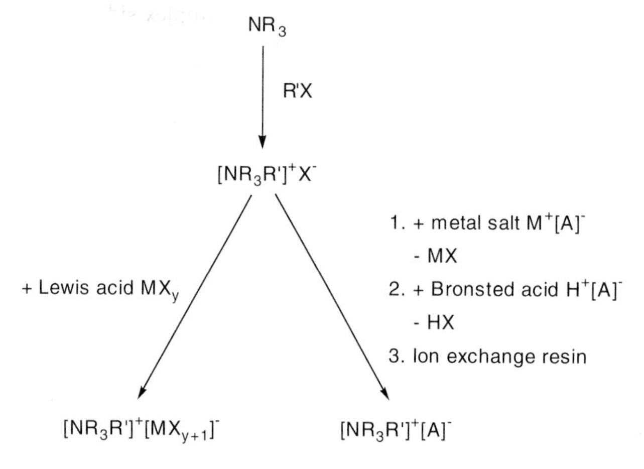

Scheme 2.1-1: Typical synthesis paths for the preparation of ionic liquids (adapted from Ref. 3c).

years. The techniques discussed in this chapter are generally applicable to the other classes of cations indicated in Figure 2.1-1, however. The original decision by Wilkes et al. to prepare 1-alkyl-3-methylimidazolium ($[RMIM]^+$) salts was prompted by the requirement for a cation with a more negative reduction potential than Al(III) [6]. The discovery that the imidazolium-based salts also generally displayed lower melting points than the 1-alkylpyridinium salts used prior to this cemented their position as the cations of choice since then. Indeed, the method reported by Wilkes et al. for the preparation of the $[RMIM]Cl/AlCl_3$-based salts remains very much that employed by most workers to this day.

2.1.2
Quaternization Reactions

The formation of the cations may be carried out either by protonation with a free acid as noted above, or by quaternization of an amine or a phosphine, most commonly with a haloalkane. The protonation reaction, as used in the formation of salts such as ethylammonium nitrate, involves the addition of 3 M nitric acid to a cooled, aqueous solution of ethylamine [7]. A slight excess of amine should be left over, and this is removed along with the water by heating to 60 °C in vacuo. The same general process may be employed for the preparation of all salts of this type, but when amines of higher molecular weight are employed, there is clearly a risk of contamination by residual amine. A similar method has been reported for the formation of low melting point, liquid crystalline, long alkyl chain-substituted 1-alkylimidazolium chloride, nitrate, and tetrafluoroborate salts [8]. For these a slight excess of acid was employed, as the products were generally crystalline at room temperature. In all cases it is recommended that addition of acid be carried out with cooling of the amine solution, as the reaction can be quite exothermic.

The alkylation process possesses the advantages that (a) a wide range of cheap haloalkanes are available, and (b) the substitution reactions generally occur smoothly at reasonable temperatures. Furthermore, the halide salts formed can easily be converted into salts with other anions. Although this section will concentrate on the reactions between simple haloalkanes and the amine, more complex side chains may be added, as discussed later in this chapter. The quaternization of amines and phosphines with haloalkanes has been known for many years, but the development of ionic liquids has resulted in several recent developments in the experimental techniques used for the reaction. In general, the reaction may be carried out with chloroalkanes, bromoalkanes, and iodoalkanes, with the reaction conditions required becoming steadily more gentle in the order Cl → Br → I, as expected for nucleophilic substitution reactions. Fluoride salts cannot be formed in this manner.

In principle, the quaternization reactions are extremely simple: the amine (or phosphine) is mixed with the desired haloalkane, and the mixture is then stirred and heated. The following section refers to the quaternization of 1-alkylimidazoles, as these are the most common starting materials. The general techniques are similar, however, for other amines such as pyridine [9], isoquinoline [10], 1,8-diazabicyclo[5,4,0]-7-undecene [11], 1-methylpyrrolidine [12], and trialkylamines [13], as well as for phosphines. The reaction temperature and time are very dependent on the haloalkane employed, chloroalkanes being the least reactive and iodoalkanes the most. The reactivity of the haloalkane also generally decreases with increasing alkyl chain length. As a general guide, in the author's laboratory it is typically found necessary to heat 1-methylimidazole with chloroalkanes to about 80 °C for 2–3 days to ensure complete reaction. The equivalent reaction with bromoalkanes is usually complete within 24 hours, and can be achieved at lower temperatures (ca. 50–60 °C). In the case of bromoalkanes, we have found that care must be taken with large-scale reactions, as a strong exotherm can occur as the reaction rate increases. Besides the obvious safety implications, the excess heat generated can result in discoloration of the final product. The reaction with iodoalkanes can often be carried out at room temperature, but the iodide salts formed are light-sensitive, requiring shielding of the reaction vessel from bright light.

A number of different methodologies have been reported, but most researchers use a simple round-bottomed flask/reflux condenser experimental setup for the quaternization reaction. If possible, the reaction should be carried out under dinitrogen or some other inert gas in order to exclude water and oxygen during the quaternization. Exclusion of oxygen is particularly important if a colorless halide salt is required. Alternatively, the haloalkane and 1-methylimidazole may be mixed in Carius tubes, degassed by freeze-pump-thaw cycles, and then sealed under vacuum and heated in an oven for the desired period. The preparation of salts with very short alkyl chain substituents, such as [EMIM]Cl, is more complex, however, as chloroethane has a boiling point of 12 °C. Such reactions are generally carried out in an autoclave, with the chloroethane cooled to below its boiling point before addition to the reaction mixture. In this case, the products should be collected at high temperature, as the halide salts are generally solids at room temperature. An autoclave may also be useful for the large-scale preparation of the quaternary salts.

In general, the most important requirement is that the reaction mixture be kept free of moisture, as the products are often extremely hygroscopic. The reaction may be carried out without the use of a solvent, as the reagents are generally liquids and mutually miscible, while the halide salt products are usually immiscible in the starting materials. A solvent is often used, however; examples include the alkyl halide itself [6], 1,1,1-trichloroethane [14], ethyl ethanoate [15], and toluene [16], although no particular advantage appears to accrue with any specific one. The unifying factor for all of these is that they are immiscible with the halide salt product, which will thus form as a separate phase. Furthermore, the halide salts are generally more dense than the solvents, so removal of excess solvent and starting material can be achieved simply by decantation. In all cases, however, after reaction is complete and the solvent is decanted, it is necessary to remove all excess solvent and starting material by heating the salt under vacuum. Care should be taken at this stage, as overheating can result in a reversal of the quaternization reaction. It not advised to heat the halide salts to temperatures greater than about 80 °C.

The halide salts are generally solids at room temperature, although some examples – such as e.g. the 1-methyl-3-octylimidazolium salts – remain viscous oils even at room temperature. Crystallization can take some time to occur, however, and many salts remain as oils even when formed in good purity. Purification of the solid salts is best achieved by recrystallisation from a mixture of dry acetonitrile and ethyl ethanoate. In cases of salts that are not solid, it is advisable to wash the oil as best as possible with an immiscible solvent such as dry ethyl ethanoate or 1,1,1-trichloroethane. If the reactions are carried out on a relatively large scale, it is generally possible to isolate product yields of >90 % even if a recrystallisation step is carried out, making this an extremely efficient reaction. A drybox is not essential, but can be extremely useful for handling the salts, as they tend to be very hygroscopic, particularly when the alkyl chain substituents are short. In the author's experience, solid 1-alkyl-3-methylimidazolium halide salts can form as extremely hard solids in round-bottomed flasks. Therefore, if a drybox is available the best approach is often to pour the hot salt into shallow trays made of aluminium foil. Once the salt cools and solidifies, it may be broken up into small pieces to aid future use.

The thermal reaction has been used in almost all reports of ionic liquids, being easily adaptable to large-scale processes, and providing high yields of products of acceptable purity with relatively simple methods. An alternative approach involving the use of microwave irradiation has recently been reported, giving high yields with very short reaction times (minutes rather than hours) [17]. The reaction was only carried out for extremely small quantities of material, however, and it is unlikely that it could be scaled up with any great feasibility.

By far the most common starting material is 1-methylimidazole. This is readily available at a reasonable cost, and provides access to the majority of cations likely to be of interest to most researchers. There is only a limited range of other N-substituted imidazoles commercially available, however, and many are relatively expensive. The synthesis of 1-alkylimidazoles may be achieved without great difficulty, though, as indicated in Scheme 2.1-2.

A wider range of C-substituted imidazoles is commercially available, and the combination of these with the reaction shown in Scheme 2.1-2 permits the formation of many different possible starting materials. In some cases, however, it may still be necessary to carry out synthesis of the heterocycle from first principles. For reasons of space, this topic is not covered here.

Relatively little has been reported regarding the determination of the purity of the halide salts other than by standard spectroscopic measurements and microanalysis. This is largely because the halide salts are rarely used as solvents themselves, but are generally simply a source of the desired cation. Also, the only impurities likely to be present in any significant quantity are unreacted starting materials and residual reaction solvents. Thus, for most applications it is sufficient to ensure that they are free of these by use of ^{1}H NMR spectroscopy.

The removal of the haloalkanes and reaction solvents is generally not a problem, especially for the relatively volatile shorter chain haloalkanes. On the other hand,

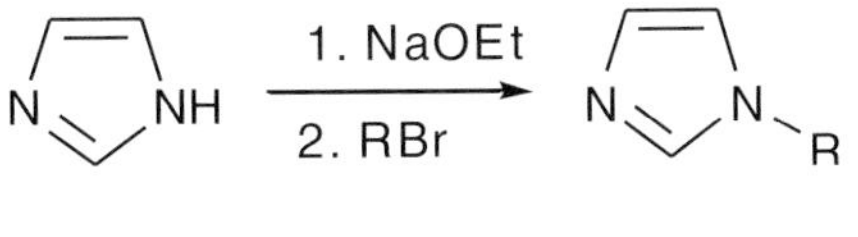

1. NaOEt
2. RBr

Scheme 2.1-2: Synthesis of alkylimidazoles.

the presence even of small quantities of unreacted 1-methylimidazole (a coordinating base) could cause problems in many applications. Furthermore, its high boiling point (198 °C) means that it can prove difficult to remove from ionic liquids. Holbrey has reported a simple colorimetric determination based on the formation of the blue $[Cu(MIM)_4]^{2+}$ ion, which is sensitive to 1-methylimidazole in the 0–3 mol% concentration range [18]. Although this does not solve the problem, it does allow samples to be checked before use, or for the progress of a reaction to be monitored.

It should be noted that it is not only halide salts that may be prepared in this manner. Quaternization reactions between 1-alkylimidazoles and methyl triflate [14], trialkylamines and methyl tosylates [19], and triphenylphosphine and octyl tosylate [20] have also been used for the direct preparation of ionic liquids, and in principle any alkyl compound containing a good leaving group may be used in this manner. The excellent leaving group abilities of the triflate and tosylate anions mean that the direct quaternization reactions can generally be carried out at ambient temperatures. It is important that these reactions be carried out under an inert atmosphere, as the alkyl triflates and tosylates are extremely sensitive to hydrolysis. This approach has the major advantage of generating the desired ionic liquid with no side products, and in particular no halide ions. At the end of the reaction it is necessary only to ensure that all remaining starting materials are removed either by washing with a suitable solvent (such as ethyl ethanoate or 1,1,1-trichloroethane) or in vacuo.

2.1.3 Anion-exchange Reactions

The anion-exchange reactions of ionic liquids can really be divided into two distinct categories: direct treatment of halide salts with Lewis acids, and the formation of ionic liquids by anion metathesis. These two approaches are dealt with separately, as quite different experimental methods are required for each.

2.1.3.1 Lewis Acid-based Ionic Liquids

The formation of ionic liquids by treatment of halide salts with Lewis acids (most notably $AlCl_3$) dominated the early years of this area of chemistry. The great breakthrough came in 1951, with the report by Hurley and Weir on the formation of a salt that was liquid at room temperature, based on the combination of 1-butylpyridinium with $AlCl_3$ in the relative molar proportions 1:2 (X = 0.66) [21].[1] More recently, the groups of Osteryoung and Wilkes have developed the technology of room temperature chloroaluminate melts based on 1-alkylpyridinium [22] and $[RMIM]^+$ cations [6]. In general terms, treatment of a quaternary halide salt Q^+X^- with a Lewis acid MX_n results in the formation of more than one anion species, depending on the relative proportions of Q^+X^- and MX_n. Such behavior can be illus-

1 Compositions of Lewis acid-based ionic liquids are generally referred to by the mole fraction (X) of monomeric acid present in the mixture.

trated for the reaction between [EMIM]Cl and $AlCl_3$ by a series of equilibria as given in Equations (2.1-1)–(2.1-3).

$$[EMIM]^+Cl^- + AlCl_3 \rightleftharpoons [EMIM]^+[AlCl_4]^- \quad (2.1\text{-}1)$$

$$[EMIM]^+[AlCl_4]^- + AlCl_3 \rightleftharpoons [EMIM]^+[Al_2Cl_7]^- \quad (2.1\text{-}2)$$

$$[EMIM]^+[Al_2Cl_7]^- + AlCl_3 \rightleftharpoons [EMIM]^+[Al_3Cl_{10}]^- \quad (2.1\text{-}3)$$

When [EMIM]Cl is present in a molar excess over $AlCl_3$, only equilibrium (2.1-1) need be considered, and the ionic liquid is basic. When a molar excess of $AlCl_3$ over [EMIM]Cl is present on the other hand, an acidic ionic liquid is formed, and equilibria (2.1-2) and (2.1-3) predominate. Further details of the anion species present may be found elsewhere [23]. The chloroaluminates are not the only ionic liquids prepared in this manner. Other Lewis acids employed have included $AlEtCl_2$ [24], BCl_3 [25], CuCl [26], and $SnCl_2$ [27]. In general, the preparative methods employed for all of these salts are similar to those indicated for $AlCl_3$-based ionic liquids as outlined below.

The most common method for the formation of such liquids is simple mixing of the Lewis acid and the halide salt, with the ionic liquid forming on contact of the two materials. The reaction is generally quite exothermic, which means that care should be taken when adding one reagent to the other. Although the salts are relatively thermally stable, the build-up of excess local heat can result in decomposition and discoloration of the ionic liquid. This may be prevented either by cooling the mixing vessel (often difficult to manage in a drybox), or else by adding one component to the other in small portions to allow the heat to dissipate. The water-sensitive nature of most of the starting materials (and ionic liquid products) means that the reaction is best carried out in a drybox. Similarly, the ionic liquids should ideally also be stored in a drybox until use. It is generally recommended, however, that only enough liquid to carry out the desired task be prepared, as decomposition by hydrolysis will inevitably occur over time unless the samples are stored in vacuum-sealed vials.

If a drybox is not available, the preparation can also be carried out by use of a dry, unreactive solvent (typically an alkane) as a "blanket" against hydrolysis. This has been suggested in the patent literature as a method for the large-scale industrial preparation of Lewis acid-based ionic liquids, as the solvent also acts as a heat-sink for the exothermic complexation reaction [28]. At the end of the reaction, the ionic liquid forms an immiscible layer beneath the protecting solvent. The ionic liquid may then either be removed by syringe, or else the solvent may be removed by distillation before use. In the former case it is likely that the ionic liquid will be contaminated with traces of the organic solvent, however.

Finally in this section, it is worth noting that some ionic liquids have been prepared by treatment of halide salts with metal halides that are not usually thought of as Lewis acids. In this case only equilibrium (2.1-1) above will apply, and the salts formed are neutral in character. Examples of these include salts of the type

$[EMIM]_2[MCl_4]$ (R = alkyl, M = Co, Ni) [29] and $[EMIM]_2[VOCl_4]$ [30]. These are formed by treatment of two equivalents of [EMIM]Cl with one equivalent of MCl_2 and $VOCl_2$, respectively.

2.1.3.2 **Anion Metathesis**

The first preparation of relatively air- and water-stable ionic liquids based on 1,3-dialkyl-methylimidazolium cations (sometimes referred to as "second generation" ionic liquids) was reported by Wilkes and Zaworotko in 1992 [31]. This preparation involved a metathesis reaction between [EMIM]I and a range of silver salts ($AgNO_3$, $AgNO_2$, $AgBF_4$, $Ag[CO_2CH_3]$, and Ag_2SO_4) in methanol or aqueous methanol solution. The very low solubility of silver iodide in these solvents allowed it to be separated simply by filtration, and removal of the reaction solvent allowed isolation of the ionic liquids in high yields and purities. This method remains the most efficient for the synthesis of water-miscible ionic liquids, but is obviously limited by the relatively high cost of silver salts, not to mention the large quantities of solid by-product produced. The first report of a water-insoluble ionic liquid was two years later, with the preparation of $[EMIM][PF_6]$ from the reaction between [EMIM]Cl and HPF_6 in aqueous solution [32]. The procedures reported in the above two papers have stood the test of time, although subsequent authors have suggested refinements of the methods employed. Most notably, many of the $[EMIM]^+$-based salts are solid at room temperature, facilitating purification, which may be achieved by recrystallisation. In many applications, however, a product that is liquid at room temperature is required, so most researchers now employ cations with 1-alkyl substituents of a chain length of four or greater, which results in a considerable lowering in melting point. Over the past few years, an enormous variety of anion exchange reactions has been reported for the preparation of ionic liquids. Table 2.1-1 gives a representative selection of both commonly used and more esoteric examples, along with references that give reasonable preparative details.

The preparative methods employed generally follow similar lines, however, and representative examples are therefore reviewed below. The main goal of all anion

Table 2.1-1: Examples of ionic liquids prepared by anion metathesis.

Salt	*Anion Source*	*Reference*
$[Cation][PF_6]$	HPF_6	9, 15, 32, 33
$[Cation][BF_4]$	HBF_4, NH_4BF_4, $NaBF_4$	31, 32, 33, 34, 35
$[Cation][(CF_3SO_2)_2N]$	$Li[(CF_3SO_2)_2N]$	14, 33
$[Cation][(CF_3SO_3)]$	$CF_3SO_3CH_3$, $NH_4[(CF_3SO_3)]$	14, 36
$[Cation][CH_3CO_2]$	$Ag[CH_3CO_2]$	31
$[Cation][CF_3CO_2]$	$Ag[CF_3CO_2]$	31
$[Cation][CF_3(CF_2)_3CO_2]$	$K[CF_3(CF_2)_3CO_2]$	14
$[Cation][NO_3]$	$AgNO_3$, $NaNO_3$	14, 33, 37
$[Cation][N(CN)_2]$	$Ag[N(CN)_2]$	38
$[Cation][CB_{11}H_{12}]$	$Ag[CB_{11}H_{12}]$	39
$[Cation][AuCl_4]$	$HAuCl_4$	40

exchange reactions is the formation of the desired ionic liquid uncontaminated with unwanted cations or anions, a task that is easier for water-immiscible ionic liquids. It should be noted, however, that low-melting salts based on symmetrical onium cations have been prepared by anion-exchange reactions for many years. For example, the preparation of tetrahexylammonium benzoate, a liquid at 25 °C, from tetrahexylammonium iodide, silver oxide, and benzoic acid was reported as early as 1967 [41]. The same authors also commented on an alternative approach involving the use of an ion-exchange resin for the conversion of the iodide salt to hydroxide, but concluded that this approach was less desirable. Low-melting salts based on cations such as tetrabutylphosphonium [42] and trimethylsulfonium [43] have also been produced by very similar synthetic methods.

To date, surprisingly few reports of the use of ion-exchange resins for large-scale preparation of ionic liquids have appeared in the open literature, to the best of the author's knowledge. One recent exception is a report by Lall et al. regarding the formation of phosphate-based ionic liquids with polyammonium cations [4]. Wasserscheid and Keim have suggested that this might be an ideal method for their preparation in high purity [3c].

As the preparation of water-immiscible ionic liquids is considerably more straightforward than that of the water-soluble analogues, these methods are considered first. The water solubility of the ionic liquids is very dependent on both the anion and cation present, and in general will decrease with increasing organic character of the cation. The most common approach for the preparation of water-immiscible ionic liquids is firstly to prepare an aqueous solution of a halide salt of the desired cation. The cation exchange is then carried out either with the free acid of the appropriate anion, or else with a metal or ammonium salt. Where available, the free acid is probably to be favored, as it leaves only HCl, HBr, or HI as the by-product, easily removable from the final product by washing with water. It is recommended that these reactions be carried out with cooling of the halide salt in an ice bath, as the metathesis reaction is often exothermic. In cases where the free acid is unavailable or inconvenient to use, however, alkali metal or ammonium salts may be substituted without major problems. It may also be preferable to avoid use of the free acid in systems where the presence of traces of acid may cause problems. A number of authors have outlined broadly similar methods for the preparation of $[PF_6]^-$ and $[(CF_3SO_2)_2N]^-$ salts that may be adapted for most purposes [14, 15].

When free acids are used, the washing should be continued until the aqueous residues are neutral, as traces of acid can cause decomposition of the ionic liquid over time. This can be a particular problem for salts based on the $[PF_6]^-$ anion, which will slowly form HF, particularly on heating if not completely acid-free. When alkali metal or ammonium salts are used, it is advisable to check for the presence of halide anions in the wash solutions, for example by testing with silver nitrate solution. The high viscosity of some ionic liquids makes efficient washing difficult, even though the presence of water results in a considerable reduction in the viscosity. As a result, a number of authors have recently recommended dissolution of these liquids in either CH_2Cl_2 or $CHCl_3$ prior to carrying out the washing step. Another advantage of this procedure is that the organic solvent/ionic liquid mixture may be placed over a drying agent such as $MgSO_4$ prior to removal of the organic solvent, thus greatly reducing the amount of water contamination of the final product.

The preparation of water-miscible ionic liquids can be a more demanding process, as separation of the desired and undesired salts may be complex. The use of silver salts described above permits the preparation of many salts in very high purity, but is clearly too expensive for large-scale use. As a result, a number of alternative methodologies that employ cheaper salts for the metathesis reaction have been developed. The most common approach is still to carry out the exchange in aqueous solution with either the free acid of the appropriate anion, the ammonium salt, or an alkali metal salt. When using this approach, it is important that the desired ionic liquid can be isolated without excess contamination from unwanted halide-containing by-products. A reasonable compromise has been suggested by Welton et al. for the preparation of [BMIM][BF_4] [35]. In this approach, which could in principle be adapted to any water-miscible system, the ionic liquid is formed by metathesis between [BMIM]Cl and HBF_4 in aqueous solution. The product is extracted into CH_2Cl_2, and the organic phase is then washed with successive small portions of deionized water until the washings are pH neutral. The presence of halide ions in the washing solutions can be detected by testing with $AgNO_3$. The CH_2Cl_2 is then removed on a rotary evaporator, and the ionic liquid then further purified by mixing with activated charcoal for 12 hours. Finally, the liquid is filtered through a short column of acidic or neutral alumina and dried by heating in vacuo. Yields of around 70 % are reported when this approach is carried out on large (~ 1 molar) scale. Although the water wash can result in a lowering of the yield, the aqueous wash solutions may ultimately be collected together, the water removed, and the crude salt added to the next batch of ionic liquid prepared. In this manner, the amount of product lost is minimized, and the purity of the ionic liquid prepared appears to be reasonable for most applications.

Alternatively, the metathesis reaction may be carried out entirely in an organic solvent such as CH_2Cl_2, as described by Cammarata et al. [33], or acetone, as described by Fuller et al. [36]. In both of these systems the starting materials are not fully soluble in the reaction solvent, so the reaction is carried out with a suspension. In the case of the CH_2Cl_2 process, it was performed by stirring the 1-alkyl-3-methylimidazolium halide salt with the desired metal salt at room temperature for 24 hours. After this, the insoluble halide by-products were removed by filtration. Although the halide by-products have limited solubility in CH_2Cl_2, they are much more soluble in the ionic liquid/CH_2Cl_2 mixture. Thus, when this method is employed it is important that the CH_2Cl_2 extracts be washed with water to minimize the halide content of the final product. This approach clearly results in a lowering of the yield of the final product, so care must be taken that the volume of water used to carry out the washing is low. Lowering of the temperature of the water to near 0 °C can also reduce the amount of ionic liquid lost. The final product was purified by stirring with activated charcoal followed by passing through an alumina column, as described in the previous paragraph. This process was reported to give final yields in the region of 70–80 %, and was used to prepare ionic liquids containing a wide variety of anions ($[PF_6]^-$, $[SbF_6]^-$, $[BF_4]^-$, $[ClO_4]^-$, $[CF_3SO_3]^-$, $[NO_3]^-$, and $[CF_3CO_2]^-$). For the acetone route, [EMIM]Cl was stirred with NH_4BF_4 or $NH_4[CF_3SO_3]$ at room temperature for 72 hours. In this case all starting materials were only slightly soluble in the reaction solvent. Once again, the insoluble NH_4Cl by-product was removed by filtration. No water wash was carried out, but trace organic impurities were removed by stirring the acetone solution with neutral alumina for two hours after removal of the metal halide salts by filtration. The salts were finally dried by heating at 120 °C for several hours, after which they were analyzed for purity by electrochemical methods, giving quoted purities of at least 99.95 %.

2.1.4
Purification of Ionic Liquids

The lack of significant vapor pressure prevents the purification of ionic liquids by distillation. The counterpoint to this is that any volatile impurity can, in principle, be separated from an ionic liquid by distillation. In general, however, it is better to remove as many impurities as possible from the starting materials, and where possible to use synthetic methods that either generate as few side products as possible, or allow their easy separation from the final ionic liquid product. This section first describes the methods employed to purify starting materials, and then moves on to methods used to remove specific impurities from the different classes of ionic liquids.

The first requirement is that all starting materials used for the preparation of the cation should be distilled prior to use. The author has found the methods described by Amarego and Perrin to be suitable in most cases [44]. In the preparation of $[RMIM]^+$ salts, for example, we routinely distil the 1-methylimidazole under vacuum from sodium hydroxide, and then immediately store any that is not used under nitrogen in the refrigerator. The haloalkanes are first washed with portions of concentrated sulfuric acid until no further color is removed into the acid layer, then neutralized with $NaHCO_3$ solution and deionized water, and finally distilled before use. All solvent used in quaternization or anion-exchange reactions should also be dried and distilled before use. If these precautions are not taken, it is often difficult to prepare colorless ionic liquids. In cases where the color of the ionic liquids is less important, the washing of the haloalkane may be unnecessary, as the quantity of colored impurity is thought to be extremely low, and thus will not affect many potential applications. It has also been observed that, in order to prepare $AlCl_3$-based ionic liquids that are colorless, it is usually necessary to sublime the $AlCl_3$ prior to use (often more than once). It is recommended that the $AlCl_3$ should be mixed with sodium chloride and aluminium wire for this process [22b].

$AlCl_3$-based ionic liquids often contain traces of oxide ion impurities, formed by the presence of small amounts of water and oxygen. These are generally referred to as $[AlOCl_2]^-$, although ^{17}O NMR measurements have indicated that a complex series of equilibria is in fact occurring [45]. It has been reported that these can be efficiently removed by bubbling phosgene ($COCl_2$) through the ionic liquid [46]. In this case the by-product of the reaction is CO_2, and thus easily removed under vacuum. This method should be approached with caution due to the high toxicity of phosgene, and an alternative approach using the less toxic triphosgene has also been reported more recently [47]. In the presence of water or other proton sources, chloroaluminate-based ionic liquids may contain protons, which will behave as a Brønsted superacid in acidic melts [48]. It has been reported that these may be removed simply by the application of high vacuum ($< 5 \times 10^{-6}$ Torr) [49].

Purification of ionic liquids formed by anion metathesis can throw up a different set of problems, as already noted in Section 2.1.3.2. In this case the most common impurities are halide anions, or unwanted cations inefficiently separated from the final product. The presence of such impurities can be extremely detrimental to the performance of the ionic liquids, particularly in applications involving transition metal-based catalysts, which are often deactivated by halide ions. In general this is much more of a problem in water-soluble ionic liquids, as water-immiscible salts can usually be purified quite efficiently by washing with water. The methods used to overcome this problem have already been covered in the previous section. The problems inherent in the preparation of water-miscible salts have been highlighted by Seddon et al. [37], who studied the Na^+ and Cl^- concentrations in a range of ionic liquids formed by treatment of [EMIM]Cl and [BMIM]Cl with Ag[BF_4], Na[BF_4], Ag[NO_3], Na[NO_3], and HNO_3. They found that the physical properties such as density and viscosity of the liquids can be radically altered by the presence of unwanted ions. The results showed that all preparations using Na^+ salts resulted in high residual concentrations of Cl^-, while the use of Ag^+ salts gave rise to much lower levels. The low solubility of NaCl in the ionic liquids, however, indicates that the impurities arise from the fact that the reaction with the Na^+ salts does not proceed to completion. Indeed, it was reported in one case that unreacted [BMIM]Cl was isolated by crystallization from [BMIM][NO_3]. A further example of the potential hazards of metal-containing impurities in ionic liquids is seen when [EMIM][CH_3CO_2] is prepared from [EMIM]Cl and Pb[CH_3CO_2]$_4$ [50]. The resulting salt has been shown to contain ca. 0.5 M residual lead [51].

In practical terms, it is suggested that, in any application where the presence of halide ions may cause problems, the concentration of these be monitored to ensure the purity of the liquids. This may be achieved either by the use of an ion-sensitive electrode, or alternatively by use of a chemical method such as the Vollhard procedure for chloride ions [52]. Seddon et al. have reported that effectively identical results were obtained with either method [37].

Most ionic liquids based on the common cations and anions should be colorless, with minimal absorbance at wavelengths greater than 300 nm. In practice, the salts often take on a yellow hue, particularly during the quaternization step. The amount of impurity causing this is generally extremely small, being undetectable by ^1H NMR or CHN microanalysis, and in many applications the discoloration may not be of any importance. This is clearly not the case, however, when the solvents are required for photochemical or UV/visible spectroscopic investigations. To date, the precise origins of these impurities have not been determined, but it seems likely that they arise from unwanted side reactions involving oligomerization or polymerization of small amounts of free amine, or else from impurities in the haloalkanes. Where it is important that the liquids are colorless, however, the color may be minimized by following a few general steps:

- All starting materials should be purified as discussed above [44].
- The presence of traces of acetone can sometimes result in discoloration during the quaternization step. Thus, all glassware used in this step should be kept free of this solvent.
- The quaternization reaction should be carried out either in a system that has been degassed and sealed under nitrogen, or else under a flow of inert gas such as nitrogen. Furthermore the reaction temperature should be kept as low as possible (no more that ca. 80 °C for Cl^- salts, and lower for Br^- and I^- salts).

If the liquids remain discolored even after these precautions, it is often possible to purify them further by first stirring with activated charcoal, followed by passing the liquid down a short column of neutral or acidic alumina as discussed in Section 2.1.3.2 [33].

Clearly, the impurity likely to be present in largest concentrations in most ionic liquids is water. The removal of other reaction solvents is generally easily achieved by heating the ionic liquid under vacuum. Water is generally one of the most problematic solvents to remove, and it is generally recommended that ionic liquids be heated to at least 70 °C for several hours with stirring to achieve an acceptably low degree of water contamination. Even water-immiscible salts such as [BMIM][PF_6] can absorb up to ca. 2 wt.% water on equilibration with the air, corresponding to a water concentration of ca. 1.1 M. Thus it is advised that all liquids be dried directly before use. If the amount of water present is of importance, it may be determined either by Karl–Fischer titration, or a less precise determination may be carried out using IR spectroscopy.

2.1.5 Conclusions

It is hoped that this section will give the reader a better appreciation of the range of ionic liquids that have already been prepared, as well as a summary of the main techniques involved and the potential pitfalls. While the basic chemistry involved is relatively straightforward, the preparation of ionic liquids of known purity may be less easily achieved, and it is hoped that the ideas given here may be of assistance to the reader. It should also be noted that many of the more widely used ionic liquids are now commercially available from a range of suppliers, including some specializing in the synthesis of ionic liquids [53].

References

1 P. Walden, *Bull. Acad. Imper. Sci. (St. Petersburg)* **1914**, 1800.
2 R.M. Pagni, *Advances in Molten Salt Chemistry 6*, Eds. G. Mamantov, C.B. Mamantov, and J. Braunstein, Elsevier: New York **1987**, pp 211–346.
3 (a) C.M. Gordon, *Appl. Cat. A, General* **2001**, *222*, 101. (b) R.A. Sheldon, *Chem. Commun.* **2001**, 2399. (c) P. Wasserscheid and W. Keim, *Angew. Chem., Int. Ed.* **2000**, *39*, 3772. (d) J.D. Holbrey and K.R. Seddon, *Clean Prod. Processes* **1999**, *1*, 223. (e) T. Welton, *Chem. Rev.* **1999**, *99*, 2071.
4 S.I. Lall, D. Mancheno, S. Castro, V. Behaj, J.L.I. Cohen, and R. Engel, *Chem. Commun.* **2000**, 2413.
5 J.H. Davis, Jr., K.J. Forrester, and T. Merrigan, *Tetrahedron Lett.* **1998**, *39*, 8955.
6 J.S. Wilkes, J.A. Levisky, R.A. Wilson, and C.L. Hussey, *Inorg. Chem.* **1982**, *21*, 1263.
7 D.F. Evans, A. Yamouchi, G.J. Wei, and V.A. Bloomfield, *J. Phys. Chem.* **1983**, *87*, 3537.
8 C.K. Lee, H.W. Huang, and I.J.B. Lin, *Chem. Commun.* **2000**, 1911.
9 C.M. Gordon, J.D. Holbrey, A.R. Kennedy, and K.R. Seddon, *J. Mater. Chem.* **1998**, *8*, 2627.
10 A.E. Visser, J.D. Holbrey, and R.D. Rogers, *Chem. Commun.* **2001**, 2484.
11 T. Kitazume, F. Zulfiqar, and G. Tanaka, *Green Chem.* **2000**, *2*, 133.
12 D.R. MacFarlane, P. Meakin, J. Sun, N. Amini, and M. Forsyth, *J. Phys. Chem. B* **1999**, *103*, 4164.
13 J. Sun, M. Forsyth, D.R. MacFarlane, *J. Phys. Chem. B* **1998**, *102*, 8858.
14 P. Bonhôte, A.-P. Dias, N. Papageorgiou, K. Kalyanasundaram, and M. Grätzel, *Inorg. Chem.* **1996**, *35*, 1168.
15 J.G. Huddleston, H.D. Willauer, R.P. Swatlowski, A.E. Visser, and R.D. Rogers, *Chem. Commun.* **1998**, 1765.
16 P. Lucas, N. El Mehdi, H.A. Ho, D. Bélanger, and L. Breau, *Synthesis* **2000**, 1253.
17 R.S. Varma and V.V. Namboodiri, *Chem. Commun.* **2001**, 643.
18 J.A. Holbrey, K.R. Seddon, and R. Wareing, *Green Chem.* **2001**, *3*, 33.
19 H. Waffenschmidt, Dissertation, RWTH Aachen, Germany **2000**.
20 N. Karodia, S. Guise, C. Newlands, and J.-A. Andersen, *Chem. Commun.* **1998**, 2341.
21 F.H. Hurley and T.P. Wier, *J. Electrochem. Soc.* **1951**, *98*, 203.
22 (a) H.L. Chum, V.R. Koch, L.L. Miller, and R.A. Osteryoung, *J. Am. Chem. Soc.* **1975**, *97*, 3264. (b) J. Robinson and R.A. Osteryoung, *J. Am. Chem. Soc.* **1979**, *101*, 323.
23 H.A. Øye, M. Jagtoyen, T. Oksefjell, J.S. Wilkes, *Mater. Sci. Forum* **1991**, *73–75*, 183.
24 (a) Y. Chauvin, S. Einloft, and H. Olivier, *Ind. Eng. Chem. Res.* **1995**, *34*, 1149. (b) B. Gilbert, Y. Chauvin, H. Olivier, F. DiMarco-van Tiggelen, *J. Chem. Soc., Dalton Trans.* **1995**, 3867.
25 S.D. Williams, J.P. Schoebrechts, J.C. Selkirk, G. Mamantov, *J. Am. Chem. Soc.* **1987**, *109*, 2218.
26 Y. Chauvin and H. Olivier-Bourbigou, *CHEMTECH* **1995**, *25*, 26.
27 G.W. Parshall, *J. Am. Chem. Soc.* **1972**, *94*, 8716.
28 S. Mori, K. Ida, H. Suzuki, S. Takahashi, I Saeki, *Eur. Pat.*, EP 90111879.4 _(Check!)._
29 P.B. Hitchcock, K.R. Seddon, and T. Welton, *J. Chem. Soc., Dalton Trans.* **1993**, 2639.
30 P.B. Hitchcock, R.J. Lewis, and T. Welton, *Polyhedron* **1993**, *12*, 2039.
31 J.S. Wilkes and M.J. Zaworotko, *Chem. Commun.* **1992**, 965.
32 J. Fuller, R.T. Carlin, H.C. DeLong, and D. Haworth, *Chem. Commun.* **1994**, 299.
33 L. Cammarata, S. Kazarian, P. Salter, and T. Welton, *Phys. Chem. Chem. Phys.* **2001**, *3*, 5192.
34 J.D. Holbrey and K.R. Seddon, *J. Chem Soc., Dalton Trans.* **1999**, 2133.
35 N. L. Lancaster, T. Welton, G.B. Young, *J. Chem. Soc., Perkin Trans. 2* **2001**, 2267.

36 J. Fuller and R.T. Carlin, *Proc. Electrochem. Soc.* **1999**, *98*, 227.
37 K.R. Seddon, A. Stark, and M.-J. Torres, *Pure Appl. Chem.* **2000**, *72*, 2275.
38 D.R. McFarlane, J. Golding, S. Forsyth, M. Forsyth, and G.B. Deacon, *Chem. Commun.* **2001**, 2133.
39 A.S. Larsen, J.D. Holbrey, F.S. Tham, and C.A. Reed, *J. Am. Chem. Soc.* **2000**, *122*, 7264.
40 M. Hasan, I.V. Kozhevnikov, M.R.H. Siddiqui, A. Steiner, and N. Winterton, *Inorg. Chem.* **1999**, *38*, 5637.
41 C.G. Swain, A. Ohno, D.K. Roe, R. Brown, and T. Maugh, II, *J. Am. Chem. Soc.* **1967**, *89*, 2648.
42 R.M. Pomaville, S.K. Poole, L.J. Davis, and C.F. Poole, *J. Chromatog.* **1988**, *438*, 1.
43 H. Matsumoto, T. Matsuda, and Y. Miyazaki, *Chem. Lett.* **2000**, 1430.
44 W.L.F. Armarego and D.D. Perrin, *Purification of laboratory chemicals, 4th Ed.*, Butterworth-Heinemann **1997**.
45 T.A. Zawodzinski and R.A. Osteryoung, *Inorg. Chem.* **1990**, *29*, 2842.
46 (a) A.K. Abdul-Sada, A.G. Avent, M.J. Parkington, T.A. Ryan, K.R. Seddon, and T. Welton, *Chem. Commun.* **1987**, 1643. (b) A.K. Abdul-Sada, A.G. Avent, M.J. Parkington, T.A. Ryan, K.R. Seddon, and T. Welton, *J. Chem. Soc., Dalton Trans.* **1993**, 3283.
47 A.J. Dent, A. Lees, R.J. Lewis, and T. Welton, *J. Chem. Soc., Dalton Trans.* **1996**, 2787.
48 (a) G.P. Smith, A.S. Dworkin, R.M. Pagni, and S.P. Zing, *J. Am. Chem. Soc.* **1989**, *111*, 525. (b) G.P. Smith, A.S. Dworkin, R.M. Pagni, and S.P. Zing, *J. Am. Chem. Soc.* **1989**, *111*, 5075. (c) S.-G. Park, P.C. Trulove, R.T. Carlin, and R.A. Osteryoung, *J. Am. Chem. Soc.* **1991**, *113*, 3334.
49 M.A.M. Noel, P.C. Trulove, and R.A. Osteryoung, *Anal. Chem.* **1991**, *63*, 2892.
50 B. Ellis, *Int. Pat.*, WO 96/18459 **1996**.
51 J.T. Hamill, C. Hardacre, M. Nieuwenhuyzen, K.R. Seddon, S.A. Thompson, and B. Ellis, *Chem. Commun.* **2000**, 1929.
52 A.I. Vogel, *A Textbook of Quantitative Inorganic Analysis*, 3rd ed., Longmans, Green and Co., London, **1961**.
53 (a) http:\\www.solvent-innovation.com. (b) http:\\www.covalentassociates.com.

2.2 Quality Aspects and Other Questions Related to Commercial Ionic Liquid Production

Claus Hilgers and Peter Wasserscheid

2.2.1 Introduction

From Section 2.1 it has become very clear that the synthesis of an ionic liquid is in general quite simple organic chemistry, while the preparation of an ionic liquid of a certain quality requires some know-how and experience. Since neither distillation nor crystallization can be used to purify ionic liquids after their synthesis (due to their nonvolatility and low melting points), maximum care has to be taken before and during the ionic liquid synthesis to obtain the desired quality.

Historically, the know-how to synthesize and handle ionic liquids has been treated somehow like a "holy grail". Up to the mid-1990s, indeed, only a small number of specialized industrial and academic research groups were able to prepare and handle the highly hygroscopic chloroaluminate ionic liquids that were the only ionic liquid systems available in larger amounts. Acidic chloroaluminate ionic liquids, for example, have to be stored in glove-boxes to prevent their contamination with traces of water. Water impurities are known to react with the anions of the melt with release of superacidic protons. These cause unwanted side reactions in many applications and possess considerable potential for corrosion (a detailed description of protic and oxidic impurities in chloroaluminate melts is given in Welton's 1999 review article [1]). This need for very special and expensive handling techniques has without doubt prevented the commercial production and distribution of chloroaluminate ionic liquids, even up to the present day.

The introduction of the more hydrolysis-stable tetrafluoroborate [2] and hexafluorophosphate systems [3], and especially the development of their synthesis by means of metathesis from alkali salts [4], can be regarded as a first key step towards commercial ionic liquid production.

However, it still took its time. When the authors founded Solvent Innovation [5] in November 1999, the commercial availability of ionic liquids was still very limited. Only a small number of systems could be purchased from Sigma–Aldrich, in quantities of up to 5 g [6].

Besides Solvent Innovation, a number of other commercial suppliers nowadays offer ionic liquids in larger quantities [7]. Moreover, the distribution of these liquids by Fluka [8], Acros Organics [9], and Wako [10] assures a certain availability of different ionic liquids on a rapid-delivery basis.

From discussions with many people now working with ionic liquids, we know that, at least for the start of their work, the ability to buy an ionic liquid was important. In fact, a synthetic chemist searching for the ideal solvent for his or her specific application usually takes solvents that are ready for use on the shelf of the laboratory. The additional effort of synthesizing a new special solvent can rarely be justified, especially in industrial research. Of course, this is not only true for ionic liquids. Very probably, nobody would use acetonitrile as a solvent in the laboratory if they had to synthesize it before use.

The commercial availability of ionic liquids is thus a key factor for the actual success of ionic liquid methodology. Apart from the matter of lowering the "activation barrier" for those synthetic chemists interested in entering the field, it allows access to ionic liquids for those communities that do not traditionally focus on synthetic work. Physical chemists, engineers, electrochemists, and scientists interested in developing new analytical tools are among those who have already developed many new exciting applications by use of ionic liquids [11].

2.2.2
Quality Aspects of Commercial Ionic Liquid Production

With ionic liquids now commercially available, it should not be forgotten that an ionic liquid is still a quite different product from traditional organic solvents, simply because it cannot be purified by distillation, due to its nonvolatile character. This, combined with the fact that small amounts of impurities can influence the ionic liquid's properties significantly [12], makes the quality of an ionic liquid quite an important consideration.

Ionic liquid synthesis in a commercial context is in many respects quite different from academic ionic liquid preparation. While, in the commercial scenario, labor-intensive steps add significantly to the price of the product (which, next to quality, is another important criterion for the customer), they can easily be justified in academia to obtain a purer material. In a commercial environment, the desire for absolute quality of the product and the need for a reasonable price have to be reconciled. This is not new, of course. If one looks into the very similar business of phase-transfer catalysts or other ionic modifiers (such as commercially available ammonium salts), one rarely finds absolutely pure materials. Sometimes the active ionic compound is only present in about 85 % purity. However, and this is a crucial point, the product is well specified, the nature of the impurities is known, and the quality of the material is absolutely reproducible from batch to batch.

From our point of view, this is exactly what commercial ionic liquid production is about. Commercial producers try to make ionic liquids in the highest quality that can be achieved at reasonable cost. For some ionic liquids they can guarantee a purity greater than 99 %, for others perhaps only 95 %. If, however, customers are offered products with stated natures and amounts of impurities, they can then decide what kind of purity grade they need, given that they do have the opportunity to purify the commercial material further themselves. Since trace analysis of impurities in ionic liquids is still a field of ongoing fundamental research, we think that anybody who really needs (or believes that they need) a purity of greater than 99.99 % should synthesize or purify the ionic liquid themselves. Moreover, they may still need to develop the methods to specify this purity.

The following subsections attempt to comment upon common impurities in commercial ionic liquid products and their significance for known ionic liquid applications. The aim is to help the reader to understand the significance of different impurities for their application. Since chloroaluminate ionic liquids are not produced or distributed commercially, we do not deal with them here.

2.2.2.1 Color

From the literature one gets the impression that ionic liquids are all colorless and look almost like water. However, most people who start ionic liquid synthesis will probably get a highly colored product at first. The chemical nature of the colored impurities in ionic liquids is still not very clear, but it is probably a mixture of traces of compounds originating from the starting materials, oxidation products, and thermal degradation products of the starting materials. Sensitivity to coloration during ionic liquid synthesis can vary significantly with the type of cation and anion of the ionic liquid. Pyridinium salts, for instance, tend to form colored impurities more easily than imidazolium salts do.

Section 2.1 excellently describes methods used to produce colorless ionic liquids. From this it has become obvious that freshly distilled starting materials and low-temperature processing during the synthesis and drying steps are key aspects for avoidance of coloration of the ionic liquid.

From a commercial point of view, it is possible to obtain colorless ionic liquids, but not on a large scale at reasonable cost. If one wants to obtain colorless material, then the labor-intensive procedures described in Section 2.1 have to be applied.

For a commercial producer, three points are important in this context:

a) The colored impurities are usually present only in trace amounts. It is impossible to detect them by NMR or by analytical techniques other than UV/VIS spectroscopy. Hence the difficulty in determining the chemical structure of the colored impurities.
b) For almost all applications involving ionic liquids, the color is not the **crucial** parameter. In catalytic applications, for example, it appears that the concentration of colored impurities is significantly lower than commonly used catalyst concentrations. Exceptions are, of course, any application in which UV spectroscopy is used for product or catalyst analysis and for all photochemical applications.
c) Prevention of coloration of the ionic liquid is not really compatible with the aim of a rational economic ionic liquid production. Additional distillative cleaning of the feedstocks consumes time and energy, and additional cleaning by chromatography after synthesis is also a time-consuming step. The most important restriction, however, is the need to perform synthesis (mainly the alkylation step) with good feedstocks at the lowest possible temperature, and thus at the slowest rate. This requires long reaction times and therefore high plant cost.

A compromise between coloration and economics in commercial ionic liquid production is therefore necessary. Since chromatographic decoloration steps are known and relatively easy to perform (see Section 2.2.3), we would not expect there to be a market for a colorless ionic liquid, if the same substance can be made in a slightly colored state, but at a much lower price.

2.2.2.2 Organic Starting Materials and other Volatiles

Volatile impurities in an ionic liquid may have different origins. They may result from solvents used in the extraction steps during the synthesis, from unreacted starting materials from the alkylation reaction (to form the ionic liquid's cation), or from any volatile organic compound previously dissolved in the ionic liquid.

In theory, volatile impurities can easily be removed from the nonvolatile ionic liquid by simple evaporation. However, this process can sometimes take a considerable time. Factors that influence the time required for the removal of all volatiles from an ionic liquid (at a given temperature and pressure) are: a) the amount of volatiles, b) their boiling points, c) their interactions with the ionic liquid, d) the viscosity of the ionic liquid, and e) the surface of the ionic liquid.

A typical example of a volatile impurity that can be found as one of the main impurities in low-quality ionic liquids with alkylmethylimidazolium cations is the methylimidazole starting material. Because of its high boiling point (198 °C) and its strong interaction with the ionic liquid, this compound is very difficult to remove from an ionic liquid even at elevated temperature and high vacuum. It is therefore important to make sure, by use of appropriate alkylation conditions, that no unreacted methylimidazole is left in the final product.

Traces of bases such as methylimidazole in the final ionic liquid product can play an unfavorable role in some common applications of ionic liquids (such as biphasic catalysis). Many electrophilic catalyst complexes will coordinate the base in an irreversible manner and be deactivated.

A number of different methods to monitor the amount of methylimidazole left in a final ionic liquid are known. NMR spectroscopy is used by most academic groups, but may have a detection limit of about 1 mol%. The photometric analysis described by Holbrey, Seddon, and Wareing has the advantage of being a relatively quick method that can be performed with standard laboratory equipment [13]. This makes it particularly suitable for monitoring of the methylimidazole content during commercial ionic liquid synthesis. The method is based on the formation and colorimetric analysis of the intensely colored complex of 1-methylimidazole with copper(II) chloride.

2.2.2.3 **Halide Impurities**

Many ionic liquids (among them the most commonly used tetrafluoroborate and hexafluorophosphate systems) are still made in two-step syntheses. In the first step, an amine or phosphine is alkylated to form the cation. For this reaction, alkyl halides are frequently used as alkylating agents, forming halide salts of the desired cation. To obtain a non-halide ionic liquid, the halide anion is exchanged in a second step. This can be achieved variously by addition of the alkali salt of the desired anion (with precipitation of the alkali halide salt), by treatment with a strong acid (with removal of the hydrohalic acid), or by use of an ion-exchange resin (for more details see Section 2.1). Alternative synthetic procedures involving the use of silver [2] or lead salts [14] are – at least from our point of view – not acceptable for commercial ionic liquid production.

All the halide exchange reactions mentioned above proceed more or less quantitatively, causing greater or lesser quantities of halide impurities in the final product. The choice of the best procedure to obtain complete exchange depends mainly on the nature of the ionic liquid that is being produced. Unfortunately, there is no general method to obtain a halide-free ionic liquid that can be used for all types of ionic liquid. This is explained in a little more detail for two defined examples: the synthesis of [BMIM][$(CF_3SO_2)_2N$] and the synthesis of [EMIM][BF_4].

[BMIM][$(CF_3SO_2)_2N$] has a miscibility gap with water (about 1.4 mass% of water dissolves in the ionic liquid [15]) and shows high stability to hydrolysis. It is therefore very easy to synthesize this ionic liquid in a halide-free state. In a procedure first described by Bônhote and Grätzel [15], [BMIM]Cl (obtained by alkylation of methylimidazole with butyl chloride) and Li[$(CF_3SO_2)_2N$] are both dissolved in water. As the aqueous solutions are mixed, the ionic liquid is formed as a second layer. After separation from the aqueous layer, the ionic liquid can easily be washed with water to a point where no traces of halide ions are detectable in the washing water (by titration with $AgNO_3$). After drying of the ionic liquid phase, an absolutely halide-free ionic liquid can be obtained (determination by ion chromatography, by titration with $AgNO_3$, or by electrochemical analysis).

The halide-free preparation of [EMIM][BF_4], however, is significantly more difficult. Since this ionic liquid is completely miscible with water and so cannot be re-extracted from aqueous solution with CH_2Cl_2 or other organic solvents, removal of the halide ions by a washing procedure with water is not an option. A metathesis reaction in water-free acetone or CH_2Cl_2 is possible, but suffers from the low solubility of Na[BF_4] in these solvents and the long reaction times. Exchange reactions in this type of suspension therefore take a long time and, when carried out on larger scales, tend to be incomplete even after long reaction times. Consequently, to synthesize [EMIM][BF_4] of completely halide-free quality, special procedures have to be applied. Two examples are synthesis with use of an ion-exchange resin [16] or the direct alkylation of ethylimidazole with Meerwein's reagent [Me_3O][BF_4].

Generally, the presence of halide impurities is not (as with the ionic liquid's color) a question of having a nice-looking ionic liquid or not. On the contrary, the halide content can seriously affect the usefulness of the material as a solvent for a given chemical reaction. Apart from the point that some physicochemical properties are highly dependent on the presence of halide impurities (as demonstrated by Seddon and al. [12]), the latter can chemically act as catalyst poisons [17], stabilizing ligands [18], nucleophiles, or reactants, depending on the chemical nature of the reaction. It is consequently necessary to have an ionic liquid free of halide impurities to investigate its properties for any given reaction, especially in catalysis, in which the amount of catalyst used can be in the range of the concentration of the halide impurities in the ionic liquid.

2.2.2.4 Protic Impurities

Protic impurities have to be taken into account for two groups of ionic liquids: those that have been produced by an exchange reaction involving a strong acid (often the case, for example, for [BMIM][PF_6]), and those that are sensitive to hydrolysis. In the latter case, the protons may originate from the hydrolysis of the anion, forming an acid that may be dissolved in the ionic liquid.

For ionic liquids that do not mix completely with water (and which display sufficient hydrolysis stability), there is an easy test for acidic impurities. The ionic liquid is added to water and a pH test of the aqueous phase is carried out. If the aqueous phase is acidic, the ionic liquid should be washed with water to the point where the washing water becomes neutral. For ionic liquids that mix completely with water we recommend a standardized, highly proton-sensitive test reaction to check for protic impurities.

Obviously, the check for protic impurities becomes crucial if the ionic liquid is to be used for applications in which protons are known to be active compounds. For some organic reactions, one has to be sure that an "ionic liquid effect" does not turn out to be a "protic impurity effect" at some later stage of the research!

2.2.2.5 Other Ionic Impurities from Incomplete Metathesis Reactions

Apart from halide and protic impurities, ionic liquids can also be contaminated with other ionic impurities from the metathesis reaction. This is especially likely if the alkali salt used in the metathesis reaction shows significant solubility in the ionic liquid formed. In this case, the ionic liquid can contain significant amounts of the alkali salt. While this may not be a problem even for some catalytic applications (since the presence of the alkali cation may not affect the catalytic cycle of a transition metal catalyst), it is of great relevance for the physicochemical properties of the melt.

In this context it is important to note that the detection of this kind of alkali cation impurity in ionic liquids is not easy with traditional methods for reaction monitoring in ionic liquid synthesis (such as conventional NMR spectroscopy). More specialized procedures are required to quantify the amount of alkali ions in the ionic liquid or the quantitative ratio of organic cation to anion. Quantitative ion chromatography is probably the most powerful tool for this kind of quality analysis.

Because of these analytical problems, we expect that some of the disagreements in the literature (mainly concerning the physicochemical data of some tetrafluoroborate ionic liquids) may have their origins in differing amounts of alkali cation impurities in the ionic liquids analyzed.

2.2.2.6 **Water**

Without special drying procedures and completely inert handling, water is omnipresent in ionic liquids. Even the apparently hydrophobic ionic liquid [BMIM][$(CF_3SO_2)_2N$] saturates with about 1.4 mass% of water [15], a significant molar amount. For more hydrophilic ionic liquids, water uptake from air can be much greater. Imidazolium halide salts in particular are known to be extremely hygroscopic, one of the reasons why it is so difficult to make completely proton-free chloroaluminate ionic liquids.

For commercial ionic liquid production, this clearly means that all products contain some greater or lesser amount of water. Depending on the production conditions and the logistics, the ionic liquids can reasonably be expected to come into some contact with traces of water.

Water in an ionic liquid may be a problem for some applications, but not for others. However, one should in all cases know the approximate amount of water present in the ionic liquid used. Moreover, one should be aware of the fact that water in the ionic liquid may not be inert and, furthermore, that the presence of water can have significant influence on the physicochemical properties of the ionic liquid, on its stability (some wet ionic liquids may undergo hydrolysis with formation of protic impurities), and on the reactivity of catalysts dissolved in the ionic liquid.

2.2.3
Upgrading of Commercial Ionic Liquids

For all research carried out with commercial ionic liquids we recommend a serious quality check of the product prior to work. As already mentioned, a good commercial ionic liquid may be colored and may contain some traces of water. However, it should be free of organic volatiles, halides (if not an halide ionic liquid), and all ionic impurities.

To remove water, commercial ionic liquids used for fundamental research purposes should be dried at 60 °C in vacuo overnight. The water content should be checked prior to use. This can be done qualitatively by infrared spectroscopy or cyclovoltametric measurements, or quantitatively by Karl–Fischer titration. If the ionic liquids cannot be dried to zero water content for any reason, the water content should always be mentioned in all descriptions and documentation of the experiments to allow proper interpretation of the results obtained.

Regarding the color, we only see a need for colorless ionic liquids in very specific applications (see above). One easy treatment that often reduces coloration quite impressively, especially of imidazolium ionic liquids, is purification by column chromatography/filtration over silica 60. For this purification method, the ionic liquid is dissolved in a volatile solvent such as CH_2Cl_2. Usually, most of the colored impurities stick to the silica, while the ionic liquid is eluted with the solvent. By repetition of the process several times, a seriously colored ionic liquid can be converted into an almost completely colorless material.

2.2.4
Scaling-up of Ionic Liquid Synthesis

For commercial ionic liquid synthesis, quality is a key factor. However, since availability and price are other important criteria for the acceptance of this new solvent concept, the scaling-up of ionic liquid production is a major research interest too.

Figure 2.2-1: One of Solvent Innovation's production plants at the Institut für Technische Chemie und Makromolekulare Chemie, University of Technology Aachen, Germany.

Figure 2.2-2: Synthesis of [BMIM]Cl in a 30 litre scale in three stages. a) start of the reaction; b) the reaction vessel after 10 min reaction time; c) some ionic liquid product at elevated temperature.

Many historical ways to make ionic liquids proved to be impractical on larger scales. Sometimes expensive starting materials are used (anion-exchange with silver salts, for example [2]), or very long reaction times are necessary for the alkylation steps, or filtration procedures are included in the synthesis, or hygroscopic solids have to be handled. All these have to be avoided for a good synthesis on larger scales.

Other important aspects to consider during the scaling-up of ionic liquid synthesis are heat management (alkylation reactions are exothermic!) and proper mass transport. For both of these the proper choice of reactor set-up is of crucial importance.

Figure 2.2-1 shows one of Solvent Innovation's production plants at the Institut für Technische Chemie und Makromolekulare Chemie, Aachen University of Technology, Germany. Figure 2.2-2 shows the synthesis of [BMIM]Cl on a 30 liter scale in three stages: a) start of the reaction, b) the reaction vessel after 10 min reaction time, and c) some ionic liquid product at elevated temperature.

2.2.5 HSE data

The production of ionic liquids on larger scales raises the question of registration of these new materials and the acquisition of HSE data. Surprisingly enough, although ionic liquids have been around for quite some years, very few data are available in this respect. Early investigations studied the effect of a basic chloroaluminate systems on the skin of rats [19]. Very recently, the acute toxicity of 1-hexyloxymethyl-3-methylimidazolium tetrafluoroborate was assessed by Pernak et al., by the Gadumm method [20]. The values were found to be LD_{50} = 1400 mg kg^{-1} for female Wistar rats and LD_{50} = 1370 mg kg^{-1} for males. The authors concluded that the tetrafluoroborate salt could be used safely.

These preliminary studies notwithstanding, much more HSE data for ionic liquids will be needed in the near future. We anticipate that commercial suppliers will play a leading role in the acquisition of these data, since considerable sums of money are involved in full HSE characterization of ionic liquids.

In the meantime, we believe that the best prediction of the toxicity of an ionic liquid of type [cation][anion] can be derived from the often well known toxicity data for the salts [cation]Cl and Na[anion]. Since almost all chemistry in nature takes place in aqueous media, the ions of the ionic liquid can be assumed to be present in dissociated form. Therefore, a reliable prediction of ionic liquids' HSE data should be possible from a combination of the known effects of the alkali metal and chloride salts. Already from these, very preliminary, studies, it is clear that HSE considerations will be an important criterion in selection and exclusion of specific ionic liquid candidates for future large-scale, technical applications.

2.2.6 Future Price of Ionic Liquids

The price of ionic liquids is determined by many parameters, such as personnel, overheads, and real production costs. One can imagine that production on a small scale would be mostly determined by the personnel cost and little by the material cost. On a large scale, the material cost should become more important and mainly determine the price of an ionic liquid. This means that the price of a large-scale commercial ionic liquid should be dictated by the price of the cation and anion source.

Table 2.2-1 shows a list of typical cations and anions ordered by their rough price on an industrial scale.

This table illustrates pretty well that the large-scale ionic liquid will probably not comprise a dialkylimidazolium cation and a $[CF_3SO_2)_2N]^-$ anion. Over a medium-term timescale, we would expect a range of ionic liquids to become commercially available for € 25–50 per liter on a ton scale. Halogen-free systems made from cheap anion sources are expected to meet this target first.

Scheme 2.2-1: Typical ions making up ionic liquids, ordered according to their rough price on an industrial scale.

Cheap		Expensive
cations:		
$[HNR_3]^+$, $[HPR_3]^+$, $[NR_4]^+$, $[PR_4]^+$	alkylpyridinium-, alkylmethylimidazolium-	dialkylimidazolium-
anions:		
$[Cl]^-$, $[MeSO_4]^-$, $[AlCl_4]^-$, [acetate], $[NO_3]^-$	$[PF_6]^-$, $[BF_4]^-$	$[SbF_6]^-$, $[CF_3SO_3]^-$, $[(CF_3SO_2)_2N]^-$

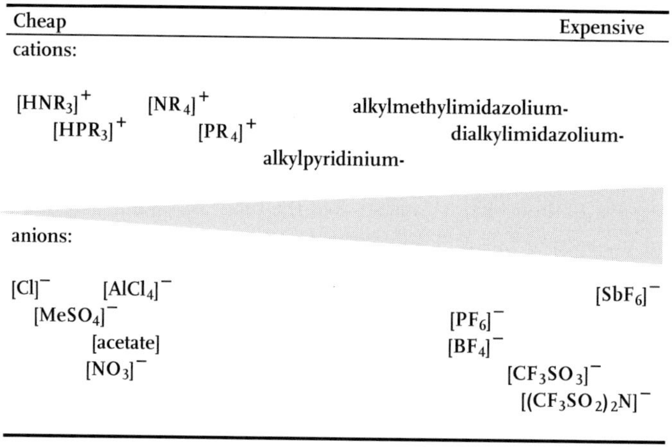

2.2.7
Intellectual Property Aspects Regarding Ionic Liquids

The future price of ionic liquids will also reflect intellectual property considerations. While the currently most frequently requested ionic liquids, the tetrafluoroborate and hexafluorophosphate ionic liquids, are all patent-free, many recently developed, new ionic liquid systems are protected by "state of matter" patents. Table 2.2-2 gives an overview of some examples published after 1999.

Table 2.2-1: Selected examples of "state of matter" patents concerning ionic liquids published since 1999.

Title	*Typical protected compound*	*Company*	*Year of pub.*	*Ref.*
Preparation of N-alkoxyalkylimidazolium salts and ionic liquids or gels containing them	1-Methyl-3-methoxyethyl imidazolium bromide	Foundation for Scientific Technology Promotion, Japan	2002	21
Ionic liquids	$[EMIM][PF_3(C_2F_5)_3]$	Merck Patent GmbH, Germany	2001	22
Ionic liquids	[EMIM] bis(1,2-oxalato-O,O')borate	Merck Patent GmbH, Germany	2001	23
Preparation of phosphonium salts as ionic liquids	$[P\mathit{i}Bu_3Et]$[tosylate]	Cytec Technology Corp., USA	2001	24
Ionic liquids derived from Lewis acids based on Ti, Nb, Sn, Sb	$[NR_3R']X/SbF_5$	Atofina, France	2001	25
Preparation of chiral ionic liquids	(*S*)-4-isopropyl-2,3-di methyloxazolinium $[BF_4]$	Solvent Innovation GmbH, Germany	2001	26
Immobilized ionic liquids	Chloroaluminate ionic liquids on inorganic supports	ICI, UK	2001	27
Preparation of ionic liquids and their use	$[NR_3R'][P(OPh\text{-}SO_3)_x(OPh)_y]$	Celanese Chemicals Europe, GmbH, Germany	2000	28
Ionic liquids prepared as low-melting salts ...	$[NR_3R']Cl/ZnCl_2$	University of Leicester, UK	2000	29
Preparation of ionic liquids for catalysis	$[BMIM][HSO_4]$, $[HNR_3][HSO_4]$	BP Chemicals, UK; Akzo Nobel NV, Netherlands; Elementis Specialities, UK	2000	30
Preparation of ionic liquids by treatment of amines with halide donors in the presence of metal halides	$[HNR_3]Cl/AlCl_3$	Akzo Nobel NV, Netherlands	2000	31

Without a doubt, tetrafluoroborate and hexafluorophosphate ionic liquids have shortcomings for larger-scale technical application. The relatively high cost of their anions, their insufficient stability to hydrolysis for long-term application in contact with water (formation of corrosive and toxic HF during hydrolysis!), and problems related to their disposal have to be mentioned here. New families of ionic liquid that should meet industrial requirements in a much better way are therefore being developed. However, these new systems will probably be protected by state of matter patents.

In this respect, there is one important statement to make. It is our belief that the owners of state of matter patents for promising new classes of ionic liquids should never stop or hinder academic research dealing with these substances. On the contrary, we think that only through fundamental academic research will we be able to gain a full understanding of a given material over time. Only this full understanding will allow the full scope and limitations of a new family of ionic liquids to be explored.

Research in ionic liquid methodology is still young and there is still a lot to explore. Prevention of fundamental research on some new families of ionic liquids by exploitation of an IP position would simply kill off a lot of future possibilities.

References

1 T. Welton, *Chem. Rev.* **1999**, *99*, 2071–2083.
2 J. S. Wilkes, M. J. Zaworotko, *J. Chem. Soc. Chem. Commun.* **1992**, 965–967.
3 J. Fuller, R. T. Carlin, H. C. de Long, D. Haworth, *J. Chem. Soc. Chem. Commun.* **1994**, 299–300.
4 P. A. Z. Suarez, J. E. L. Dullius, S. Einloft, R. F. de Souza, J. Dupont, *Polyhedron* **1996**, *15(7)*, 1217–1219.
5 Solvent Innovation GmbH, Cologne (www.solvent-innovation.com).
6 M. Tinkl, *Chem. Rundschau* **1999**, *2*, 59.
7 a) Covalent Associates (www.covalent-associates.com), b) Sachem Inc. (www.sacheminc.com).
8 Fluka (www.fluka.com)
9 Acros Organics (www. acros.com); ionic liquids are offered in collaboration with QUILL (quill.qub.ac.uk).
10 Wako (www.wako-chem.co.jp).
11 For example: a) D. W. Armstrong, L. He, Y.-S. Liu, *Anal. Chem.* **1999**, *71*, 3873–3876; b) D. W. Armstrong, *Anal. Chem.* **2001**, *73*, 3679–3686; c) F. Endres, *Phys. Chem. Chem. Phys.* **2001**, *3*, 3165; d) C. Ye, W. Liu, Y. Chen, L. Yu, *Chem. Commun.* **2001**, 2244–2245.
12 K. R. Seddon, A. Stark, M. J. Torres, *Pure Appl. Chem.* **2000**, *72*, 2275–2287.
13 J. D. Holbrey, K. R. Seddon, R. Wareing, *Green Chem.* **2001**, *3*, 33–36.
14 B. Ellis, WO 9618459 (to BP Chemicals Limited, UK) **1996** [Chem. Abstr. **1996**, *125*, 114635].
15 P. Bonhôte, A.-P. Dias, N. Papageorgiou, K. Kalyanasundaram, M. Grätzel, *Inorg. Chem.* **1996**, *35*, 1168–1178.
16 H. Waffenschmidt, dissertation, RWTH Aachen, **2000**.
17 For example: Y. Chauvin, L. Mußmann, H. Olivier, *Angew. Chem. Int. Ed. Engl.* **1995**, *34*, 2698–2700.
18 For example: C. J. Mathews, P. J. Smith, T. Welton, A. J. P. White, D. J. Williams, *Organometallics* **2001**, *20(18)*, 3848–3850.
19 W. J. Mehm, J. B. Nold, R. C. Zernach, *Aviat. Space Environ. Med.* **1986**, *57*, 362.
20 J. Pernak, A. Czepukowicz, R. Pozniak, *Ind. Eng. Chem. Res.* **2001**, *40*, 2379–2283.

21 N. Imizuka, T. Nakashima, JP 2002003478 (to Foundation for Scientific Technology Promotion, Japan), **2002** [Chem. Abstr. **2002**, *136*, 85811].
22 M. Schmidt, U. Heider, W. Geissler, N. Ignatyev, V. Hilarius, EP 1162204 (to Merck Patent GmbH, Germany), **2001** [Chem. Abstr. **2001**, *136*, 20157].
23 V. Hilarius, U. Heider, M. Schmidt EP 1160249 (to Merck Patent GmbH, Germany) **2001** [Chem. Abstr. **2001**, *136*, 6139].
24 A. J. Robertson, WO 0187900 (to Cytec Technology Corp., USA), **2001** [Chem. Abstr. **2001**, *135*, 371866].
25 P. Bonnet, E. Lacroix, J.-P. Schirmann, WO 0181353 [to Atofina, France) **2001** [Chem. Abstr. **2001**, *135*, 338483].
26 P. Wasserscheid, W. Keim, C. Bolm, A. Boesmann, WO 0155060 (to Solvent Innovation GmbH, Germany) **2001** [Chem. Abstr. **2001**, *135*, 152789].
27 M. H. Valkenberg, E. Sauvage, C. P. De Castro-Moreira, W. F. Hölderich, WO 0132308 **2001** [Chem. Abstr. **2001**, *134*, 342374].
28 H. Bahrmann, H. Bohnen, DE 19919494 (to Celanese Chemicals Europe GmbH, Germany) **2000** [Chem. Abstr. **2000**, *133*, 321998].
29 A. P. Abbott, D. L. Davies, WO 0056700 (to University of Leicester, UK) **2000** [Chem. Abstr. **2000**, *133*, 269058].
30 W. Keim, W. Korth, P. Wasserscheid WO 0016902 (to BP Chemicals Limited, UK; Akzo Nobel NV; Elementis UK Limited) **2000** [Chem. Abstr. **2000**, *132*, 238691].
31 C. P. M. Lacroix, F. H. M. Dekker, A. G. Talma, J. W. F. Seetz EP 989134 (to Akzo Nobel N. V., Netherlands) **2000** [Chem. Abstr. **2000**, *132*, 238691].

2.3 Synthesis of Task-specific Ionic Liquids

James H. Davis, Jr.

Early studies probing the feasibility of conducting electrophilic reactions in chloroaluminate ionic liquids (ILs) demonstrated that the ionic liquid could act both as solvent and catalyst for the reaction [1–3]. The success of these efforts hinged upon the capacity of the salt itself to manifest the catalytic activity necessary to promote the reaction. Specifically, it was found that the capacity of the liquid to function as an electrophilic catalyst could be adjusted by varying the $Cl^-/AlCl_3$ ratio of the complex anion. Anions that were even marginally rich in $AlCl_3$ catalyzed the reaction.

Despite the utility of chloroaluminate systems as combinations of solvent and catalysts in electrophilic reactions, subsequent research on the development of newer ionic liquid compositions focused largely on the creation of liquid salts that were water-stable [4]. To this end, new ionic liquids that incorporated tetrafluoroborate, hexafluorophosphate, and bis(trifluoromethyl)sulfonamide anions were introduced. While these new anions generally imparted a high degree of water-stability to the ionic liquid, the functional capacity inherent in the IL due to the chloroaluminate anion was lost. Nevertheless, it is these water-stable ionic liquids that have become the de rigueur choices as solvents for contemporary studies of reactions and processes in these media [5].

A 1998 report on the formation of ionic liquids by relatively large, structurally complex ions derived from the antifungal drug miconazole reemphasized the possibilities for the formulation of salts that remain liquids at low temperatures, even with incorporation of functional groups in the ion structure [6]. This prompted the introduction of the concept of "task-specific" ionic liquids [7]. Task-specific ionic liquids (TSILs) may be defined as ionic liquids in which a functional group is covalently tethered to the cation or anion (or both) of the IL. Further, the incorporation of this functionality should imbue the salt with a capacity to behave not only as a reaction medium but also as a reagent or catalyst in some reaction or process. The definition of TSILs also extends to "conventional" ionic liquids to which are added ionic solutes that introduce a functional group into the liquid. Logically, when added to a "conventional" ionic liquid, these solutes become integral elements of the overall "ion soup" and must then be regarded as an element of the ionic liquid as a whole, making the resulting material a TSIL.

Viewed in conjunction with the solid-like, nonvolatile nature of ionic liquids, it is apparent that TSILs can be thought of as liquid versions of solid-supported reagents. Unlike solid-supported reagents, however, TSILs possess the added advantages of kinetic mobility of the grafted functionality and an enormous operational surface area (Figure 2.3-1). It is this combination of features that makes TSILs an aspect of ionic liquids chemistry that is poised for explosive growth.

Conceptually, the functionalized ion of a TSIL can be regarded as possessing two elements. The first element is a core that bears the ionic charge and serves as the locus for the second element, the substituent group. Save for the well documented chloroaluminate ionic liquids, established TSILs are largely species in which the functional group is cation-tethered. Consequently, discussion of TSIL synthesis from this point will stress the synthesis of salts possessing functionalized cations, though the general principles outlined are pertinent to the synthesis of functionalized anions as well.

The incorporation of functionality into an ion slated for use in formulation of an ionic liquid is a usually a multi-step process. Consequently, a number of issues must be considered in planning the synthesis of the ion. The first of these is the choice of the cationic core. The core of a TSIL cation may be as simple as a single

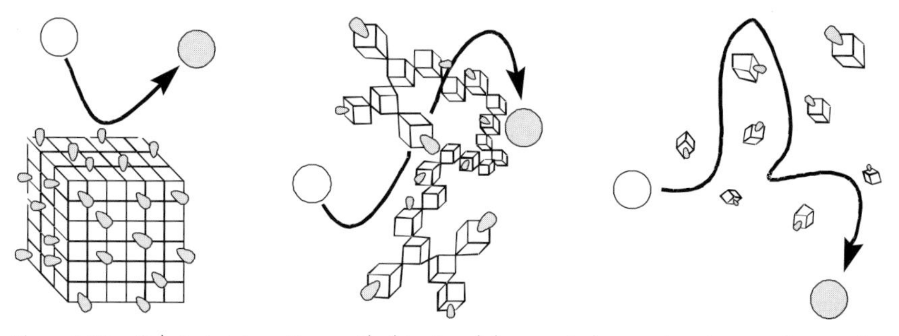

Figure 2.3-1: Substrate interactions with (l to r): solid-supported reagent, polymer gel supported reagent, task-specific ionic liquid.

atom such as N, P, or S, as found in ammonium, phosphonium, or sulfonium ions, respectively. Alternatively, the core of the ion may be (and frequently is) a heterocycle such as imidazole or pyridine. The choices made in this regard will play a large role in determining both the chemical and physical properties of the resulting salt. For example, ionic liquids incorporating phosphonium cations generally exhibit the greatest thermal stability, but also commonly possess melting points higher than those of salts of other cations [8]. Thus, if the desired ionic liquid is to be used in a process intended to be conducted at 0 °C, it may prove especially challenging to build the cation core around a phosphonium ion. If the ionic liquid is to be used in a metal-catalyzed reaction, the use of an imidazolium-based ionic liquid might be critical, especially in view of the possible involvement in some reactions of imidazolylidene carbenes originating with the IL solvent [9].

The second element of general importance in the synthesis of a task-specific ionic liquid is the source of the functional group that is to be incorporated. Key to success here is the identification of a substrate containing two functional groups with different reactivities, one of which allows the attachment of the substrate to the core, and the other of which either is the functional group of interest or is modifiable to the group of interest. Functionalized alkyl halides are commonly used in this capacity, although the triflate esters of functionalized alcohols work as well.

The choice of reaction solvent is also of concern in the synthesis of new TSILs. Toluene and acetonitrile are the most widely used solvents, the choice in any given synthesis being dictated by the relative solubilities of the starting materials and products. The use of volatile organic solvents in the synthesis of ionic liquids is decidedly the least "green" aspect of their chemistry. Notably, recent developments in the area of the solventless synthesis of ionic liquids promise to improve this situation [10].

The choice of the anion ultimately intended to be an element of the ionic liquid is of particular importance. Perhaps more than any other single factor, it appears that the anion of the ionic liquid exercises a significant degree of control over the molecular solvents (water, ether, etc.) with which the IL will form two-phase systems. Nitrate salts, for example, are typically water-miscible while those of hexafluorophosphate are not; those of tetrafluoroborate may or may not be, depending on the nature of the cation. Certain anions such as hexafluorophosphate are subject to hydrolysis at higher temperatures, while those such as bis(trifluoromethane)sulfonamide are not, but are extremely expensive. Additionally, the cation of the salt used to perform any anion metathesis is important. While salts of potassium, sodium, and silver are routinely used for this purpose, the use of ammonium salts in acetone is frequently the most convenient and least expensive approach.

Although the first ionic liquid expressly categorized as being "task-specific" featured the incorporation of the function within the cation core, subsequent research has focused on the incorporation of functionality into a branch appended to the cation [11]. In this fashion, a number of task-specific ionic liquids built up from 1-methyl- and 1-butylimidazole have been prepared, produced by means of the reaction between these imidazoles and haloalkanes also incorporating a desired functional group (Scheme 2.3-1). Bazureau has used this approach to prepare imida-

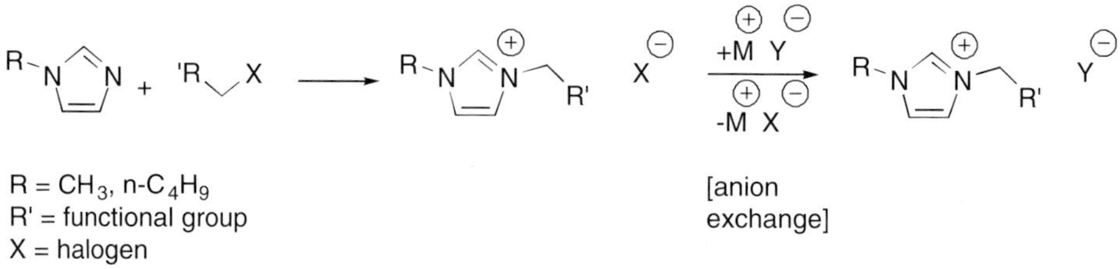

R = CH_3, n-C_4H_9
R' = functional group
X = halogen

[anion exchange]

Scheme 2.3-1: General synthesis of task-specific ionic liquids from 1-alkylimidazoles. The preparation of functionalized pyridinium, phosphonium, etc. cations may be accomplished in like fashion.

zolium ions with appended carboxylic acid groups, which have been used as replacements for solid polymer supports in the heterogeneous-phase synthesis of small organic molecules by means of Knoevenagel and 1, 3-dipolar cycloaddition reactions [12].

Another commercially available imidazole "scaffold" upon which a number of other functionalized cations have been constructed is 1-(3-aminopropyl)imidazole. The appended amino group in this material is a versatile reactive site that lends itself to conversion into a variety of derivative functionalities (Scheme 2.3-2).

Treatment of 1-(3-aminopropyl)imidazole with isocyanates and isothiocyanates gives urea and thiourea derivatives [13]. These elaborated imidazoles can then be quaternized at the ring nitrogen by treatment with alkyl iodides to produce the corresponding N(3)-alkylimidazolium salts. Because of a competing side reaction arising from the interaction of the alkylating species with the urea or thiourea groups, the reactions must be conducted within relatively narrow temperature and solvent parameters (below reflux in acetonitrile). Similar care must be exercised in the synthesis of IL cations with appended acetamide and formamide groups.

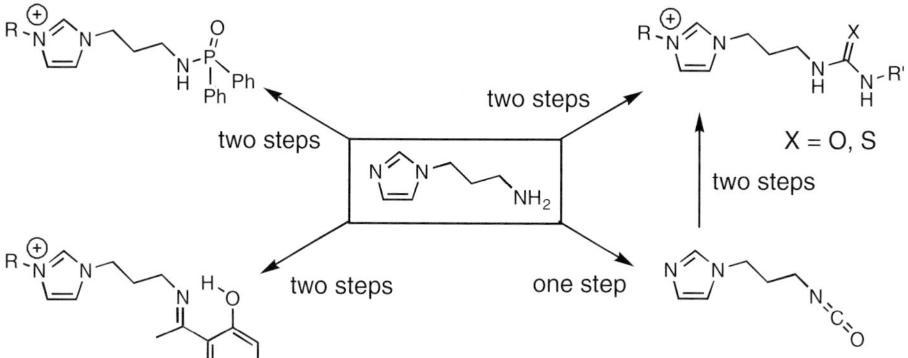

two steps
two steps
X = O, S
two steps
two steps
one step

Scheme 2.3-2: Representative syntheses of task-specific ionic liquids beginning with 1-(3-aminopropyl)imidazole. Step one of the synthetic transformations is the conversion of the primary amine moiety into the functional group of interest. Step two of the process is the quaternization of the imidazole ring by alkylation at N(3).

A variation on this overall synthetic approach allows the formation of related TSIL ureas by initial conversion of 1-(3-aminopropyl)imidazole into an isocyanate, followed by treatment with an amine and alkylating agent. This approach has been used to append both amino acids and nucleic acids onto the imidazolium cation skeleton [14].

The incorporation of more "inorganic" appendages into TSIL cations has also been achieved through the use of 1-(3-aminopropyl)imidazole. Phosphoramide groups are readily synthesized by treatment of phosphorous(V) oxyhalides with primary or secondary amines. In just such an approach, 1-(3-aminopropyl)imidazole was allowed to react with commercially available $(C_6H_5)_2POCl_2$ in dichloromethane. After isolation, the resulting phosphoramide was then quaternized at the imidazole N(3) position by treatment with ethyl iodide (Scheme 2.3-2). The viscous, oily product was found to mix readily with more conventional ionic liquids such as $[HMIM][PF_6]$, yielding a more tractable material. This particular TSIL has been used to extract a number of actinide elements from water. Similarly, the thiourea-appended TSILs discussed earlier have been used for the extraction of Hg^{2+} and Cd^{2+} from IL-immiscible aqueous phases.

While certain TSILs have been developed to pull metals into the IL phase, others have been developed to keep metals in an IL phase. The use of metal complexes dissolved in IL for catalytic reactions has been one of the most fruitful areas of IL research to date. However, these systems still have a tendency to leach dissolved catalyst into the co-solvents used to extract the product of the reaction from the ionic liquid. Consequently, Wasserscheid et al. have pioneered the use of TSILs based upon the dissolution into a "conventional" IL of metal complexes that incorporate charged phosphine ligands in their structures [16–18]. These metal complex ions become an integral part of the ionic medium, and remain there when the reaction products arising from their use are extracted into a co-solvent. Certain of the charged phosphine ions that form the basis of this chemistry (e.g., $P(m\text{-}C_6H_4SO_3^-Na^+)_3$) are commercially available, while others may be prepared by established phosphine synthetic procedures.

An example of this approach to TSIL formulation is the synthesis from 1-vinylimidazole of a series of imidazolium cations with appended tertiary phosphine groups [Scheme 2.3-3]. The resulting phosphines are then coordinated to a Rh(I) organometallic and dissolved in the conventional IL $[BMIM][PF_6]$, the mixture constituting a TSIL. The resulting system is active for the hydroformylation of 1-octene, with no observable leaching of catalyst [17].

Task-specific ionic liquids designed for the binding of metal ions need not be only monodentate in nature. Taking a hint from classical coordination chemistry, a bidentate TSIL has been prepared and used in the extraction of Ni^{2+} from an aqueous solution. This salt is readily prepared in a two-step process. Firstly, 1-(3-aminopropyl)imidazole is condensed with 2-salicylaldehyde under Dean–Stark conditions, giving the corresponding Schiff base. This species is readily alkylated in acetonitrile to form the imidazolium salt. Mixed as the $[PF_6]^-$ salt in a 1:1 (v/v) fashion with $[HMIM][PF_6]$, this new TSIL quickly decolorizes green, aqueous solutions containing Ni^{2+} when it comes into contact with them, the color moving completely into the IL phase (Scheme 2.3-4).

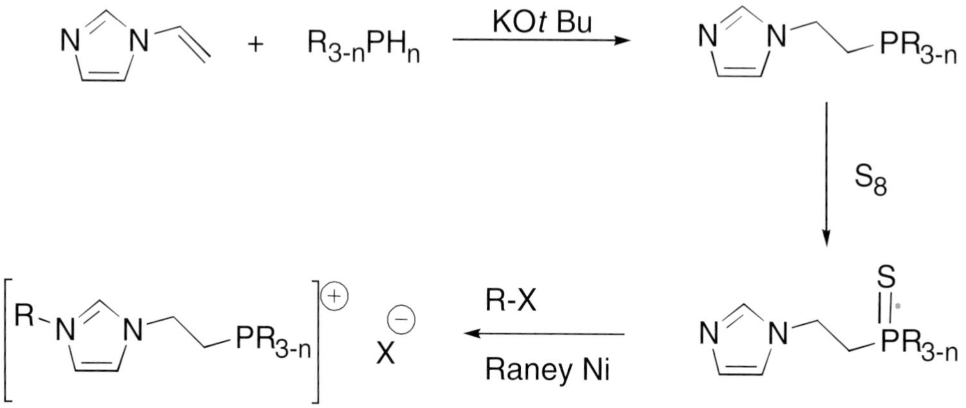

N N + $R_{3-n}PH_n$ KO*t* Bu → N N PR_{3-n}

S_8

$[R-N N PR_{3-n}]^{+}$ X^{-} ← R-X / Raney Ni — N N S PR_{3-n}

Scheme 2.3-3: Synthesis of phosphine-appended imidazolium salts. Combination of these species with the conventional IL $[BMIM]PF_6$ and Rh(I) gives rise to a task-specific ionic liquid active for the hydroformylation of 1-octene.

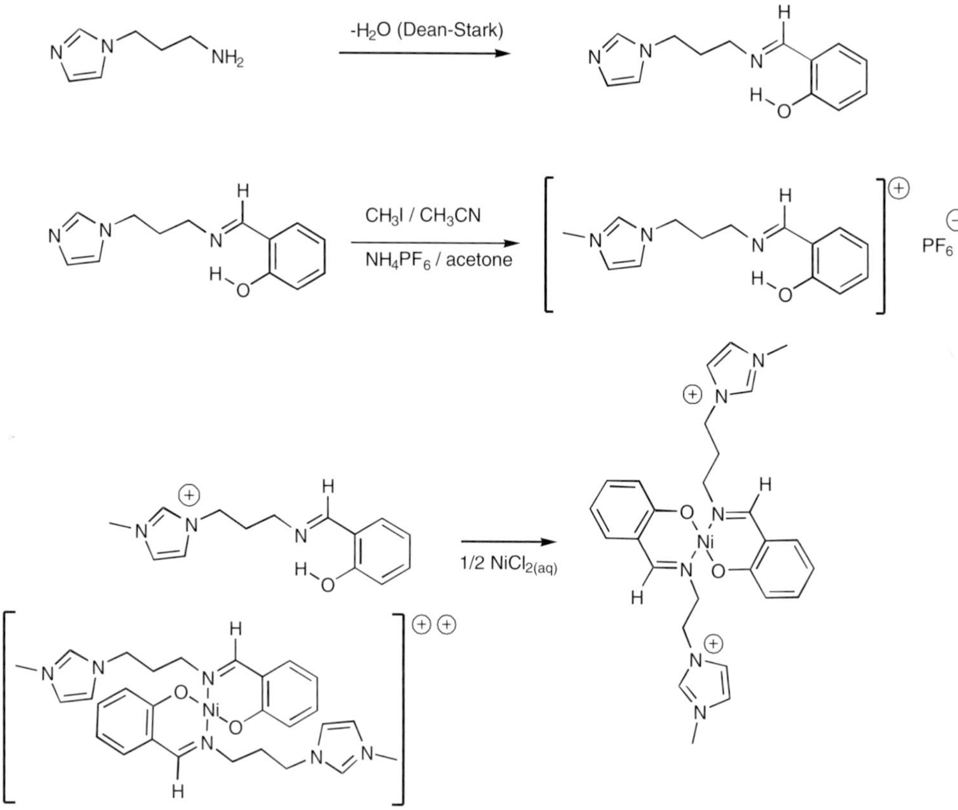

N N NH_2 —$-H_2O$ (Dean-Stark)→ N N N H H O

N N N H H O —CH_3I / CH_3CN, NH_4PF_6 / acetone→ [N N N H H O]$^{+}$ PF_6^{-}

N N N H H O (+) —1/2 $NiCl_{2(aq)}$→ N N (+) O N Ni N O H H N N (+)

[N N N H O Ni O N H N N]$^{+ +}$

Scheme 2.3-4: Synthesis of a chelating task-specific ionic liquid and its utilization for metal extraction.

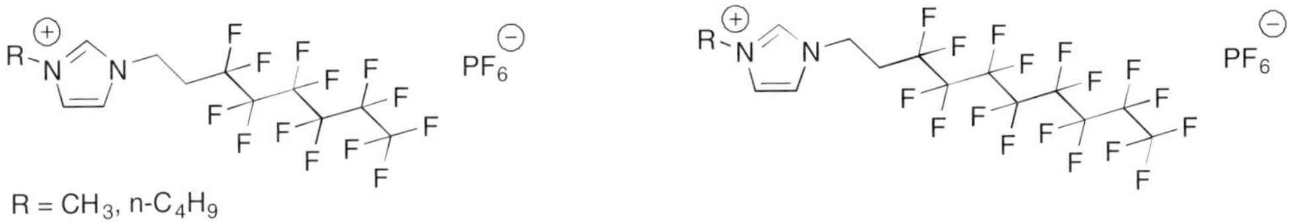

$R = CH_3$, $n\text{-}C_4H_9$

Figure 2.3-5: Imidazolium-based task-specific ionic liquids with cation-appended fluorous tails.

The types of functional groups incorporated into TSILs need not be limited to those based upon nitrogen, oxygen, or phosphorus. Ionic liquids containing imidazolium cations with long, appended fluorous tails have been reported, for example, (Figure 2.3-2). Although these species are not liquids at room temperature (melting in the 60–150 °C range), they nevertheless exhibit interesting chemistry when "alloyed" with conventional ILs. While their solubility in conventional ionic liquids is rather limited (saturation concentrations of about 5 mM), the TSILs apparently form fluorous micelles in the ILs. Thus, when a conventional ionic liquid doped with a fluorous TSIL is mixed with perfluorocarbons, extremely stable emulsions can be formed. These may be of use in the development of two-phase fluorous/ionic liquid reaction systems [19]. As with many other TSILs reported so far, these compounds are prepared by direct treatment of 1-alkylimidazoles with a (polyfluoro)alkyl iodide, followed by anion metathesis.

While the overwhelming bulk of research on and with TSILs has been done on imidazolium-based systems, there is little obvious reason for this to remain the case. Rather, because of the relatively high cost of commercial imidazole starting materials, economic considerations would suggest that future research place more emphasis on the less costly ammonium- and phosphonium-based systems. Indeed, it is notable that a huge number of functionalized phosphonium salts (mostly halides) are in the literature, having been synthesized over the past forty-odd years as Wittig reagent precursors [20]. Many of these compounds will probably be found to give rise to ionic liquids when the cations are paired with an appropriate anion. In similar fashion, large numbers of known natural products are quaternized (or quarternizable) ammonium species that incorporate other, useable functional groups elsewhere in their structure. Many of these molecules are optically active, and could form the basis of entirely new TSIL systems for use in catalysis and chiral separations. Clearly, the potential for development of new TSILs is limited only by the imaginations of the chemists working to do so.

References

1 J. A. Boon, J. A. Levisky, J. L. Pflug, and J. S. Wilkes, *J. Org. Chem.* **1986**, *51*, 480.

2 Y. Chauvin, A. Hirschauer, and H. Olivier, *J. Mol. Catal.* **1994**, *92*, 155.

3 J. A. Boon, S. W. Lander, Jr., J. A. Leviski, J. L. Pflug, L. M. Skrzynecki-Cooke, and J. S. Wilkes, *Proc. Electrochem. Soc. 87-7 (Proc. Jt. Int. Symp. Molten Salts)* **1987**, 979.

4 J. S. Wilkes and M. J. Zaworotko, *J. Chem. Soc. , Chem. Commun.* **1990**, 965.

5 T. Welton, *Chem. Rev.* **1999**, *99*, 2071.

6 K. J. Forrester, T. L. Merrigan, and J. H. Davis, Jr., *Tetrahedron Letters* **1998**, *39*, 8955.

7 A. Wierzbicki, and J. H. Davis, Jr., *Proceedings of the Symposium on Advances in Solvent Selection and Substitution for Extraction,* March 5–9, **2000**, Atlanta, Georgia. AIChE, New York, **2000**.

8 N Karodia, S. Guise, C. Newlands, and J. -A. Andersen, *Chem. Commun.* **1998**, 2341.

9 C. J. Mathews, P. J. Smith, T. Welton, A. J. P. White, and D. J. Williams, *Organometallics* **2001** *20*, 3848.

10 R. S. Varma, and V. V. Namboodiri, *Chem. Commun.* **2001**, 643.

11 K. J. Forrester, and J. H. Davis, Jr., *Tetrahedron Letters* **1999**, *40*, 1621.

12 J. Fraga-Dubreuil and J. P. Bazureau, *Tetrahedron Letters* **2001**, *42*, 6097.

13 A. E. Visser, R. P. Swatloski, W. M. Reichert, R. Mayton, S Sheff, A. Wierzbicki, J. H. Davis, Jr., and R. D. Rogers, *Chemical Communications* **2001**, 135.

14 J. H. Davis, Jr. and E. D. Bates, unpublished results.

15 J. H. Davis, Jr., "Working Salts: Syntheses and Uses of Ionic Liquids Containing Functionalised Ions," *ACS Symp. Ser.* **2002**, *in press.*

16 C. C. Brasse, U. Englert, A. Salzer, H. Waffenschmidt, and P. Wasserscheid, *Organometallics* **2000**, *19*, 3818.

17 K. W. Kottsieper, O. Stelzer, and P. Wasserscheid, *J. Mol. Catal. A.* **2001**, *175*, 285.

18 D. J. Brauer, K. W. Kottsieper, C. Liek, O. Stelzer, H. Waffenschmidt, and P. Wasserscheid, *J. Organomet. Chem.* **2001**, *630*, 177.

19 T. L. Merrigan, E. D. Bates, S. C. Dorman, and J. H. Davis, Jr., *Chemical Communications* **2000**, 2051–2052.

20 A. W. Johnson, *Ylides and Imines of Phosphorous,* Wiley-Interscience, New York, **1993**.

3
Physicochemical Properties of Ionic Liquids

Jennifer L. Anthony, Joan F. Brennecke, John D. Holbrey, Edward J. Maginn, Rob A. Mantz, Robin D. Rogers, Paul C. Trulove, Ann E. Visser, and Tom Welton

3.1
Melting Points and Phase Diagrams

John D. Holbrey and Robin D. Rogers

3.1.1
Introduction

What constitutes an *ionic liquid*, as distinct from a *molten salt*? It is generally accepted that ionic liquids have relatively low melting points, ideally below ambient temperature [1, 2]. The distinction is arbitrarily based on the salt exhibiting liquidity at or below a given temperature, often conveniently taken to be 100 °C. However, it is clear from observation that the principle distinction between the materials of interest today as ionic liquids (and more as specifically room-temperature ionic liquids) and conventional molten salts is that ionic liquids contain organic cations rather than inorganic ones. This allows a convenient differentiation without concern that some 'molten salts' may have lower melting points than some 'ionic liquids'.

It should also be noted that terms such as 'high temperature' and 'low temperature' are also subjective, and depend to a great extent on experimental context. If we exclusively consider ionic liquids to incorporate an organic cation, and further limit the selection of salts to those that are liquid below 100 °C, a large range of materials are still available for consideration.

The utility of ionic liquids can primarily be traced to the pioneering work by Osteryoung et al. [3] on *N*–butylpyridinium-containing, and by Wilkes and Hussey [4–6] on 1-ethyl-3-methylimidazolium-containing ionic liquids for electrochemical studies. These studies have strongly influenced the choice of ionic liquids for subsequent research [7]. The vast majority of work published to date on room-temperature ionic liquids relates to *N*–butylpyridinium and 1-ethyl-3-methylimidazolium [EMIM] tetrachloroaluminate(III) systems. The large variety of available ion combinations (and composition variation in mixtures) gives rise to extensive ranges of salts and salt mixtures, with solidification points ranging from –90 °C upwards.

However, ionic liquids containing other classes of organic cations are known. Room-temperature ionic liquids containing organic cations including quaternary ammonium, phosphonium, pyridinium, and – in particular – imidazolium salts are currently available in combination with a variety of anions (Figure 3.1-1 provides some common examples) and have been studied for applications in electrochemistry [7, 8] and in synthesis [9–11].

It should be emphasized that ionic liquids are simply organic salts that happen to have the characteristic of a low melting point. Many ionic liquids have been widely investigated with regard to applications other than as liquid materials: as electrolytes, phase-transfer reagents [12], surfactants [13], and fungicides and biocides [14, 15], for example.

The wide liquid ranges exhibited by ionic liquids, combined with their low melting points and potential for tailoring size, shape, and functionality, offer opportunities for control over reactivity unobtainable with molecular solvents. It is worth noting that quaternary ammonium, phosphonium, and related salts are being widely reinvestigated [16–18] as the best ionic liquid choices for particular applications, particularly in synthetic chemistry, are reevaluated. Changes in ion types, substitution, and composition produce new ionic liquid systems, each with a unique set of properties that can be explored and hopefully applied to the issues. With the potential large matrix of both anions and cations, it becomes clear that it will be impossible to screen any particular reaction in all the ionic liquids, or even all within a subset containing only a single anion or cation. Work is clearly needed to determine how the properties of ionic liquids vary as functions of anion/cation/substitution patterns etc., and to establish which, if any, properties change in systematic (that is, predictable) ways.

The most simple ionic liquids consist of a single cation and single anion. More complex examples can also be considered, by combining of greater numbers of

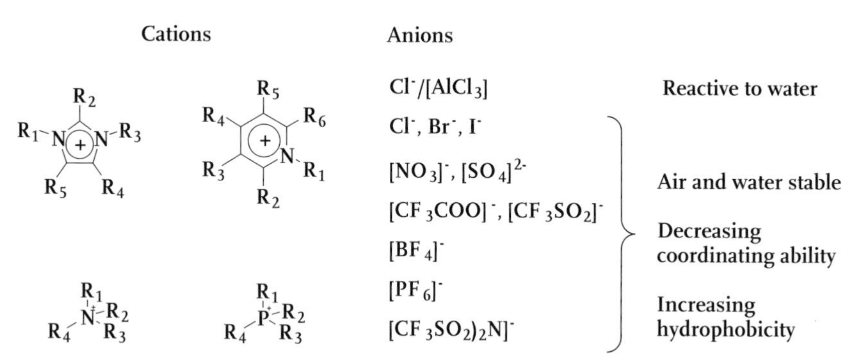

Figure 3.1-1: Examples of common cation and anion pairs used in the formation of ionic liquids, and general progression of changes in IL properties with anion type.

cations and/or anions, or when complex anions are formed as the result of equilibrium processes, as can be seen in Equation (3.1-1):

$$Cl^- + AlCl_3 \rightleftharpoons [AlCl_4]^- + AlCl_3 \rightleftharpoons [Al_2Cl_7]^- \qquad (3.1\text{-}1)$$

Chloroaluminate(III) ionic liquid systems are perhaps the best established and have been most extensively studied in the development of low-melting organic ionic liquids with particular emphasis on electrochemical and electrodeposition applications, transition metal coordination chemistry, and in applications as liquid Lewis acid catalysts in organic synthesis. Variable and tunable acidity, from basic through neutral to acidic, allows for some very subtle changes in transition metal coordination chemistry. The melting points of [EMIM]Cl/$AlCl_3$ mixtures can be as low as –90 °C, and the upper liquid limit almost 300 °C [4, 6].

The following discussion concerns the thermal liquidus ranges available in different ionic liquids, as functions of cation and anion structure and composition. In particular, those structural features of cation and anion that promote these properties (while providing other desirable, and sometimes conflicting characteristics of the liquid, such as low viscosity, chemical stability, etc.) and variations in liquidus ranges and stabilities are the focus of this chapter.

The general observations made regarding structural influences on melting points are transferable across cation type, and apply in each case. The primary focus is on 1-alkyl-3-methylimidazolium cations, coupled with simple organic and inorganic anions. Complex anions, such as mixed X^-/MX_n systems, are mentioned, as are other series of cations (including some examples of tetraalkylammonium salts).

3.1.2 Determination of Liquidus Ranges

The liquidus ranges exhibited by ionic liquids can be much greater that those found in common molecular solvents. Water, for example, has a liquidus range of 100 °C (0 to 100 °C), and dichloromethane has one of 145 °C (–95 to 40 °C). The lower temperature limit, solidification (either as crystallization or glassification), is governed by the structure and interactions between the ions. Ionic liquids, consisting of totally ionized components and displaying relatively weak ion–ion pairing (in comparison to molten salts), have little or no measurable vapor pressure. In contrast to molecular solvents, the upper liquidus limit for ionic liquids is usually that of thermal decomposition rather than vaporization.

3.1.2.1 Melting points

The solid–liquid transition temperatures of ionic liquids can (ideally) be below ambient and as low as –100 °C. The most efficient method for measuring the transition temperatures is differential scanning calorimetry (DSC). Other methods that have been used include cold-stage polarizing microscopy, NMR, and X-ray scattering.

The thermal behavior of many ionic liquids is relatively complex. For a typical IL, cooling from the liquid state causes glass formation at low temperatures; solidification kinetics are slow. On cooling from the liquid, the low-temperature region is not usually bounded by the phase diagram liquidus line, but rather is extended down to a lower temperature limit imposed by the glass transition temperature [19]. This tendency is enhanced by addition of lattice-destabilizing additives, including organic solutes, and by mixing salts. Solidification (glass) temperatures recorded on cooling are not true measures either of heating T_g values or of melting points, and represent kinetic transitions. Thermodynamic data must be collected in heating mode to obtain reproducible results. Hence, in order to obtain reliable transition data, long equilibration times, with small samples that allow rapid cooling, are needed to quench non-equilibrium states in mixtures. Formation of metastable glasses is common in molten salts. In many cases, the glass transition temperatures are low: for 1-alkyl-3-methylimidazolium salts, glass transition temperatures recorded are typically in the region between –70 and –90 °C. In many cases, heating from the glassy state yields an exothermic transition associated with sample crystallization, followed by subsequent melting.

In some cases there is evidence of multiple solid–solid transitions, either crystal–crystal polymorphism (seen for Cl^- salts [20]) or, more often, formation of plastic crystal phases – indicated by solid–solid transitions that consume a large fraction of the enthalpy of melting [21], which also results in low-energy melting transitions. The overall enthalpy of the salt can be dispersed into a large number of fluxional modes (vibration and rotation) of the organic cation, rather than into enthalpy of fusion. Thus, energetically, crystallization is often not overly favored.

3.1.2.2 **Upper limit decomposition temperature**

The upper limit of the liquidus range is usually bounded by the thermal decomposition temperature of the ionic liquid, since most ionic liquids are nonvolatile. In contrast to molten salts, which form tight ion-pairs in the vapor phase, the reduced Coulombic interactions between ions energetically restricts the ion-pair formation required for volatilization of salts, producing low vapor pressures. This gives rise to high upper temperature limits, defined by decomposition of the IL rather than by vaporization. The nature of the ionic liquids, containing organic cations, restricts upper stability temperatures, pyrolysis generally occurs between 350–450 °C if no other lower temperature decomposition pathways are accessible [22]. In most cases, decomposition occurs with complete mass loss and volatilization of the component fragments. Grimmett et al. have studied the decomposition of imidazolium halides [22] and identified the degradation pathway as E2 elimination of the *N*-substituent, essentially the reverse of the S_N2 substitution reaction to form the ionic liquid.

If then decomposition temperatures for a range of ionic liquids with differing anions are compared, the stability of the ionic liquid is inversely proportional to the tendency to form a stable alkyl-X species. As can be seen from TGA decomposition data for a range of $[RMIM]^+$ salts (Figure 3.1-2) collected by anion type, the decomposition temperatures vary with anion type and follow the general stability order, $Cl^- < [BF_4]^- \sim [PF_6]^- < [NTf_2]^-$, so that ionic liquids containing weakly coordinating anions are most stable to high-temperature decomposition [23–27].

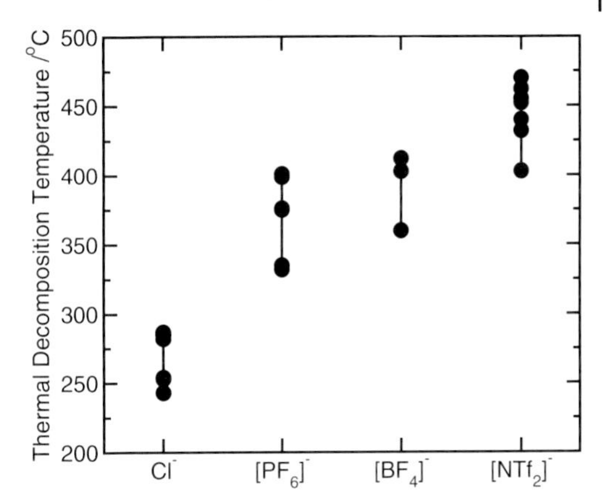

Figure 3.1-2: Thermal decomposition temperature ranges (in °C) for ionic liquids containing 1-alkyl-3-methylimidazolium cations. The thermal stability of the ionic liquids depends on the nucleophilicity of the anion.

Ngo et al. [24] have shown that the thermal decomposition of ionic liquids, measured by TGA, varies depending on the sample pans used. Increased stabilization of up to 50 °C was obtained in some cases on changing from aluminium to alumina sample pans.

3.1.3 Effect of Ion Sizes on Salt Melting Points

It is well known that the characteristic properties of ionic liquids vary with the choice of anion and cation. The structure of an ionic liquid directly impacts upon its properties, in particular the melting point and liquidus ranges. The underlying principles behind the drive to reduce the melting points (and thus operational range limits) for battery electrolytes have been described elsewhere [4]. Exploitation of the changes in these characteristics enables ionic liquids with a wide range of properties to be designed.

The charge, size and distribution of charge on the respective ions are the main factors that influence the melting points of the salts, as generic classes. Within a similar series of salts, however, small changes in the shape of uncharged, covalent regions of the ions can have an important influence on the melting points of the salts.

The dominant force in ionic liquids is Coulombic attraction between ions. The Coulombic attraction term is given by Equation (3.1-2):

$$E_c = MZ^+Z^-/4\pi\varepsilon_0 r \tag{3.1-2}$$

where Z^+ and Z^- are the ion charges, and r is the inter-ion separation.

The overall lattice energies of ionic solids, as treated by the Born–Landé or Kaputinskii equations, thus depends on (i) the product of the net ion charges, (ii) ion–ion separation, and (iii) packing efficiency of the ions (reflected in the Madelung constant, M, in the Coulombic energy term). Thus, low-melting salts should be most preferred when the charges on the ions are ±1 and when the sizes of the ions are large, thus ensuring that the inter-ion separation (*r*) is also large. In addition, large ions permit charge delocalization, further reducing overall charge density.

This can be illustrated for a series of sodium salts, shown in Table 3.1-1, in which the size of the anion is varied.

As the size of the anion increases, the melting point of the salt decreases, reflecting the weaker Coulombic interactions in the crystal lattice. With increasing thermochemical radius of the anion, from Cl^- to $[BF_4]^-$ to $[PF_6]^-$ to $[AlCl_4]^-$, the melting points of the sodium salts decrease from 801 to 185 °C. The results from the sodium salts can be roughly extrapolated to room temperature, and indicate that in order to obtain a salt that would melt at room temperature, the anion would be required to have a radius in excess of about 3.4–4 Å [28]. Large anions are, in general, non-spherical and have significant associated covalency. A similar increase is observed with increasing cation size, on moving down a group in the periodic table, for example. Lithium salts tend to be higher melting than their sodium or cesium analogues. If the charge on the ion can also be delocalized or if the charge-bearing regions can be effectively isolated in the interior of the ionic moiety, then Coulombic terms are further reduced.

Reduction in melting points can, simplistically, be achieved by increasing the size of the anion, or that of the cation. Ionic liquids contain organic cations that are large in comparison to the thermodynamic radii of inorganic cations. This results in significant reductions in the melting points for the organic salts, as illustrated by the $[EMIM]^+$ examples in Table 3.1-1. The Coulombic attraction terms for ionic liquids are of comparable magnitude to the intermolecular interactions in molecular liquids.

3.1.3.1 Anion size

As shown above, increases in anion size give rise to reductions in the melting points of salts through reduction of the Coulombic attraction contributions to the lattice energy of the crystal and increasing covalency of the ions. In ionic liquids generally, increasing anion size results in lower melting points, as can be seen for a selection of [EMIM]X salts in Table 3.1-2.

Ionic liquids containing carborane anions, described by Reed et al. [35], contain large, near-spherical anions with highly delocalized charge distributions. These ionic liquids have low melting points relative to the corresponding lithium and ammonium salts, but these melting points are higher than might be anticipated

X^-	*r.*	*Melting point*	
		NaX	*[EMIM]X*
Cl^-	1.7	801	87
$[BF_4]^-$	2.2	384	6
$[PF_6]^-$	2.4	> 200	60
$[AlCl_4]^-$	2.8	185	7

Table 3.1-1: Melting points (°C) and thermochemical radii of the anions (Å) for Na^+ and $[EMIM]^+$ salts. The ionic radii of the cations are 1.2 Å (Na^+) and 2 x 2.7 Å ($[EMIM]^+$, non-spherical).

Table 3.1-2: [EMIM]X salts and melting points, illustrating anion effects.

Anion [X]	*Melting point (°C)*	*Reference*
Cl^-	87	4
Br^-	81	22
I^-	79–81	22
$[BF_4]^-$	15	6, 29, 26
$[AlCl_4]^-$	7	4, 6
$[GaCl_4]^-$	47	30
$[AuCl_3]^-$	58	31
$[PF_6]^-$	62	32
$[AsF_6]^-$	53	
$[NO_3]^-$	38	33
$[NO_2]^-$	55	33
$[CH_3CO_2]^-$	ca. 45	33
$[SO_4]\cdot 2H_2O^{2-}$	70	33
$[CF_3SO_3]^-$	−9	25
$[CF_3CO_2]^-$	−14	25
$[N(SO_2CF_3)_2]^-$	−3	25
$[N(CN)_2]^-$	−21	34
$[CB_{11}H_{12}]^-$	122	35
$[CB_{11}H_6Cl_6]^-$	114	35
$[CH_3CB_{11}H_{11}]^-$	59	35
$[C_2H_5CB_{11}H_{11}]^-$	64	35

from a simplistic model based solely on comparison of the size of the anions. Similarly, it should be noted that [EMIM][PF_6] [33] appears to have a higher melting point than would be anticipated. In other large anions – tetraphenylborate, for example [39] – additional attractive interactions such as aromatic π–π stacking can give rise to increased melting points.

Anion and cation contributions cannot be taken in isolation; induced dipoles can increase melting points through hydrogen bonding interactions, seen in the crystal structures of [EMIM]X (X = Cl, Br, I) salts [36] and absent from the structure of [EMIM][PF_6] [32]. In addition to increasing ion–ion separations, larger (and in general, more complex) anions can allow greater charge delocalization. For salts with the anion $[(CF_3SO_2)_2N]^-$ [25, 37, 38], this is effected by $-SO_2CF_3$ groups, which effectively provide a steric block, isolating the delocalized $[S–N–S]^-$ charged region in the center of the anion.

3.1.3.2 **Mixtures of anions**

Complex anions, formed when halide salts are combined with Lewis acids (e.g., $AlCl_3$) produce ionic liquids with reduced melting points through the formation of eutectic compositions [8]. The molar ratio of the two reactants can influence the melting point of the resultant mixed salt system through speciation equilibria. For [EMIM]Cl/$AlCl_3$, an apparently simple phase diagram for a binary mixture forming a 1:1 compound and exhibiting two eutectic minima is formed, with a characteristic W-shape to the melting point transition [4, 6] (Figure 3.1-3). Polyanionic species

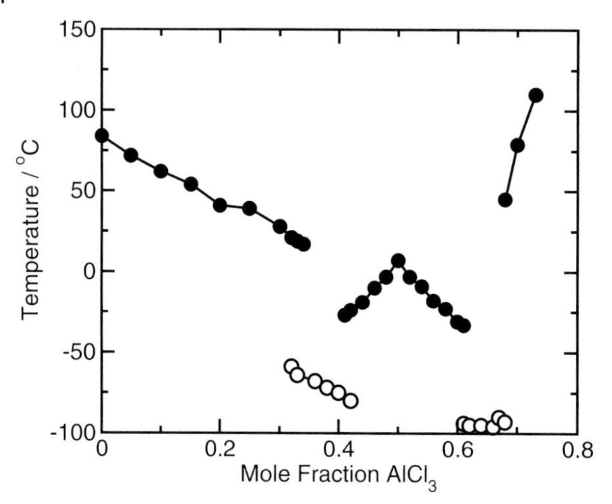

Figure 3.1-3: Phase diagram for [EMIM]Cl/$AlCl_3$: (○) melting and freezing points; (•) glass transition points.

including $[Al_2Cl_7]^-$ and $[Al_3Cl_{10}]^-$ have been identified. Only at 50 % composition is the compound [EMIM][$AlCl_4$] formed.

The two eutectic minima, corresponding to 1:2 and 2:1 compositions, result in liquids with very low solidification temperatures; the glass transition temperature for [EMIM]Cl/$AlCl_3$ (0.33:0.66) is –96 °C. Similar compositional variation should be anticipated in the phase diagrams of other metal halide ionic liquids. The phase behavior of [EMIM]Cl/CuCl [40, 41] and [EMIM]Cl/$FeCl_3$ [42], for example, is similar. For the [EMIM]Cl/CuCl system, the lower liquidus temperatures in the basic and acidic regions are –40 and –65 °C, respectively. More complex phase diagrams have been reported; one example is [HPy]Cl/$ZnCl_2$ [19], in which a range of multinuclear zinc halide anions can exist.

The presence of several anions in these ionic liquids has the effect of significantly decreasing the melting point. Considering that the formation of eutectic mixtures of molten salts is widely used to obtain lower melting points, it is surprising that little effort has been put into identifying the effects of mixtures of cations or anions on the physical properties of other ionic liquids [17].

3.1.3.3 **Cation size**

The sizes and shapes of cations in ionic liquids are important in determining the melting points of the salts. On a simple basis, large ions tend to produce reductions in the melting points. Tetraalkylammonium and phosphonium salts are examples of salts containing large cations with delocalized, or more correctly alkyl-shielded charge. The cation radius, *r*, is large, and the salts correspondingly display melting points lower than those of their Group 1 analogues. The reduction in melting point over a series of tetraalkylammonium bromide salts with increasing cation size, for example, is shown in Table 3.1-3.

Additionally, the salts contain linear alkyl substituents, which have many rotational degrees of freedom, allowing the alkyl chains to 'melt' at temperatures below

Table 3.1-3: Changes in melting points for symmetric tetraalkylammonium bromide salts with increasing size of alkyl substituents.

Cation	*Melting point (°C)*
$[NMe_4]$	> 300
$[NEt_4]$	284
$[NBu_4]$	124–128
$[NHex_4]$	99–100
$[NOct_4]$	95–98

the melting point and resulting in solid–solid polymorphic transitions. For example, tetrabutylammonium thiocyanate ionic liquid (mp 49.5 °C) has a number of solid–solid transitions associated with changes in alkyl chain conformation [43], which change the density of the solid below the melting point.

In these salts, the interactions in the liquid phase are Coulombic terms also present in the ionic crystalline phase. The ionic attractions are relatively small in comparison to those in analogous inorganic salts, and they are dispersed by the large, hydrocarbon-rich cations. The liquid–solid transition is largely caused by a catastrophic change in the rotational and vibrational freedom of these ions. Since the charge terms are dominant, substituents only appear to contribute to symmetry (rotational freedom) and to dispersal of the charge–charge interactions (large charge–charge separation and distortion from cubic salt-like packing), until sufficient hydrocarbon groups are introduced that van der Waals interactions start to contribute to the crystal ordering.

3.1.3.4 Cation symmetry

Melting points of organic salts have an important relationship to the symmetry of the ions: increasing symmetry in the ions increases melting points, by permitting more efficient ion–ion packing in the crystal cell. Conversely, a reduction in the symmetry of the cations causes a distortion from ideal close-packing of the ionic charges in the solid state lattice, a reduction in the lattice energy, and depression of melting points. A change from spherical or high-symmetry ions such as Na^+ or $[NMe_4]^+$ to lower-symmetry ions such as imidazolium cations distorts the Coulombic charge distribution. In addition, cations such as the imidazolium cations contain alkyl groups that do not participate in charge delocalization.

Reduction in cation symmetry (ideally to C1) lowers the freezing point and markedly expands the range of room-temperature liquid salts. Table 3.1-4 shows the effect of symmetry for a series of $[NR_4]X$ salts, in which all the cations contain 20 carbon atoms in the alkyl substituents [44].

Room-temperature liquids are obtained for the salts $[N_{6554}]Br$, $[N_{10,811}]Br$, $[N_{6644}]Br$, $[N_{8543}][ClO_4]$, $[N_{10,811}][ClO_4]$, $[N_{9551}][ClO_4]$, and $[N_{8651}][ClO_4]$, whereas the salts containing cations with high symmetry have much higher melting points. It can be seen that the melting points of these isomeric salts vary by over 200 °C depending on the symmetry of the cation.

Cation ($[N_{nmop}]^+$)	Br^-	$[ClO_4]^-$	$[BPh_4]^-$
5555	101.3	117.7	203.3
6554	83.4		
6644	83.0		
8444	67.3		
8543	*l*	109.5	
6662	46.5		
7733	*l*	45–58	138.8
8663	*l*	*l*	110.2
7751	*l*	104	
8651	*l*		
9551	*l*		
9641	*l*	*l*	
11333	67–68	65.5	
11432	*l*		
8822	62		
9821	*l*		
13331	71–72	52–53	
9911	*l*		
10811	*l*	*l*	
14222	170	152	
16211	180	155	
17111	210	205	

Table 3.1-4: Effects of cation symmetry on the melting points of isomeric tetraalkylammonium salts. In each case the cation (designated $[N_{nmop}]^+$) has four linear alkyl substituents, together containing a total of 20 carbons. Salts that are liquid at room temperature are indicated by *l*.

3.1.4.1 Imidazolium salts

Changes in the ring substitution patterns can have significant effects on the melting points of imidazolium salts, beyond those anticipated by simple changes in symmetry or H-bonding interactions (i.e., substitution at the C(2,4,5)-positions on an imidazolium ring affects packing and space-filling of the imidazolium cations). Substitution at the C(2)-position of the imidazolium ring, for example, increases the melting points of the salts. This is not necessarily an obvious or straightforward result, but may be caused by changes in the cation structure that can induce aromatic stacking or methyl–π interactions between cations. The introduction of other functionalities around the periphery of the ions can also change the interactions between ions. In most cases, additional functions, such as ether groups, increase the number of interactions, and thus increase melting points.

3.1.4.2 Imidazolium substituent alkyl chain length

The data in Table 3.1-4 illustrate the changes in melting points that can be achieved by changing the symmetry of the cation. $[RMIM]^+$ salts with asymmetric *N*-substitution have no rotation or reflection symmetry operations. A change in the alkyl chain substitution on one of the ring heteroatoms does not change the symmetry of the cation. However, manipulation of the alkyl chain can produce major changes in the melting points, and also in the tendency of the ionic liquids to form glasses rather than crystalline solids on cooling, by changing the efficiency of ion packing.

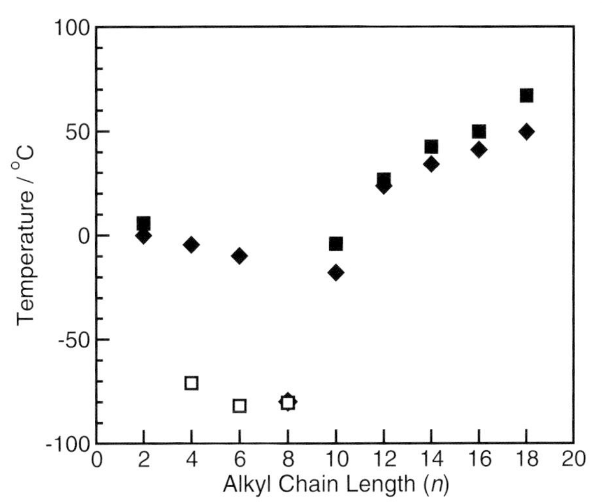

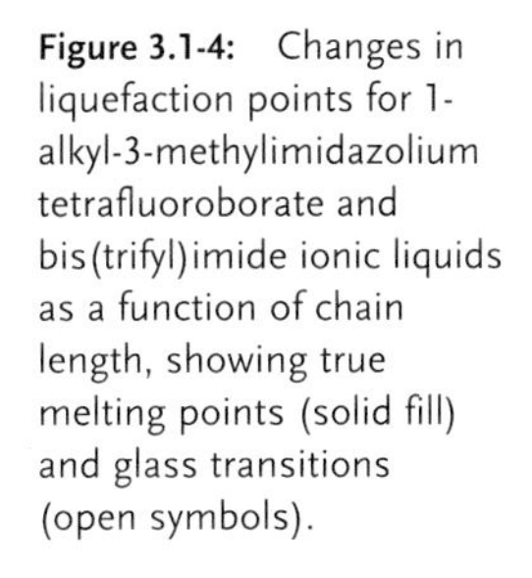
Figure 3.1-4: Changes in liquefaction points for 1-alkyl-3-methylimidazolium tetrafluoroborate and bis(trifyl)imide ionic liquids as a function of chain length, showing true melting points (solid fill) and glass transitions (open symbols).

Figure 3.1-4 shows the changes in liquefaction points (either melting points or glass transitions) for a series of 1-alkyl-3-methylimidazolium tetrafluoroborate [26] and bis(trifyl)imide [45] ionic liquids with changing length of the linear alkyl-substituent on the *N*(3)-position.

It is immediately noticeable that an increase in the substituent length initially reduces the melting point of the IL, with the major trend towards glass formation on cooling for $n = 4–10$. On extending the alkyl chain lengths beyond a certain point (around 8–10 carbons for alkyl-methylimidazolium salts), the melting points of the salts start to increase again with increasing chain length, as van der Waals interactions between the long hydrocarbon chains contribute to local structure by induction of microphase separation between the covalent, hydrophobic alkyl chains and charged ionic regions of the molecules.

Consideration of the changes in molecular structure and the underlying effects that this will have in both the liquid and crystal phases helps to explain changes in melting points with substitution. The crystalline phases of the ionic liquid are dominated by Coulombic ion–ion interactions, comparable to those in typical salt crystals, although since the ions are larger, the Coulombic interactions are weaker (decreasing with r^2). An effect of this is that many organic salts (including 'ionic liquids') crystallize with simple, salt-like packing of the anions and cations.

The reported transition temperatures for a range of $[RMIM]^+$ ionic liquids [6, 23–26, 46] are shown in Figure 3.1-5, with varying anion and alkyl chain substituent length.

The melting transition varies by up to 100 °C with changes in anion (common cation) and almost 250 °C with changes in cation. The phase diagram shows a number of salts that are liquid at or substantially below room temperature. A steady decrease in melting point with increasing chain length, up to a minimum around $n = 6–8$, is followed by a progressive increase in melting points with increasing chain length for the longer chain homologues, which form ordered, lamellar ionic

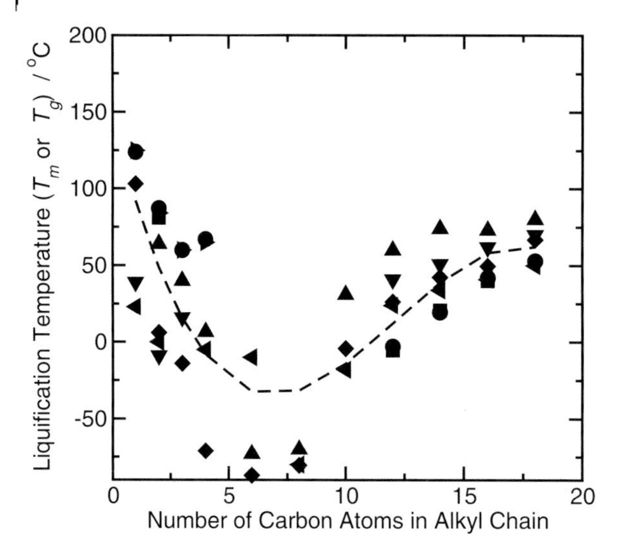

Figure 3.1-5: Variation in melting point with alkyl chain length for ionic liquids containing 1-alkyl-3-methylimidazolium cations: chloride (circle), bromide (square), tetrafluoroborate (diamond), hexafluorophosphate (triangle), bis(trifyl)imide (left triangle), triflate (down triangle), and tetrachloroaluminate (right triangle). The data show the general trend for decreasing melting point with increasing *n* up to *n* = 8, followed by an increase in melting point with *n*.

liquid crystalline phases on melting. Most of the salts that are ionic liquids at room temperature form glasses rather than crystalline phases on cooling. The glass transition temperatures all tend towards –90 °C, which is comparable with the T_g observed for the [EMIM]Cl/$AlCl_3$ (0.33/0.66) ionic liquid system.

Increased asymmetric substitution on 1-alkyl-3-methylimidazolium salts increases the asymmetric disruption and distortion of the Coulombic packing of ions, giving rise to substantial decreases in the melting point as the efficiency of packing and crystallization is reduced. This results in (i) melting point reduction and (ii) a pronounced tendency towards glass formation rather than crystallization on cooling, on extending the alkyl substituents. This is indicative of inefficient packing within the crystal structures, which is a function of the low-symmetry cations employed. Increasing alkyl chain substitution can also introduce other rheological changes in the ionic liquids, including increased viscosity, reduced density, and increased lipophilicity, which must also be taken into account.

The incorporation of alkyl substituents of increasing chain length in a non-symmetrical arrangement on the ions results in the introduction of 'bulk' into the crystalline lattice, which disrupts the attractive charge–charge lattice. Relatively short alkyl chains act as buffers in this manner, and do not pack well into the available space in the crystalline lattice; high rotational freedom results in low occupation densities over a relatively large volume of space. This free rotation volume probably gives rise to the 'void-space' considered by Brennecke [48] to explain the extraordinary propensity for *sc*-CO_2 to dissolve in ILs without substantially changing the volume of the liquid phase.

3.1.4.3 **Branching**

Table 3.1-5 provides data for a series of ionic liquids in which the only difference is the degree of branching within the alkyl chain at the imidazolium ring 3-position [24, 47].

Table 3.1-5: Melting points and heats of fusion for isomeric $[BMIM][PF_6]$ and $[PMIM][PF_6]$ ionic liquids, showing melting point and crystal stability increasing with the degree of branching in the alkyl substituent.

N(1)-Substitution	*Melting point (°C)*	ΔH_{fusion} *(kJ mol*$^{-1}$*)*
n-Butyl	6.4	31
sec-Butyl	83.3	72
tert-Butyl	159.7	83
n-Propyl	40	
Isopropyl	102	

The melting points and enthalpies of the three isomeric 1-butyl-3-methylimidazolium hexafluorophosphate salts $[BMIM][PF_6]$ [47] increase with the degree of chain branching, reflecting the changes in efficiency of the crystal packing as free-rotation volume decreases and atom density is increased. The same effects are also observed for the two isomers of 1-propyl-3-methylimidazolium hexafluorophosphate $[PMIM][PF_6]$ [24].

3.1.5 Summary

Liquid structure is defined by short-range ordering, with long-range disorder. The short-range (near neighbor) structuring of the liquids is a combination of dominant Coulombic charge–charge attractions balanced against rotational and vibrational freedom of the ions. Changes in the degrees of freedom and increases in nonparticipating portions of the cation that do not contribute to Coulombic stabilization of the crystal in salt-like lattices result in decreases in melting points and heats of formation. At longer chain lengths, amphiphilic nature is manifested, resulting in hydrophobic van der Waals contributions and formation of bilayer lattices.

The effects of cation symmetry are relatively clear: the melting points of symmetrically substituted 1,3-dialkylimidazolium cations are higher than those of the unsymmetrical cations, and continue to decrease with increasing alkyl substitution up to a critical point around 8–10 carbons, then increase with increasing additional substitution. Both alkyl substitution and ion asymmetry interfere with efficient packing of ions into a crystalline lattice based on Coulombic attractions. However, there appears to be no simple correlation with hydrogen-bonding ability. The absence of strong H-bonding is certainly a major contributor to low melting points, but ionic liquids containing strongly H-bonding anions (such as $[CH_3COO]^-$) have melting points similar to those of ionic liquids incorporating anions that are highly delocalized and unable to H-bond (such as $[(CF_3SO_2)_2N]^-$). Similarly, C(2)-substitution might be expected to suppress melting points, by suppressing hydrogen-bonding. This does not appear to be the case, with significant increases in melting points occurring with C(2)-substitution. This implies that the effects of van der Waals interactions through the methyl group, or methyl-π interactions, etc., are more important than the electrostatic interactions through the C(2)-hydrogen.

Hagiwara and Ito [49] and Bonhôte et al. [25] have indicated that there appears to be no overall correlation, based on non-systematic changes in cation substitution and anion types, between the composition of an ionic liquid and its melting point. Ngo et al. [24] indicate that the melting points decrease with incorporation of larger, more asymmetrical cations. Ionic liquids containing highly fluorinated anions – $[BF_4]^-$, $[PF_6]^-$, $[(CF_3SO_2)_2N]^-$, $[CF_3COO]^-$, etc. – are generally liquid down to low temperatures, forming glasses on solidification (slow crystallization prior to melting is often observed on heating). However, Katritzky et al. [50, 51] have started to show that the physical properties of imidazolium and pyridinium salts (including ionic liquids) can be modeled by QSPR and CODESSA computational methods, allowing melting points to be predicted with reasonable confidence.

It is important that the forces and interactions that govern the melting points of ionic liquids are not considered in isolation; these interactions also control the dissolution and solubility of other components in the ionic liquids. If, for example, there is a requirement for an ionic liquid to have strong H-bond accepting character (in the anion), then it should be anticipated that this will also give rise to hydrogen-bonding interactions between ions, resulting in greater attractive forces and elevated melting points.

References

1 Seddon, K. R., *J. Chem. Tech. Biotech.* **1997**, *68*, 351.
2 Rogers, R. D., *Green Chem.* **2000**, *5*, G94.
3 Osteryoung, R. A., Gale, R. J., Robinson, J., Linga, H., Cheek, G., *J. Electrochem. Soc.* **1981**, *128*, 79.
4 Wilkes, J. S., Levisky, J. A., Wilson, R. A., Hussey, C. L., *Inorg. Chem.* **1982**, *21*, 1263.
5 Fannin, A. A., King, L. A., Levisky, J. A., Wilkes, J. S. *J. Phys. Chem.* **1984**, *88*, 2609.
6 Fannin, A. A., Floreani, D. A., King, L. A., Landers, J. S., Piersma, B. J., Stech, D. J., Vaughn, R. L., Wilkes, J. S., Williams, J. L., *J. Phys. Chem.* **1984**, *88*, 2614.
7 Hussey, C. L. *Adv. Molten Salt Chem.* **1983**, *5*, 185.
8 Cooper, E. I., Sullivan, E. S. M., *Eighth International Symposium on Molten Salts*, Vol. 92-16; The Electrochemical Society: Pennington, NJ, **1992**, 386.
9 Welton, T., *Chem. Rev.* **1999**, *99*, 2071.
10 Holbrey, J. D., Seddon, K. R., *Clean Prod. Proc.* **1999**, *1*, 233.
11 Keim, W., Wasserscheid, P., *Angew. Chem. Int. Ed.* **2000**, *39*, 3772.
12 Albanese, D., Landini, D., Maia, A., Penso, M., *J. Mol. Catal. A* **1999**, *150*, 113.
13 Blackmore, E. S., Tiddy, G. J. T., *J. Chem. Soc., Faraday Trans. 2* **1988**, *84*, 1115.
14 Pernak, J., Krysinski, J., Skrzypczak, A., *Pharmazie* **1985**, *40*, 570.
15 Pernak, J., Czepukowicz, A., Pozniak, R., *Ind. Eng. Chem. Res.* **2001**, *40*, 2379.
16 Sun, J., Forsyth, M., MacFarlane, D. R., *J. Phys. Chem. B* **1998**, *102*, 8858.
17 MacFarlane, D. R., Meakin, P., Sun, J., Amini, N., Forsyth, M., *J. Phys. Chem. B* **1999**, *103*, 4164.
18 Abdallah, D. J., Robertson, A., Hsu, H.-F., Weiss, R. G., *J. Amer. Chem. Soc.* **2000**, *122*, 3053.

19 Easteal, E. J., Angell, C. A., *J. Phys. Chem.* **1970**, *74*, 3987.
20 Holbrey, J. D., Seddon, K. R., Rogers, R. D., unpublished data.
21 Hardacre, C., Holbrey, J. D., McCormac, P. B., McMath, S. E. J., Nieuwenhuyzen, M., Seddon, K. R., *J. Mat. Chem.* **2001**, *11*, 346.
22 Chan, B. K. M., Chang, N.-H., Grimmett, M. R., *Aust. J. Chem.* **1977**, *30*, 2005.
23 Huddleston, J. G., Visser, A. E., Reichert, W. M., Willauer, H. D., Broker, G. A., Rogers, R. D., *Green Chem.* **2001**, 156.
24 Ngo, H. L., LeCompte, K., Hargens, L., McEwan, A. B., *Thermochim. Acta.* **2000**, *357-358*, 97.
25 Bonhôte, P., Dias, A. P., Papageorgiou, N., Kalyanasundaram, K., Grätzel, M., *Inorg. Chem.* **1996**, *35*, 1168.
26 Holbrey, J. D., Seddon, K. R., *J. Chem. Soc., Dalton Trans.* **1999**, 2133.
27 Takahashi, S., Koura, N., Kohara, S., Saboungi, M.-L., Curtiss, L. A., *Plasmas & Ions* **1999**, *2*, 91.
28 Rooney, D. W., Seddon, K. R., 'Ionic Liquids' in *Handbook of Solvents* (Wypych, G., ed.), ChemTec, Toronto, **2001**, p. 1459.
29 Fuller, J., Carlin, R. T., Osteryoung, R. A., *J. Electrochem. Soc.* **1997**, *144*, 3881.
30 Wicelinski, S. P., Gale, R. J., Wilkes, J. S., *J. Electrochem. Soc.* **1987**, *134*, 262.
31 Hasan, M., Kozhevnikov, I. V., Siddiqui, M. R. H., Steiner, A., Winterton, N., *Inorg. Chem.* **1999**, *38*, 5637.
32 Fuller, J., Carlin, R. T., De Long, H. C., Haworth, D., *Chem. Commun.* **1994**, 229.
33 Wilkes, J. S., Zaworotko, M. J., *J. Chem. Soc., Chem. Commun.* **1992**, 965.
34 MacFarlane, D. R., Golding, J., Forsyth, S., Forsyth, M., Deacon, G. B., *Chem. Commun.* **2001**, 1430.
35 Larsen, A. S., Holbrey, J. D., Tham, F. S., Reed, C. A., *J. Amer. Chem. Soc.* **2000**, *122*, 7264.
36 Elaiwi, A., Hitchcock, P. B., Seddon, K. R., Srinivasan, N., Tan, Y. M., Welton, T., Zora, J. A., *J. Chem. Soc., Dalton Trans.* **1995**, 3467.
37 Golding, J. J., MacFarlane, D. R., Spicca, L., Forsyth, M., Skelton, B. W., White, A. H., *Chem. Commun.* **1998**, 1593.
38 Noda, A., Hayamizu, K., Watanabe, M., *J. Phys. Chem. B* **2001**, *105*, 4603.
39 Suarez, P. A. Z., Dupont, J., Souza, R. F., Burrow, R. A., Kintzinger, J.-P., *Chem. Eur. J.* **2000**, 2377.
40 Bolkan, S. A., Yoke, J. T., *J. Chem. Eng. Data.* **1986**, *31*, 194.
41 Bolkan, S. A., Yoke, J. T. *Inorg. Chem.*, **1986**, *25*, 3587.
42 Sitze, M. S., Schreiter, E. R., Patterson, E. V., Freeman, R. G., *Inorg. Chem.* **2001**, *40*, 2298.
43 Coker, T. G., Wunderlich, B., Janz, G. J., *Trans. Faraday Soc.* **1969**, *65*, 3361.
44 Gordon, J. E., SubbaRao, G. N., *J. Amer. Chem. Soc.* **1978**, *100*, 7445.
45 Holbrey, J. D., Johnston, S., Rogers, G., Rooney, D. W., Seddon, K. R., unpublished data.
46 Gordon, C. M., Holbrey, J. D., Kennedy, A., Seddon, K. R., *J. Mater. Chem.* **1998**, *8*, 2627.
47 Carmichael, A. J., Hardacre, C., Holbrey, J. D., Nieuwenhuyzen, M., Seddon, K. R., *Eleventh International Symposium on Molten Salts*, Vol. 99-41 (Truelove, P. C., De Long, H. C., Stafford, G. R., Deki, S. eds.), The Electrochemical Society, Pennington, NJ, **1999**, 209.
48 Blanchard, L. A., Gu, Z., Brennecke, J. F., *J. Phys. Chem. B* **2001**, 2437.
49 Hagiwara, R., Ito, Y., *J. Fluor. Chem.* **2000**, *105*, 221.
50 Katritzky, A. R., Lomaka, A., Petrukhin, R., Jain, R., Karelson, M., Visser, A. E., Rogers, R. D., *J. Chem. Inf. Comp. Sci.* **2002**, *42*, 71.
51 Katritzky, A. R., Jain, R., Lomaka, A., Petrukhin, R., Karelson, M., Visser, A. E., Rogers, R. D., *J. Chem. Inf. Comp. Sci.* **2002**, *42*, 225.

3.2
Viscosity and Density of Ionic Liquids

Robert A. Mantz and Paul C. Trulove

3.2.1
Viscosity of Ionic Liquids

The viscosity of a fluid arises from the internal friction of the fluid, and it manifests itself externally as the resistance of the fluid to flow. With respect to viscosity there are two broad classes of fluids: Newtonian and non-Newtonian. Newtonian fluids have a constant viscosity regardless of strain rate. Low-molecular-weight pure liquids are examples of Newtonian fluids. Non-Newtonian fluids do not have a constant viscosity and will either thicken or thin when strain is applied. Polymers, colloidal suspensions, and emulsions are examples of non-Newtonian fluids [1]. To date, researchers have treated ionic liquids as Newtonian fluids, and no data indicating that there are non-Newtonian ionic liquids have so far been published. However, no research effort has yet been specifically directed towards investigation of potential non-Newtonian behavior in these systems.

Experimentally determined viscosities are generally reported either as absolute viscosity (η) or as kinematic viscosity (υ). Kinematic viscosity is simply the absolute viscosity normalized by the density of the fluid. The relationship between absolute viscosity (η), density (ρ), and kinematic viscosity (υ) is given by Equation 3.2-1.

$$\eta / \rho = \upsilon \qquad (3.2\text{-}1)$$

The unit of absolute viscosity is the Poise (P, g $cm^{-1}s^{-1}$ or mPa s), while the unit for kinematic viscosity is the Stoke (St, cm^2s^{-1}). Because of the large size of these viscosity units, absolute viscosities for ionic liquids are usually reported in centipoises (cP) and kinematic viscosities reported in centistokes (cSt).

3.2.1.1 Viscosity measurement methods

The viscosities of ionic liquids have normally been measured by one of three methods: falling ball, capillary, or rotational. Falling ball viscometers can easily be constructed from a graduated cylinder and appropriately sized ball bearings. The ball bearing material and the diameter can be varied. The experiment is conducted by filling the graduated cylinder with the fluid to be investigated and carefully dropping the ball through the fluid. After the ball has reached steady state, the velocity is measured. The absolute viscosity can then be calculated by Stokes' law (Equation 3.2-2) [1]:

$$\eta = \left(\frac{2}{9}\right)\frac{(\rho_s - \rho)gR^2}{v} \qquad (3.2\text{-}2)$$

where η is the absolute viscosity, ρ_S is the density of the ball, ρ is the density of the fluid, g is the gravity constant (980 cm s^{-2}), R is the radius of the ball, and υ is the steady-state velocity of the ball. A falling ball viscometer is commonly calibrated with a standard fluid similar in viscosity to the fluid of interest, and an instrument constant (k) is then determined. Comparisons between the standard fluid and the unknown fluid can then be made by means of Equation 3.2-3

$$\upsilon = k(\rho_s - \rho)\theta \tag{3.2-3}$$

where θ is the time of fall between two fiducial marks on the viscometer tube. This technique does have several limitations: the fluid must be Newtonian, the density of the fluid must be known, and the downward velocity of the ball should not exceed ~1 cm s^{-1} to aid in time measurement. The falling ball method is generally used to measure absolute viscosities from 10^{-3} to 10^{7} P [2].

Capillary viscometers are simple and inexpensive. They are normally constructed from glass and resemble a U-tube with a capillary section between two bulbs. The initial design originated with Ostwald and is shown as part A in Figure 3.2-1. The Cannon–Fenske type, a popular modification of the Ostwald design that moves the bulbs into the same vertical axis, is shown as part B in Figure 3.2-1.

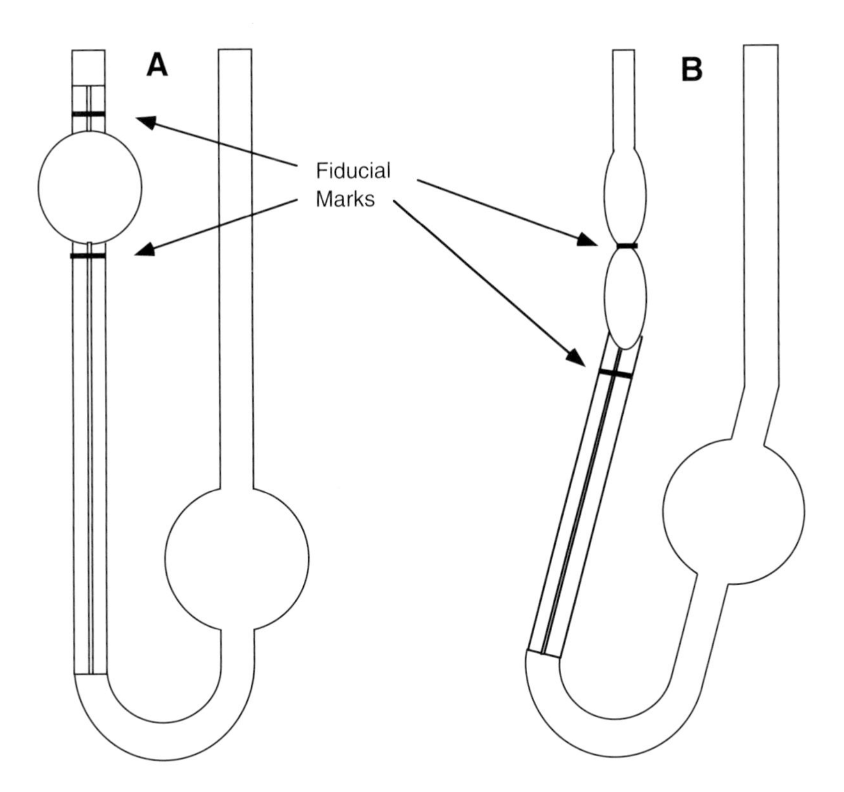

Figure 3.2-1: Diagrams of (A) Ostwald and (B) Cannon–Fenske capillary viscometers.

Capillary viscometers are normally immersed in a constant-temperature bath, to regulate the sample temperature precisely during the experiment. To determine the viscosity, the fluid in the viscometer is drawn into the upper bulb by vacuum. The vacuum is released, and the time for the fluid to fall past the marks above and below the bulb is measured. The main driving force for flow in this type of viscometer is gravity, although pressure can be applied to one side of the viscometer to provide an additional driving force (increased head pressure) [1]. Since the driving pressure is governed by the difference in heights of the liquid in the viscometer, it is important always to use the same volume of liquid in each experiment. The kinematic viscosity can be calculated by Equation 3.2-4 [2];

$$\nu = \frac{\left[\pi g\left(z_1 - z_2\right)D^4\right]}{128LV_o}\Delta t \tag{3.2-4}$$

where $(z_1 - z_2)$ is the difference in height, D is the capillary inner diameter, L is the length of the capillary, and V_o is the volume between the fiducial marks. This equation only holds as long as the liquid behaves as a Newtonian fluid and the length-to-diameter ratio of the capillary tube is large.

Capillary viscometers measure the kinematic viscosity directly, because the head pressure is generated by the weight of the fluid. In order to convert to absolute viscosity, the kinematic viscosity must be multiplied by the fluid density [Equation (3.2-1)]. Obviously this requires additional experiments to determine fluid density so that the absolute viscosity can be calculated. The capillary type viscometer is normally used to measure kinematic viscosities spanning the range from 4×10^{-3} to 1.6×10^{2} Stokes, with experimental times ranging from 200 to 800 seconds [1]. This range of kinematic viscosities corresponds to absolute viscosities of 6×10^{-3} to 2.4×10^{2} P, assuming an average ionic liquid density of 1.5 g cm^{-3}.

The last type of widely used viscometer is the rotational viscometer. These can adopt a variety of geometries, including concentric cylinders, cone and plate, and parallel disks. Of the three geometries, the concentric cylinders is the most common, because it is well suited for low-viscosity fluids [2]. Rotational viscometers consist of two main elements: a rotating element and a fixed element. The liquid to be measured is placed in the space between the two elements. The viscosity is determined by measurement of the torque transferred between the two elements by the liquid. For the concentric cylinder geometry, the outer cylinder is often rotated at a fixed speed and the torque is measured on the fixed center cylinder immersed in the liquid. By measuring the angular speed of the rotating cylinder and the torque on the fixed cylinder, the fluid viscosity can be calculated by Equation 3.2-5 [2]:

$$\eta = \left\{\frac{\left[\beta^2 - 1\right]}{\left[4\pi R_2^2 L_e\right]}\right\}\frac{T}{\omega_2} \tag{3.2-5}$$

where β is the ratio of the cylinder radii, R_2 is the radius of the outer cylinder, L_e is the effective length of the cylinder, T is the torque applied to the rotating cylinder, and ω_2 is the rotational speed of the outer cylinder [2]. The effective length of the cylinder (L_e) consists of the immersion depth of the center cylinder plus an end effect correction. This equation requires β to be less than 1.2.

All three methods discussed above appear to provide equally high quality ionic liquid viscosity data. However, the rotational viscometer could potentially provide additional information concerning the Newtonian behavior of the ionic liquids. The capillary method has been by far the most commonly used to generate the ionic liquid viscosity data found in the literature. This is probably due to its low cost and relative ease of use.

3.2.1.2 **Ionic liquid viscosities**

As a group, ionic liquids are more viscous than most common molecular solvents. Ionic liquid viscosities at room temperature range from a low of around 10 cP to values in excess of 500 cP. For comparative purposes, the viscosities of water, ethylene glycol, and glycerol at room temperature are 0.890, 16.1, and 934 cP, respectively [3]. The room-temperature viscosity data (also conductivity and density data) for a wide variety of ionic liquids are listed in Tables 3.2-1, 3.2-2, and 3.2-3. These tables are organized by the general type of ionic liquid. Table 3.2-1 contains data for non-haloaluminate alkylimidazolium ionic liquids, Table 3.2-2 for the haloaluminate ionic liquids, and Table 3.2-3 for other types of ionic liquids. There are multiple listings for several of the ionic liquids in Tables 3.2-1–3.2-3. These represent measurements by different researchers and have been included to help emphasize the significant variability in the viscosity data found in the literature.

The viscosities of many ionic liquids are strongly dependent upon temperature. For example, the viscosity of 1-butyl-3-methylimidazolium hexafluorophosphate increases 27 % with a 5 degree change between 298 and 293 K [16]. Clearly some of the apparent variability in the literature data seen in Tables 3.2-1–3.2-3 may have resulted from errors associated with control of experimental temperature. However, much of this variability is probably the result of impurities in the ionic liquids. Recent work with non-haloaluminate alkylimidazolium ionic liquids has established the ubiquitous nature of impurities in these ionic liquids, and has demonstrated the dramatic impact relatively that small amounts of impurities can have on ionic liquid viscosity [28]. In this study, a series of ionic liquids were prepared and purified by a variety of techniques. They were then analyzed for impurities and their physical properties evaluated. Chloride concentrations of up to 6 wt. % were found for some of the preparative methods. Residual chloride concentrations of between 1.5 and 6 wt. % increased the observed viscosity by between 30 and 600 % [28]. This work also showed the strong propensity of the non-haloaluminate alkylimidazolium ionic liquids to absorb water from laboratory air, and the significant solubility of water in these same ionic liquids (up to 14 wt. % for one ionic liquid). Surprisingly, as little as 2 wt. % (20 mol %) water (as well as other co-solvents, *vide infra*) reduced the observed viscosity of [BMIM][BF_4] by more than 50 %. Given this information, it is highly likely that many of the ionic liquids listed in Table 3.2-1 (and by analogy Tables 3.2-2 and 3.2-3) contained significant concentrations of impurities (especially water). This, in turn, complicates the evaluation of the literature data, and any conclusions drawn below must consequently be used with care.

Table 3.2-1: Specific conductivity data for non-haloaluminate alkylimidazolium-based ionic liquids.

Cation	*Anion*	*Temperature (K)*	*Conductivity (κ), mS cm^{-1}*	*Conductivity method*	*Viscosity (υ), cP*	*Viscosity method*	*Density (ρ), g cm^{-3}*	*Density method*	*Molar Conductivity (Λ), cm$^2\Omega^{-1}$mol^{-1}*	*Walden product ($\Lambda\eta$)*	*Ref.*
$[MMIM]^+$	$[(CF_3SO_2)_2N]^-$	293	8.4	B	44	R	1.559	U	2.03	89.44	4
$[EMIM]^+$	$[BF_4]^-$	295	12	I							5
$[EMIM]^+$	$[BF_4]^-$	298	14	B	34	U	1.240	U	2.24	76.22	6
$[EMIM]^+$	$[BF_4]^-$	298	14	I	32	R	1.279	I	2.17	68.91	7[a]
$[EMIM]^+$	$[BF_4]^-$	299	13	B	43	R					8
$[EMIM]^+$	$[BF_4]^-$	303	20	I							9
$[EMIM]^+$	$[PF_6]^-$	299	5.2	B							8
$[EMIM]^+$	$[CH_3CO_2]^-$	293	2.8	B	162	R					4
$[EMIM]^+$	$[CF_3CO_2]^-$	293	9.6	B	35	R	1.285	U	1.67	58.62	4
$[EMIM]^+$	$[C_3F_7CO_2]^-$	293	2.7	B	105	R	1.450	U	0.60	63.39	4
$[EMIM]^+$	$[CH_3SO_3]^-$	298	2.7	B	160	C	1.240	V	0.45	71.86	10
$[EMIM]^+$	$[CF_3SO_3]^-$	293	8.6	B	45	R	1.390	U	1.61	72.45	4
$[EMIM]^+$	$[CF_3SO_3]^-$	298	9.2	B	43	C	1.380	V	1.73	74.08	10
$[EMIM]^+$	$[CF_3SO_3]^-$	303	8.2	B							11[b]
$[EMIM]^+$	$[(CF_3SO_2)_2N]^-$	293	8.8	B	34	R	1.520	U	2.27	77.03	4
$[EMIM]^+$	$[(CF_3SO_2)_2N]^-$	298	5.7	I	31	R	1.518	I	1.47	44.82	7[c]
$[EMIM]^+$	$[(CF_3SO_2)_2N]^-$	303	10	I							11[b]
$[EMIM]^+$	$[(CF_3SO_2)_2N]^-$	298	9.2	B	34	U	1.510	U	2.38	81.06	12
$[EMIM]^+$	$[(CF_3SO_2)_2N]^-$	299	8.4	B	28	R					8
$[EMIM]^+$	$[(C_2F_5SO_2)_2N]^-$	299	3.4	B	61	R					8
$[EMMIM]^+$	$[(CF_3SO_2)_2N]^-$	293	3.2	B	88	R	1.495	U	0.87	76.35	4
$[EMM(5)IM]^+$	$[CF_3SO_3]^-$	293	6.4	B	51	R	1.334	U	1.32	67.11	4
$[EMM(5)IM]^+$	$[(CF_3SO_2)_2N]^-$	293	6.6	B	37	R	1.470	U	1.82	67.34	4
$[PMIM]^+$	$[PF_6]^-$	293					1.333	V			13
$[PMMIM]^+$	$[BF_4]^-$	295	5.9	I							14
$[PMMIM]^+$	$[PF_6]^-$	308	0.5	B							8
$[PMMIM]^+$	$[(CF_3SO_2)_2N]^-$	299	3.0	B	60	R					8

$[BMIM]^+$	$[PF_6]^-$	295	1.8	I						15
$[BMIM]^+$	$[PF_6]^-$	293					1.363	V		13
$[BMIM]^+$	$[PF6]^-$	298			207	R				16
$[BMIM]^+$	$[CF_3CO_2]^-$	293	3.2	B	73	R	1.209	U	0.67	48.74	4
$[BMIM]^+$	$[C_3F_7CO_2]^-$	293	1.0	B	182	R	1.333	U	0.26	48.09	4
$[BMIM]^+$	$[CF_3SO_3]^-$	293	3.7	B	90	R	1.290	U	0.83	74.42	4
$[BMIM]^+$	$[C_4F_9SO_3]^-$	293	0.45	B	373	R	1.427	U	0.14	51.56	4
$[BMIM]^+$	$[(CF_3SO_2)_2N]^-$	293	3.9	B	52	R	1.429	U	1.14	59.52	4
$[iBMIM]^+$	$[(CF_3SO_2)_2N]^-$	293	2.6	B	83	R	1.428	U	0.76	63.37	4
$[BMMIM]^+$	$[BF_4]^-$	295	0.23	I							5
$[BMMIM]^+$	$[PF_6]^-$	295	0.77	I							5
$[HMIM]^+$	$[PF_6]^-$	293					1.307	V			13
$[OMIM]^+$	$[PF_6]^-$	293					1.237	V			13
$[EEIM]^+$	$[CF_3CO_2]^-$	293	7.4	B	43	R	1.250	U	1.41	60.64	4
$[EEIM]^+$	$[CF_3SO_3]^-$	293	7.5	B	53	R	1.330	U	1.55	81.97	4
$[EEIM]^+$	$[(CF_3SO_2)_2N]^-$	293	8.5	B	35	R	1.452	U	2.37	83.05	4
$[EEM(5)IM]^+$	$[(CF_3SO_2)_2N]^-$	293	6.2	B	36	R	1.432	U	1.82	65.36	4
$[BEIM]^+$	$[CF_3CO_2]^-$	293	2.5	B	89	R	1.183	U	0.56	50.08	4
$[BEIM]^+$	$[CH_3SO_3]^-$	298	0.55	B			1.140	V	0.12		10
$[BEIM]^+$	$[CF_3SO_3]^-$	298	2.7	B			1.270	V	0.64		10
$[BEIM]^+$	$[C_4F_9SO_3]^-$	293	0.53	B	323	R	1.427	U	0.17	54.26	4
$[BEIM]^+$	$[(CF_3SO_2)_2N]^-$	293	4.1	B	48	R	1.404	U	1.27	60.75	4
$[DEIM]^+$	$[CF_3SO_3]^-$	298					1.10	V			10
$[MeOEtMIM]^+$	$[CF_3SO_3]^-$	293	3.6	B	74	R	1.364	U	0.77	56.69	4
$[MeOEtMIM]^+$	$[(CF_3SO_2)_2N]^-$	293	4.2	B	54	R	1.496	U	1.18	63.88	4
$[CF_3CH_2MIM]^+$	$[(CF_3SO_2)_2N]^-$	293	0.98	B	248	R	1.656	U	0.25	62.56	4

I = complex impedance, B = conductivity bridge, C = capillary viscometer, P = pycnometer or dilatometer, V = volumetric glassware, I = instrument, U = method unknown (not provided in the reference) [a] Conductivity at 298K Calculated from VTF Parameters given in reference. [b] Conductivity estimated from graphical data provided in the reference. [c] Density estimated from graphical data provided in the reference.

Table 3.2-2: Specific conductivity data for binary haloaluminate ionic liquids.

Ionic liquid System	*Cation*	*Anion(s)*	*Temperature, (K)*	*Conductivity (κ), mS cm⁻¹*	*Conductivity method*	*Viscosity (η), cP*	*Viscosity method*	*Density (ρ), g cm⁻³*	*Density method*	*Molar conductivity (Λ), cm²Ω⁻¹mol⁻¹*	*Walden product (Λη)*	*Ref.*
34.0–66.0 mol% [MMIM]Cl/$AlCl_3$	$[MMIM]^+$	$[Al_2Cl_7]^-$	298	15.0	B	17	C	1.404	P	4.26	72.07	17[a]
34.0–66.0 mol% [EMIM]Cl/$AlCl_3$	$[EMIM]^+$	$[Al_2Cl_7]^-$	298	15.0	B	14	C	1.389	P	4.46	62.95	17[a]
50.0–50.0 mol% [EMIM]Cl/$AlCl_3$	$[EMIM]^+$	$[AlCl_4]^-$	298	23.0	B	18	C	1.294	P	4.98	89.07	17[a]
60.0–40.0 mol% [EMIM]Cl/$AlCl_3$	$[EMIM]^+$	Cl^-, $[AlCl_4]^-$	298	6.5	B	47	C	1.256	P	1.22	57.77	17[a]
34.0–66.0 mol% [EMIM]Br/$AlBr_3$	$[EMIM]^+$	$[Al_2Br_7]^-$	298	5.8	B	32	C	2.219	P	1.89	59.64	18[a,b]
60.0–40.0 mol% [EMIM]Br/$AlBr_3$	$[EMIM]^+$	Br^-, $[AlBr_4]^-$	298	5.7	B	67	C	1.828	P	1.15	76.72	18[a,b]
40.0–60.0 mol% [PMIM]Cl/$AlCl_3$	$[PMIM]^+$	$[AlCl_4]^-$, $[Al_2Cl_7]^-$	298	11.0	B	18	C	1.351	P	2.94	53.44	17[a]
50.0–50.0 mol% [PMIM]Cl/$AlCl_3$	$[PMIM]^+$	$[AlCl_4]^-$	298	12.0	B	27	C	1.262	P	2.79	76.29	17[a]
60.0–40.0 mol% [PMIM]Cl/$AlCl_3$	$[PMIM]^+$	Cl^-, $[AlCl_4]^-$	298	3.3	B		C		P			17[a]
34.0–66.0 mol% [BMIM]Cl/$AlCl_3$	$[BMIM]^+$	$[Al_2Cl_7]^-$	298	9.2	B	19	C	1.334	P	3.04	58.45	17[a]
50.0–50.0 mol% [BMIM]Cl/$AlCl_3$	$[BMIM]^+$	$[AlCl_4]^-$	298	10.0	B	27	C	1.238	P	2.49	67.42	17[a]
34.0–66.0 mol% [BBIM]Cl/$AlCl_3$	$[BBIM]^+$	$[Al_2Cl_7]^-$	298	6.0	B	24	C	1.252	P	2.32	55.36	17[a]
50.0–50.0 mol% [BBIM]Cl/$AlCl_3$	$[BBIM]^+$	$[AlCl_4]^-$	298	5.0	B	38	C	1.164	P	1.50	56.83	17[a]
33.3–66.7 mol% [MP]Cl/$AlCl_3$	$[MP]^+$	$[Al_2Cl_7]^-$	298	8.1	B	21	C	1.441	P	2.23	46.12	19[a]
33.3–66.7 mol% [EP]Cl/$AlCl_3$	$[EP]^+$	$[Al_2Cl_7]^-$	298	10.0	B	18	C	1.408	P	2.91	51.29	19[a]
33.3–66.7 mol% [EP]Br/$AlCl_3$	$[EP]^+$	$[Al_2Cl_xBr_{7-x}]^-$	298	8.4	B	22	C	1.524	P			19[a]
33.3–66.7 mol% [EP]Br/$AlCl_3$	$[EP]^+$	$[Al_2Cl_xBr_{7-x}]^-$	298	17.0	B	25						20
33.3–66.7 mol% [EP]Br/$AlBr_3$	$[EP]^+$	$[Al_2Br_7]^-$	298			50	C	2.20	V			104
33.3–66.7 mol% [PP]Cl/$AlCl_3$	$[PP]^+$	$[Al_2Cl_7]^-$	298	8.0	B	18	C	1.375	P	2.47	44.93	19[b]
33.3–66.7 mol% [BP]Cl/$AlCl_3$	$[BP]^+$	$[Al_2Cl_7]^-$	298	6.7	B	21	C	1.346	P	2.18	45.81	19[b]

I = complex impedance, B = conductivity bridge, C = capillary viscometer, P = pycnometer or dilatometer, V = volumetric glassware, I = instrument, U = method unknown (not provided in the reference). [a] Conductivity at 298K calculated from least-squares-fitted parameters given in reference. [b] Conductivity estimated from graphical data provided in the reference.

Table 3.2-3: Specific conductivity data for other room-temperature ionic liquids

	Cation	Anion	Temperature, (K)	Conductivity (κ), mS cm^{-1}	Conductivity method	Viscosity (η), cP	Viscosity method	Density (ρ), g cm^{-3}	Density method	Molar conductivity (Λ), $cm^2 \Omega^{-1} mol^{-1}$	Walden product ($\Lambda\eta$)	Ref
Ammonium												
	$[(CH_3)_2(C_2H_5)(CH_3OC_2H_4)N]^+$	$[BF_4]^-$	298	1.7	B							22
	$[(n\text{-}C_3H_7)(CH_3)_3N]^+$	$[(CF_3SO_2)_2N]^-$	298	3.3	B	72	U	1.440	U	0.88	63.09	12
	$[(n\text{-}C_6H_{13})(C_2H_5)_3N]^+$	$[(CF_3SO_2)_2N]^-$	298	0.67	I	167	C	1.270	V	0.25	41.10	23
	$[(n\text{-}C_8H_{17})(C_2H_5)_3N]^+$	$[(CF_3SO_2)_2N]^-$	298	0.33	I	202	C	1.250	V	0.13	26.37	23
	$[(n\text{-}C_8H_{17})(C_4H_9)_3N]^+$	$[(CF_3SO_2)_2N]^-$	298	0.13	I	574	C	1.120	V	0.07	38.56	23
	$[(CH_3)_3(CH_3OCH_2)N]^+$	$[(CF_3SO_2)_2N]^-$	298	4.7	B	50	U	1.510	U	1.20	59.81	12
	1-methyl-1-propyl-pyrrolidinium	$[(CF_3SO_2)_2N]^-$	298	1.4	B	63	C	1.45	V			24
	1-butyl-1-methyl-pyrrolidinium	$[(CF_3SO_2)_2N]^-$	298	2.2	B	85	C	1.41	V			24
Pyrazolium												
	1,2-dimethyl-4-fluoropyrazolium	$[BF_4]^-$	298	1.3	B							25
Pyridinium												
	$[BP]^+$	$[BF_4]^-$	298	1.9	I	103	R	1.220	I	0.35	35.77	7[a]
	$[BP]^+$	$[BF_4]^-$	303	3.0	I							9
	$[BP]^+$	$[(CF_3SO_2)_2N]^-$	298	2.2	I	57	R	1.449	I	0.63	35.91	7[a]
Sulfonium												
	$[(CH_3)_3S]^+$	$[HBr_2]^-$	298	34	B	20.5	C	1.74	P	4.62	95.33	21
	$[(CH_3)_3S]^{+b}$	$[HBr_2]^-$, $[H_2Br_3]^-$	298	56	B	8.3	C	1.79	P	8.41	69.80	21
	$[(CH_3)_3S]^+$	$[Al_2Cl_7]^-$	298	5.5	B							26
	$[(CH_3)_3S]^+$	$[Al_2Cl_7]^-$	298	5.5	B	39.3	C	1.40	V	1.49	58.56	27
	$[(CH_3)_3S]^+$	$[Al_2Cl_6Br]^-$	298	4.21	B	54.9	C	1.59	V	1.12	61.60	27
	$[(CH_3)_3S]^+$	$[Al_2Br_7]^-$	298	1.44	B	138	C	2.40	V	0.41	57.17	27
	$[(CH_3)_3S]^+$	$[(CF_3SO_2)_2N]^-$	318	8.2	B	44	U	1.580	U	1.85	81.59	27
	$[(C_2H_5)_3S]^+$	$[(CF_3SO_2)_2N]^-$	298	7.1	B	30	U	1.460	U	1.94	58.27	27
	$[(n\text{-}C_4H_9)_3S]^+$	$[(CF_3SO_2)_2N]^-$	298	1.4	B	75	U	1.290	U	0.52	39.36	27
Thiazolium												
	1-ethylthiazolium	$[CF_3SO_3]^-$	298	4.2	B			1.50	V			10

I = complex impedance, B = conductivity bridge, C = capillary viscometer, P = pycnometer or dilatometer, V = volumetric glassware, I = instrument, U = method unknown (not provided in the reference). [a] Conductivity at 298K calculated from VTF Parameters given in reference. [b] Binary composition of 42.0–58.0 mol % $[(CH_3)_3S]Br–HBr$.

Within a series of non-haloaluminate ionic liquids containing the same cation, a change in the anion clearly affects the viscosity (Tables 3.2-1 and 3.2-3). The general order of increasing viscosity with respect to the anion is: $[(CF_3SO_2)_2N]^- \leq [BF_4]^- \leq [CF_3CO_2]^- \leq [CF_3SO_3]^- < [(C_2H_5SO_2)_2N]^- < [C_3F_7CO_2]^- < [CH_3CO_2]^- \leq [CH_3SO_3]^- < [C_4F_9SO_3]^-$. Obviously, this trend does not exactly correlate with anion size. This may be due to the effect of other anion properties on the viscosity, such as their ability to form weak hydrogen bonds with the cation.

The viscosities of non-haloaluminate ionic liquids are also affected by the identity of the organic cation. For ionic liquids with the same anion, the trend is that larger alkyl substituents on the imidazolium cation give rise to more viscous fluids. For instance, the non-haloaluminate ionic liquids composed of substituted imidazolium cations and the bis-trifyl imide anion exhibit increasing viscosity from $[EMIM]^+$, $[EEIM]^+$, $[EMM(5)IM]^+$, $[BEIM]^+$, $[BMIM]^+$, $[PMMIM]^+$, to $[EMMIM]^+$ (Table 3.2-1). Were the size of the cations the sole criteria, the $[BEIM]^+$ and $[BMIM]^+$ cations from this series would appear to be transposed and the $[EMMIM]^+$ would be expected much earlier in the series. Given the limited data set, potential problems with impurities, and experimental differences between laboratories, we are unable to propose an explanation for the observed disparities.

The haloaluminate ionic liquids are prepared by mixing two solids, an organic chloride and an aluminium halide (e.g., [EMIM]Cl and $AlCl_3$). These two solids react to form ionic liquids with a single cation and a mix of anions, the anion composition depending strongly on the relative molar amounts of the two ingredients used in the preparation. The effect of anionic composition on the viscosity of haloaluminate ionic liquids has long been recognized. Figure 3.2-2 shows the absolute viscosities of the [EMIM]Cl/$AlCl_3$ ionic liquids at 303 K over a range of compositions.

When the amount of [EMIM]Cl is below 50 mol %, the viscosity is relatively constant, only varying from 14 to 18 cP. However, when the [EMIM]Cl exceeds 50 mol %, the absolute viscosity begins to increase, eventually rising to over 190 cP at 67 mol % [EMIM]Cl [17]. This dramatic increase in viscosity is strongly correlated to the corresponding growth in chloride ion concentration as the [EMIM]Cl mol % increases, and appears to be the result of hydrogen bonding between the Cl^- anions and the hydrogen atoms on the imidazolium cation ring [29–32].

The size of the cation in the chloroaluminate ionic liquids also appears to have an impact on the viscosity. For ionic liquids with the same anion(s) and compositions, the trend is for greater viscosity with larger cation size (Table 3.2-2). An additional contributing factor to the effect of the cation on viscosity is the asymmetry of the alkyl substitution. Highly asymmetric substitution has been identified as important for obtaining low viscosities [17].

The addition of co-solvents to ionic liquids can result in dramatic reductions in the viscosity without alteration of the cations or anions in the system. The haloaluminate ionic liquids present a challenge, due to the reactivity of the ionic liquid. Nonetheless, several compatible co-solvents including benzene, dichloromethane, and acetonitrile have been investigated [33–37]. The addition of as little as 5 wt. % acetonitrile or 15 wt. % benzene or methylene chloride was able to reduce the

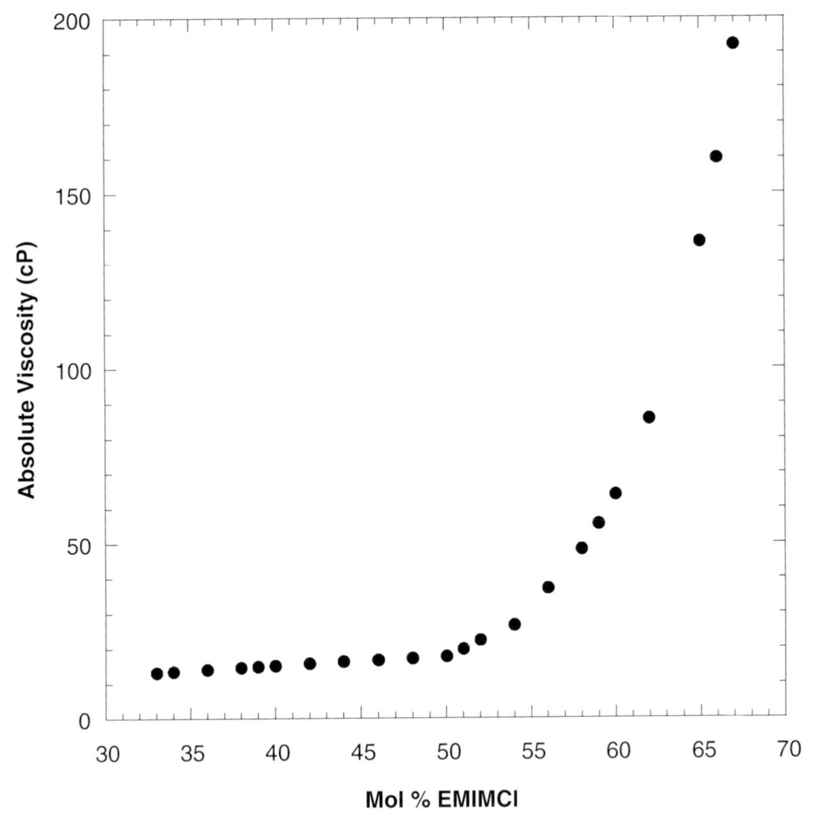

Figure 3.2-2: Change in the absolute viscosity (cP) as a function of the [EMIM]Cl mol % in an [EMIM]Cl/AlCl3 ionic liquid at 303 K.

absolute viscosity by 50 % for [EMIM]Cl/$AlCl_3$ ionic liquids with less than 50 mol % $AlCl_3$ [33]. Non-haloaluminate ionic liquids have also been studied with a range of co-solvents including water, toluene, and acetonitrile. The ionic liquid response is similar to that observed in the haloaluminate ionic liquids. The addition of as little as 20 mol % co-solvent reduced the viscosity of a [BMIM][BF_4] melt by 50 % [28].

3.2.2
Density of Ionic Liquids

Densities are perhaps the most straightforwardly determined and unambiguous physical property of ionic liquids. Given a quality analytical balance and good volumetric glassware the density of an ionic liquid can be measured gravimetrically (i.e., the sample can be weighed).

3.2.2.1 **Density measurement**

To measure density properly with a minimal amount of sample, a pycnometer should be employed. A pycnometer removes the ambiguity of measuring the bottom of the meniscus in a piece of glassware calibrated with aqueous solutions that have potentially very different surface tensions. The most common types of pycnometers are the Ostwald–Sprengel and the Weld or stopper pycnometer. These devices are generally constructed of glass and consist of a reservoir connected to a capillary or capillaries with fiducial marks. The pycnometer is weighed while empty, filled with the fluid of interest, and allowed to equilibrate thermally. The fluid above the fiducial marks is removed and the pycnometer is weighed [38, 39]. Pycnometers must be calibrated prior to use to determine the exact volume. The density is then calculated by dividing the mass of the fluid by the pycnometer volume.

3.2.2.2 **Ionic liquid densities**

The reported densities of ionic liquids vary between 1.12 g cm^{-3} for $[(n\text{-}C_8H_{17})(C_4H_9)_3N][(CF_3SO_2)_2N]$ and 2.4 g cm^{-3} for a 34–66 mol% $[(CH_3)_3S]Br/AlBr_3$ ionic liquid [21, 23]. The densities of ionic liquid appear to be the physical property least sensitive to variations in temperature. For example, a 5 degree change in temperature from 298 to 303 K results in only a 0.3 % decrease in the density for a 50.0:50.0 mol % $[EMIM]Cl/AlCl_3$ [17]. In addition, the impact of impurities appears to be far less dramatic than in the case of viscosity. Recent work indicates that the densities of ionic liquids vary linearly with wt. % of impurities. For example, 20 wt. % water (75 mol %) in $[BMIM][BF_4]$ results in only a 4 % decrease in density [33].

In the binary haloaluminate ionic liquids, an increase in the mole percent of the imidazolium salt decreases the density of the liquid (see Table 3.2-2). The bromoaluminate ionic liquids are substantially denser than their chloroaluminate counterparts, being between 0.57 g cm^{-3} and 0.83 g cm^{-3} denser than the analogous chloroaluminate ionic liquids (see Table 3.2-2). Variation of the substituents on the imidazolium cation in the chloroaluminate ionic liquids has been shown to affect the density on the basis of the cation size [17].

Within a series of non-haloaluminate ionic liquids containing the same cation species, increasing anion mass corresponds to increasing ionic liquid density (see Tables 3.2-1 and 3.2-3). Generally, the order of increasing density for ionic liquids composed of a single cation is: $[CH_3SO_3]^- \approx [BF_4]^- < [CF_3CO_2]^- < [CF_3SO_3]^- < [C_3F_7CO_2] < [(CF_3SO_2)_2N]$.

The densities of the non-haloaluminate ionic liquids are also affected by the identity of the organic cation. As in the haloaluminate ionic liquids, the density decreases as the size of the cation increases. In non-haloaluminate ionic liquids composed of substituted imidazolium cations and the triflate anion, for instance, the density decreases from 1.390 g cm^{-3} for $[EMIM]^+$ to 1.334 g cm^{-3} for $[EMM(5)IM]^+$, to 1.330 g cm^{-3} for $[EEIM]^+$, to 1.290 g cm^{-3} for $[BMIM]^+$, and 1.270 g cm^{-3} for $[BEIM]^+$ (see Table 3.2-3).

References

1 J. R. Van Wazer, J. W. Lyons, K. Y. Kim, R. E. Colwell, *Viscosity and Flow Measurement: A Laboratory Handbook of Rheology*, Interscience Publishers-John Wiley, New York, **1963**, chapters 2, 4, and 5.

2 G. E. Leblanc, R. A. Secco, M. Kostic, in *Mechanical Variables Measurement: Solid, Fluid, and Thermal* (J. G. Webster, ed.), CRC Press: Boca Raton, **2000**, chapter 11.

3 Handbook of Chemistry and Physics, 82nd Edition, D. R. Linde, Ed., CRC Press, New York, **2001**, pp. 6-182 – 6-186.

4 P. Bonhôte, A.-P. Dias, N. Papageorgiou, K. Kalyanasundaram, M. Grätzel, *Inorg. Chem.* **1996**, *35*, 1168.

5 T. E. Sutto, H. C. De Long, P. C. Trulove, in *Progress in Molten Salt Chemistry 1* (R. W. Berg, H. A. Hjuler, eds.), Elsevier: Paris, **2000**, 511.

6 a. J. Fuller, R. T. Carlin, R. A. Osteryoung, *J. Electrochem. Soc.*, **1997** *144*, 3881. b. J. Fuller, R. A. Osteryoung, R. T. Carlin, *Abstracts of Papers*, 187th Meeting of The Electrochemical Society, Reno, NV, **1995**, Vol. 95-1, p. 27.

7 a. A. Noda, K. Hayamizu, M. Watanabe, *J. Phys. Chem. B* **2001**, *105*, 4603; b. A. Noda, M. Watanabe, in *Proceedings of the Twelfth International Symposium on Molten Salts* (P. C. Trulove, H. C. De Long, G. R. Stafford, S. Deki, eds.), The Electrochemical Society: Pennington NJ, **2000**, Vol. 99-41, pp. 202–208.

8 A. B. McEwen, H. L. Ngo, K. LeCompte, J. L. Goldman, *J. Electrochem. Soc.* **1999**, *146*, 1687.

9 A. Noda, M. Watanabe, *Electrochem. Acta* **2000**, *45*, 1265.

10 E. I. Cooper, E. J. M. O'Sullivan, in *Proceedings of the Eighth International Symposium on Molten Salts* (R. J. Gale, G. Blomgren, eds.), The Electrochemical Society: Pennington NJ, **2000**, Vol. 92-16, pp. 386–396.

11 H. Every, A.G. Bishop, M. Forsyth, D. R. MacFarlane, *Electrochim. Acta* **2000**, *45*, 1279.

12 H. Matsumoto, M. Yanagida, K. Tanimoto, M. Nomura, Y. Kitagawa, Y. Miyazaki, *Chem. Lett.* **2000**, 922.

13 S. Chun, S. V. Dzyuba, R. A. Bartsch, *Anal. Chem.* **2001**, *73*, 3737.

14 T. E. Sutto, H. C. De Long, P. C. Trulove, in *Progress in Molten Salt Chemistry 1* (R. W. Berg, H. A. Hjuler, eds.), Elsevier, Paris, **2000**, 511.

15 J. Fuller, A. C. Breda, R. T. Carlin, *J. Electroanal. Chem.* **1998**, *459*, 29.

16 S. N. Baker, G. A. Baker, M. A. Kane, F. V. Bright, *J. Phys. Chem. B* **2001**, *105*, 9663.

17 a. J. S. Wilkes, J. A. Levisky, R. A. Wilson, C. L. Hussey, *Inorg. Chem.* **1982**, *21*, 1263. b. A. A. Fannin Jr., D. A. Floreani, L. A. King, J. S. Landers, B. J. Piersma, D. J. Stech, R. J. Vaughn, J. S. Wilkes, J. L. Williams, *J. Phys. Chem.* **1984**, *88*, 2614.

18 a. J. R. Sanders, E. H. Ward, C. L. Hussey, in *Proceedings of the Fifth International Symposium on Molten Salts* (M.-L. Saboungi, K. Johnson, D. S. Newman, D. Inman, Eds.), The Electrochemical Society: Pennington NJ **1986**, Vol. 86-1, pp.307–316. b. J. R. Sanders, E. H. Ward, C. L. Hussey, *J. Electrochem. Soc.*, **1986**, *133*, 325.

19 R. A. Carpio, L. A. King, R. E. Lindstrom, J. C. Nardi, C. L. Hussey, *J. Electrochem. Soc.* **1979**, *126*, 1644.

20 V. R. Koch, L. L. Miller, R. A. Osteryoung, *J. Am. Chem. Soc.* **1976**, *98*, 5277.

21 M. Ma, K. E. Johnson, in *Proceedings of the Ninth International Symposium on Molten Salts* (C. L. Hussey, D. S. Newman, G. Mamantov, Y. Ito, eds.), The Electrochemical Society: Pennington NJ, **1994**, Vol. 94-13, pp. 179–186.

22 E. I. Cooper, C. A. Angell, *Solid State Ionics* **1983**, *9–10*, 617.

23 a. J. Sun, M. Forsyth, D. R. MacFarlane, *Molten Salt Forum*, **1998**, *5–6*, 585. b. J. Sun, M. Forsyth, D. R. MacFarlane, *J. Phys. Chem. B.* **1998**, *102*, 8858.

24 D. R. MacFarlane, P. Meakin, J. Sun, N. Amini, M. Forsyth, *J. Phys. Chem. B.* **1999**, *103*, 4164.

25 J. Caja, T. D. J. Dunstan, D. M. Ryan, V. Katovic, in *Proceedings of the Twelfth International Symposium on Molten Salts* (P. C. Trulove, H. C. De Long, G. R. Stafford, S. Deki, eds.), The Electrochemical Society: Pennington NJ, **2000**, Vol. 99-41, pp. 150–161.

26 S. D. Jones, G. E. Blomgren, in *Proceedings of the Seventh International Symposium on Molten Salts* (C. L. Hussey, S. N. Flengas, J. S. Wilkes, Y. Ito, eds.), The Electrochemical Society: Pennington NJ, **1990**, Vol. 90-17, pp. 273–280.

27 H. Matsumoto, T. Matsuda, Y. Miyazaki, *Chem. Lett.* **2000**, 1430.

28 K. R. Seddon, A. Stark, M.-J. Torres, *Pure Appl. Chem.* **2000**, *72*, 2275.

29 C. J. Dymek, D. A. Grossie, A. V. Fratini, W. W. Adams, *J. Mol. Struct.* **1989**, *213*, 25.

30 C. J. Dymek, J. J. Stewart, *Inorg. Chem.* **1989**, *28*, 1472.

31 A. G. Avent, P. A. Chaloner, M. P. Day, K. R. Seddon, T. Welton, in *Proceedings of the Seventh International Symposium on Molten Salts* (C. L. Hussey, J. S. Wilkes, S. N. Flengas, Y. Ito, eds.), The Electrochemical Society: Pennington NJ, **1990**, Vol. 90-17, pp. 98–133.

32 A. G. Avent, P. A. Chaloner, M. P. Day, K. R. Seddon, T. Welton, *J. Chem. Soc. Dalton Trans.* **1994**, 3405.

33 R. L. Perry, K. M. Jones, W. D. Scott, Q. Liao, C. L. Hussey, *J. Chem. Eng. Data* **1995**, *40*, 615.

34 Q. Liao, C. L. Hussey, *J. Chem. Eng. Data* **1996**, *41*, 1126.

35 N. Papageorgiou, Y. Athanassov, M. Armand, P. Bonhôte, H. Pattersson, A. Azam, M. Grätzel, *J. Electrochem. Soc.* **1996**, *143*, 3099.

36 R. Moy, R.-P. Emmenegger, *Electrochimica Acta* **1992**, *37*, 1061.

37 J. Robinson, R. C. Bugle, H. L. Chum, D. Koran, R. A. Osteryoung, *J. Am. Chem. Soc.* **1979**, *101*, 3776.

38 D. P. Shoemaker, C. W. Garland, J. I. Steinfeld, J. W. Nibler, *Experiments in Physical Chemistry*, 4th ed., McGraw-Hill: New York, **1981**, exp. 11.

39 H. Eren, in *Mechanical Variables Measurement: Solid, Fluid, and Thermal* (J. G. Webster, ed.), CRC Press: Boca Raton, **2000**, chapter 2.

3.3 Solubility and Solvation in Ionic Liquids

John D. Holbrey, Ann E. Visser, and Robin D. Rogers

3.3.1 Introduction

Interest in using ionic liquid (IL) media as alternatives to traditional organic solvents in synthesis [1–4], in liquid/liquid separations from aqueous solutions [5–9], and as liquid electrolytes for electrochemical processes, including electrosynthesis, primarily focus on the unique combination of properties exhibited by ILs that differentiate them from molecular solvents.

ILs are considered to be polar solvents, but can be non-coordinating (mainly depending on the IL's anion). Solvatochromatic studies indicate that ILs have polarities similar to those of short-chain alcohols and other polar, aprotic solvents

(DMSO, DMF, etc.) [10–14]. That is, the polarity of many ILs is intermediate between water and chlorinated organic solvents and varies, depending on the nature of the IL components (for more details see Section 3.5).

By changing the nature of the ions present in an IL, it is possible to change the resulting properties of the IL. For example, the miscibility with water can be varied from complete miscibility to almost total immiscibility, by changing the anion from, for example, Cl^- to $[PF_6]^-$. Similarly, the lipophilicity of an IL is modified by the degree of cation substitution. Primary solvent features of ILs include the capability for H-bond donation from the cation to polar or dipolar solutes, H-bond accepting functionality in the anion (this is variable – Cl^- is a good H-bond acceptor, for example, whereas $[PF_6]^-$ is poor), and π–π or C–H...π interactions (which enhance aromatic solubility) (for more details see Section 3.5). ILs tend to be immiscible with alkanes and other non-polar organic solvents and hence can be used in two-phase systems. Similarly, it is possible to design ILs that are hydrophobic and can be used in aqueous/IL biphasic systems.

The solubilities both of organic compounds and of metal salts in ILs are important with regard to stoichiometric chemical synthesis and catalytic processes. Not only must reagents and catalysts be sufficiently soluble in the solvent, but differential solubilities of reagents, products, and catalysts are required in order to enable effective separation and isolation of products. As well as requiring a knowledge of solute solubility in ILs, to assess the relative merits of a particular IL for chemical or separations processes, relative solubility and partitioning information about the preference of the solutes for IL phases relative to extractants is needed in order to design systems in which both reactions and extractions can be performed efficiently. However, only limited systematic data on these properties are available in the literature. In many cases, solutes and solvents are described as immiscible in a particular IL on the basis of the observation that two phases are formed, rather than compositional analysis to determine the limits of solubility or co-miscibility.

ILs have been investigated as alternatives to traditional organic solvents in liquid/liquid separations. Reports highlighting separations based on ILs [5, 15–18] for implementation into industrial separations systems demonstrate the design principles, and have identified hydrophobic ILs as replacements for VOCs in aqueous/IL biphase separations schemes. Other work on novel solvent media has shown how scH_2O [19, 20], $scCO_2$ [15], and fluorous phases [21] can be used in efforts to broaden the scope of possibilities available for more environmentally responsible processes. From the synthetic perspective, desirable features of an IL are: (i) catalyst solubility, enabling high catalyst capacity, immobilization to extraction processes, (ii) reagent solubility, ideally in high concentration, and (iii) product extractability.

Extractions and separations in two-phase systems require knowledge of the miscibilities and immiscibilities of ILs with other solvents compatible with the process. These are most usually IL/aqueous biphase systems in which the IL is the less polar phase and organic/IL systems in which the IL is used as the polar phase. In these two-phase systems, extraction both to and from the IL phase is important.

3.3.2 Metal Salt Solubility

Transition metal catalysis in liquid/liquid biphasic systems principally requires sufficient solubility and immobilization of the catalysts in the IL phase relative to the extraction phase. Solubilization of metal ions in ILs can be separated into processes, involving the dissolution of simple metal salts (often through coordination with anions from the ionic liquid) and the dissolution of metal coordination complexes, in which the metal coordination sphere remains intact.

3.3.2.1 Halometalate salts

The formation of halometalate ionic liquids by the use of the equilibrium reactions of organic halide salts with metal halide compounds is well established and has been reviewed by Hussey [22]. A wide range of metalates have been prepared and investigated, primarily as liquid electrolytes for electrochemistry and battery applications, and for electroplating, electrowinning, and as Lewis acid catalysts for chemical synthesis. In particular, acidic tetrachloroaluminate(III) ILs have been used in place of solid $AlCl_3$, with the IL acting as a liquid catalyst and with product separation from the IL encouraged by differential solubility of the reagents and products.

Many simple metal compounds (ionic salts) are dissolved in 'basic' ionic liquids, containing coordinating Lewis base ligands, by complexation mechanisms; most metal halides can be dissolved in chloride-rich ILs as chloro-containing metalate species. This is the basis for the formation of chlorometalate ionic liquids, containing metal complex anions. Among the more novel recent examples of metal-containing ILs is the gold-containing [EMIM][$AuCl_4$] [23]. Seddon and Hussey have investigated the dissolution of many transition metals in halometalate ILs (see, for example, references [24–26]).

Metal halide compounds can be dissolved in basic tetrachloroaluminate ionic liquids, but can in many cases be precipitated from acidic ILs. For example, crystals of $[EMIM]_2[PdCl_4]$ were obtained by dissolving $PdCl_2$ in acidic [EMIM]Cl/$AlCl_3$ IL [27]. This reflects changes in ligating ability of the predominant anions present in tetrachloroaluminate ILs on changing from the basic regime (Cl^-) through to acidic ($[AlCl_4]^-$). Simple metal salts can also be dissolved in other ionic liquids, containing coordinating anions such as nitrates.

3.3.2.2 Metal complexes

Examination of metal complex solubility in ILs has mainly stemmed from requirements for metal complex solubility for transition metal catalysis. The most effective method is through selective solubility and immobilization of the catalyst in the reacting phase, allowing product separation (with no catalyst leaching) into a second, extracting phase. In the context of emerging separations and extraction investigations and homogeneous catalysis, efficient recycling of metal catalysts is an absolute necessity. Systematic studies of metal complex solubility in ILs have yet to be reported and warrant investigation.

As a set of general observations:
(i) Simple ionic *compounds* are generally poorly soluble in ILs.
(ii) Ionic *complexes* are more soluble.
(iii) Compounds are solubilized by complexation.
(iv) The peripheral environments of the ligands are important in affecting solubility, and can be modified to provide better solubility.

Solubility depends on the nature of the IL and on solvation or complex formation. Most metal ions display preferential partitioning into water in IL aqueous systems and are hence less soluble in the IL than in water.

Simple metal compounds are poorly soluble in non-coordinating ILs, but the solubility of metal ions in an IL can be increased by addition of lipophilic ligands. However, enhancement of lipophilicity also increases the tendency for the metal complex to leach into less polar organic phases.

Ionic complexes tend to be more soluble than neutral complexes in ILs. Representative examples of transition metal salts and complexes that have been used as homogeneous catalysts in IL systems include $[LNiCH_2CH_3][AlCl_4]$, used in the Difasol olefin oligomerization process, $[Rh(nbd)(PPh_3)_2][PF_6]$ [28], $[Rh(cod_2)][BF_4]_2$ [29], and $[H_4Ru(\pi^6\text{-}C_6H_6)_4][BF_4]_2$ [30] complexes, which have been described as catalysts for hydrogenation reactions. In catalytic hydrogenation studies, Chauvin has noted that neutral catalysts, such as $Rh(CO)_2(acac)$, are leached into the organic phase whereas charged species are maintained in the IL phase [31] (see Chapter 5.2 for more detail).

Precipitation of neutral complexes from solution, or extraction into a secondary phase, has enormous implications in the design of two-phase catalytic systems (to eliminate catalyst leaching) and in extractions (where selective extraction from either aqueous or organic phases is required, followed by controlled stripping of metals from the IL phase for recovery). Metal ion solubility in ILs can be increased by changing the complexing ligands present, for example by the use of soluble organic complexants such as crown ethers, or by modifying the ligands to increase solubility in the IL.

Chauvin showed that sulfonated triphenylphosphine ligands (e.g., tppts and tppms (iv) in Figure 3.3-1) prevented leaching of neutral Rh hydrogenation catalysts from ILs [28], although Cole-Hamilton and co-workers [32] have noted that the solubility of Rh-tppts complexes in ILs is low. Wasserscheid and co-workers [33] and Olivier-Bourbigou and co-workers [34] have demonstrated that addition of cationic functionality to the periphery of otherwise neutral ligands can be used to increase solubility and stability of metal complexes in the IL phase relative to leaching into an organic extractant phase ((i), (ii), (iv), and (v) in Figure 3.3-1). This approach mimics that taken to confer greater water solubility on metal complexes for aqueous-biphasic catalysis, and is equivalent to the TSIL approach of Davis, Rogers and co-workers [6] for enhanced metal transfer and binding in IL phases for extractions ((iii) in Figure 3.3-1).

ILs have also been used as inert additives to stabilize transition metal catalysts during evaporative workup of reactions in organic solvent systems [35,36]. The non-

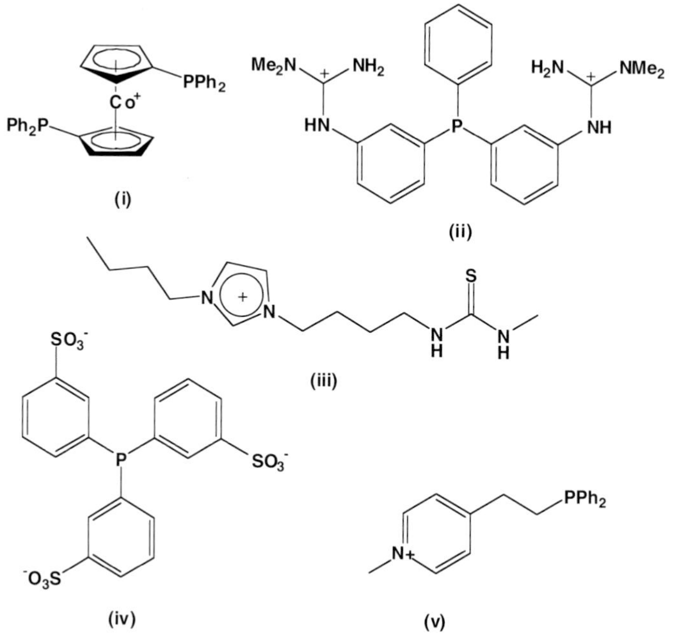

Figure 3.3-1: Incorporation of groups with high affinities for ILs (such as cobaltacenium (i), guanadinium (ii), sulfonate (iv), and pyridinium (v)) or even groups that are themselves ionic liquid moieties (such as imidazolium (iii)) as peripheral functionalities on coordinating ligands increases the solubility of transition metal complexes in ILs.

volatile IL component solubilizes the catalyst upon concentration and removal of organic solvent and products, thereby preventing catalyst decomposition and enabling catalysts to be recycled and reused in batch processes.

3.3.3
Extraction and Separations

Studies of extractions and separations provide information on the relative solubilities of solutes between two phases, such as partitioning data, required in order to design systems in which a solute is either selectively extracted from, or immobilized in one phase. Liquid/liquid separation studies of metal ions are principally concerned with aqueous/organic two-phase systems, with relevance for extraction and concentration of metal ions in the organic phase. In terms of IL/aqueous partitioning, there is considerable interest in the replacement of organic extracting phases with ILs for recovery of metals from waste water in mining, in nuclear fuel and waste reprocessing, and in immobilization of transition metal catalysts. The hydrated natures of most metal ions lower their affinity for the less-polar extracting phases; this is the case in IL systems, where hydrated metal ions do not partition into the IL from water, except for the most hydrophobic cations [8]. The affinity of metal ions for less polar phases can be enhanced by changing the hydration environments of the metal ions either by using organic ligands [37–39], which provide more hydrophobic regions around the metal, or with inorganic anions [40] that form softer, more extractable anionic complexes with the metal.

Ideally, to ensure the complete removal of the metal ions from the aqueous phase, the complexant and the metal complex should remain in the hydrophobic phase. Thus, the challenges for separations include the identification of extractants that quantitatively partition into the IL phase and can still readily complex target metal ions, and also the identification of conditions under which specific metal ion species can be selectively extracted from aqueous streams containing inorganic complexing ions.

3.3.3.1 Anionic extractants

Most hydrated metal ions are more soluble in water than in ILs. The distribution ratios of some metal ions between aqueous and IL phases may be enhanced in the presence of coordinating anions, such as halides or pseudohalides, capable of modifying the metal complex hydrophobicity, increasing partitioning from water [41]. The effect of halide, cyanate, cyanide, and thiocyanate ions on the partitioning of Hg^{2+} in [BMIM][PF_6]/aqueous systems (Figure 3.3-2) has been studied [8]. The results indicate that the metal ion transfer to the IL phase depends on the relative hydrophobicity of the metal complex. Hg-I complexes have the highest formation constants, decreasing to those of Hg-F [42]. Results from pseudohalides, however, suggest a more complex partitioning mechanism, since Hg-CN complexes have even higher formation constants [42], but display the lowest distribution ratios.

3.3.3.2 Organic extractants

Macrocyclic ligands such as crown ethers have been widely used for metal ion extraction, the basis for metal ion selectivity being the structure and cavity size of the crown ether. The hydrophobicity of the ligand can be adjusted by attachment of alkyl or aromatic ligands to the crown. Impressive results have been obtained with dicyclohexano-18-crown-6 as an extractant for Sr^{2+} in [RMIM][$(CF_3SO_2)_2N$] IL/aque-

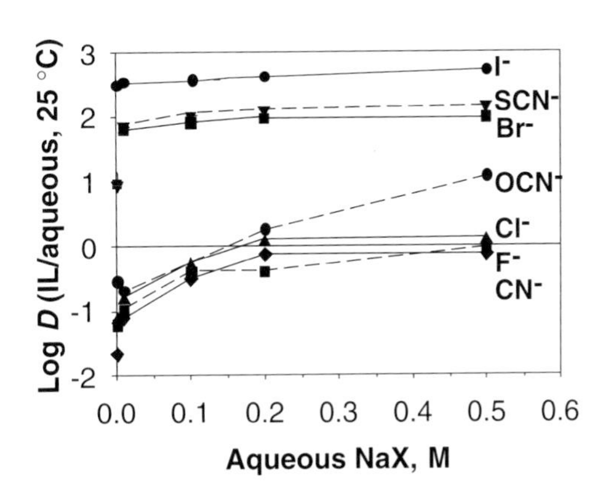

Figure 3.3-2: Hg^{2+} distribution ratios with increasing aqueous halide concentrations in [BMIM][PF_6]/aqueous systems. From reference [8].

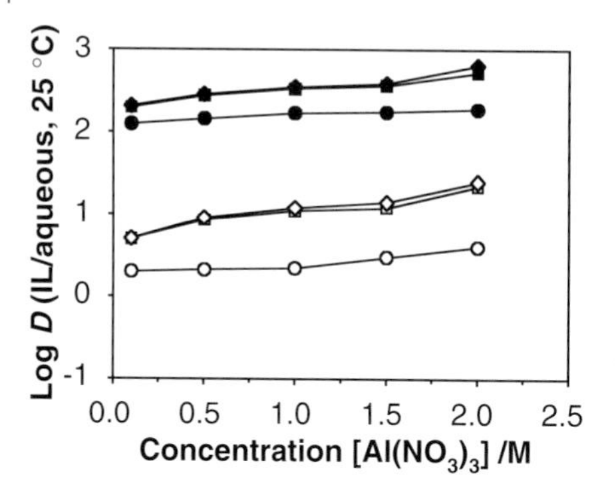

Figure 3.3-3: Distribution ratios for Sr^{2+} (closed symbol) and Cs^{+} (open symbol) with dibenzo-18-crown-6 (0.1 M) as extractant in IL/aqueous systems ([BMIM][PF_6] (●), [HMIM][PF_6] (■), [OMIM][PF_6] (◆)) as a function of increasing aqueous phase concentrations of [$Al(NO_3)_3$] (M). From reference [7].

ous systems [17] and with 18-crown-6, dicyclohexano-18-crown-6, and 4,4'-(5')-bis-(*tert*-butylcyclohexano)-18-crown-6 in [RMIM][PF_6]/aqueous systems as extractants for Sr^{2+}, Cs^{+}, and Na^{+} (Figure 3.3-3) [7]. Metal ion extraction into the IL is greatest at high anion concentration in the aqueous phase, but decreases with increasing acid concentration, in contrast to typical solvent extraction, where the higher distribution ratios are obtained with increasing acid concentrations. Results indicate that metal ion partitioning is very complex in IL-based liquid/liquid systems, and that other factors such as aqueous phase composition and water content of the IL have a dramatic effect on the metal ion extraction and the stability of the IL.

Organic extractants can be used to complex metal ions and to increase lipophilicity. The traditional metal extractants 1-(2-pyridylazo)naphthol (PAN) and 1-(2-thiazolyl)-2-naphthol (TAN) have been used in polymer-based aqueous biphasic systems [43] and traditional solvent extraction systems [44]. These are conventional metal extractants widely used in solvent extraction applications. When the aqueous phase is basic, both molecules are ionized, yet they quantitatively partition into [HMIM][PF_6] over the pH range 1–13. The distribution ratios for Fe^{3+}, Co^{2+}, and Cd^{2+} (Figure 3.3-4) show that the coordinating and complexing abilities of the extractants are dependent on pH and that metal ions can be extracted from the

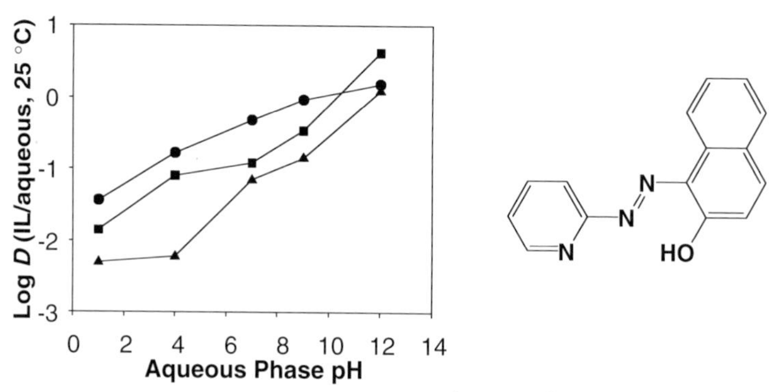

Figure 3.3-4: Metal ion distribution ratios for Fe^{3+} (●), Cd^{2+} (▲), and Co^{2+} (■) with 0.1 mM PAN in [HMIM][PF_6]/aqueous systems as a function of aqueous phase pH. From reference [8].

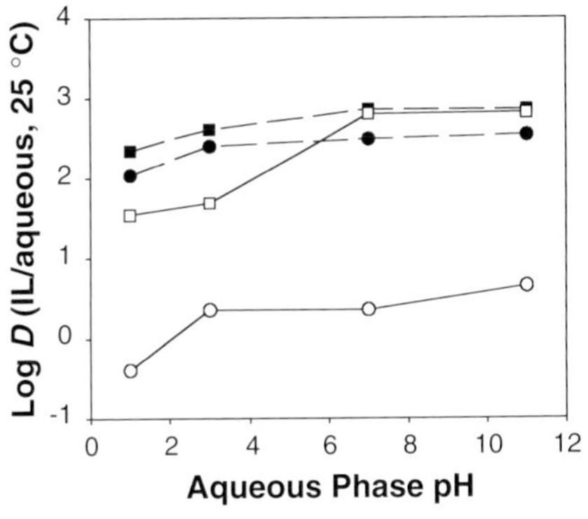

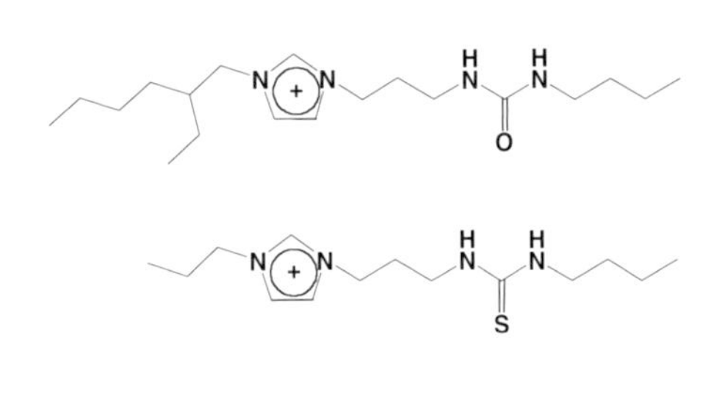

Figure 3.3-5: Hg^{2+} (■) and Cd^{2+} (○) distribution ratios between IL and aqueous phase with 1:1 [BMIM][PF_6] and urea-appended (dashed lines) or thiourea-appended (solid lines) TSILs as the extracting phase.

aqueous phase at basic pH and stripped from the IL under acidic conditions, as is the case when more conventional organic extracting phases are utilized.

Davis first introduced the term Task-specific Ionic Liquid (TSIL) to describe ILs prepared using the concept of increasing the IL affinity of the extractants through incorporation of the complexing functionality as an integral part of the IL [9]. It is thus possible to prepare ILs with built in extracting capability, and to achieve differentiation from both aqueous and organic phases through modification of the IL co-miscibility. These TSILs can be used as IL extracting phases, or may be mixed with a second, more conventional IL to modify rheological properties. Metal ion-ligating groups are incorporated into the cationic moiety of the IL by tethering to the imidazolium cation; thioether, urea, and thiourea-derivatized imidazolium ILs have been investigated as Hg^{2+} and Cd^{2+} extractants (see Figure 3.3-5) [9]. The distribution ratios are typically higher for Hg^{2+} and a change in the aqueous phase pH has only a slight effect on the partitioning.

In liquid/liquid systems that use ILs as alternatives to organic solvents, the tendency for metal ions to remain in the aqueous phase can be offset by the presence in the system either of organic or inorganic extractants or of TSILs. These extractants serve to modify the hydration environment of metal ions through complexation with ligating functional groups, increasing metal ion partitioning to the IL phase. In the design of an IL, fine-tuning of the properties can be achieved by changing the cation substituent groups and/or anion identity, or by mixing two types of IL with differing, but defined characteristics.

3.3.4 Organic Compounds

In general, ILs behave as moderately polar organic solvents with respect to organic solutes. Unlike the organic solvents to which they are commonly compared, however, they are poorly solvating and are rarely found as solvates in crystal structures.

Armstrong and co-workers [45] have investigated the interactions of solutes with ILs by using the ILs as stationary phases for gas-liquid chromatography (GLC), and have shown that ILs appear to act as low-polarity phases in their interactions with non-polar compounds and that the solubility increases with increasing lipophilicity (alkyl chain length etc.). However, polar molecules or those containing strong proton donor functionality (such as phenols, carboxylic acids, diols) also interact strongly with ILs. Compounds with weak proton donor/acceptor functions (such as aromatic and aliphatic ketones, aldehydes, and esters) appear to interact with the ionic liquids through induced ion dipole, or weak van der Waals interactions.

Bonhôte and co-workers [10] reported that ILs containing triflate, perfluorocarboxylate, and bistrifylimide anions were miscible with liquids of medium to high dielectric constant (ε), including short-chain alcohols, ketones, dichloromethane, and THF, while being immiscible with low dielectric constant materials such as alkanes, dioxane, toluene, and diethyl ether. It was noted that ethyl acetate ($\varepsilon = 6.04$) is miscible with the 'less-polar' bistrifylimide and triflate ILs, and only partially miscible with more polar ILs containing carboxylate anions. Brennecke [15] has described miscibility measurements for a series of organic solvents with ILs with complementary results based on bulk properties.

We have shown that ILs in general display partitioning properties similar to those of dipolar aprotic solvents or short chain alcohols. The relationship between octanol/water partitioning and IL/water partitioning [46] (Figure 3.3-6), despite the clear polarity differences between the solvents, allows the solubility or partitioning of organic solutes with ILs to be predicted from the relative polarities of the materials (by use of solvatochromatic scales etc.). Complex organic molecules such as cyclodextrins, glycolipids [45], and antibiotics [16] can be dissolved in ILs; the solubility of these complex molecules increases in the more polar ILs. The interactions are greatest when the ILs have H-bond acceptor capability (chloride-containing ILs, for example). The miscibility of ILs with water varies with cation substitution and with anion types; coordinating anions generally produce water-soluble ILs, whereas the presence of large, non-coordinating, charge-diffuse anions generates hydrophobic ILs.

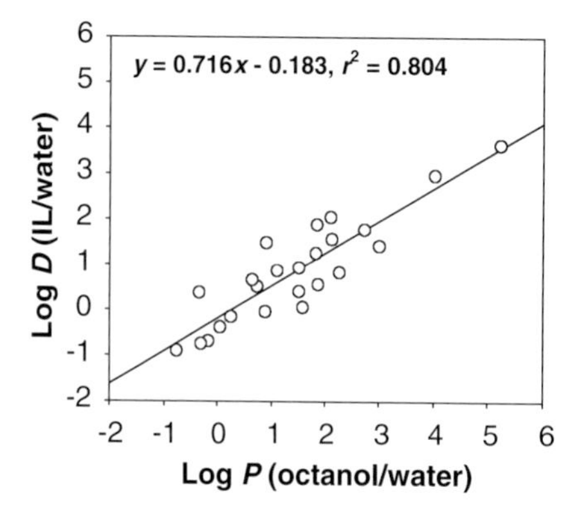

Figure 3.3-6: The distribution ratios between $[BMIM][PF_6]$ and water (neutral pH) for organic solutes correlate with literature partition functions of the solutes between octanol and water (log *P*).

From empirical observation, ILs tend to be immiscible with non-polar solvents. They can therefore be washed or brought into contact with diethyl ether or hexane to extract non-polar reaction products. Among solvents of greater polarity, esters (ethyl acetate, for example) exhibit variable solubility with ILs, depending on the nature of the IL. Polar or dipolar solvents (including chloroform, acetonitrile, and methanol) appear to be totally miscible with all ILs (excepting tetrachloroaluminate IL and the like, which react). Among notable exceptions, [EMIM]Cl and [BMIM]Cl are insoluble in dry acetone.

Although hydrocarbons are poorly soluble in most ILs, they are not insoluble. The relative solubility of short-chain hydrocarbons over oligomers and polymers is the basis for the efficient separation of products described for alkylation, oligomerization, and hydrogenation reactions using IL catalyst systems. The lipophilicity of the IL and the solubility of non-polar solutes can be increased by adding additional non-polar alkyl-functionality to the IL, thus reducing further Coulombic ion–ion interactions. Alkanes, for example, are essentially insoluble in all ILs, while the solubility of linear alkenes [34] is low, but increases with increased alkyl chain substitution in the IL cations and with delocalization of charge in the anion. Aromatic compounds are more soluble in ILs: benzene can be dissolved in [EMIM]Cl/$AlCl_3$, [BMIM][PF_6], and [EMIM][$(CF_3SO_2)_2N$] at up to ca. 1:1 ratios, reflecting the importance of CH...π and π–π stacking interactions and formation of liquid clathrate structures. Similarly, [BMIM][PF_6], [EMIM][BF_4], and [BMIM][PF_6] form liquid clathrates in chloroform, readily observable by ^{1}H NMR spectroscopy.

We have recently shown that the hydrophobic hexafluorophosphate ILs can in fact be made totally miscible with water by addition of alcohols [47, 48]; the ternary phase diagram for [BMIM][PF_6]/water/ethanol (left part of Figure 3.3-7) shows the

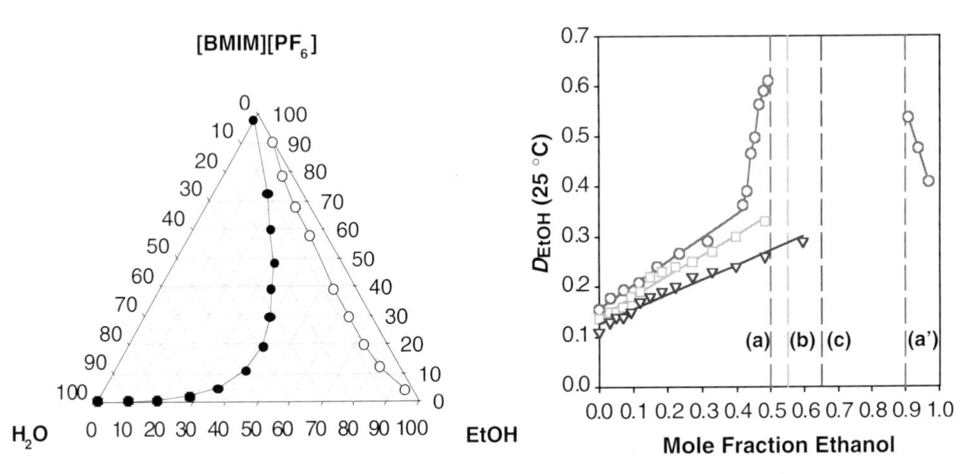

Figure 3.3-7: Ethanol/water/[BMIM][PF_6] ternary phase diagram (a, left) and solute distribution in EtOH/water/IL mixtures (b, right) for [BMIM][PF_6] (○), [HMIM][PF_6] (□), and [OMIM][PF_6] (∇) as a function of initial mole fraction of ethanol in the aqueous phase, measured at 25 °C. From references [47, 48].

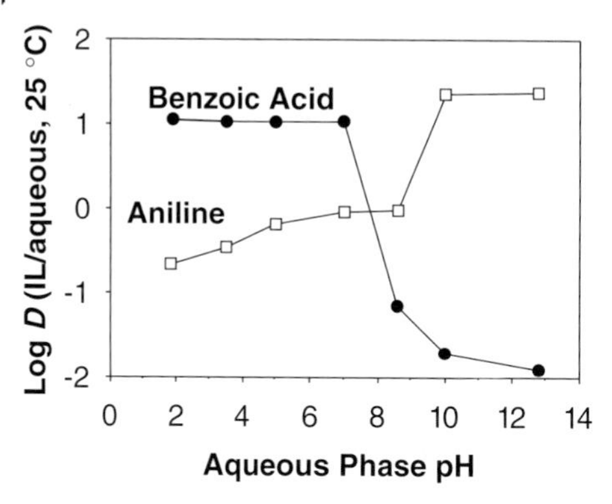

Figure 3.3-8: Distribution ratios for aniline ($pK_b = 9.42$) and benzoic acid ($pK_a = 4.19$) in [BMIM][PF_6]/aqueous systems as a function of the pH of the aqueous phase.

large co-miscibility region of the three components. In fact, ethanol forms biphasic mixtures with [BMIM][PF_6], [HMIM][PF_6], and [OMIM][PF_6], the degree of miscibility depending on temperature and on the water composition of the mixtures (right part of Figure 7). In each case, increasing the water content of the IL increases ethanol solubility. This process has many potential uses for washing and removal of ionic liquids from products and reactors or catalyst supports and has important implications in the design of IL/aqueous two-phase extraction systems.

With charged or ionizable solutes, changes in the pH of the aqueous phase resulted in certain ionizable solutes exhibiting pH-dependent partitioning such that their affinity for the IL decreased upon ionization [6]. Solute ionization effects, as demonstrated for aniline and benzoic acid (Figure 3.3-8), can modify solubility and partitioning of solutes into an IL by several orders of magnitude difference in the partitioning.

The pH-dependent partitioning of the ionizable, cationic dye thymol blue has also been investigated [6]. In its neutral, zwitterionic, and monoanionic forms, the dye preferentially partitions into the IL phase (from acidic solution), the partition coefficient to the IL increasing with increasing IL hydrophobicity. Under basic conditions, the dye is in the dianionic form and partitions into water (Figure 3.3-9).

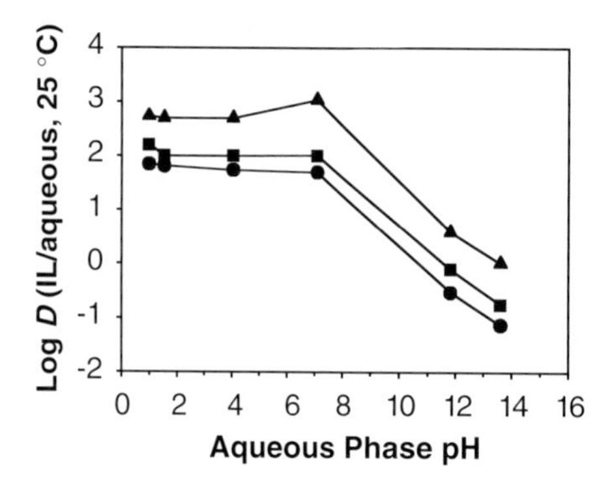

Figure 3.3-9: pH-switchable partitioning of the ionic dye thymol blue in [BMIM][PF_6] (●), [HMIM][PF_6] (■), [OMIM][PF_6] (▲)/aqueous biphasic systems. From reference [6].

Partitioning and solubility of specific organic compounds in ILs depends on the solubilizing interactions between the solute and IL components. As observed, ionized or polar compounds have high affinities for IL phases. It has been recognized that aromatic compounds have relatively high solubilities in ILs, even benzene being soluble to high concentrations in a wide range of ILs, presumably through C-H...π and π–π aromatic interactions with the IL cations. Atwood has shown that the high solubility of aromatics can be attributed to clathrate formation [49] and has described model extractions of toluene from toluene/heptane mixtures with $[EMIM][I_3]$ and $[BMIM][I_3]$ [50].

ILs containing 1-alkylisoquinolinium cations combined with the bis(perfluoroethylsulfonyl)imide anion ($[N(SO_2CF_2CF_3)_2]^-$, $[BETI]^-$) have been reported [51] and tested for organic partitioning in aqueous/IL two-phase systems. The large, extended aromatic cores in the cations of these ILs were expected to exhibit greater affinities for aromatic solutes in IL/aqueous partitioning experiments, and it was found, in particular, that the distribution ratio for 1,2,4-trichlorobenzene in $[C_{14}isoq][BETI]$ was much greater than that in $[BMIM][PF_6]$ [5, 46]. It is not yet clear whether interactions of aromatic solutes with the extended aromatic region of the isoquinolinium cations or increased lipophilicity factors were responsible for the increased aromatic partitioning.

3.3.5 Conclusions

Ionic liquids are similar to dipolar, aprotic solvents and short-chain alcohols in their solvent characteristics. These vary with anion (from very 'ionic' Cl^- to more 'covalent' $[BETI]^-$). ILs become more lipophilic with increasing alkyl substitution, resulting in increasing solubility of hydrocarbons and non-polar organics.

For separations, challenges lie in reconciling the partitioning results with those from more traditional systems. Continued study of organic solute behavior in IL-based liquid/liquid separations may facilitate a molecular level understanding of the partitioning mechanisms for neutral and ionic solutes, ultimately providing a predictive tool for their behavior. Exploration of the driving forces for organic solute partitioning should contribute to the understanding of metal ion extractants and partitioning mechanisms in IL systems. Incorporation of new concepts such as task-specific ionic liquids for separations should broaden both understanding and applicability.

References

1 R. Sheldon, *Chem. Commun.* **2001**, 2399.

2 W. Keim, P. Wasserscheid, *Angew. Chem. Int. Ed.* **2000**, *39*, 3772.

3 T. Welton, *Chem. Rev.* **1999**, *99*, 2071.

4 J. D. Holbrey, K. R. Seddon, *Clean Prod. Proc.* **1999**, *1*, 233.

5 J. G. Huddleston, A. E. Visser, W. M. Reichert, H. D. Willauer, G. A. Broker, R. D. Rogers, *Green Chem.* **2001**, *3*, 156.

6 A. E. Visser, R. P. Swatloski, R. D. Rogers, *Green Chem.* **2000**, *2*, 1.

7 A. E. Visser, R. P. Swatloski, W. M. Reichert, S. T. Griffin, R. D. Rogers, *Ind. Eng. Chem. Res.* **2000**, *39*, 3596.

8 A. E. Visser, R. P. Swatloski, S. T. Griffin, D. H. Hartman, R. D. Rogers, *Sep. Sci. Technol.* **2001**, *36*, 785.

9 A. E. Visser, R. P. Swatloski, W. M. Reichert, R. D. Rogers, R. Mayton, S. Sheff, A. Wierzbicki, J. H. Davis, Jr., *Chem. Commun.* **2001**, 135.

10 P. Bonhôte, A.-P. Dias, N. Papageorgiou, K. Kalyanasundaram, M. Grätzel, *Inorg. Chem.* **1996**, *35*, 1168.

11 A. J. Carmichael, K. R. Seddon, *J. Phys. Org. Chem.* **2000**, *13*, 591.

12 M. J. Muldoon, C. M. Gordon, I. R. Dunkin, *J. Chem. Soc., Perkin Trans. 2* **2001**, 433

13 S. N. V. K. Aki, J. F. Brennecke, A. Samanta, *Chem. Commun.* **2001**, 413.

14 D. Behar, C. Gonzalez, P. Neta, *J. Phys. Chem. A* **2001**, *105*, 7607.

15 L. A. Blanchard, J. F. Brennecke, *Ind. Eng. Chem. Res.* **2001**, *40*, 287.

16 S. G. Cull, J. D. Holbrey, V. Vargas-Mora, K. R. Seddon, G. J. Lye, *Biotech. Bioeng.* **2000**, *69*, 227.

17 S. Dai, Y. H. Ju, C. E. Barnes, *J. Chem. Soc. Dalton Trans.* **1999**, 1201.

18 W. R. Pitner, D. W. Rooney, K. R. Seddon, R. C. Thied, *WO 99/41752*, **2000**.

19 Z. Fang, S. Xu, J. A. Kozinski, *Ind. Eng. Chem. Res.*, **2000**, *39*, 4536.

20 M. Sasaki, Z. Fang, Y. Fukuskima, T. Adschiri, K. Arai, *Ind. Eng. Chem. Res.* **2000**, *39*, 2883.

21 C. Ohrenberg, W. E. Geiger, *Inorg. Chem.* **2000**, *39*, 2948.

22 C. L. Hussey, *Pure Appl. Chem.* **1988**, *60*, 1763.

23 M. Hasan, I. V. Kozhevnikov, M. R. H. Siddiqui, A. Steiner, N. Winterton, *Inorg. Chem.*, **1999**, *38*, 5637.

24 D. Appleby, P. B. Hitchcock, K. R. Seddon, J. E. Turp, J. A. Zora, C. L. Hussey, J. R. Sanders, T. A. Ryan, *J. Chem. Soc., Dalton Trans.* **1990**, *6*, 1879.

25 D. Appleby, R. I. Crisp, P. B. Hitchcock, C. L. Hussey, T. A. Ryan, J. R. Sanders, K. R. Seddon, J. E. Turp, J. A. Zora, *J. Chem. Soc., Chem. Commun.* **1986**, 483.

26 D. Appleby, C. L. Hussey, K. R. Seddon, J. E. Turp, *Nature* **1986**, *323*, 614.

27 M. Ortwerth, M. J. Wyzlic, R. Baughman, *Acta. Crystallogr.* **1998**, *C54*, 1594.

28 Y. Chauvin, L. Mussmann, H. Olivier, *Angew. Chem. Int. Ed.* **1996**, *34*, 2698.

29 S. Einloft, F. K. Dietrich, R. F. Desouza, J. Dupont, *Polyhedron* **1996**, *15*, 3257.

30 P. J. Dyson, D. J. Ellis, D. G. Parker, T. Welton, *Chem. Commun.* **1996**, 25.

31 Y. Chauvin, H. Olivier-Bourbigiou, *CHEMTECH* **1995**, *25*, 26.

32 M. F. Sellin, P. B. Webb, D. J. Cole-Hamilton, *Chem. Commun.* **2000**, 781.

33 P. Wasserscheid, H. Waffenschmidt, P. Machnitzki, K. W. Kottseiper, O. Stetzler, *Chem. Commun.* **2001**, 451.

34 F. Favre, H. Olivier-Bourbigou, D. Commereuc, L. Saussine, *Chem. Commun.* **2001**, 1360.

35 S. V. Ley, C. Ramarao, M. D. Smith, *Chem. Commun.* **2001**, 2278.

36 C. E. Song, E. J. Roh, *Chem. Commun.* **2000**, 837.

37 E. P. Horwitz, A. C. Muscatello, D. G. Kalina, L. Kaplan, *Sep. Sci. Technol.* **1981**, *16*, 1127.

38 E. P. Horwitz, K. A. Martin, H. Diamond, L. Kaplan, *Solv. Extr. Ion Exch.* **1986**, *4*, 449.

39 R. A. Sachleben, Y. Deng, D. R. Bailey, B. A. Moyer, *Solv. Extr. Ion Exch.*, **1997**, *14*, 995.

40 B. A. Moyer, P. V. Bonnesen, in *Physical Factors in Anion Separations* (A. Bianchi, K. Bowman-James, K. Garcia-Espana eds.), Wiley, New York, **1997**, 1–38.

41 R. D. Rogers and S. T. Griffin, *J. Chrom. B* **1998**, *711*, 277.

42 NIST Database 46: Critically Selected Stability Constants of Metal Complexes Database, U.S. Department of Commerce, Gaithersburg, MD, **1998**, ver. 5.0.

43 A. E. Visser, S. T. Griffin, D. H. Hartman, R. D. Rogers, *J. Chromatogr. B* **2000**, *743*, 107.

44 J. Gao, G. Hu, J. Kang, G. Bai, *Talanta* **1993**, *40*, 195.

45 D. W. Armstrong, L. He, Y. S. Lui, *Anal. Chem.* **1999**, *71*, 3873.

46 J. G. Huddleston, H. D. Willauer, R. P. Swatloski, A. E. Visser, R. D. Rogers, *Chem. Commun.* **1998**, 1765.

47 R. P. Swatloski, A. E. Visser, W. M. Reichert, G. A. Broker, L. M. Farina, J. D. Holbrey, R. D. Rogers, *Chem. Commun.* **2001**, 2070.

48 R. P. Swatloski, A. E. Visser, W. M. Reichert, G. A. Broker, L. M. Farina, J. D. Holbrey, R. D. Rogers, *Green Chem.* **2002**, *4*, 81.

49 J. L. Atwood, 'Liquid Clathrates', in *Inclusion Compounds* (J. L. Atwood, J. E. D. Davies, D. D. MacNicol eds.), Academic Press: London, **1984**, Vol. 1.

50 M. S. Selvan, M. D. McKinley, R. H. Dubois, J. L. Atwood, *J. Chem. Eng. Data.* **2000**, *45*, 841.

51 A. E. Visser, J. D. Holbrey, R. D. Rogers, *Chem. Commun.* **2001**, 2484.

3.4 Gas Solubilities in Ionic Liquids

Joan F. Brennecke, Jennifer L. Anthony, and Edward J. Maginn

3.4.1 Introduction

A wide variety of physical properties are important in the evaluation of ionic liquids (ILs) for potential use in industrial processes. These include pure component properties such as density, isothermal compressibility, volume expansivity, viscosity, heat capacity, and thermal conductivity. However, a wide variety of *mixture* properties are also important, the most vital of these being the phase behavior of ionic liquids with other compounds. Knowledge of the phase behavior of ionic liquids with gases, liquids, and solids is necessary to assess the feasibility of their use for reactions, separations, and materials processing. Even from the limited data currently available, it is clear that the cation, the substituents on the cation, and the anion can be chosen to enhance or suppress the solubility of ionic liquids in other compounds and the solubility of other compounds in the ionic liquids. For instance, an increase in alkyl chain length decreases the mutual solubility with water, but some anions ($[BF_4]^-$, for example) can increase mutual solubility with water (compared to $[PF_6]^-$, for instance) [1–3]. While many mixture properties and many types of phase behavior are important, we focus here on the solubility of gases in room temperature ILs.

A primary motivation for understanding gas solubilities in ILs is to be found in the many successful demonstrations of the use of ILs as solvents for reactions [4–6]. Some of these reactions, such as hydrogenations, oxidations, and hydroformylations, involve treatment of substrates in the ionic liquid solution with permanent and condensable gases. If a gas has limited solubility in the IL, then significant efforts will have to be made to increase interfacial area and enhance mass transfer, and/or high-pressure operation will be required. This may limit the ability of ILs to compete with conventional solvents, unless there are other significant chemical processing advantages to the IL. Conversely, high solubility or selective solubility of the desired gases might make ILs quite attractive.

A second motivation for understanding gas solubilities in ILs is the possibility of using ILs to separate gases. Because they are not volatile [7, 8] and would not contaminate the gas stream in even small amounts, ILs have an automatic advantage over conventional absorption solvents for the performance of gas separations. In addition, their high thermal stabilities mean that they could be used to perform gas separations at higher temperatures than is possible with conventional absorption solvents. Whether used in a conventional absorber arrangement or as a supported liquid membrane, the important physical properties for this application (besides low volatility) are the solubility and diffusivity of the gases of interest in the ILs.

A third motivation for studying gas solubilities in ILs is the potential to use compressed gases or supercritical fluids to separate species from an IL mixture. As an example, we have shown that it is possible to recover a wide variety of solutes from ILs by supercritical CO_2 extraction [9]. An advantage of this technology is that the solutes can be removed quantitatively without any cross-contamination of the CO_2 with the IL. Such separations should be possible with a wide variety of other compressed gases, such as C_2H_6, C_2H_4, and SF_6. Clearly, the phase behavior of the gas in question with the IL is important for this application.

Finally, a fourth motivation for exploring gas solubilities in ILs is that they can act as probes of the molecular interactions with the ILs. Information can be discerned on the importance of specific chemical interactions such as hydrogen bonding, as well as dipole–dipole, dipole–induced dipole, and dispersion forces. Of course, this information can be determined from the solubility of a series of carefully chosen liquids, as well. However, gases tend to be of the smallest size, and therefore the simplest molecules with which to probe molecular interactions.

In this section, we first discuss various experimental techniques that can be used to measure gas solubilities and related thermodynamic properties in ILs. We then describe the somewhat limited data currently available on gas solubilities in ILs. Finally, we discuss the impact that gas solubilities in ILs have on the applications described above (reactions, gas separations, separation of solutes from ILs) and draw some conclusions.

3.4.2 Experimental Techniques

In this section we describe some of the various experimental techniques that can be used to measure gas solubilities and related thermodynamic properties.

3.4.2.1 Gas solubilities and related thermodynamic properties

In general, gas solubilities are measured at constant temperature as a function of pressure. Permanent gases (gases with critical temperatures below room temperature) will not condense to form an additional liquid phase no matter how high the applied pressure. However, condensable gases (those with critical temperatures above room temperature) will condense to form a liquid phase when the vapor pressure is reached. The solubilities of many gases in normal liquids are quite low and can be adequately described at ambient pressure or below by Henry's law. The Henry's law constant is defined as

$$H_1(T,P) \equiv \lim_{x_1 \to 0} \frac{f_1^L}{x_1} \tag{3.4-1}$$

where x_1 is the mole fraction of gas in the liquid and $f_1{}^L$ the fugacity of the gas (species 1) in the liquid phase. If the gas phase behaves ideally (i.e., the fugacity coefficient is close to 1), then the fugacity is equal to the pressure of gas above the IL sample. This is because there is essentially no IL in the vapor phase, due to its nonvolatility. Experimentally, the Henry's law constant can be determined from the limiting slope of the solubility as a function of pressure. A large Henry's law constant indicates a low gas solubility, and a small Henry's law constant indicates a high gas solubility. One might also choose to express the limiting gas solubility (especially for condensable gases) in terms of an infinite dilution activity coefficient, where the standard state is pure condensed liquid at the temperature of the experiment $\left(P_1^{sat}\right)$. In this case, the infinite dilution activity coefficient, γ_1^∞ can be related to the Henry's law constant simply by $\gamma_1^\infty = H_1(T,P)P_1^{sat}(T)$

Also of importance is the effect of temperature on the gas solubility. From this information it is possible to determine the enthalpy and entropy change experienced by the gas when it changes from the ideal gas state $\left(h_1^{ig} \text{ and } s_1^{ig}\right)$ to the mixed liquid state $\left(\bar{h}_1 \text{ and } \bar{s}_1\right)$.

$$\Delta h_1 = \bar{h}_1 - h_1^{ig} = R\left(\frac{\partial \ln P}{\partial (1/T)}\right)_{x_1} \tag{3.4-2}$$

$$\Delta s_1 = \bar{s}_1 - s_1^{ig} = -R\left(\frac{\partial \ln P}{\partial \ln T}\right)_{x_1} \tag{3.4-3}$$

Thus, Δh_1 and Δs_1 can be obtained by determining the pressure required to achieve a specified solubility at several different temperatures and constant composition, x_1. In the Henry's law region, Δh_1 and Δs_1 can be found directly from the temperature dependence of the Henry's law constant, as given by the familiar van't Hoff equations:

$$\Delta h_1 = R\left(\frac{\partial \ln H_1}{\partial (1/T)}\right)_P \tag{3.4-4}$$

$$\Delta s_1 = -R\left(\frac{\partial \ln H_1}{\partial \ln T}\right)_P \tag{3.4-5}$$

The enthalpy and entropy of gas dissolution in the IL provide information about the strength of the interaction between the IL and the gas, and about the ordering that takes place in the gas/IL mixture, respectively.

Of course, a primary concern for any physical property measurement, including gas solubility, is the purity of the sample. Since impurities in ILs have been shown to affect pure component properties such as viscosity [10], one would anticipate that impurities might affect gas solubilities as well, at least to some extent. Since ILs are hygroscopic, a common impurity is water. There might also be residual impurities, such as chloride, present from the synthesis procedure. Surprisingly though, we found that even as much as 1400 ppm residual chloride in 1-*n*-octyl-3-methylimidazolium hexafluorophosphate and tetrafluoroborate ([OMIM][PF_6] and [OMIM][BF_4]) did not appear to have any detectable effect on water vapor solubility [1].

3.4.2.2 Stoichiometric technique

The simplest method to measure gas solubilities is what we will call the stoichiometric technique. It can be done either at constant pressure or with a constant volume of gas. For the constant pressure technique, a given mass of IL is brought into contact with the gas at a fixed pressure. The liquid is stirred vigorously to enhance mass transfer and to allow approach to equilibrium. The total volume of gas delivered to the system (minus the vapor space) is used to determine the solubility. If the experiments are performed at pressures sufficiently high that the ideal gas law does not apply, then accurate equations of state can be employed to convert the volume of gas into moles. For the constant volume technique, a known volume of gas is brought into contact with the stirred ionic liquid sample. Once equilibrium is reached, the pressure is noted, and the solubility is determined as before. The effect of temperature (and thus enthalpies and entropies) can be determined by repetition of the experiment at multiple temperatures.

The advantage of the stoichiometric technique is that it is extremely simple. Care has to be taken to remove all gases dissolved in the IL sample initially, but this is easily accomplished because one does not have to worry about volatilization of the IL sample when the sample chamber is evacuated. The disadvantage of this technique is that it requires relatively large amounts of ILs to obtain accurate measurements for gases that are only sparingly soluble. At ambient temperature and pressure, for instance, 10 cm^3 of 1-*n*-butyl-3-methylimidazolium hexafluorophosphate ([BMIM][PF_6]) would take up only 0.2 cm^3 of a gas with a Henry's law constant of

5000 bar. Also, small temperature variations can cause large uncertainties. For instance, for 50 cm^3 of gas, a temperature fluctuation of just 1 °C would also cause about a 0.2 cm^3 volume change. For some metal apparatus of this type, gas adsorption on the metal surfaces can be an additional source of error. Thus, stoichiometric measurements are best for high-solubility gases and, in general, require excellent temperature and pressure control and measurement, as well as relatively large samples.

3.4.2.3 Gravimetric technique

An alternative technique to the stoichiometric method for measuring gas solubilities has evolved as a result of the development of extremely accurate microbalances. The gravimetric technique involves the measurement of the weight gain of an IL sample when gases are introduced into a sample chamber at a given pressure. There are various commercial apparatus (e.g., Hiden Analytical, Cahn, Rubotherm) well suited for this purpose. The gravimetric technique was originally designed for gas uptake by solids (such as zeolites), but it is well suited for ILs. Even the powerful vacuum (~ 10^{-9} bar) used to evacuate the system prior to gas introduction does not evaporate the ionic liquid.

The main advantage of the gravimetric technique is that it requires a much smaller sample than the stoichiometric technique. In many cases, samples as small as 70 mg are sufficient. Accurate temperature and pressure control and measurement are still required, but gas adsorption on the metal walls of the equipment is no longer a concern because it is only the weight gain of the sample that is measured.

There are two main disadvantages to this technique. Firstly, the sample is placed in a static sample "bucket", and so there is no possibility of stirring. Equilibrium is thus reached solely by diffusion of the gas into the IL sample. For the more viscous samples this can require equilibration times of as much as several hours. Secondly, the weight gain must be corrected for the buoyancy of the sample in order to determine the actual gas solubility. While the mass is measured accurately, the density of the sample must also be known accurately for the buoyancy correction. This is a particularly important problem for low-solubility gases, where the buoyancy correction is a large percentage of the weight gain. For example, the density of an IL must be known to at least $\pm$ 0.5 % if one wishes to measure the solubility of gases with Henry's law constants greater than 2000 bar accurately. A detailed description of a Hiden Analytical (IGA003) microbalance and its use for the measurement of gas solubilities in ILs can be found elsewhere [1].

3.4.2.4 Gas chromatography

Another method to determine infinite dilution activity coefficients (or the equivalent Henry's law coefficients) is gas chromatography [11, 12]. In this method, the chromatographic column is coated with the liquid solvent (e.g., the IL). The solute (the gas) is introduced with a carrier gas and the retention time of the solute is a measure of the strength of interaction (i.e., the infinite dilution activity coefficient, γ_1^∞) of the solute in the liquid. For the steady-state method, γ_1^∞ is given by [11, 12]:

$$F^0(t - t_{ref})J_3^2 = \frac{RTn_2^1}{\gamma_1^\infty P_1^{sat}} \qquad (3.4\text{-}6)$$

where F^0 is the flow rate of the carrier gas, t and t_{ref} are the retention times of the solute of interest and a solute that is not retained, respectively, J_3^2 is a correction factor for the pressure drop across the column, R is the gas constant, T is the temperature, n_2^1 is the mass of solvent on the column, and P_1^{sat} is the vapor pressure of the solute. Additional corrections are required for a high pressure, non-ideal gas phase. This technique has been used to measure the infinite dilution activity coefficients of a wide variety of liquid solutes in ionic liquids [13] and could conceivably be used to determine the infinite dilution activity coefficients of condensable gases as well. This technique would work for condensable gases that were retained by the IL more strongly than the carrier gas (usually helium), which may well be the case for many of the alkanes and alkenes of interest, as shown below.

3.4.3
Gas Solubilities

Although the solubilities of gases in ILs are extremely important, at the time of this writing the number of published studies are limited. Some measurements were presented in oral and poster presentations at a five day symposium dedicated to ionic liquid research at the American Chemical Society national meeting in San Diego in April, 2001. Scovazzo et al. [14], for instance, presented preliminary results for CO_2 and N_2 solubility in [BMIM][PF_6], and Rooney et al. [15] presented the solubility of several gases in several different ILs as determined by the stoichiometric technique. A recent manuscript [16] presented Henry's law constants for H_2 in two ILs. Given the lack of availability of other data, we concentrate below on the data collected in our laboratories.

3.4.3.1 Water vapor

The solubility of water vapor in ionic liquids is of interest because ionic liquids are extremely hygroscopic. In addition, the solubility of water vapor in ILs is an excellent test of the strength of molecular interactions in these fluids. By using the gravi-

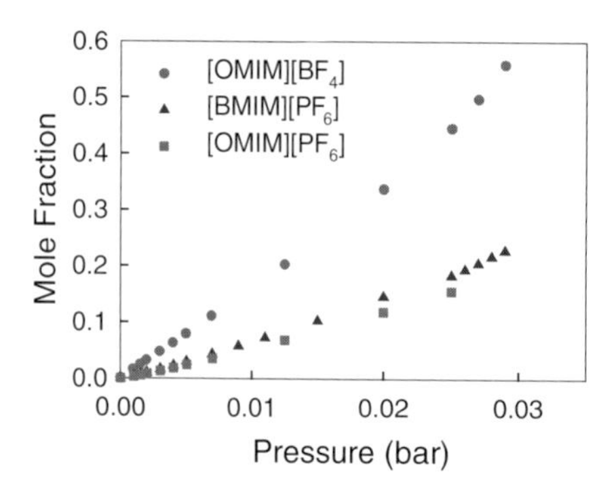

Figure 3.4-1: Solubility of water vapor in [BMIM][PF_6], [OMIM][PF_6], and [OMIM][BF_4] at 25 °C.

Table 3.4-1: Henry's Law Constants, H_1, for water in [OMIM][BF_4], [BMIM][PF_6], and [OMIM][PF_6].

T (°C)	P_{sat} *(bar)*	*[OMIM][BF_4]*	*[BMIM][PF_6]*	*[OMIM][PF_6]*
		H_1 (bar)	H_1 (bar)	H_1 (bar)
10	0.012	0.033 ± 0.014	0.09 ± 0.02	0.11 ± 0.03
25	0.031	0.055 ± 0.006	0.17 ± 0.02	0.20 ± 0.03
35	0.055	0.118 ± 0.014	0.25 ± 0.04	0.30 ± 0.02
50	0.122	–	0.45 ± 0.05	–

metric technique (Hiden Analytical IGA003) we have measured the solubility of water vapor in [BMIM][PF_6], [OMIM][PF_6], and [OMIM][BF_4] at temperatures between 10 and 50 °C and pressures up to about 80 % of the saturation pressure at each temperature [1]. The data for the three compounds at 25 °C, at which the saturation pressure is just 0.031 bar, are shown in Figure 3.4-1.

The solubility of water vapor is much greater in the IL containing $[BF_4]^-$ as the anion, perhaps due to the greater charge density [17] or simply to there being more space for the water molecules [18]. Increasing the length of the alkyl chain on the imidazolium ring decreases the solubility of water, as would be expected. The Henry's law constants are shown in Table 3.4-1. The small values indicate extremely high water vapor solubility in all the ILs.

The Δh_1 and Δs_1 values for the absorption of water into the three ILs are most similar to water absorption into polar compounds such as 2-propanol. The enthalpy change when water vapor is dissolved in the ILs is compared to dissolution in various organics solvents in Figure 3.4-2 [19, 20]. The enthalpies for absorption of water vapor into polar and protic solvents are much greater than for its absorption into non-polar solvents, indicating much stronger molecular interactions between water and the polar solvents (including opportunities for hydrogen-bonding). In fact, the enthalpy for the absorption of water vapor into the ILs is almost as great as the value for water condensation, which is –44 kJ/mol. Thus, the interactions between water and the ILs are quite strong and probably involve hydrogen bonding.

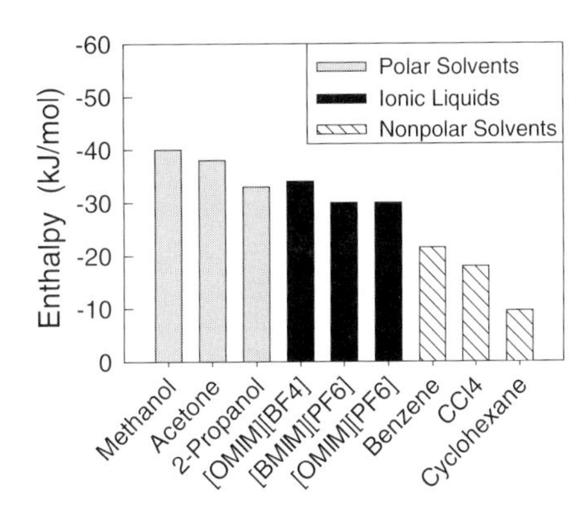

Figure 3.4-2: Enthalpy of absorption for water vapor in various solvents at 25 °C.

3.4.3.2 Other gases

We have also used the gravimetric technique to measure the solubilities of CO_2, C_2H_6, C_2H_4, CH_4, O_2, Ar, H_2, N_2, and CO in [BMIM][PF_6] at temperatures between 10 and 50 °C and pressures to 13 bar [21, 22]. In complementary work, we used a constant pressure stoichiometric technique to measure the solubility of CO_2 in [BMIM][PF_6], [OMIM][PF_6], [OMIM][BF_4], 1-*n*-butyl-3-methylimidazolium nitrate ([BMIM][NO_3]), 1-ethyl-3-methylimidazolium ethylsulfate ([EMIM][$EtSO_4$]), and *N*-butylpyridinium tetrafluoroborate ([BP][BF_4]) at 40, 50, and 60 °C and pressures to 95 bar [23].

CO_2 has by far the highest solubility of all of the gases tested (with the exception of water vapor), as can be seen from Table 3.4-2, which gives the Henry's law constants of the various gases in [BMIM][PF_6] at 25 °C. In the high-pressure measurements, CO_2 solubilities reached as high as 70 mole % [23]. Although there were some differences (e.g., at a given pressure the CO_2 solubility in [EMIM][$EtSO_4$] was about half that in [BMIM][PF_6]), the solubility of CO_2 was remarkably high in all of the ILs tested. However, CO_2/IL mixtures did not become single-phase at higher pressures and there was no detectable IL dissolved in the CO_2-rich gas phase. Even more interestingly, the dissolution of the CO_2 in the IL was accompanied by very little volume increase.

Of the remaining gases, C_2H_4, C_2H_6, and CH_4 are the next most soluble, in that order. O_2 and Ar display some measurable solubility, but the solubilities of H_2, N_2, and CO are below the detection limit of our equipment. The estimates of these detection limits for each gas, which depend on the molecular weight of the gas, are listed in Table 3.4-2. H_2 is particularly difficult to measure by the gravimetric techniques, due to its low molecular weight. For the low-solubility gases, the greatest source of uncertainty in the measurements is the uncertainty in the IL density needed for the buoyancy correction. The solubilities of the non-polar gases, apart from CO_2, which has a large quadrupole moment, correlate reasonably well with the polarizabilities of the gases. On the basis of this correlation, we would predict Henry's law constants for H_2, N_2, and CO of 20,000 bar, 6000 bar, and 5000 bar, respectively, at 25 °C. This would suggest that we should have been able to measure the Henry's law constants for N_2 and CO, but this was not the case. Similar trends are observed at the other temperatures [22].

Gas	*Henry's Law Constant (Bar)*
CO_2	53.4 ± 0.3
C_2H_4	173 ± 17
C_2H_6	355 ± 36
CH_4	1690 ± 180
O_2	8000 ± 5400
Ar	8000 ± 3800
H_2	Non-detectable (>1500)
N_2	Non-detectable (>20000)
CO	Non-detectable (>20000)

Table 3.4-2: Henry's Law Constants, H_1, for various gases in [BMIM][PF_6] at 25 °C.

Berger and coworkers [16] presented measurements for the Henry's law constant (defined as M = K*P, where M is the gas solubility in mol L^{-1}, K is the constant in mol $L^{-1}atm^{-1}$, and P is the gas partial pressure in atm.) of H_2 in [BMIM][PF_6] and [BMIM][BF_4] of 3.0×10^{-3} mol L^{-1} atm^{-1} and 8.8×10^{-4} mol L^{-1} atm^{-1}, respectively. The value for [BMIM][PF_6] corresponds to about 5700 bar, which is consistent with our measurements. However, it should be noted that Berger et al. [16] used a constant volume stoichiometric technique with a 50 cm^3 vessel, pressurized to 50 atm. and containing 10 cm^3 of IL. The resulting pressure drop when the gas is absorbed into the liquid would only be on the order of 0.005 atm. The authors do not report the uncertainty of their Henry's constants, nor the accuracy of their pressure gauge. Unless a highly accurate differential pressure transducer was employed, it is likely that these values are good order of magnitude estimates only.

Not surprisingly, the enthalpy of adsorption of CO_2 is the highest of the gases listed in Table 3.4-2, indicating strong interactions between the CO_2 and [BMIM][PF_6]. However, it is significantly less than that for the dissolution of H_2O vapor in [BMIM][PF_6]. The enthalpies decrease in magnitude in the order of decreasing solubility. Interestingly, O_2 and Ar exhibit positive enthalpies of absorption, indicating that there are no measurable attractive forces and that their dissolution in the IL is entirely entropically driven. Thus, gas solubilities are a useful "probe" of interactions in ILs.

3.4.4 Applications

The solubilities, discussed above, of the various gases in the ionic liquids have important implications for applications of ILs. The impact of gas solubilities on reactions, gas separations and the use of compressed gases or supercritical fluids to separate solutes from ILs are discussed below.

3.4.4.1 Reactions involving gases

In general, the rates of most reactions depend on the concentration of the reacting species, by some positive power. Thus, it is generally desirable to have high solubilities of reactants. For reactions in ILs involving gases, this means that one would desire high gas solubilities. This is true in all cases except for instances in which the inherent reaction rate is extremely slow (i.e., slower than the natural diffusion of the gas into the IL). Otherwise, one would have to resort to high-pressure operation or vigorous stirring in an attempt to increase interfacial area and to promote faster mass transfer. In this case, the reaction rate is limited by the rate at which the gas is transferred into the liquid, rather than by the inherent reaction kinetics. This could be a particular challenge when using ILs as solvents, since they tend to be more viscous than conventional solvents [10, 24]. High viscosity means lower diffusivities and more difficulties in attempts to promote adequate mass transfer.

Some types of reactions involving gases that have been studied in ILs are hydrogenations [16, 25–37], oxidations [38, 39], and hydroformylations [25, 40–45]. In addition, some dimerizations and alkylations may involve the dissolution of condensable gases (e.g., ethylene, propylene, isobutene) in the IL solvent [46–50].

Surprisingly, no effort was made to determine the influence of gas solubility in most of these reaction studies, nor whether the reaction was, in fact, mass-transfer limited. Because of the low solubilities of H_2, O_2, and CO in [BMIM][PF_6], one of the most common IL solvents used in the above studies, however, most of the reactions in these studies are likely to have been mass transfer-limited. Suarez et al. [27], for example, noted that in the absence of stirring the consumption of hydrogen in their hydrogenation reaction stopped completely. This same group [16] is also responsible for the estimate of the Henry's law constant of hydrogen in [BMIM][PF_6] and [BMIM][BF_4], as discussed above. They correctly note that, on comparing reaction results in different ILs, the important physical parameter to hold constant is the gas solubility rather than the gas partial pressure. Moreover, for mass transfer-limited reactions, it is not possible to compare results of reaction rates for reactions performed by different research groups. This is because the rates depend solely on the quality of the interphase mass transfer, and all the researchers perform the reactions with different mixing conditions.

As mentioned above, the solubilities of the gases needed for hydrogenation reactions (H_2), oxidation reactions (O_2), and hydroformylation reactions (H_2 and CO) in [BMIM][PF_6] are extremely low. In fact, the solubilities are generally lower than in conventional organic solvents [22]. The solubility of these gases may be higher in other ILs; for instance, Berger et al. [16] estimate that H_2 solubility is four times greater in [BMIM][BF_4] than in [BMIM][PF_6]. However, one would not expect orders of magnitude changes in solubility when the anion or substituents on the cation were changed slightly. The primary substrate in the hydroformylation reactions is an alkene, the lightest example of which would be ethylene. Since the ethylene solubility is quite high, the limiting factors are likely to be the CO and H_2 solubilities, rather than the solubility of the alkene. Thus, overall, [BMIM][PF_6] does not appear to be a good industrial solvent for reactions involving permanent gases such as H_2, O_2, or CO, because these reactions would have to be carried out at extremely high pressures or would be limited by interphase mass transfer. While there are certainly examples of commercial biphasic reactions that are limited by interphase mass transfer [51], that type of reactor configuration is generally not preferred. There would have to be significant independent factors favoring the IL solvent (such as higher selectivity) to warrant the use of the IL solvent.

3.4.4.2 **Gas separations**

ILs, on the other hand, are uniquely suited for use as solvents for gas separations. Since they are non-volatile, they cannot evaporate to cause contamination of the gas stream. This is important when selective solvents are used in conventional absorbers, or when they are used in supported liquid membranes. For conventional absorbers, the ability to separate one gas from another depends entirely on the relative solubilities (ratio of Henry's law constants) of the gases. In addition, ILs are particularly promising for supported liquid membranes, because they have the potential to be incredibly stable. Supported liquid membranes that incorporate conventional liquids eventually deteriorate because the liquid slowly evaporates. Moreover, this finite evaporation rate limits how thin one can make the membrane. This means that the net flux through the membrane is decreased. These problems could be eliminated with a non-volatile liquid. In the absence of facilitated transport (such as complexation of CO_2 with amines to form carbamates), permeability of gases through supported liquid membranes depends both on the solubility and on the diffusivity. The flux of one gas relative to the other can be estimated by a simplified solution-diffusion model:

$$\frac{N_A}{N_B} = \frac{H_B D_A}{H_A D_B} \tag{3.4-7}$$

where N is the flux, H is the Henry's law constant, and D is the diffusivity. Thus, the ratio of Henry's law constants is critical in determining the performance of an IL-based supported liquid membrane.

The solubilities of the various gases in [BMIM][PF_6] suggests that this IL should be an excellent candidate for a wide variety of industrially important gas separations. There is also the possibility of performing higher-temperature gas separations, thanks to the high thermal stability of the ILs. For supported liquid membranes this would require the use of ceramic or metallic membranes rather than polymeric ones. Both water vapor and CO_2 should be removed easily from natural gas since the ratios of Henry's law constants at 25 °C are ~9950 and 32, respectively. It should be possible to scrub CO_2 from stack gases composed of N_2 and O_2. Since we know of no measurements of H_2S, SO_x, or NO_x solubility in [BMIM][PF_6], we do not know if it would be possible to remove these contaminants as well. Nonetheless, there appears to be ample opportunity for use of ILs for gas separations on the basis of the widely varying gas solubilities measured thus far.

3.4.4.3 Extraction of solutes from ionic liquids with compressed gases or supercritical fluids

We have shown that it is possible to extract a wide variety of solutes from ILs with supercritical CO_2 [9, 52]. The advantage of this technique is that it combines two potentially environmentally benign solvents to perform the reaction and separation steps. Subsequently, several groups have shown that this strategy can be combined with reaction operations [36, 37, 45]. The key to this separation is the phase behavior of ILs with CO_2. Although large amounts of CO_2 dissolve in the IL, no measurable IL dissolves in the CO_2. Conceivably, it should be possible to extract solutes from IL mixtures with other supercritical fluids or compressed gases. For instance, supercritical ethane and propane are excellent solvents for a wide variety of nonpolar and aromatic species. These are not as environmentally benign as CO_2 but may be options if it is shown that the ILs have negligible solubility in them, as well.

3.4.5 Summary

The solubility of various gases in ionic liquids is extremely important in evaluating ILs as solvents for reactions, separations, and materials processing. There are a number of viable techniques for measuring gas solubilities in ILs, including the stoichiometric method, gravimetric methods, or even gas chromatography. In general, the measurement of these solubilities is facilitated by the non-volatility of the solvent. For [BMIM][PF_6], we have shown very large solubility differences between the relatively high-solubility gases (CO_2, C_2H_4, C_2H_6, and CH_4) and the low-solubility gases (CO, H_2, O_2, Ar, and N_2). Unfortunately, many of the gases of interest for reactions (H_2, O_2 and CO) are only sparingly soluble in the IL. Conversely, the large differences in the solubilities of different gases in the IL suggest that ILs may be ideal solvents for performing gas separations. Limited measurements of CO_2 solubility in different ILs suggest that the trends found for [BMIM][PF_6] may be representative. However, the wide variety of choice of cations, anions, and substituents make the possibility of tailoring ILs for specific gas separations or reactions involving gases an exciting option.

References

1. J. L. Anthony, E. J. Maginn, J. F. Brennecke, *J. Phys. Chem. B* **2001**, *105*, 10,942–10,949.
2. A. G. Fadeev, M. M. Meagher, *Chem. Commun.* **2001**, 295–296.
3. K. R. Seddon, A. Stark, M. J. Torres, *Pure Appl. Chem.* **2000**, *72*, 2275–2287.
4. J. D. Holbrey, K. R. Seddon, *Clean Products and Processes* **1999**, *1*, 233–236.
5. T. Welton, *Chem. Rev.* **1999**, *99*, 2071–2083.
6. P. Wasserscheid, W. Keim, *Angew. Chem. Int. Ed.* **2000**, *39*, 3772–3789.
7. C. L. Hussey, *Pure Appl. Chem.* **1988**, *60*, 1763–1772.
8. K. R. Seddon, *Kinet. Catal.* **1996**, *37*, 693–697.
9. L. A. Blanchard, J. F. Brennecke, *Ind. Eng. Chem. Res.* **2001**, *40*, 287–292.
10. K. R. Seddon, A. Stark, M. J. Torres, *ACS Symp. Ser.* (L. Moens, M. Abraham eds.), *ACS Symp. Ser. 819* (L. Moens, M. Abraham eds.), **2002**, p. 34–49. The title of the symposium series is “Clean Solvents”.
11. R. J. Laub, R. L. Pecsok, *Physicochemical Applications of Gas Chromatography*, **1978**, Wiley, New York.
12. H. Orbey, S. I. Sandler, *Ind. Eng. Chem. Res.* **1991**, *30*, 2006–2011.
13. A. Heintz, D. V. Kulikov, S. P. Verevkin, *J. Chem. Eng. Data* **2001**, 46, 1526–1529.
14. P. Scovazzo, A. E. Visser, J. H. Davis, R. D. Rogers, R. D. Noble, C. Koval, *Abstracts of Papers of the American Chemical Society* **2001**, *221*, 28-IEC.
15. D. W. Rooney, C. Hardacre, S. Kadare, *Abstracts of Papers of the American Chemical Society* **2001**, *221*, 56-IEC.
16. A. Berger, R. F. de Souza, M. R. Delgado, J. Dupont, *Tetrahedron: Asymmetry* **2001**, *12*, 1825–1828.
17. A. B. McEwen, H. L. Ngo, K. LeCompte, J. L. Goldman, *J. Electrochem. Soc.* **1999**, *146*, 1687–1695.
18. I. Jureviciute, S. Bruckenstein, A. R. Hillman, *J. Electroanal. Chem.* **2000**, *488*, 73–81.
19. S. Goldman, *Can. J. Chem.* **1974**, *88*, 1668.
20. J. Gmehling, J. Menke, M. Schiller, *Activity Coefficients at Infinite Dilution*; DECHEMA, Frankfurt, **1994**.
21. J. L. Anthony, E. J. Maginn, J. F. Brennecke, *ACS Symp. Series 818* (R. D. Rogers, K. R. Seddon, Eds.), **2002**, p. 260–270.
22. J. L. Anthony, E. J. Maginn, J. F. Brennecke, *J. Phys. Chem. B*, **2002**, *106*, 7315–7320.

23 L. A. Blanchard, Z. Gu, J. F. Brennecke, *J. Phys. Chem. B* **2001**, *105*, 2437–2444.

24 P. Bonhôte, A.-P. Dias, N. Papageorgiou, K. Kalyanasundaram, M. Grätzel, *Inorg. Chem.* **1996**, *35*, 1168–1178.

25 Y. Chauvin, L. Mussmann, H. Olivier, *Angew. Chem. Int. Ed. Engl.* **1995**, *34*, 2698–2700.

26 P. A. Z. Suarez, J. E. L. Dullius, S. Einloft, R. F. de Souza, J. Dupont, *Polyhedron* **1996**, *15*, 1217–1219.

27 P. A. Z. Suarez, J. E. L. Dullius, S. Einloft, R. F. de Souza, J. Dupont, *Inorganica Chimica Acta* **1997**, *255*, 207–209.

28 A. L. Monteiro, F. K. Zinn, R. F. de Souza, J. Dupont, *Tetrahedron: Asymmetry* **1997**, *8*, 177–179.

29 R. T. Carlin, J. Fuller, *Chem. Commun.* **1997**, 1345–1346.

30 T. H. Cho, J. Fuller, R. T. Carlin, *High Temp. Material Processes* **1998**, *2*, 543–558.

31 L. A. Müller, J. Dupont, R. F. de Souza, *Macromol. Rapid Commun.* **1998**, *19*, 409–411.

32 P. J. Dyson, D. J. Ellis, D. G. Parker, T. Welton, *Chem. Commun*, **1999**, 25–26.

33 D. J. Ellis, P. J. Dyson, D. G. Parker, T. Welton, *J. Mol. Cat. A* **1999**, *150*, 71–75.

34 S. Steines, P. Wasserscheid, B. Drießen-Hölscher, *J. Prakt. Chem.* **2000**, *342*, 348–354.

35 S. Guernik, A. Wolfson, M. Herskowitz, N. Greenspoon, S. Geresh, *Chem. Commun.* **2001**, 2314–2315.

36 R. A. Brown, P. Pollet, E. McKoon, C. A. Eckert, C. L. Liotta, P. G. Jessop, *J. Am. Chem. Soc.* **2001**, *123*, 1254–1255.

37 F. Liu, M. B. Abrams, R. T. Baker, W. Tumas, *Chem. Commun.* **2001**, 433–434.

38 J. Howarth, *Tetrahedron Lett.* **2000**, *41*, 6627–6629.

39 L. Gaillon, F.Bedioni, *Chem. Commun.* **2001**, 1458–1459.

40 N. Karodia, S. Guise, C. Newlands, J. Anderson, *Chem. Commun.* **1998**, 2341–2342.

41 W. Keim, D. Vogt, H. Waffenschmidt, P. Wasserscheid, *J. Catalysis* **1999**, *186*, 481–484.

42 C. C. Brasse, U. Englert, A. Salzer, H. Waffenschmidt, P. Wasserscheid, *Organometallics* **2000**, *19*, 3818–3823.

43 F. Favre, H. Olivier-Bourbigou, D. Commereuc, L. Saussine, *Chem. Commun.* **2001**, 1360–1361.

44 D. J. Brauer, K. W. Lottsieper, C. Like, O. Stelzer, H. Waffenschmidt, P. Wasserscheid, *J. Organometallic Chem.* **2001**, *630*, 177–184.

45 M. F. Sellin, P. B. Webb, D. J. Cole-Hamilton, *Chem. Commun.* **2001**, 781–782.

46 Y. Chauvin, B. Gilbert, I. Guibard, *J. Chem. Soc., Chem. Commun.* **1990**, 1715–1716.

47 Y. Chauvin, A. Hirschauer, H. Olivier, *J. Mol. Cat.* **1994**, *92*, 155–165.

48 Y. Chauvin, S. Einloft, H. Olivier, *Ind. Eng. Chem. Res.* **1995**, *34*, 1149–1155.

49 S. Einloft, F. K. Dietrich, R. F. de Souza, J. Dupont, *Polyhedron*, **1996**, *15*, 1257–1259.

50 Y. Chauvin, H. Olivier, C. N. Wyrvalski, L. C. Simon, R. F. de Souza, *J. Catalysis* **1997**, *165*. 275–278.

51 E. G. Kuntz, *CHEMTECH* **1987**, 570.

52 L. A. Blanchard, D. Hancu, E. J. Beckman, J. F. Brennecke, *Nature* **1999**, *399*, 28–29.

3.5
Polarity

Tom Welton

It is well known that the choice of solvent can have a dramatic effect upon a chemical reaction [1]. As early as 1862 the ability of solvents to decelerate the reaction between acetic acid and ethanol had been noted [2]. Thirty years later the influence of solvents on reaction equilibria was demonstrated for the first time [3].

Once such effects had been noted, it became necessary to interpret the observed results and to classify the solvents. The earliest attempts at this were by Stobbe, who reviewed the effects of solvents on keto-enol tautomers [4]. Since then many attempts have been used to explain solvent effects, some based on observations of chemical reactions, others on physical properties of the solvents, and yet others on spectroscopic probes. All of these have their advantages and disadvantages and no one approach can be thought of as exclusively "right". This review is organized by type of measurement, and the available information is then summarized at the end.

Most modern discussions of solvent effects rely on the concept of solvent polarity. Qualitative ideas of polarity are based on observations such as "like dissolves like" and are well accepted. However, quantification of polarity has proven to be extraordinarily difficult. Since the macroscopic property polarity arises from a myriad of possible microscopic interactions, this is perhaps unsurprising. Hence, it is important that care is taken when measuring the "polarity" of any liquid to ensure that it is clearly understood what is actually being measured.

The most common measure of polarity used by chemists in general is that of dielectric constant. It has been measured for most molecular liquids and is widely available in reference texts. However, direct measurement, which requires a nonconducting medium, is not available for ionic liquids. Other methods to determine the "polarities" of ionic liquids have been used and are the subject of this chapter. However, these are early days and little has been reported on ionic liquids themselves. I have therefore included the literature on higher melting point organic salts, which has proven to be very informative.

3.5.1
Chromatographic Measurements

In a series of papers published throughout the 1980s, Colin Poole and his co-workers investigated the solvation properties of a wide range of alkylammonium and, to a lesser extent, phosphonium salts. Parameters such as McReynolds' phase constants were calculated by using the ionic liquids as stationary phases for gas chromatography and analysis of the retention of a variety of probe compounds. However, these analyses were found to be unsatisfactory and were abandoned in favour of an analysis that used Abraham's solvation parameter model [5].

Abraham's model was constructed to describe solute behavior, but it recognizes the intimacy of the solute–solvent relationship and so provides a useful model of solvent properties [6]. The model is based on solvation occurring in two steps. First, a cavity is generated in the solvent. This process is endoergic, as the self-association of the solvent is overcome. Then the solute is incorporated in the cavity. This step is exoergic, as the solvent–solute interactions are formed. By working with G.C., and therefore gaseous solutes, the self-association of the solute can be ignored. The form of the model used to investigate ionic liquid properties is given in Equation (3.5-1) [5]:

$$\log K_L = c + rR_2 + s\pi_2{}^H + a\alpha_2{}^H + b\beta_2{}^H + l \log L^{16} \quad (3.5\text{-}1)$$

Where K_L is the solute gas–liquid partition coefficient, r is the tendency of the solvent to interact through π- and n-electron pairs (Lewis basicity), s the contribution from dipole–dipole and dipole-induced dipole interactions (in molecular solvents), a is the hydrogen bond basicity of the solvent, b is its hydrogen bond acidity and l is how well the solvent will separate members of a homologous series, with contributions from solvent cavity formation and dispersion interactions.

This model has been applied to 38 different organic salts, primarily tetraalkylammonium halides and substituted alkanesulfonates [5]. For the range of salts included in the study, it was found that they were strong hydrogen-bond bases, with the basicity being diminished by fluorination of the anion. The organic salts had large s values, which were interpreted as a significant capacity for dipole–dipole and dipole–induced dipole interactions, but it is in the s value that any Coulombic effect would be expected to show itself.

Some exceptions to this general observation were found: halide and nitrite salts have unusually high hydrogen-bond basicities (as would be expected), while pentacyanopropionide, picrate, triflate, and perfluorobezenesulfonate salts not only had unusually low hydrogen-bond basicities (also as would be expected) but also lower s values, perhaps due to the weakening of the Coulombic interactions by delocalization of the charge on the anions.

It was noted that, on going from tetrabutylammonium to tetrabutylphosphonium, salts with a common anion displayed identical solvation properties. Hence, with these simple cations, the solvent properties are dominated by the choice of anion. It is possible that, had cations with acidic protons, such as trialkylammonium and trialkylphosphonium, been included in the study, these may then have also had an influence.

One interesting point of note is that the component arising from the solvent's ability to form a cavity and its dispersion interactions is unusually high – in comparison to those for non-ionic polar solvents – for most of the ionic liquids with poorly associating anions and increases as the cation becomes bulkier. With the hydrogen bond base anions, this ability weakened. High values for l are usually associated with non-polar solvents. This observation agrees with the often repeated statement that ionic liquids have unusual mixing properties (see Sections 3.3 and 3.4).

Finally, none of the ionic liquids were found to be hydrogen bond acids [5], although this may well be a consequence of the salts selected, none of which had a cation that would be expected to act as a hydrogen bond donor. Earlier qualitative measurements on ionic liquid stationary phases of mono-, di-, and trialkylammonium salts suggest that hydrogen bond donation can be important where a potentially acidic proton is available [7–9]. More recent work, with $[BMIM]^+$ salts, also indicates that these ionic liquids should be considered to be hydrogen bond donor solvents [10]. However, this has yet to be quantified.

3.5.2
Absorption Spectra

The longest-wavelength absorption band of Reichardt's dye (2,4,6-triphenylpyridinium-*N*-4-(2,6-diphenylphenoxide) betaine, Figure 3.5-1) shows one of the largest solvatochromic shifts known (375 nm between diphenyl ether and water) [11]. It can register effects arising from solvent dipolarity, hydrogen bonding, and Lewis acidity, with the greatest contribution coming from the hydrogen bond donor property of the solvent [12]. The E_T^N values of a small number of alkylammonium nitrate, thiocyanate, and sulfonate salts [13, 14] have been recorded, as have those of some substituted imidazolium tetrafluoroborate, hexafluorophosphate, triflate, and trifluoromethanesulfonylimide salts [15]. It has been noted that these measurements are highly sensitive to the preparation of the ionic liquids used [15], so it is unfortunate that there are no examples that appear in all studies.

Table 3.5-1 lists the E_T^N values for the alkylammonium thiocyanates and nitrates and the substituted imidazolium salts. It can be seen that the values are dominated by the nature of the cation. For instance, values for monoalkylammonium nitrates and thiocyanates are ca. 0.95–1.01, whereas the two tetraalkylammonium salts have values of ca. 0.42–0.46. The substituted imidazolium salts lie between these two extremes, with those with a proton at the 2-position of the ring having higher values than those with this position methylated. This is entirely consistent with the expected hydrogen bond donor properties of these cations.

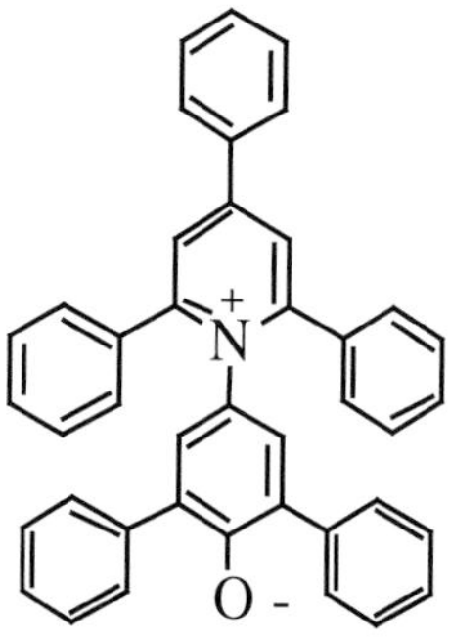

Figure 3.5-1: Reichardt's Dye.

Table 3.5-1: Solvent polarity measurements for some ionic liquids.

Salt	E_T^N	π^*	α	β	Ω
$[Pr_2NH_2][SCN]$	1.006	1.16	0.97	0.39	
$[sec\text{-}BuNH_3][SCN]$	1.006	1.28	0.91		
Water	1.000	1.09	1.17	0.47	0.869
$[EtNH_3][NO_3]$	0.954	1.24	0.85	0.46	0.82
$[BuNH_3][SCN]$	0.948	1.23	0.92		
$[PrNH_3][NO_3]$	0.923	1.17	0.88	0.52	
$[Bu_3NH][NO_3]$	0.802	0.97	0.84		
$[BMIM][ClO_4]$	0.684				0.67
$[BMIM][BF_4]$	0.673	1.09	0.73	0.72	0.66
[BMIM][TfO]	0.667				0.65
$[BMIM][PF_6]$	0.667	0.91	0.77	0.41	0.68
Ethanol	0.654	0.54	0.75	0.75	0.718
$[BMIM][Tf_2N]$	0.642				
[BMIM]Cl		1.17	0.41	0.95	
$[EtNH_3]Cl$	0.636				
$[OMIM][PF_6]$	0.633	0.88	0.58	0.46	
$[OMIM][Tf_2N]$	0.630				
[OMIM]Cl		1.09	0.33	0.90	
$[Pr_4N][CHES]$*	0.62	1.08	0.34	0.80	
$[BMIM][CF_3CO_2]$	0.620				
$[Bu_4N][CHES]$*	0.62	1.01	0.34	0.98	
$[Pe_4N][CHES]$*	0.58	1.00	0.15	0.91	
$[Bu_4N][BES]$*	0.53	1.07	0.14	0.81	
$[BMMIM][Tf_2N]$	0.525				
$[Bu_4N][MOPSO]$*	0.49	1.07	0.03	0.74	
$[OMMIM][BF_4]$	0.543				
$[OMMIM][Tf_2N]$	0.525				
$[Et_4N][NO_3]$	0.460				
Acetonitrile	0.460	0.75	0.19	0.31	0.692
$[Et_4N]Cl$	0.454				
$[Hx_4N][PhCO_2]$	0.420				
Diethyl ether	0.117	0.27	0.00	0.47	0.466
Cyclohexane	0.009	0.00	0.00	0.00	0.595

*CHES is 2-(cyclohexylamino)ethanesulfonate, BES is 2-{bis(2-hydroxoethyl)amino}ethanesulfonate, MOPSO is 2-hydroxo-4-morpholinepropanesulfonate.

The role of the anion is less clear-cut. It can be seen that ionic liquids with the same cation but different anions have different E_T^N values. However, the difference in the values for $[Et_4N]Cl$ and $[Et_4N][NO_3]$ is only 0.006, whereas the difference between $[EtNH_3]Cl$ and $[EtNH_3][NO_3]$ is 0.318. Less dramatically, the difference in the values for $[OMMIM][BF_4]$ and $[OMMIM][Tf_2N]$ is only 0.018, whereas the difference between $[BMIM][BF_4]$ and $[BMIM][Tf_2N]$ is 0.031. Hence, it is clear that the effect of changing the anion depends on the nature of the cation.

If the cation has been unchanged, its ability to act as a hydrogen-bond donor has been unchanged, so why is an effect seen at all? I propose that there is competition between the anion and the Reichardt's dye solute for the proton. Thus, the E_T^N values of the ionic liquids are controlled by the ability of the liquid to act as a hydrogen bond donor (cation effect) moderated by its hydrogen bond acceptor ability (anion effect). This may be described in terms of two competing equilibria. The cation can hydrogen bond to the anion [Equation (3.5-2)]:

$$[BMIM]^+ + A^- \rightleftharpoons [BMIM]...A$$

$$K'_{eqm} = \frac{[[BMIM]...A]}{[[BMIM]^+][A^-]} \qquad (3.5\text{-}2)$$

The cation can hydrogen bond to the solute (Reichardt's dye in this case) [Equation (3.5-3)]:

$$[BMIM]^+ + \text{solute} \rightleftharpoons [BMIM]...\text{solute}$$

$$K''_{eqm} = \frac{[[BMIM]...\text{solute}]}{[[BMIM]^+][\text{solute}]} \qquad (3.5\text{-}3)$$

It can easily be shown that the value of K'' is inversely proportional to the value of K' and that K' is dependent on both the cation and the anion of the ionic liquid. Hence, it is entirely consistent with this model that the difference made by changing the anion should depend on the hydrogen bond acidity of the cation.

Attempts have also been made to separate non-specific effects of the local electrical field from hydrogen-bonding effects for a small group of ionic liquids through the use of the π^* scale of dipolarity/polarizability, the α scale of hydrogen bond donor acidity, and the β scale of hydrogen bond basicity (see Table 3.5-1) [13, 16].

The π^* values were high for all of the ionic liquids investigated (0.97–1.28) when compared to molecular solvents. The π^* values result from measuring the ability of the solvent to induce a dipole in a variety of solute species, and they will incorporate the Coulombic interactions from the ions as well as dipole–dipole and polarizability effects. This explains the consistently high values for all of the salts in the studies. The values for quaternary ammonium salts are lower than those for the monoalkylammonium salts. This probably arises from the ability of the charge center on the cation to approach the solute more closely for the monoalkylammonium salts. The values for the imidazolium salts are lower still, probably reflecting the delocalization of the charge in the cation.

The difference in the hydrogen bond acidities and basicities was far more marked. The α value is largely determined by the availability of hydrogen bond donor sites on the cation. Values range from 0.8–0.9 for the monoalkylammonium salts, and are slightly lower (0.3–0.8) for the imidazolium salts. In the absence of a hydrogen bond donor, cation values are lower still: in the 0.1–0.3 range. It appears that more basic anions give lower values of α with a common cation. This is further evidence for the idea of competing equilibria detailed above for Reichardt's dye [Equations (3.5-2) and (3.5-3)].

At first glance the hydrogen bond basicity β is controlled solely by the anions, with basicity decreasing in the order Cl^- > $[RSO_3]^-$ >$[BF_4]^-$ > $[PF_6]^-$ > $[NO_3]^-$ > $[SCN]^-$. However, while the general trend is clear, this is not the order that one would have expected, and the cations are obviously playing a role. Again, this may be a consequence of competition for the basic site (anion) between the test solute and the acidic site (cation) of the ionic liquid. It is unfortunate that no study to date has used a common anion across all possible cations.

3.5.3 Fluorescence Spectra

A number of workers have attempted to study the polarity of ionic liquids with the aid of the fluorescence spectra of polycyclic aromatic hydrocarbons. Of these, the most commonly applied has been that of pyrene [17–19]. The measurements are of the ratio of the intensities of the first and third vibronic bands in the π–π* emission spectrum of monomer pyrene (I_1/I_3). The increase in I_1/I_3 values in more polar solvents has been attributed to a reduction in local symmetry [20], but the mechanism for this is poorly understood, although some contribution from solvent hydrogen bond acidity has been noted [21]. It is hence difficult to know what the measurements are telling us about the ionic liquids in anything other than the most general terms. These measurements have generally placed the ionic liquids in the polarity range of moderately polar solvents, with monoalkylammonium thiocyanates displaying values in the 1.01–1.23 range, $[EMIM][(CF_3SO_2)_2N]$ having a value of 0.85, and $[BMIM][PF_6]$ a particularly high value of 2.08 (water = 1.87, acetonitrile = 1.79, methanol = 1.35). It should be noted that the spectrum of pyrene would be expected to be sensitive to HF, which could well be present in these $[PF_6]^-$ ionic liquids, and would give a high value of I_1/I_3. Other fluorescence probes that have been used give broadly similar results [17–22].

3.5.4 Refractive Index

The refractive index of a medium is the ratio of the speed of light in a vacuum to its speed in the medium, and is the square root of the relative permittivity of the medium at that frequency. When measured with visible light, the refractive index is related to the electronic polarizability of the medium. Solvents with high refractive indexes, such as aromatic solvents, should be capable of strong dispersion interactions. Unlike the other measures described here, the refractive index is a property of the pure liquid without the perturbation generated by the addition of a probe species.

Although the measurement of the refractive index of a liquid is relatively straightforward, few have been recorded for ionic liquids to date. Monoalkylammonium nitrate and thiocyanates give moderate values, higher for thiocyanates than for nitrates with a common cation, and the refractive index increases with the chain length of the cation [23]. However, little more than that can be said at this stage, since there have been too few measurements reported to date.

3.5.5 Organic Reactions

An alternative avenue for the exploration of the polarity of a solvent is by investigation of its effect on a chemical reaction. Since the purpose of this book is to review the potential application of ionic liquids in synthesis, this could be the most productive way of discussing ionic liquid polarity. Again, the field is in its infancy, but some interesting results are beginning to appear.

3.5.5.1 Alkylation of sodium 2-naphthoxide

The C- vs. O-alkylation of sodium 2-naphthoxide in simple molten phosphonium and ammonium halides (which are molten at temperatures of ca. 110 °C) has been studied in order to compare their properties with those of conventional organic solvents [24]. The regioselectivity of the reaction is dependent on the nature of the counterion of the 2-naphthol salt and the solvent. In dipolar aprotic solvents, O-alkylation is favoured. The use of [*n*-Bu_4P]Br, [*n*-Bu_4N]Br, [EMIM]Br, and [*n*-Bu_4P]Cl as solvents resulted in every case in high regioselectivity for the O-alkylation product (between 93 % and 97 %), showing the polar nature of the ionic liquids. Analysis by ^1H and ^{31}P NMR showed that the ionic liquids were unaffected by the reaction and could be reused to achieve the same results [24]. Almost identical findings were obtained for the alkylation of 2-naphthol in the ionic liquid [BMIM][PF_6] [25]. This would suggest that, for this reaction and with this range of ionic liquids, the specific ions that compose the ionic liquid are less important than the ionic nature of the medium in determining the outcome of the reaction.

3.5.5.2 Diels–Alder reactions

One of the earliest solvent polarity scales is Berson's Ω scale. This scale is based on the *endo*/*exo* ratio of the Diels–Alder reaction between cyclopentadiene and methyl acrylate (Figure 3.5-2, $\Omega = \log_{10}$ *endo*/*exo*). This reaction has been conducted in a number of ionic liquids, giving values in the 0.46–0.83 range [26].

Berson postulated that the transition state giving rise to the *endo* adduct should be more polar than that giving rise to the *exo* adduct, and so polar solvents should

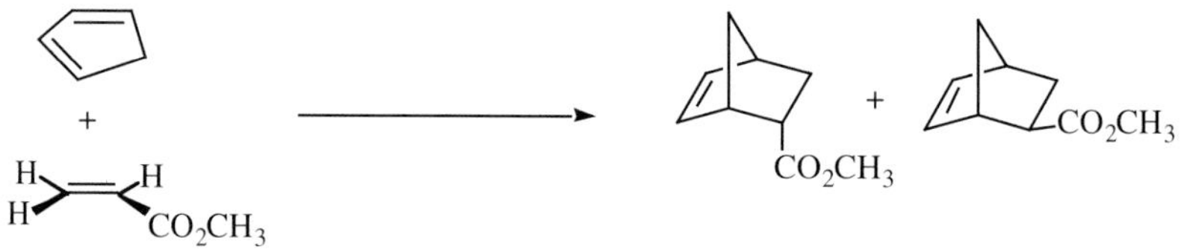

Figure 3.5-2: Diels–Alder cycloaddition between cyclopentadiene and methyl acrylate.

Figure 3.5-3: The hydrogen bond (Lewis acid) interaction of an imidazolium cation with the carbonyl oxygen of methyl acrylate in the activated complex of the Diels-Alder reaction.

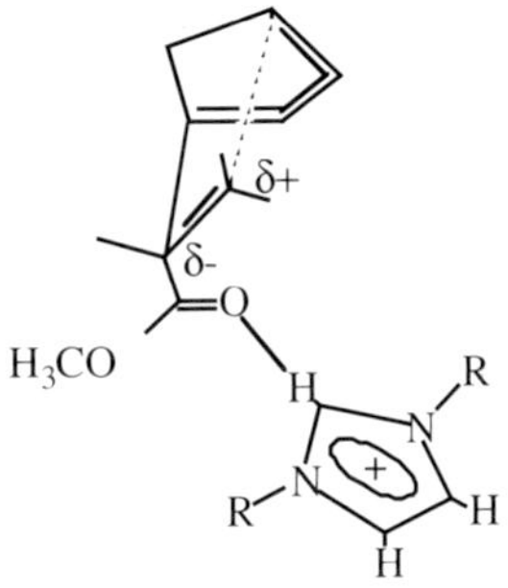

favour the formation of the *endo* adduct in a kinetically controlled reaction. Alternative explanations, based on solvophobic interactions, have been advanced to explain the behavior of this reaction in water [27]. However, it has been shown that, at least in ionic liquids, the *endo/exo* ratio is controlled by the ability of the solvent to hydrogen-bond to the solute (Figure 3.5-3). Again, it was shown that the effect results from competition between the hydrogen bond acceptor sites of the methyl acrylate and the ionic liquid (anion) for the hydrogen bond donor site of the ionic liquid (cation). Clearly, in this case, the chemical natures of the ions of the ionic liquid are dominant in determining the outcome of the reaction, rather than the ionic nature of the medium itself.

3.5.5.3 **Photochemical reactions**

Radiolysis of solutions of CCl_4 and O_2 in [BMIM][PF_6] results in the formation of $CCl_3O_2^{\bullet}$ radicals [28]. These have then been allowed to react with the organic reductant chlorpromazine (ClPz) according to Equation (3.5-3):

$$CCl_3O_2^{\bullet} + ClPz \longrightarrow CCl_3O_2^{-} + ClPz^{\bullet+} \qquad (3.5\text{-}3)$$

This reaction has been investigated in a range of molecular solvents, and the rate constants have been correlated with the Hildebrand solubility parameter (δ_H). The δ_H parameter is derived from the molar energy of vaporization of the solvent (clearly it is not possible to measure an energy of vaporization for a nonvolatile ionic liquid) and is a measure of the work required to create a cavity of the size of the solvent molecules in the solvent, with contributions from all of the possible nonspecific interactions between the solvent molecules. It does not take into account any interaction between the solvent and any potential solute. However, in these fast reactions, the reacting species would not be in equilibrium with their surroundings throughout the reaction process. Hence, direct solute–solvent interactions might be expected to play a secondary role in controlling the rate of the reaction and the ability of the solvent to reorganize itself to accommodate the activated complex of the reaction would become predominant. The second order rate constant for the reaction was found to lie between those of isopropanol and *tert*-butanol.

3.5.6
General Conclusions

To date, most studies of ionic liquids have used a small set of ionic liquids and have been based on the idea that, if the response of a particular probe molecule or reaction is like that in some known molecular solvent, then it can be said that the polarities of the ionic liquid and the molecular solvent are the same. This may not necessarily be the case. Only systematic investigations will show whether this is true, and only when a wide range of ionic liquids with a wide range of different solvent polarity probes have been studied will we be able to make any truly general statements about the polarity of ionic liquids. Indeed, in our attempts to understand the nature of solvent effects in ionic liquids, we will probably have to refine our notion of polarity itself. However, it is possible to draw some tentative general conclusions.

Not all ionic liquids are the same, different combinations of anions and cations produce solvents with different polarities. No ionic liquids have shown themselves to be "super-polar"; regardless of the method used to assess their polarities, ionic liquids come within the range of molecular solvents. Most general measures of overall polarity place ionic liquids in the range of the short- to medium-chain alcohols.

It becomes more interesting when the solvent properties are broken down into their component parts. Ionic liquids can act as hydrogen bond acids and/or hydrogen bond bases, or as neither. Generally, the hydrogen bond basicity is determined by the anion and the hydrogen bond acidity is determined by the cation. There is no obvious unique "ionic effect" to be seen in the available data, but this may yet be found after further study and might explain the consistently high π^* values (spectroscopic) and *s* values (chromatographic), where they have been measured [13].

References

1 C. Reichardt, "Solvents and Solvent Effects in Organic Chemistry", 2nd ed., VCH, Weinheim, **1990**.

2 M. Berthelot, L. Péan de Saint-Giles, *Ann. Chim. Et Phys.*, 3 Ser. **1862**, *65*, 385; **1862** *66*, 5; **1863**, *68*, 255.

3 L. Claisen, *Liebigs Ann. Chem.* **1896**, *291*, 25; W. Wislicenus, *Liebigs Ann. Chem.* **1896**, *291*, 147; L. Knorr, *Liebigs Ann. Chem.* **1896**, *293*, 70.

4 H. Stobbe, *Liebigs Ann. Chem.* **1903**, *326*, 347.

5 S. K. Poole, C. F. Poole, *Analyst* **1995**, *120*, 289 and references therein.

6 M. H. Abraham, *Chem. Soc. Rev.* **1993**, 73.

7 F. Pacholec, H. Butler, C. F. Poole, *Anal. Chem.*, **1982**, *54*, 1938.

8 C. F. Poole, K. G. Furton, B. R. Kersten, *J. Chromatogr. Sci.*, **1986**, *24*, 400.

9 M. E. Coddens, K. G. Furton, C . F. Poole, *J. Chromatogr.*, **1986**, *356*, 59.

10 D. W. Armstrong, L. He, Y.-S. Liu, *Anal. Chem.*, **1999**, *71*, 3873.

11 C. Reichardt, *Chem. Soc. Rev.* **1992**, 147.

12 R. W. Taft, M. J. Kamlet, *J. Am. Chem. Soc.* **1976**, *98*, 2886.

13 S. K. Poole, P. H. Shetty, C. F. Poole, *Anal. Chim. Acta* **1989**, *218*, 241.
14 I. M. Herfort, H Schneider, *Liebigs Ann. Chem.* **1991**, 27.
15 M. J. Muldoon, C. M. Gordon, I. R. Dunkin, *J. Chem. Soc., Perkin Trans. 2* **2001**, 433.
16 J. G. Huddleston, G. A. Broker, H. D. Willauer, R. D. Rogers, in "Green (or greener) industrial applications of ionic liquids.", ACS Symposium Series, in press.
17 K. W. Street, Jr., W. E. Acree, Jr., J. C. Fetzer, P. H. Shetty, C. F. Poole, *Appl. Spectrosc.*, **1989**, *43*, 1149.
18 P. Bonhôte, A.-P. Dias, N. Papageorgiou, K. Kalyanasundaram, M. Grätzel, *Inorg. Chem.*, **1996**, *35*, 1168.
19 S. N. Baker, G. A. Baker, M. A. Kane, F. V. Bright, *J. Phys. Chem. B*, **2001**, *105*, 9663.
20 D. C. Dong, M. A. Winnik, *Can. J. Chem.* **1984**, *62*, 2560.
21 J. Catalán, *J. Org. Chem.*, **1997**, *62*, 8231.
22 S. N. V. K. Aki, J. F. Brennecke, A. Samanta, *Chem. Commun.* **2001**, 413.
23 C. F. Poole, B. R. Kersten, S. S. J. Ho, M. E. Coddens, K. G. Furton, *J. Chromatogr.* **1986**, *352*, 407.
24 M. Badri, J.-J. Brunet, R. Perron, *Tetrahedron Lett.* **1992**, *33*, 4435.
25 M. J. Earle, P. B. McCormac, K. R. Seddon, *J. Chem. Soc., Chem. Commun.* **1998**, 2245.
26 (a) A. Sethi, T. Welton, J. Wolff, *Tetrahedron Lett.*, **1999**, *40*, 793; (b) A. Sethi, T. Welton, in "Green (or greener) industrial applications of ionic liquids.", ACS Symposium Series, in press.
27 R. Breslow, *Acc. Chem. Res.*, **1991**, *24*, 159.
28 D. Behar, C. Gonzalez, P. Neta, *J. Phys. Chem. A* ***2001***, *105*, 7607.

3.6 Electrochemical Properties of Ionic Liquids

Paul C. Trulove and Robert A. Mantz

The early history of ionic liquid research was dominated by their application as electrochemical solvents. One of the first recognized uses of ionic liquids was as a solvent system for the room-temperature electrodeposition of aluminium [1]. In addition, much of the initial development of ionic liquids was focused on their use as electrolytes for battery and capacitor applications. Electrochemical studies in the ionic liquids have until recently been dominated by work in the room-temperature haloaluminate molten salts. This work has been extensively reviewed [2–9]. Development of non-haloaluminate ionic liquids over the past ten years has resulted in an explosion of research in these systems. However, recent reviews have provided only a cursory look at the application of these "new" ionic liquids as electrochemical solvents [10, 11].

Ionic liquids possess a variety of properties that make them desirable as solvents for investigation of electrochemical processes. They often have wide electrochemical potential windows, they have reasonably good electrical conductivity and solvent transport properties, they have wide liquid ranges, and they are able to solvate a wide variety of inorganic, organic, and organometallic species. The liquid ranges of ionic liquids have been discussed in Section 3.1 and their solubility and solvation in

Section 3.3. In this section we deal specifically with the electrochemical properties of ionic liquids (electrochemical windows, conductivity, and transport properties); we will discuss the techniques involved in measuring these properties, summarize the relevant literature data, and discuss the effects of ionic liquid components and purity on their electrochemical properties.

3.6.1
Electrochemical Potential Windows

A key criterion for selection of a solvent for electrochemical studies is the electrochemical stability of the solvent [12]. This is most clearly manifested by the range of voltages over which the solvent is electrochemically inert. This useful electrochemical potential "window" depends on the oxidative and reductive stability of the solvent. In the case of ionic liquids, the potential window depends primarily on the resistance of the cation to reduction and the resistance of the anion to oxidation. (A notable exception to this is in the acidic chloroaluminate ionic liquids, where the reduction of the heptachloroaluminate species $[Al_2Cl_7]^-$ is the limiting cathodic process). In addition, the presence of impurities can play an important role in limiting the potential windows of ionic liquids.

The most common method used for determining the potential window of an ionic liquid is cyclic voltammetry (or its digital analogue, cyclic staircase voltammetry). In a three-electrode system, the potential of an inert working electrode is scanned out to successively greater positive (anodic) and negative (cathodic) potentials until background currents rise dramatically due to oxidation and reduction of the ionic liquid, respectively. The oxidative and reductive potential limits are assigned when the background current reaches a threshold value. The electrochemical potential window is the difference between these anodic and cathodic potential limits. Since the choice of the threshold currents is somewhat subjective, the potential limits and corresponding electrochemical window have a significant uncertainty associated with them. Normally this is in the range of ± 0.2 V.

It must be noted that impurities in the ionic liquids can have a profound impact on the potential limits and the corresponding electrochemical window. During the synthesis of many of the non-haloaluminate ionic liquids, residual halide and water may remain in the final product [13]. Halide ions (Cl^-, Br^-, I^-) are more easily oxidized than the fluorine-containing anions used in most non-haloaluminate ionic liquids. Consequently, the observed anodic potential limit can be appreciably reduced if significant concentrations of halide ions are present. Contamination of an ionic liquid with significant amounts of water can affect both the anodic and the cathodic potential limits, as water can be both reduced and oxidized in the potential limits of many ionic liquids. Recent work by Schröder et al. demonstrated considerable reduction in both the anodic and cathodic limits of several ionic liquids upon the addition of 3 % water (by weight) [14]. For example, the electrochemical window of 'dry' $[BMIM][BF_4]$ was found to be 4.10 V, while that for the ionic liquid with 3 % water by weight was reduced to 1.95 V. In addition to its electrochemistry, water can react with the ionic liquid components (especially anions) to produce products that are electroactive in the electrochemical potential window. This has been well documented in the chloroaluminate ionic liquids, in which water will react to produce electroactive proton-containing species (e.g., HCl and $[HCl_2]^-$) [4, 9]. In addition, water appears to react with some of the anions commonly used in the non-haloaluminate ionic liquids [15]. The $[PF_6]^-$ anion, for example, is known to react with water to form HF [16, 17].

Glassy carbon (GC), platinum (Pt), and tungsten (W) are the most common working electrodes used to evaluate electrochemical windows in ionic liquids. The choice of the working electrode has some impact on the overall electrochemical window measured. This is due to the effect of the electrode material on the irreversible electrode reactions that take place at the oxidative and reductive limits. For example, W gives a 0.1 to 0.2 V greater oxidative limit for $[EMIM]Cl/AlCl_3$ ionic liquids than Pt, due to a greater overpotential for the oxidation of the chloroaluminate anions [18]. In addition, GC (and to a lesser extent W) exhibits a large overpotential for proton reduction. Under normal circumstances, the electrochemistry of protonic impurities (i.e., water) will not be observed in the ionic liquid electrochemical window with GC. Pt, on the other hand, exhibits good electrochemical behavior for proton. Consequently, protonic impurities will give rise to a reduction wave(s) at Pt positive of the cathodic potential limit. Interestingly, comparison of the background electrochemical behavior of an ionic liquid at both Pt and GC working electrodes can be an excellent qualitative tool for determining if significant amounts of protonic impurities are present.

Figure 3.6-1 shows the electrochemical window of a 76–24 mol % $[BMMIM][(CF_3SO_2)_2N]/Li[(CF_3SO_2)_2N]$ ionic liquid at both GC and Pt working electrodes [15]. For the purposes of assessing the electrochemical window, the current threshold for both the anodic and cathodic limits was set at an absolute value of 100 $\mu A\ cm^{-2}$.

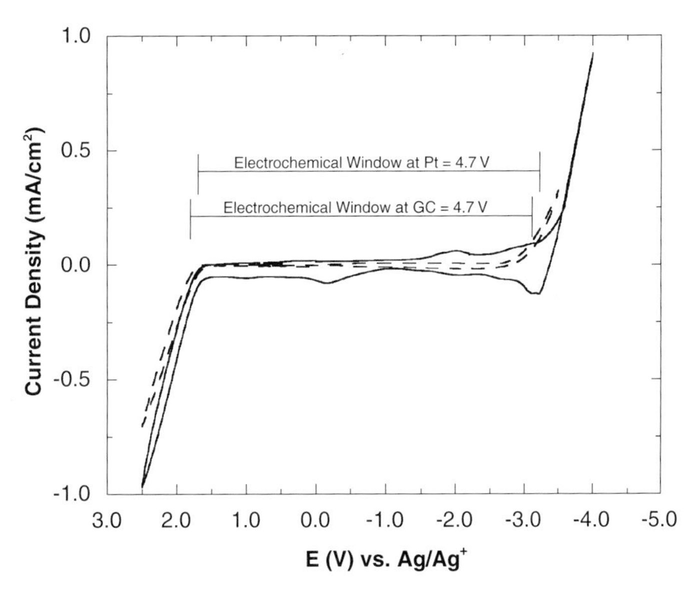

Figure 3.6-1: The electrochemical window of 76–24 mol % $[BMMIM][(CF_3SO_2)_2N]/Li[(CF_3SO_2)_2N]$ binary melt at: a) a platinum working electrode (solid line), and b) a glassy carbon working electrode (dashed line). Electrochemical window set at a threshold of 0.1 mA cm^{-2}. The reference electrode was a silver wire immersed in 0.01 M $AgBF_4$ in $[EMIM][BF_4]$ in a compartment separated by a Vicor frit, and the counter-electrode was a graphite rod.

Table 3.6-1: The room-temperature electrochemical potential windows for non-haloaluminate ionic liquids.

Cation	*Anion(s)*	*Working electrode*[c]	*Window (V)*	*Ref.*
Ammonium				
$[(n\text{-}C_3H_7)(CH_3)_3N]^+$	$[(CF_3SO_2)_2N]^-$	GC	5.7	19[a]
$[(n\text{-}C_6H_{13})(C_2H_5)_3N]^+$	$[(CF_3SO_2)_2N]^-$	GC	4.5[b]	20
$[(n\text{-}C_8H_{17})(C_2H_5)_3N]^+$	$[(CF_3SO_2)_2N]^-$	GC	5.0	20
$[(n\text{-}C_8H_{17})(C_4H_9)_3N]^+$	$[(CF_3SO_2)_2N]^-$	GC	5.0	20
$[(CH_3)_3(CH_3OCH_2)N]^+$	$[(CF_3SO_2)_2N]^-$	GC	5.2	19[a]
1-butyl-1-methyl-pyrrolidinium	$[(CF_3SO_2)_2N]^-$	GC	5.5	21
Imidazolium				
$[EMIM]^+$	F^-	Pt	3.1	22[a]
$[EMIM]^+$	$[BF_4]^-$	Pt	4.3	23[a]
$[EMIM]^+$	$[BF_4]^-$	Pt	4.5	24
$[EMIM]^+$	$[CH_3CO_2]^-$	Pt	3.6	25[a]
$[EMIM]^+$	$[CF_3CO_2]^-$	Pt	3.8[b]	26[a]
$[EMIM]^+$	$[CF_3SO_3]^-$	Pt	4.1	26[a]
$[EMIM]^+$	$[CF_3SO_3]^-$	Pt	4.3	25[a]
$[EMIM]^+$	$[(CF_3SO_2)_2N]^-$	GC	4.1	27
$[EMIM]^+$	$[(CF_3SO_2)_2N]^-$	Pt	4.5	26[a]
$[EMIM]^+$	$[(CF_3SO_2)_2N]^-$	GC	4.5	19[a]
$[EMIM]^+$	$(C_2F_5SO_2)_2N^-$	GC	4.1	27
$[BMIM]^+$	$[BF_4]^-$	Pt	4.1	14
$[BMIM]^+$	$[PF_6]^-$	Pt	4.2	14
$[EMMIM]^+$	$[(CF_3SO_2)_2N]^-$	Pt	4.7	26[a]
$[PMMIM]^+$	$[(CF_3SO_2)_2N]^-$	GC	4.3	27
$[PMMIM]^+$	$[(CF_3SO_2)_2N]^-$	GC	5.2	28
$[PMMIM]^+$	$(CF_3SO_2)_3C^-$	GC	5.4	28
$[PMMIM]^+$	$[PF_6]^-$	GC	4.3[d]	28[e]
$[PMMIM]^+$	$[AsF_6]^-$	GC	4.4[d]	28[e]
Pyrazolium				
1,2-dimethyl-4-fluoropyrazolium	$[BF_4]^-$	GC	4.1	29
Pyridinium				
$[BP]^+$	$[BF_4]^-$	Pt	3.4	23
Sulfonium				
$[(C_2H_5)_3S]^+$	$[(CF_3SO_2)_2N]^-$	GC	4.7	30
$[(n\text{-}C_4H_9)_3S]^+$	$[(CF_3SO_2)_2N]^-$	GC	4.8	30

[a] Voltage window estimated from cyclic voltammograms contained in the reference.
[b] Voltage window may be limited by impurities.
[c] Working electrode, Pt = platinum, GC = glassy carbon, W = tungsten.
[d] Voltage window at 80 °C.
[e] Voltage window determined assuming cathodic limit of 0.63 V vs. Li/Li^+ reference.

As shown in Figure 3.6-1, GC and Pt exhibit anodic and cathodic potential limits that differ by several tenths of volts. However, somewhat fortuitously, the electrochemical potential windows for both electrodes in this ionic liquid come out to be 4.7 V. What is also apparent from Figure 3.6-1 is that the GC electrode exhibits no significant background currents until the anodic and cathodic potential limits are reached, while the Pt working electrode shows several significant electrochemical processes prior to the potential limits. This observed difference is most probably due to trace amounts of water in the ionic liquid, which is electrochemically active on Pt but not on GC (vide supra).

Tables 3.6-1 and 3.6-2 contain electrochemical potential windows for a wide variety of ionic liquids. Only limited information concerning the purity of the ionic liquids listed in Tables 3.6-1 and 3.6-2 was available, so these electrochemical potential windows must be treated with caution, as it is likely that many of the ionic liquids would have had residual halides and water present.

Ideally, one would prefer to compare anodic and cathodic potential limits instead of the overall ionic liquid electrochemical window, because difference sets of anodic and cathodic limits can give rise to the same value of electrochemical window (see Figure 3.6-1). However, the lack of a standard reference electrode system within and between ionic liquid systems precludes this possibility. Consequently, significant care must be taken when evaluating the impact of changes in the cation or anion on the overall ionic liquid electrochemical window.

As indicated by the data in Tables 3.6-1 and 3.6-2, the trend in the electrochemical stabilities of the types of ionic liquid cations is: pyridinium $<$ pyrazolium $\leq$ imidazolium $\leq$ sulfonium $\leq$ ammonium. Overall, the quaternary ammonium-based ionic liquids are the potential window champs [36], the $[(n\text{-}C_3H_7)(CH_3)_3N][(CF_3SO_2)_2N]$ ionic liquid exhibiting the largest electrochemical window, of 5.7 V, at GC [19]. (For comparison, one of the best non-aqueous electrolyte systems, acetonitrile/tetrabutylammonium hexafluorophosphate, exhibits a potential window of 6.3 V [12]). Because of uncertainties in the purity of the quaternary ammonium-based ionic liquids listed in Table 3.6-1, it is impossible to determine from the data listed what effect changes in the alkyl substituents have on the electrochemical stability of the cation. However, within the group of imidazolium-based ionic liquids there is a clear increase in cation stability when the 2-position on the imidazolium ring is capped by an alkyl substituent, as in $[EMMIM]^+$. It has been proposed that the cathodic limiting reactions of imidazolium cations proceed initially by the reduction of ring protons to molecular hydrogen [37]. Since the 2-position on the imidazolium ring is the most acidic hydrogen [38], it is reasonable to conclude that substitution of an alkyl substituent at that position would result in an improvement in the reductive stability of the imidazolium cation. From the data in Tables 3.6-1 and 3.6-2, the anion stabilities towards oxidation appear to follow the order: halides (Cl^-, F^-, Br^-) $<$ chloroaluminates ($[AlCl_4]^-$, $[Al_2Cl_7]^-$) $\leq$ fluorinated ions ($[BF_4]^-$, $[PF_6]^-$, $[AsF_6]^-$) $\leq$ triflate/triflyl ions ($[CF_3SO_3]^-$, $[(CF_3SO_2)_2N]^-$, $[(C_2F_5SO_2)_2N]^-$, $[(CF_3SO_2)_3C]^-$).

The electrochemical windows exhibited by the chloroaluminates tend to fall into three ranges that correspond to the types of chloroaluminate ionic liquids: basic,

Table 3.6-2: The room-temperature electrochemical potential windows for binary and ternary chloroaluminate and related ionic liquids.

Ionic liquid system	*Cation(s)*	*Anion(s)*	*Working electrode*[e]	*Window (V)*	*Ref.*
60.0–40.0 mol % $[EMIM]Cl/AlCl_3$	$[EMIM]^+$	$[AlCl_4]^-/Cl^-$	W	2.8	31[a]
50.0–50.0 mol % $[EMIM]Cl/AlCl_3$	$[EMIM]^+$	$[AlCl_4]^-$	W	4.4	31[a]
45.0–55.0 mol % $[EMIM]Cl/AlCl_3$	$[EMIM]^+$	$[Al_2Cl_7]^-/[AlCl_4]^-$	W	2.9	31[a]
45.0–55.0 mol % $[EMIM]Cl/AlCl_3$	$[EMIM]^+$	$[Al_2Cl_7]^-/[AlCl_4]^-$	W	2.9	31[a]
60.0–40.0 mol % $[PMMIM]Cl/AlCl_3$	$[PMMIM]^+$	$[AlCl_4]^-/Cl^-$	GC	3.1	15[a]
50.0–50.0 mol % $[PMMIM]Cl/AlCl_3$	$[PMMIM]^+$	$[AlCl_4]^-$	GC	4.6	15[a]
40.0–60.0 mol % $[PMMIM]Cl/AlCl_3$	$[PMMIM]^+$	$[Al_2Cl_7]^-/[AlCl_4]^-$	GC	2.9	15[a]
45.5–50.0–4.5 mol % $[EMIM]Cl/AlCl_3/LiCl$	$[EMIM]^+/Li^+$	$[AlCl_4]^-$	W	4.3	32[a,c]
45.5–50.0–4.5 mol % $[EMIM]Cl/AlCl_3/LiCl$	$[EMIM]^+/Li^+$	$[AlCl_4]^-$	W	4.6	32[a,c,d]
47.6–50.0–2.4 mol % $[EMIM]Cl/AlCl_3/NaCl$	$[EMIM]^+/Na^+$	$[AlCl_4]^-$	W	4.5	32[a,c]
47.6–50.0–2.4 mol % $[EMIM]Cl/AlCl_3/NaCl$	$[EMIM]^+/Na^+$	$[AlCl_4]^-$	W	4.6	32[a,c]
45.5–50.0–4.5 mol % $[PMMIM]Cl/AlCl_3/NaCl$	$[PMMIM]^+/Na^+$	$[AlCl_4]^-$	W	4.6	32[a,c]
45.5–50.0–4.5 mol % $[PMMIM]Cl/AlCl_3/NaCl$	$[PMMIM]^+/Na^+$	$[AlCl_4]^-$	W	4.7	32[a,c,d]
50.0–50.0 mol % $[BP]Cl/AlCl_3$	$[BP]^+$	$[AlCl_4]^-$	W	3.6	31[a]
52.0–48.0 mol % $[EMIM]Cl/GaCl_3$	$[EMIM]^+$	$[GaCl_4]^-/Cl^-$ [b]	W	2.4	33[a]
50.0–50.0 mol % $[EMIM]Cl/GaCl_3$	$[EMIM]^+$	$[GaCl_4]^-$ [b]	W	4.0	33[a]
49.0–51.0 mol % $[EMIM]Cl/GaCl_3$	$[EMIM]^+$	$[Ga_2Cl_7]^-/[GaCl_4]^-$ [b]	W	2.2	33[a]
52.0–48.0 mol % $[BP]Cl/GaCl_3$	$[BP]^+$	$[GaCl_4]^-/Cl^-$ [b]	W	2.2	33[a]
50.0–50.0 mol % $[BP]Cl/GaCl_3$	$[BP]^+$	$[GaCl_4]^-$ [b]	W	3.7	33[a]
49.0–51.0 mol % $[BP]Cl/GaCl_3$	$[BP]^+$	$[Ga_2Cl_7]^-/[GaCl_4]^-$ [b]	W	2.2	33[a]
Basic $[(CH_3)_2(C_2H_5)(C_2H_5OCH_2)N]Cl/AlCl_3$	$[(CH_3)_2(C_2H_5)(C_2H_5OCH_2)N]^+$	$[AlCl_4]^-/Cl^-$	Pt	3.5	34[a]
33.0–67.0 mol % $[(CH_3)_3S]Cl/AlCl_3$	$[(CH_3)_3S]^+$	$[Al_2Cl_7]^-/[AlCl_4]^-$	GC	2.5	35[a]

[a] Voltage window estimated from cyclic voltammograms contained in the reference.

[b] The exact nature of the anions is unknown; anions listed are those that would be expected if the system behaved similarly to the chloroaluminates.

[c] Voltage window determined assuming anodic limit of 2.4 V vs. Al/Al(III) reference.

[d] Small amount of $[EMIM][HCl]_2$ added.

[e] Working electrode, Pt = platinum, GC = glassy carbon, W = tungsten.

neutral, and acidic. Basic ionic liquids contain an excess of the organic chloride salt (> 50 mol %), resulting in the presence of free chloride ion (a Lewis base); this, in turn, significantly restricts the anodic limit of the basic ionic liquids. Acidic ionic liquids are prepared with an excess of the aluminium chloride (organic chloride < 50 mol %), and contain two chloroaluminate species: $[AlCl_4]^-$ and $[Al_2Cl_7]^-$ (a Lewis acid). Both anions are significantly more stable towards oxidation than chloride ion. Furthermore, the $[Al_2Cl_7]^-$ is more readily reduced than the organic cation. The acidic ionic liquids thus have a limited cathodic range, but an extended anodic potential range. In the special case in which the organic chloride salt and aluminium chloride are present in equal amounts (50 mol %), these ionic liquids are termed neutral, because they contain only the organic cation and $[AlCl_4]^-$. The neutral chloroaluminate ionic liquids possess the widest electrochemical windows, but they are difficult to prepare and to maintain at the exact neutral composition. A solution to this problem has been developed through the introduction of a third component, an alkali halide, to the chloroaluminate ionic liquids [39]. When an excess of alkali halide (e.g., LiCl, NaCl) is added to an acidic chloroaluminate ionic liquid it dissolves to the extent that it reacts with $[Al_2Cl_7]^-$ ion to produce $[AlCl_4]^-$ and the alkali metal cation; this results in an ionic liquid that is essentially neutral, and at that point the alkali halide is no longer soluble. This neutral ionic liquid is "buffered" to the addition either of more $[Al_2Cl_7]^-$ or of organic chloride. Consequently, the buffered neutral ionic liquids possess wide, and stable, electrochemical windows. However, the cathodic limits of the imidazolium-based buffered neutral ionic liquids are not sufficient to obtain reversible alkali metal deposition and stripping. Interestingly, addition of small amounts of proton to the buffered neutral ionic liquids shifts the reduction of the imidazolium cation sufficiently negative such that reversible lithium and sodium deposition and stripping can be obtained [32, 37].

3.6.2 Ionic Conductivity

The ionic conductivity of a solvent is of critical importance in its selection for an electrochemical application. There are a variety of DC and AC methods available for the measurement of ionic conductivity. In the case of ionic liquids, however, the vast majority of data in the literature have been collected by one of two AC techniques: the impedance bridge method or the complex impedance method [40]. Both of these methods employ simple two-electrode cells to measure the impedance of the ionic liquid (Z). This impedance arises from resistive (R) and capacitive contributions (C), and can be described by Equation (3.6-1):

$$Z = \sqrt{\left(\frac{1}{\omega C}\right)^2 + R^2} \tag{3.6-1}$$

where ω is the frequency of the AC modulation. One can see from Equation 3.6-1 that as the AC frequency increases the capacitive contribution to the impedance becomes vanishingly small and Equation 3.6-1 reduces to $Z = R$, the resistance of the ionic liquid in the impedance cell. Under these conditions the conductivity (κ) of the ionic liquid may be obtained from the measured resistance by Equation 3.6-2:

$$\kappa = \frac{l}{AR} \tag{3.6-2}$$

where l is the distance between the two electrodes in the impedance cell and A is the area of the electrodes. The term in l/A is often referred to as the cell constant and it is normally determined by measuring the conductivity of a standard solution (usually aqueous KCl).

The impedance bridge method employs the AC version of a Wheatstone bridge (i.e., an Impedance Bridge) to measure the unknown cell impedance. Impedance measurements are carried out at a relatively high fixed frequency, normally in the range of a few kHz, in order to minimize the impact of capacitive contribution on the cell impedance. This contribution is often further reduced by increasing the electrode surface area, and correspondingly increasing its capacitance, with a fine deposit of platinum black. The complex impedance method involves the measurement of the cell impedance at frequencies ranging from a few Hz up to several MHz. The impedance data is collected with standard electrochemical impedance hardware (potentiostat/impedance analyzer) and is separated out into its real and imaginary components. These data are then plotted in the form of a Nyquist Plot (imaginary vs. real impedance), and the ionic liquid resistance is taken as the point at which the data crosses the real axis at high frequency.

In general, there appears to be no significant difference between the data collected by either method. There is some evidence that data collected by the bridge method, at lower frequencies, may provide an underestimation of the true conductivity [20, 24], but there is no indication that this error is endemic to the impedance bridge method. The instrumentation for the impedance bridge method, although somewhat specialized, is generally less costly than the instrumentation required by the complex impedance method. However, the complex impedance method has gained popularity in recent years, most probably due to the increased availability of electrochemical impedance hardware.

The conductivity of an electrolyte is a measure of the available charge carriers and their mobility. Superficially, one would expect ionic liquids to possess very high conductivities because they are composed entirely of ions. Unfortunately this is not the case. As a class, ionic liquids possess reasonably good ionic conductivities, comparable to the best non-aqueous solvent/electrolyte systems (up to ~10 mS cm^{-1}). However, they are significantly less conductive than concentrated aqueous electrolytes. The smaller than expected conductivity of ionic liquids can be attributed to the reduction of available charge carriers due to ion pairing and/or ion aggregation, and to the reduced ion mobility resulting from the large ion size found in many ionic liquids.

The conductivity of ionic liquids often exhibits classical linear Arrhenius behavior above room temperature. However, as the temperatures of these ionic liquids approach their glass transition temperatures (T_gs), the conductivity displays significant negative deviation from linear behavior. The observed temperature-dependent conductivity behavior is consistent with glass-forming liquids, and is often best described by the empirical Vogel–Tammann–Fulcher (VTF) equation (3.6-3):

$$\kappa = AT^{-1/2}\exp[-B/(T-T_o)] \tag{3.6-3}$$

where A and B are constants, and T_o is the temperature at which the conductivity (κ) goes to zero [60]. Examples of Arrhenius plots of temperature-dependent conductivity data for three ionic liquids are shown in Figure 3.6-2 [15, 27, 41].

The data in Figure 3.6-2 are also fit to the VTF equation. As can be seen from these data, the change in conductivity with temperature clearly varies depending on the ionic liquid. The conductivity of [EMIM][BF_4], for example, decreases by a factor of 10 over the 375 to 275 K temperature range, while the conductivity of [PMMIM][$(CF_3SO_2)_2N$] decreases by a factor of 30 over the same range of temperatures (Figure 3.6-2). The temperature dependence of conductivity of an ionic liquid involves a complex interplay of short- and long-range forces that is strongly impacted by the type and character of the cation and anion. At our current level of understanding it is not possible to predict accurately how the conductivity of a given ionic liquid will vary with temperature [42].

The room temperature conductivity data for a wide variety of ionic liquids are listed in Tables 3.6-3, 3.6-4, and 3.6-5. These tables are organized by the general type of ionic liquid. Table 3.6-3 contains data for imidazolium-based non-haloaluminate alkylimidazolium ionic liquids, Table 3.6-4 data for the haloaluminate ionic liquids, and Table 3.6-5 data for other types of ionic liquids. There are multiple listings for several of the ionic liquids in Tables 3.6-3–3.6-5. These represent measurements by different researchers and have been included to help emphasize the significant vari-

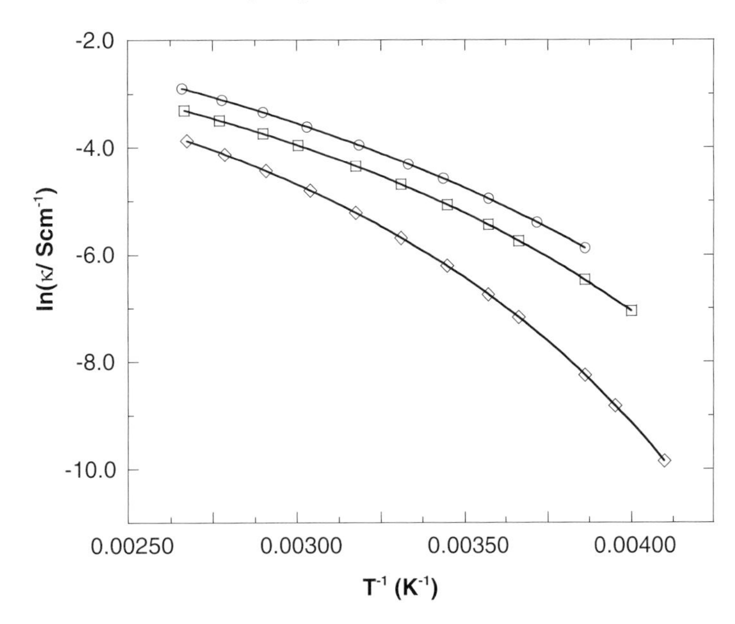

Figure 3.6-2: Examples of Arrhenius plots of temperature-dependent conductivity for [EMIM][BF_4] (○), [EMIM][$(CF_3SO_2)_2N$] (□), and [PMMIM][$(CF_3SO_2)_2N$] (◇). The solid lines through each set of data represents the best fit of the data to the VTF equation.

Table 3.6-3: Specific conductivity data for non-haloaluminate alkylimidazolium-based ionic liquids.

Cation	*Anion*	*Temperature (K)*	*Conductivity (κ), mS cm^{-1}*	*Conductivity method*	*Viscosity (υ), cP*	*Viscosity method*	*Density (ρ), g cm^{-3}*	*Density method*	*Molar Conductivity (Λ), cm$^2\Omega^{-1}$mol^{-1}*	*Walden product ($\Lambda\eta$)*	*Ref.*
$[MMIM]^+$	$[(CF_3SO_2)_2N]^-$	293	8.4	B	44	R	1.559	U	2.03	89.44	26
$[EMIM]^+$	$[BF_4]^-$	295	12	I							43
$[EMIM]^+$	$[BF_4]^-$	298	14	B	34	U	1.240	U	2.24	76.22	24
$[EMIM]^+$	$[BF_4]^-$	298	14	I	32	R	1.279	I	2.17	68.91	41[a]
$[EMIM]^+$	$[BF_4]^-$	299	13	B	43	R					27
$[EMIM]^+$	$[BF_4]^-$	303	20	I							23
$[EMIM]^+$	$[PF_6]^-$	299	5.2	B							27
$[EMIM]^+$	$[CH_3CO_2]^-$	293	2.8	B	162	R					26
$[EMIM]^+$	$[CF_3CO_2]^-$	293	9.6	B	35	R	1.285	U	1.67	58.62	26
$[EMIM]^+$	$[C_3F_7CO_2]^-$	293	2.7	B	105	R	1.450	U	0.60	63.39	26
$[EMIM]^+$	$[CH_3SO_3]^-$	298	2.7	B	160	C	1.240	V	0.45	71.86	25
$[EMIM]^+$	$[CF_3SO_3]^-$	293	8.6	B	45	R	1.390	U	1.61	72.45	26
$[EMIM]^+$	$[CF_3SO_3]^-$	298	9.2	B	43	C	1.380	V	1.73	74.08	25
$[EMIM]^+$	$[CF_3SO_3]^-$	303	8.2	B							44[b]
$[EMIM]^+$	$[(CF_3SO_2)_2N]^-$	293	8.8	B	34	R	1.520	U	2.27	77.03	26
$[EMIM]^+$	$[(CF_3SO_2)_2N]^-$	298	5.7	I	31	R	1.518	I	1.47	44.82	41[a]
$[EMIM]^+$	$[(CF_3SO_2)_2N]^-$	303	10	I							44[a]
$[EMIM]^+$	$[(CF_3SO_2)_2N]^-$	298	9.2	B	34	U	1.510	U	2.38	81.06	19
$[EMIM]^+$	$[(CF_3SO_2)_2N]^-$	299	8.4	B	28	R					27
$[EMIM]^+$	$[(C_2F_5SO_2)_2N]^-$	299	3.4	B	61	R					27
$[EMMIM]^+$	$[(CF_3SO_2)_2N]^-$	293	3.2	B	88	R	1.495	U	0.87	76.35	26
$[EMM(5)IM]^+$	$[CF_3SO_3]^-$	293	6.4	B	51	R	1.334	U	1.32	67.11	26
$[EMM(5)IM]^+$	$[(CF_3SO_2)_2N]^-$	293	6.6	B	37	R	1.470	U	1.82	67.34	26
$[PMIM]^+$	$[PF_6]^-$	293					1.333	V			45
$[PMMIM]^+$	$[BF_4]^-$	295	5.9	I							43
$[PMMIM]^+$	$[PF_6]^-$	308	0.5	B							27
$[PMMIM]^+$	$[(CF_3SO_2)_2N]^-$	299	3.0	B	60	R					27

$[BMIM]^+$	$[PF_6]^-$	295	1.8	I							47
$[BMIM]^+$	$[PF_6]^-$	293					1.363	V			45
$[BMIM]^+$	$[PF_6]^-$	298			207	R					46
$[BMIM]^+$	$[CF_3CO_2]^-$	293	3.2	B	73	R	1.209	U	0.67	48.74	26
$[BMIM]^+$	$[C_3F_7CO_2]^-$	293	1.0	B	182	R	1.333	U	0.26	48.09	26
$[BMIM]^+$	$[CF_3SO_3]^-$	293	3.7	B	90	R	1.290	U	0.83	74.42	26
$[BMIM]^+$	$[C_4F_9SO_3]^-$	293	0.45	B	373	R	1.427	U	0.14	51.56	26
$[BMIM]^+$	$[(CF_3SO_2)_2N]^-$	293	3.9	B	52	R	1.429	U	1.14	59.52	26
$[iBMIM]^+$	$[(CF_3SO_2)_2N]^-$	293	2.6	B	83	R	1.428	U	0.76	63.37	26
$[BMMIM]^+$	$[BF_4]^-$	295	0.23	I							43
$[BMMIM]^+$	$[PF_6]^-$	295	0.77	I							43
$[HMIM]^+$	$[PF_6]^-$	293					1.307	V			45
$[OMIM]^+$	$[PF_6]^-$	293					1.237	V			45
$[EEIM]^+$	$[CF_3CO_2]^-$	293	7.4	B	43	R	1.250	U	1.41	60.64	26
$[EEIM]^+$	$[CF_3SO_3]^-$	293	7.5	B	53	R	1.330	U	1.55	81.97	26
$[EEIM]^+$	$[(CF_3SO_2)_2N]^-$	293	8.5	B	35	R	1.452	U	2.37	83.05	26
$[EEM(5)IM]^+$	$[(CF_3SO_2)_2N]^-$	293	6.2	B	36	R	1.432	U	1.82	65.36	26
$[BEIM]^+$	$[CF_3CO_2]^-$	293	2.5	B	89	R	1.183	U	0.56	50.08	26
$[BEIM]^+$	$[CH_3SO_3]^-$	298	0.55	B			1.140	V	0.12		25
$[BEIM]^+$	$[CF_3SO_3]^-$	298	2.7	B			1.270	V	0.64		25
$[BEIM]^+$	$[C_4F_9SO_3]^-$	293	0.53	B	323	R	1.427	U	0.17	54.26	26
$[BEIM]^+$	$[(CF_3SO_2)_2N]^-$	293	4.1	B	48	R	1.404	U	1.27	60.75	26
$[DEIM]^+$	$[CF_3SO_3]^-$	298					1.10	V			25
$[MeOEtMIM]^+$	$[CF_3SO_3]^-$	293	3.6	B	74	R	1.364	U	0.77	56.69	26
$[MeOEtMIM]^+$	$[(CF_3SO_2)_2N]^-$	293	4.2	B	54	R	1.496	U	1.18	63.88	26
$[CF_3CH_2MIM]^+$	$[(CF_3SO_2)_2N]^-$	293	0.98	B	248	R	1.656	U	0.25	62.56	26

I = complex impedance, B = conductivity bridge, C = capillary viscometer, P = pycnometer or dilatometer, V = volumetric glassware, I = instrument, U = method unknown (not provided in the reference) [a] Conductivity at 298 K calculated from VTF parameters given in reference. [b] Conductivity estimated from graphical data provided in the reference. [c] Density estimated from graphical data provided in the reference.

ability in the conductivity data found in the literature. For example, there are five separate listings for the $[EMIM][(CF_3SO_2)_2N]$ ionic liquid in Table 3.6-3, with conductivity values ranging from 5.7 to 10.0 mS cm^{-1}. Some of these differences may be accountable for by slight differences in experimental temperature, or they could result from measurement error. However, most of this variability is undoubtedly due to impurities in the ionic liquids. Recent work has shown, for example, that contamination with chloride ion increases ionic liquid viscosity, while contamination with water decreases the viscosity [13]. This work has also shown the significant solubility of water in many so-called "hydrophobic" ionic liquids and the strong propensity of these same ionic liquids to absorb water from laboratory air. As can be seen in the discussion below, ionic liquid viscosity is strongly coupled to ionic conductivity. The likelihood that many of the ionic liquids listed in Tables 3.6-3–3.6-5 contained significant concentrations of impurities (especially water) makes evaluation of the literature data difficult. Consequently, any conclusions drawn below must be used with caution. Ionic liquid conductivity appears to be only weakly correlated with the size and type of the cation (Tables 3.6-3–3.6-5). Increasing cation size tends to give rise to lower conductivity, most probably due to the lower mobility of the larger cations.

The overall trend in conductivity with respect to cation type follows the order: imidazolium $\geq$ sulfonium $>$ ammonium $\geq$ pyridinium. Interestingly, the correlation between the anion type or size and the ionic liquid conductivity is very limited. Other than the higher conductivities observed for ionic liquids with the $[BF_4]^-$ anion, there appears to be no clear relationship between anion size and conductivity. Ionic liquids with large anions such as $[(CF_3SO_2)_2N]^-$, for example, often exhibit higher conductivities than those with smaller anions, such as $[CH_3CO_2]^-$.

The conductivity and viscosity of an ionic liquid is often combined into what is termed Walden's rule [Equation (3.6-4)] [54],

$$\Lambda\eta = \text{constant} \tag{3.6-4}$$

where Λ is the molar conductivity of the ionic liquid, and it is given by Equation (3.6-5)

$$\Lambda = \kappa M/\rho \tag{3.6-5}$$

where M is the equivalent weight (molecular weight) of the ionic liquid and ρ is the ionic liquid density. Ideally, the Walden Product ($\Lambda\eta$) remains constant for a given ionic liquid regardless of temperature. The magnitude of the Walden Product for different ionic liquids has been shown to vary inversely with ion size [27, 54]. This inverse relationship between ion size and the magnitude of $\Lambda\eta$ is generally followed for the cations in Tables 3.6-3–3.6-5. The clearest example of this can be seen for the sulfonium ionic liquids, in which increasing cation size from $[(CH_3)_3S]^+$, $[(C_2H_5)_3S]^+$, and $[(n\text{-}C_4H_9)_3S]^+$ results in Walden products of 81.59, 58.27, and 39.36, respectively. As was the case with conductivity, the size of the anions in Tables 3.6-3–3.6-5 exhibits no clear correlation to the magnitude of the Walden product.

Table 3.6-4: Specific conductivity data for binary haloaluminate ionic liquids.

Ionic liquid system	*Cation*	*Anion(s)*	*Temperature (K)*	*Conductivity (κ), mS cm^{-1}*	*Conductivity method*	*Viscosity (η), cP*	*Viscosity method*	*Density (ρ), g cm^{-3}*	*Density method*	*Molar Conductivity (Λ), cm^2 Ω^{-1}mol^{-1}*	*Walden product (Λη)*	*Ref.*
34.0–66.0 mol % [MMIM]Cl/$AlCl_3$	$[MMIM]^+$	$[Al_2Cl_7]^-$	298	15.0	B	17	C	1.404	P	4.26	72.07	48[a]
34.0–66.0 mol % [EMIM]Cl/$AlCl_3$	$[EMIM]^+$	$[Al_2Cl_7]^-$	298	15.0	B	14	C	1.389	P	4.46	62.95	48[a]
50.0–50.0 mol % [EMIM]Cl/$AlCl_3$	$[EMIM]^+$	$[AlCl_4]^-$	298	23.0	B	18	C	1.294	P	4.98	89.07	48[a]
60.0–40.0 mol % [EMIM]Cl/$AlCl_3$	$[EMIM]^+$	Cl^-, $[AlCl_4]^-$	298	6.5	B	47	C	1.256	P	1.22	57.77	48[a]
34.0–66.0 mol % [EMIM]Br/$AlBr_3$	$[EMIM]^+$	$[Al_2Br_7]^-$	298	5.8	B	32	C	2.219	P	1.89	59.64	49[a,b]
60.0–40.0 mol % [EMIM]Br/$AlBr_3$	$[EMIM]^+$	Br^-, $[AlBr_4]^-$	298	5.7	B	67	C	1.828	P	1.15	76.72	49[a,b]
40.0–60.0 mol % [PMIM]Cl/$AlCl_3$	$[PMIM]^+$	$[AlCl_4]^-$, $[Al_2Cl_7]^-$	298	11.0	B	18	C	1.351	P	2.94	53.44	48[a]
50.0–50.0 mol % [PMIM]Cl/$AlCl_3$	$[PMIM]^+$	$[AlCl_4]^-$	298	12.0	B	27	C	1.262	P	2.79	76.29	48[a]
60.0–40.0 mol % [PMIM]Cl/$AlCl_3$	$[PMIM]^+$	Cl^-, $[AlCl_4]^-$	298	3.3	B		C		P			48[a]
34.0–66.0 mol % [BMIM]Cl/$AlCl_3$	$[BMIM]^+$	$[Al_2Cl_7]^-$	298	9.2	B	19	C	1.334	P	3.04	58.45	48[a]
50.0–50.0 mol % [BMIM]Cl/$AlCl_3$	$[BMIM]^+$	$[AlCl_4]^-$	298	10.0	B	27	C	1.238	P	2.49	67.42	48[a]
34.0–66.0 mol % [BBIM]Cl/$AlCl_3$	$[BBIM]^+$	$[Al_2Cl_7]^-$	298	6.0	B	24	C	1.252	P	2.32	55.36	48[a]
50.0–50.0 mol % [BBIM]Cl/$AlCl_3$	$[BBIM]^+$	$[AlCl_4]^-$	298	5.0	B	38	C	1.164	P	1.50	56.83	48[a]
33.3–66.7 mol % [MP]Cl/$AlCl_3$	$[MP]^+$	$[Al_2Cl_7]^-$	298	8.1	B	21	C	1.441	P	2.23	46.12	50[a]
33.3–66.7 mol % [EP]Cl/$AlCl_3$	$[EP]^+$	$[Al_2Cl_7]^-$	298	10.0	B	18	C	1.408	P	2.91	51.29	50[a]
33.3–66.7 mol % [EP]Br/$AlCl_3$	$[EP]^+$	$[Al_2Cl_xBr_{7-x}]^-$	298	8.4	B	22	C	1.524	P			50[a]
33.3–66.7 mol % [EP]Br/$AlCl_3$	$[EP]^+$	$[Al_2Cl_xBr_{7-x}]^-$	298	17.0	B	25						51
33.3–66.7 mol % [EP]Br/$AlBr_3$	$[EP]^+$	$[Al_2Br_7]^-$	298			50	C	2.20	V			52
33.3–66.7 mol % [PP]Cl/$AlCl_3$	$[PP]^+$	$[Al_2Cl_7]^-$	298	8.0	B	18	C	1.375	P	2.47	44.93	50[b]
33.3–66.7 mol % [BP]Cl/$AlCl_3$	$[BP]^+$	$[Al_2Cl_7]^-$	298	6.7	B	21	C	1.346	P	2.18	45.81	50[b]

I = complex impedance, B = conductivity bridge, C = capillary viscometer, P = pycnometer or dilatometer, V = volumetric glassware, I = instrument, U = method unknown (not provided in the reference).

[a] Conductivity at 298K calculated from least-squares-fitted parameters given in reference.

[b] Conductivity estimated from graphical data provided in the reference.

Table 3.6-5: Specific conductivity data for other room-temperature ionic liquids.

Cation	Anion	Temperature (K)	Conductivity (κ), mS cm^{-1}	Conductivity method	Viscosity (η), cP	Viscosity method	Density (ρ), g cm^{-3}	Density method	Molar Conductivity (Λ), cm$^2\Omega^{-1}$mol^{-1}	Walden product ($\Lambda\eta$)	Ref.
Ammonium											
$[(CH_3)_2(C_2H_5)(CH_3OC_2H_4)N]^+$	$[BF_4]^-$	298	1.7	B							53
$[(n\text{–}C_3H_7)(CH_3)_3N]^+$	$[(CF_3SO_2)_2N]^-$	298	3.3	B	72	U	1.440	U	0.88	63.09	19
$[(n\text{-}C_6H_{13})(C_2H_5)_3N]^+$	$[(CF_3SO_2)_2N]^-$	298	0.67	I	167	C	1.270	V	0.25	41.10	20
$[(n\text{-}C_8H_{17})(C_2H_5)_3N]^+$	$[(CF_3SO_2)_2N]^-$	298	0.33	I	202	C	1.250	V	0.13	26.37	20
$[(n\text{-}C_8H_{17})(C_4H_9)_3N]^+$	$[(CF_3SO_2)_2N]^-$	298	0.13	I	574	C	1.120	V	0.07	38.56	20
$[(CH_3)_3(CH_3OCH_2)N]^+$	$[(CF_3SO_2)_2N]^-$	298	4.7	B	50	U	1.510	U	1.20	59.81	19
1-propyl-1-methyl-pyrrolidinium	$[(CF_3SO_2)_2N]^-$	298	1.4	B	63	C	1.45	V			21
1-butyl-1-methyl-pyrrolidinium	$[(CF_3SO_2)_2N]^-$	298	2.2	B	85	C	1.41	V			21
Pyrazolium											
1,2-dimethyl-4-fluoropyrazolium	$[BF_4]^-$	298	1.3	B							29
Pyridinium											
$[BP]^+$	$[BF_4]^-$	298	1.9	I	103	R	1.220	I	0.35	35.77	41[a]
$[BP]^+$	$[BF_4]^-$	303	3.0	I							23
$[BP]^+$	$[(CF_3SO_2)_2N]^-$	298	2.2	I	57	R	1.449	I	0.63	35.91	41[a]
Sulfonium											
$[(CH_3)_3S]^+$	$[HBr_2]^-$	298	34	B	20.5	C	1.74	P	4.62	95.33	52
$[(CH_3)_3S]^{+b}$	$[HBr_2]^-$, $[H_2Br_3]^-$	298	56	B	8.3	C	1.79	P	8.41	69.80	52
$[(CH_3)_3S]^+$	$[Al_2Cl_7]^-$	298	5.5	B							35
$[(CH_3)_3S]^+$	$[Al_2Cl_7]^-$	298	5.5	B	39.3	C	1.40	V	1.49	58.56	52
$[(CH_3)_3S]^+$	$[Al_2Cl_6Br]^-$	298	4.21	B	54.9	C	1.59	V	1.12	61.60	52
$[(CH_3)_3S]^+$	$[Al_2Br_7]^-$	298	1.44	B	138	C	2.40	V	0.41	57.17	52
$[(CH_3)_3S]^+$	$[(CF_3SO_2)_2N]^-$	318	8.2	B	44	U	1.580	U	1.85	81.59	30
$[(C_2H_5)_3S]^+$	$[(CF_3SO_2)_2N]^-$	298	7.1	B	30	U	1.460	U	1.94	58.27	30
$[(n\text{-}C_4H_9)_3S]^+$	$[(CF_3SO_2)_2N]^-$	298	1.4	B	75	U	1.290	U	0.52	39.36	30
Thiazolium											
1-ethylthiazolium	$[CF_3SO_3]^-$	298	4.2	B			1.50	V			25

I = complex impedance, B = conductivity bridge, C = capillary viscometer, P = pycnometer or dilatometer, V = volumetric glassware, I = instrument, U = method unknown (not provided in the reference). [a] Conductivity at 298 K calculated from VTF parameters given in reference. [b] Binary composition of 42.0–58.0 mol % $[(CH_3)_3S]Br\text{–}HBr$.

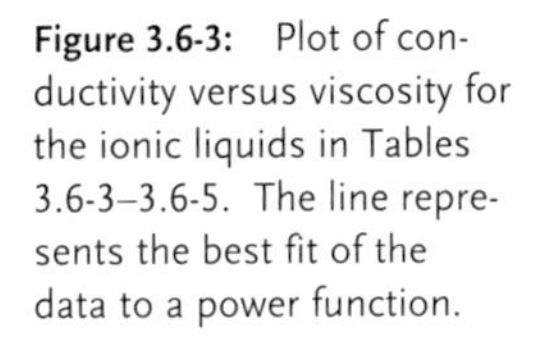

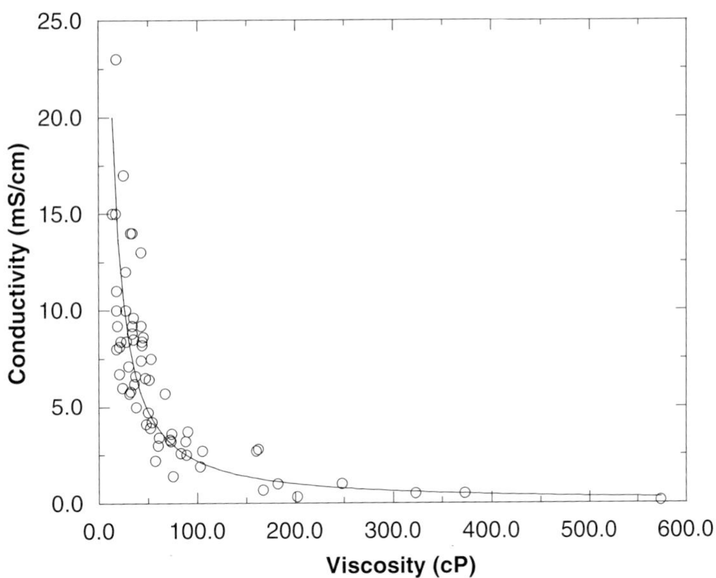

Figure 3.6-3: Plot of conductivity versus viscosity for the ionic liquids in Tables 3.6-3–3.6-5. The line represents the best fit of the data to a power function.

Ionic liquid conductivity appears to be most strongly correlated with viscosity (η). Figure 3.6-3 shows a plot of conductivity versus viscosity for the data in Tables 3.6-3–3.6-5. This figure clearly demonstrates an inverse relationship between conductivity and viscosity.

The data in Figure 3.6-3 were fitted to a simple power function to give the resulting equation, $\kappa = 390\eta^{-1.125}$, $R^2 = 0.83$. A potentially more informative way to look at the relationship between conductivity and viscosity is through the use of a Walden Plot (log Λ versus log η^{-1}) [54]. Plotting the molar conductivity (Λ) instead of the absolute conductivity (κ) normalizes to some extent for the effects of molar concentration and density on the conductivity, and thus gives a better indication of the number of mobile charge carriers in an ionic liquid. Figure 3.6-4 shows the Walden Plot for the data in Tables 3.6-3–3.6-5.

A linear regression was performed on the data, giving a slope of 1.08, an intercept of 1.922, and $R^2 = 0.94$. The fit of the data to the linear relationship is surprisingly good when one considers the wide variety of ionic liquids and the unknown errors in the literature data. This linear behavior in the Walden Plot clearly indicates that the number of mobile charge carriers in an ionic liquid and its viscosity are strongly coupled.

The physical properties of ionic liquids can often be considerably improved through the judicious addition of co-solvents [55–58]. However, surprisingly, this approach has been relatively underutilized. Hussey and co-workers investigated the effect of co-solvents on the physical properties of [EMIM]Cl/$AlCl_3$ ionic liquids [55, 56]. They found significant increases in ionic conductivity upon the addition of a variety of co-solvents. Figure 3.6-5 displays representative data from this work. The magnitude of the conductivity increase depends both on the type and amount of the co-solvent [55, 56].

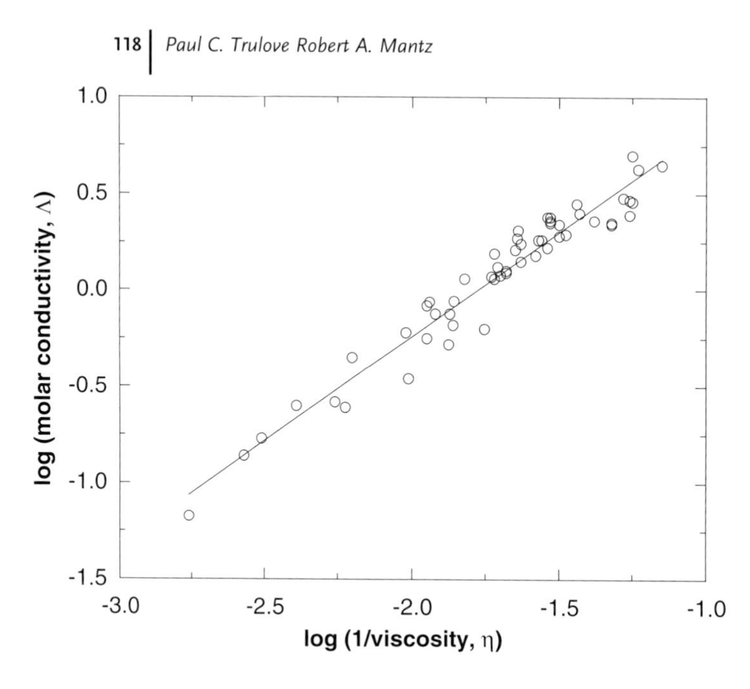

Figure 3.6-4: The Walden plot of the molar conductivity and viscosity data in Tables 3.6-3–3.6-5. The line represents the linear regression fit of the data.

The observed effect of co-solvent addition has been explained in terms of the solvation of the constituent ions of the ionic liquid by the co-solvent. This solvation, in turn, reduces ion-pairing or ion aggregation in the ionic liquid, resulting in an increase in the number of available charge carriers and an increase in the mobility of these charge carriers. Counteracting this solvating effect is the dilution of the number of free ions as the mole fraction of the co-solvent increases. These counteracting effects help to explain the observed maximum in conductivity for benzene added to a 40.00–60.00 mol % [EMIM]Cl/$AlCl_3$ ionic liquid, shown in Figure 3.6-5.

3.6.3
Transport Properties

The behavior of ionic liquids as electrolytes is strongly influenced by the transport properties of their ionic constituents. These transport properties relate to the rate of ion movement and to the manner in which the ions move (as individual ions, ion-pairs, or ion aggregates). Conductivity, for example, depends on the number and mobility of charge carriers. If an ionic liquid is dominated by highly mobile but neutral ion-pairs it will have a small number of available charge carriers and thus a low conductivity. The two quantities often used to evaluate the transport properties of electrolytes are the ion-diffusion coefficients and the ion-transport numbers. The diffusion coefficient is a measure of the rate of movement of an ion in a solution, and the transport number is a measure of the fraction of charge carried by that ion in the presence of an electric field.

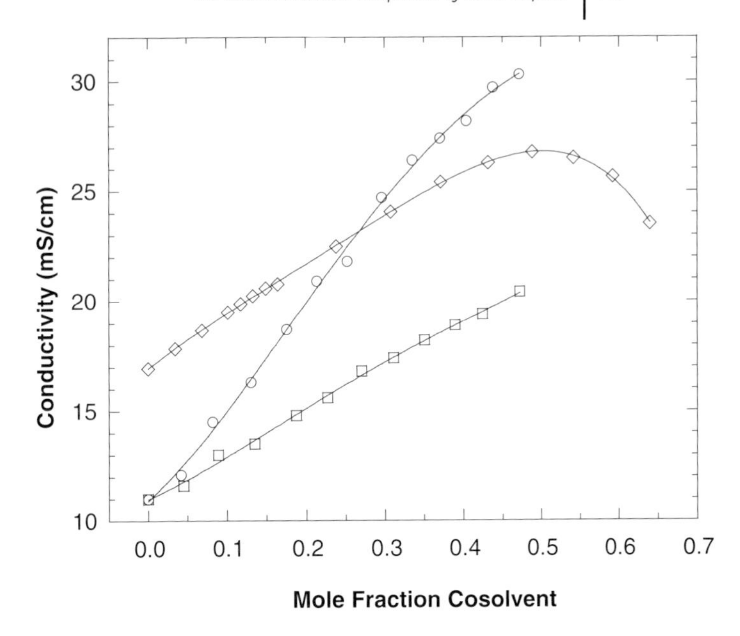

Figure 3.6-5: Change in the conductivity of [EMIM]Cl/$AlCl_3$ ionic liquids with the mole fraction of co-solvent: (□) benzene or (○) CH_2Cl_2 added to a 55.56–44.44 mol % [EMIM]Cl/$AlCl_3$ ionic liquid, and (◇) benzene added to a 40.00–60.00 mol % [EMIM]Cl/$AlCl_3$ ionic liquid.

The diffusion coefficients of the constituent ions in ionic liquids have most commonly been measured either by electrochemical or by NMR methods. These two methods in fact measure slightly different diffusional properties. The electrochemical methods measure the diffusion coefficient of an ion in the presence of a concentration gradient (Fick diffusion) [59], while the NMR methods measure the diffusion coefficient of an ion in the absence of any concentration gradients (self-diffusion) [60]. Fortunately, under most circumstances these two types of diffusion coefficients are roughly equivalent.

There are a number of NMR methods available for evaluation of self-diffusion coefficients, all of which use the same basic measurement principle [60]. Namely, they are all based on the application of the spin-echo technique under conditions of either a static or a pulsed magnetic field gradient. Essentially, a spin-echo pulse sequence is applied to a nucleus in the ion of interest while at the same time a constant or pulsed field gradient is applied to the nucleus. The spin echo of this nucleus is then measured and its attenuation due to the diffusion of the nucleus in the field gradient is used to determine its self-diffusion coefficient. The self-diffusion coefficient data for a variety of ionic liquids are given in Table 3.6-6.

Electrochemically generated diffusion coefficients are normally determined from the steady state voltammetric limiting current arising from the reduction or oxidation of the ion of interest. In the case of ionic liquids this requires that the potential of a working electrode be scanned into the cathodic and anodic potential limits in the hopes of obtaining clear limiting current plateaus for the reduction of the cation and the oxidation of the anion, respectively. This process is fraught with difficulty and has met with limited success. The very large limiting currents encountered scanning beyond the normal potential limits result in significant migration effects, and the irreversible electrode reactions (especially for the cation reduction) often foul the working electrode surface. The one successful application for these electrochemical methods in ionic liquids has been in the evaluation of the diffusion coefficient of the chloride ion in basic [EMIM]Cl/$AlCl_3$ ionic liquids [63, 64]. In slightly basic ionic liquids the chloride ion concentration is reasonably low and its oxidation well separated from that of the other anion present ([$AlCl_4$]$^-$). These diffusion coefficient data are given in Table 3.6-6.

Table 3.6-6: Diffusion coefficients for ionic liquids.

Ionic liquid system	*Cation*	D_{R+} (10^{11} m^2s^{-1})	*Anion(s)*	D_{X^-} (10^{11} m^2s^{-1})	*Temperature (K)*	*Method*[a]	*Ref*
[EMIM][BF_4]	[EMIM]$^+$	5.0	[BF_4]$^-$	4.2	298	PNMR	41[b]
[EMIM][BF_4]	[EMIM]$^+$	3.0	[BF_4]$^-$	–	300	PNMR	61[c]
[EMIM][CF_3SO_2]	[EMIM]$^+$	5	[CF_3SO_2]$^-$	3	298	FNMR	44[e]
[EMIM][$(CF_3SO_2)_2N$]	[EMIM]$^+$	6.3	[$(CF_3SO_2)_2N$]$^-$	3.5	298	PNMR	41[b]
[EMIM][$(CF_3SO_2)_2N$]	[EMIM]$^+$	5	[$(CF_3SO_2)_2N$]$^-$	4	298	FNMR	44[e]
[BP][BF_4]	[BP]$^+$	0.91	[BF_4]$^-$	1.0	298	PNMR	41[b]
[BP][$(CF_3SO_2)_2N$]	[BP]$^+$	2.4	[BF_4]$^-$	2.0	298	PNMR	41[b]
33.0–67.0 mol % [EMIM]Cl/$AlCl_3$	[EMIM]$^+$	14.4	[Al_2Cl_7]$^-$ / [$AlCl_4$]$^-$	–	303	DNMR	62
50.0–50.0 mol % [EMIM]Cl/$AlCl_3$	[EMIM]$^+$	18	[$AlCl_4$]$^-$	–	298	PNMR	65[c]
50.0–50.0 mol % [EMIM]Cl/$AlCl_3$	[EMIM]$^+$	12.7	[$AlCl_4$]$^-$	–	303	DNMR	62
55.0–45.0 mol % [EMIM]Cl/$AlCl_3$	[EMIM]$^+$	10	Cl^-/[$AlCl_4$]$^-$	–	298	PNMR	65[c]
60.0–40.0 mol % [EMIM]Cl/$AlCl_3$	[EMIM]$^+$	4.3	Cl^-/[$AlCl_4$]$^-$	–	298	PNMR	65[c]
65.0–35.0 mol % [EMIM]Cl/$AlCl_3$	[EMIM]$^+$	2.2	Cl^-/[$AlCl_4$]$^-$	–	298	PNMR	65[c]
70.0–30.0 mol % [EMIM]Cl/$AlCl_3$	[EMIM]$^+$	1.2	Cl^-/[$AlCl_4$]$^-$	–	298	PNMR	65[c]
56.0–44.0 mol % [EMIM]Cl/$AlCl_3$	[EMIM]$^+$	–	Cl^-/[$AlCl_4$]$^-$	6.1[d]	299	Elec	63
50.5–49.5 mol % [EMIM]Cl/$AlCl_3$	[EMIM]$^+$	–	Cl^-/[$AlCl_4$]$^-$	5.7[d]	299	Elec	64
51.0–49.0 mol % [EMIM]Cl/$AlCl_3$	[EMIM]$^+$	–	Cl^-/[$AlCl_4$]$^-$	5.3[d]	299	Elec	64
51.5–48.5 mol % [EMIM]Cl/$AlCl_3$	[EMIM]$^+$	–	Cl^-/[$AlCl_4$]$^-$	4.6[d]	299	Elec	64

[a] PNMR = Pulse Gradient-Field Spin Echo (PGSE) NMR, DNMR = ^1H Diffusion Ordered Spectroscopy (DOSY) NMR, Elec = electrochemistry, FNMR = fringe field NMR. [b] Diffusion coefficients at 298 K Calculated from VTF Parameters given in reference. [c] Only cation diffusion coefficients were determined. [d] Diffusion coefficient for the chloride ion. [e] Diffusion coefficients estimated from graphical data provided in the reference.

The cation diffusion coefficient data in Table 3.6-6 correlate well with the corresponding ionic conductivity data given in Tables 3.6-3 and 3.6-4. As would be expected, the cation diffusion coefficient increases with increasing conductivity. The more limited anion diffusion coefficient data, on the other hand, do not show any significant correlation to conductivity. The cation diffusion coefficients for the [EMIM]Cl/$AlCl_3$ ionic liquids decrease significantly as the mol % of $AlCl_3$ decreases below 50 %. This "basic" composition regime is characterized by increasing chloride as the mole percent of $AlCl_3$ decreases. The decline in cation diffusion coefficients is consistent with the observation of significant hydrogen bonding between chloride ion and the $[EMIM]^+$ cation [4], which would be expected to cause reduced cation mobility.

Transport numbers are intended to measure the fraction of the total ionic current carried by an ion in an electrolyte as it migrates under the influence of an applied electric field. In essence, transport numbers are an indication of the relative ability of an ion to carry charge. The classical way to measure transport numbers is to pass a current between two electrodes contained in separate compartments of a two-compartment cell. These two compartments are separated by a barrier that only allows the passage of ions. After a known amount of charge has passed, the composition and/or mass of the electrolytes in the two compartments are analyzed. From these data the fraction of the charge transported by the cation and the anion can be calculated. Transport numbers obtained by this method are measured with respect to an external reference point (i.e., the separator), and, therefore, are often referred to as external transport numbers. Two variations of the above method, the Moving Boundary method [66] and the Hittorff method [66–69], have been used to measure cation (t_{R+}) and anion (t_{X-}) transport numbers in ionic liquids, and these data are listed in Table 3.6-7.

The measurement of transport numbers by the above electrochemical methods entails a significant amount of experimental effort to generate high-quality data. In addition, the methods do not appear applicable to many of the newer non-haloaluminate ionic liquid systems. An interesting alternative to the above method utilizes the NMR-generated self-diffusion coefficient data discussed above. If both the cation (D_{R+}) and anion (D_{X-}) self-diffusion coefficients are measured, then both the cation (t_{R+}) and anion (t_{X-}) transport numbers can be determined by using the following Equations (3.6-6) and (3.6-7) [41, 44]:

$$t_{R^+} = \frac{D_{R^+}}{D_{R^+} + D_{X^-}} \tag{3.6-6}$$

$$t_{X^-} = \frac{D_{X^-}}{D_{R^+} + D_{X^-}} \tag{3.6-7}$$

Transport numbers for several non-haloaluminate ionic liquids generated from ionic liquid self-diffusion coefficients are listed in Table 3.6-7. The interesting, and still open, question is whether the NMR-generated transport numbers provide the same measure of the fraction of current carried by an ion as the electrochemically generated transport numbers. The electrochemical experiment measures the relative movement of charge carriers in the presence of an applied field, while the NMR experiment measures the movement of all ions regardless of whether they are free ions or neutral ion-pairs. If an ion spends a significant portion of its time as part of a neutral association of ions, will its NMR-generated transport number differ significantly from the electrochemical transport number? Clearly further work is needed to resolve this issue.

Table 3.6-7: External ion transport numbers in ionic liquids.

Ionic liquid system	*cation*	t_{R+}	*Anion(s)*	t_{X^-}	*Temperature (K)*	*Method*[a]	*Ref*
$[EMIM][BF_4]$	$[EMIM]^+$	0.54	$[BF_4]^-$	0.46	298	PNMR	41[b]
$[EMIM][(CF_3SO_2)_2N]$	$[EMIM]^+$	0.64	$[(CF_3SO_2)_2N]^-$	0.36	298	PNMR	41[b]
$[BP][BF_4]$	$[BP]^+$	0.48	$[BF_4]^-$	0.52	298	PNMR	41[b]
$[BP][(CF_3SO_2)_2N]$	$[BP]^+$	0.55	$[(CF_3SO_2)_2N]^-$	0.45	298	PNMR	41[b]
45.0–55.0 mol % $[EMIM]Cl/AlCl_3$	$[EMIM]^+$	0.71	$[AlCl_4]^-$	0.23	303	MH	68[c]
			$[Al_2Cl_7]^-$	0.06	303	MH	68[c]
50.0–50.0 mol % $[EMIM]Cl/AlCl_3$	$[EMIM]^+$	0.70	$[AlCl_4]^-$	–	303	MH	66
50.0–50.0 mol % $[EMIM]Cl/AlCl_3$	$[EMIM]^+$	–	$[AlCl_4]^-$	0.30	366	MB	66
60.0–40.0 mol % $[EMIM]Cl/AlCl_3$	$[EMIM]^+$	0.71	$[AlCl_4]^-$	0.19	298	MH	67[c]
			Cl^-	0.10	298	MH	67[c]
70.0–30.0 mol % $[EMIM]Cl/AlCl_3$	$[EMIM]^+$	0.71	$[AlCl_4]^-$	0.12	298	MH	67[c]
			Cl^-	0.17	298	MH	67[c]
45.0–55.0 mol % $[EMIM]Br/AlBr_3$	$[EMIM]^+$	0.76	$[AlBr_4]^-$	0.22	333	MH	68[c]
			$[Al_2Br_7]^-$	0.02	333	MH	68[c]
50.0–50.0 mol % $[EMIM]Br/AlBr_3$	$[EMIM]^+$	0.76	$[AlBr_4]^-$	0.24	333	MH	69
60.0–40.0 mol % $[EMIM]Br/AlBr_3$	$[EMIM]^+$	0.76	$[AlBr_4]^-$	0.22	333	MH	69[c]
			Br^-	0.02	333	MH	67[c]
70.0–30.0 mol % $[EMIM]Br/AlBr_3$	$[EMIM]^+$	0.76	$[AlBr_4]^-$	0.16	333	MH	67[c]
			Br^-	0.08	333	MH	67[c]

[a] MB = Moving boundary, MH = modified Hittorf, PNMR = Pulse Gradient-Field Spin Echo (PGSE) NMR. [b] Transport numbers at 298 K determined from self-diffusion data provided in the reference. [c] Anion transport numbers calculated from formulas provided in the reference.

As one can see from the data in Table 3.6-7, the $[EMIM]^+$ cation carries the major portion of the charge (t_{R+} 0.70) for all the haloaluminate ionic liquids measured. This result is very surprising in view of the large size of the cation compared with that of the extant anions and the fact that the relative fraction of the charge carried by the anions remains essentially constant even with significant changes in the anion composition occurring with changes in ionic liquid composition. It has been proposed that these observations result from the fact that the smaller anions are more structurally constrained in the charge-transport process [67]. However, this explanation seems overly simplistic. The NMR-generated transport numbers in Table 3.6-7 indicate that, in general, more charge is carried by the cation. However, the relative fraction of this charge is significantly less than that observed in the electrochemical transport data for the haloaluminate ionic liquids.

It is unclear at this time whether this difference is due to the different anions present in the non-haloaluminate ionic liquids or due to differences in the two types of transport number measurements. The apparent greater importance of the cation to the movement of charge demonstrated by the transport numbers (Table 3.6-7) is consistent with the observations made from the diffusion and conductivity data above. Indeed, these data taken in total may indicate that the cation tends to be the majority charge carrier for all ionic liquids, especially the alkylimidazoliums. However, a greater quantity of transport number measurements, performed on a wider variety of ionic liquids, will be needed to ascertain whether this is indeed the case.

References

1 a) F. H. Hurley, T. P. Wier, *J. Electrochem. Soc.*, **1951**, *98*, 203; b) F. H. Hurley, T. P. Wier, *J. Electrochem. Soc.*, **1951**, *98*, 207.

2 H. L. Chum, R. A. Osteryoung, in *Ionic Liquids* (D. Inman, D. G. Lovering eds.), Plenum Press, New York, **1981**, pp. 407–423.

3 R. J. Gale, R. A. Osteryoung, in *Molten Salt Techniques*, Vol. 1 (D. G. Lovering, R. J. Gale eds), Plenum Press, New York, **1983**, pp. 55–78.

4 C. L. Hussey, in *Advances in Molten Salt Chemistry* (G. Mamantov, C. B. Mamantov eds.), Elsevier, Amsterdam, **1983**, pp. 185–230.

5 G. Mamantov, in *Molten Salt Chemistry* (G. Mamantov, R. Marassi eds.), D. Reidel, New York, **1987**, pp. 259–270.

6 R. A. Osteryoung, in *Molten Salt Chemistry* (G. Mamantov, R. Marassi eds.), D. Reidel, New York, **1987**, pp. 329–364.

7 C. L. Hussey, *Pure Appl. Chem.*, **1988**, *60*, 1763.

8 C. L. Hussey, in *Chemistry of Nonaqueous Solutions – Current Progress* (G. Mamantov, A. I. Popov eds.), VCH, New York, **1994**, pp. 227–275.

9 R. T. Carlin, J. S. Wilkes, in *Chemistry of Nonaqueous Solutions – Current Progress* (G. Mamantov, A. I. Popov eds.), VCH, New York, **1994**, pp. 277–306.

10 T. Welton, *Chem. Rev.*, **1999**, *99*, 2071.

11 Y. Ito, T. Nohira, *Electrochim. Acta*, **2000**, *45*, 2611.

12 A. Fry, W. E. Britton, in *Laboratory Techniques in Electroanalytical Chemistry* (P. T. Kissinger, W. R. Heineman eds.), Marcel Dekker, New York, **1984**, chapter 13.

13 K. R. Seddon, A. Stark, M.-J. Torres, *Pure Appl. Chem.*, **2000**, *72*, 2275.

14 U. Schröder, J. D. Wadhawan, R. G. Compton, F. Marken, P. A. Z. Suarez, C. S. Consorti, R. F. de Souza, J. Dupont, *New J. Chem.*, **2000**, *24*, 1009.

15 T. E. Sutto, H. C. De Long, P. C. Trulove, unpublished results.

16 A. E. Visser, R. P. Swatloski, W. M. Reichert, S. T. Griffin, R. D. Rogers, *Ind. Eng. Chem. Res.*, **2000**, *39*, 3596.

17 J. G. Huddleston, A. E. Visser, W. M. Reichert, H. D. Willauer, G. A. Broker, R. D. Rogers, *Green Chem.*, **2001**, *3*, 156.

18 R. T. Carlin, T. Sullivan, *J. Electrochem. Soc.*, **1992**, *139*, 144.

19 H. Matsumoto, M. Yanagida, K. Tanimoto, M. Nomura, Y. Kitagawa, Y. Miyazaki, *Chem. Lett.*, **2000**, 922.

20 a) J. Sun, M. Forsyth, D. R. MacFarlane, *Molten Salt Forum*, **1998**, *5-6*, 585; b) J. Sun, M. Forsyth, D. R. MacFarlane, *J. Phys. Chem. B.*, **1998**, *102*, 8858.

21 D. R. MacFarlane, P. Meakin, J. Sun, N. Amini, M. Forsyth, *J. Phys. Chem. B.*, **1999**, *103*, 4164.

22 a) R. Hagiwara, T. Hirashige, T. Tsuda, Y. Ito, *J. Fluorine Chem.*, **1999**, *99*, 1. b) R. Hagiwara, T. Hirashige, T. Tsuda, Y. Ito, *J. Electrochem. Soc.*, **2002**, *149*, D1.

23 A. Noda, M. Watanabe, *Electrochem. Acta*, **2000**, *45*, 1265.

24 a) J. Fuller, R. T. Carlin, R. A. Osteryoung, *J. Electrochem. Soc.*, **1997**, *144*, 3881. b) J. Fuller, R. A. Osteryoung, R. T. Carlin, *Abstracts of Papers*, 187th Meeting of The Electrochemical Society, Reno, NV, **1995**, Vol. 95-1, p. 27.

25 E. I. Cooper, E. J. M. O'Sullivan, in *Proceedings of the Eighth International Symposium on Molten Salts* (R. J. Gale, G. Blomgren eds.), The Electrochemical Society: Pennington NJ, **2000**, Vol. 92-16, pp. 386–396.

26 P. Bonhôte, A.-P. Dias, N. Papageorgiou, K. Kalyanasundaram, M. Grätzel, *Inorg. Chem.*, **1996**, *35*, 1168.

27 A. B. McEwen, H. L. Ngo, K. LeCompte, J. L. Goldman, *J. Electrochem. Soc.*, **1999**, *146*, 1687.

28 V. R. Koch, L. A. Dominey, C. Najundiah, M. J. Ondrechen, *J. Electrochem. Soc.*, **1996**, *143*, 798.

29 J. Caja, T. D. J. Dunstan, D. M. Ryan, V. Katovic, in *Proceedings of the Twelfth International Symposium on Molten Salts* (P. C. Trulove, H. C. De Long, G. R. Stafford, S. Deki eds.), The Electrochemical Society: Pennington NJ, **2000**, Vol. 99-41, pp.150–161.

30 H. Matsumoto, T. Matsuda, Y. Miyazaki, *Chem. Lett.*, **2000**, 1430.

31 M. Lipsztajn, R. A. Osteryoung, *J. Electrochem. Soc.*, **1983**, *130*, 1968.

32 a) T. L. Riechel, J. S. Wilkes, *J. Electrochem. Soc.*, **1992**, *139*, 977; b) C. Scordilis-Kelly, R. T. Carlin, *J. Electrochem. Soc.*, **1993**, *140*, 1606; c) C. Scordilis-Kelly, R. T. Carlin, *J. Electrochem. Soc.*, **1994**, *141*, 873.

33 S. P. Wicelinski, R. J. Gale, J. S. Wilkes, *J. Electrochem. Soc.*, **1987**, *134*, 262.

34 J. R. Stuff, S. W. Lander Jr., J. W. Rovang, J. S. Wilkes, *J. Electrochem. Soc.*, **1990**, *137*, 1492.

35 S. D. Jones, G. E. Blomgren, in *Proceedings of the Seventh International Symposium on Molten Salts* (C. L. Hussey, S. N. Flengas, J. S. Wilkes, Y. Ito eds.), The Electrochemical Society: Pennington NJ, **1990**, Vol. 90-17, pp. 273–280.

36 In the publication, P. A. Z. Suarez, V. M. Selbach, J. E. L. Dullius, S. Einloft, C. M. S. Piatnecki, D. S. Azambuja, R. F. de Souza, J. Dupont, *Electrochim. Acta*, **1997**, *42*, 2533, the authors claim to have observed electrochemical windows for [BMIM][BF_4] and [BMIM][PF_6] of 6.1 V and 7.1 V, respectively, at a tungsten working electrode, while at a Pt working electrode the windows were 4.6 V and 5.7 V, respectively. The significant disparity between the windows given by the two working electrodes, plus the disagreement with values at Pt (4.1 V and

4.2 V, respectively) generated by this same group in a later publication [13], cause us to treat these impressive electrochemical windows as suspect. Consequently, they have not been included in Table 3.6-1.

37 G. E. Gray, J. Winnick, P. A. Kohl, *J. Electrochem. Soc.*, **1996**, *143*, 3820.

38 J. D. Vaugh, A. Mughrabi, E. C. Wu, *J. Org. Chem.*, **1970**, *35*, 1141.

39 T. J. Melton, J. Joyce, J. T. Maloy, J. A. Boon, J. S. Wilkes, *J. Electrochem. Soc.*, **1990**, *137*, 3865.

40 F. J. Holler, C. G. Enke, in *Laboratory Techniques in Electroanalytical Chemistry* (P. T. Kissinger, W. R. Heineman eds.), Marcel Dekker, New York, **1984**, chapter 8.

41 a) A. Noda, K. Hayamizu, M. Watanabe, *J. Phys. Chem. B*, **2001**, *105*, 4603; b) A. Noda, M. Watanabe, in *Proceedings of the Twelfth International Symposium on Molten Salts* (P. C. Trulove, H. C. De Long, G. R. Stafford, S. Deki eds.), The Electrochemical Society: Pennington NJ, **2000**, Vol. 99-41, pp. 202–208.

42 C. A. Angell, in *Molten Salts: From Fundamentals to Applications* (M. Gaune-Escard, Ed.), Kluwer Academic Publishers, London, **2002**, pp. 305–320.

43 T. E. Sutto, H. C. De Long, P. C. Trulove, in *Progress in Molten Salt Chemistry 1* (R. W. Berg, H. A. Hjuler eds.), Elsevier: Paris, **2000**, 511.

44 H. Every, A.G. Bishop, M. Forsyth, D. R. MacFarlane, *Electrochim. Acta*, **2000**, *45*, 1279.

45 S. Chun, S. V. Dzyuba, R. A. Bartsch, *Anal. Chem.*, **2001**, *73*, 3737.

46 S. N. Baker, G. A. Baker, M. A. Kane, F. V. Bright, *J. Phys. Chem. B*, **2001**, *105*, 9663.

47 J. Fuller, A. C. Breda, R. T. Carlin, *J. Electroanal. Chem.*, **1998**, *459*, 29.

48 a) J. S. Wilkes, J. A. Levisky, R. A. Wilson, C. L. Hussey, *Inorg. Chem.*, **1982**, *21*, 1263; b) A. A. Fannin Jr., D. A. Floreani, L. A. King, J. S. Landers, B. J. Piersma, D. J. Stech, R. J. Vaughn, J. S. Wilkes, J. L. Williams, *J. Phys. Chem.*, **1984**, *88*, 2614.

49 a) J. R. Sanders, E. H. Ward, C. L. Hussey, in *Proceedings of the Fifth International Symposium on Molten Salts* (M.-L. Saboungi, K. Johnson, D. S. Newman, D. Inman eds.), The Electrochemical Society: Pennington NJ, **1986**, Vol. 86-1, pp.307–316; b) J. R. Sanders, E. H. Ward, C. L. Hussey, *J. Electrochem. Soc.*, **1986**, *133*, 325.

50 R. A. Carpio, L. A. King, R. E. Lindstrom, J. C. Nardi, C. L. Hussey, *J. Electrochem. Soc.*, **1979**, *126*, 1644.

51 V. R. Koch, L. L. Miller, R. A. Osteryoung, *J. Am. Chem. Soc.*, **1976**, *98*, 5277.

52 M. Ma, K. E. Johnson, in *Proceedings of the Ninth International Symposium on Molten Salts* (C. L. Hussey, D. S. Newman, G. Mamantov, Y. Ito eds.), The Electrochemical Society: Pennington NJ, **1994**, Vol. 94-13, pp.179–186.

53 E. I. Cooper, C. A. Angell, *Solid State Ionics*, **1983**, *9-10*, 617.

54 S. I. Smedley, *The Interpretation of Ionic Conductivity in Liquids*, Plenum, New York, **1980**, chapter 3.

55 R. L. Perry, K. M. Jones, W. D. Scott, Q. Liao, C. L. Hussey, *J. Chem. Eng. Data*, **1995**, *40*, 615.

56 Q. Liao, C. L. Hussey, *J. Chem. Eng. Data*, **1996**, *41*, 1126.

57 R. Moy, R.-P. Emmenegger, *Electrochimica Acta*, **1992**, *37*, 1061.

58 J. Robinson, R. C. Bugle, H. L. Chum, D. Koran, R. A. Osteryoung, *J. Am. Chem. Soc.*, **1979**, *101*, 3776.

59 A. J. Bard, L. R. Faulkner, Electrochemical Methods – Fundamentals and Applications, 2nd Ed., Wiley, New York, **2001**, chap. 4.

60 P. Stilbs, *Progress in NMR Spectroscopy*, **1987**, *19*, 1.

61 J.-F. Huang, P.-Y. Chen, I.-W. Sun, S. P. Wang, *Inorg. Chim. Acta*, **2001**, *320*, 7.

62 W. R. Carper, G. J. Mains, B. J. Piersma, S. L. Mansfield, C. K. Larive, *J. Phys. Chem.*, **1996**, *100*, 4724.

63 R. T. Carlin, R. A. Osteryoung, *J. Electroanal. Chem.*, **1988**, *252*, 81.

64 L. R. Simonsen, F. M. Donahue, *Electrochim. Acta*, **1990**, *35*, 89.

65 R. A. Mantz, H. C. De Long, R. A. Osteryoung, P. C. Trulove, in *Proceedings of the Twelfth International Symposium on Molten Salts* (P. C. Trulove, H. C. De Long, G. R. Stafford, S. Deki eds.), The Electrochemical Society: Pennington NJ, *2000*, Vol. 99-41, pp.169–176.

66 C. J. Dymek, L. A. King, *J. Electrochem. Soc.*, **1985**, *132*, 1375.

67 C. L. Hussey, J. R. Sanders, H. A. Øye, *J. Electrochem. Soc.*, **1985**, *132*, 2156.

68 C. L. Hussey, H. A. Øye, *J. Electrochem. Soc.*, **1984**, *131*, 1623.

69 C. L. Hussey, J. R. Sanders, *J. Electrochem. Soc.*, **1987**, *134*, 1977.

4
Molecular Structure and Dynamics

W. Robert Carper, Andreas Dölle, Christof G. Hanke, Chris Hardacre, Axel Leuchter, Ruth M. Lynden-Bell, Zhizhong Meng, Günter Palmer, and Joachim Richter

4.1
Order in the Liquid State and Structure

Chris Hardacre

The structure of liquids has been studied for many years. These investigations have, in general, been focussed on the arrangements in molecular solvents such as water, *t*-butanol, and simple chlorinated solvents. The field of molten salts and their structures is much less studied, and within this field the study of the structure of room-temperature ionic liquids is in its infancy. A variety of techniques have been used to investigate liquid structure, including neutron diffraction, X-ray scattering, and extended X-ray absorption fine structure. This chapter summarizes some of the techniques used, including practical details, and shows examples of where they have been employed previously. The examples given are not meant to be exhaustive and are provided for illustration only. Where possible, examples relating to the more recent air- and moisture-stable ionic liquids have been included.

4.1.1
Neutron Diffraction

Neutron diffraction is one of the most widely used techniques for the study of liquid structure. In the experiment, neutrons are elastically scattered off the nuclei in the sample and are detected at different scattering angles, typically 3° to 40°, for the purpose of measuring intermolecular structure whilst minimizing inelasticity corrections. The resultant scattering profile is then analyzed to provide structural information.

The data taken is normally presented as the total structure factor, $F(Q)$. This is related to the neutron scattering lengths b_i, the concentrations c_i, and the partial structure factor $S_{ij}(Q)$ for each pair of atoms i and j in the sample, by Equation 4.1-1:

$$F(Q) = \sum_{i,j} c_i c_j b_i b_j \left(S_{ij}(Q) - 1\right) \quad (4.1\text{-}1)$$

where Q is the scattering vector and is dependent on the scattering angle θ, and the wavelength λ of the neutrons used.

$$Q = \frac{4\pi \sin\theta}{\lambda} \quad (4.1\text{-}2)$$

The real space pair distributions $g_{ij}(r)$ is the inverse Fourier transform of $(S_{ij}(Q)\text{-}1)$, that is:

$$g_{ij}(r) = 1 + \frac{1}{2\pi^2 \rho} \int_0^{\infty} Q^2 \frac{\sin(Qr)}{Qr} \left(S_{ij}(Q) - 1\right) dQ \quad (4.1\text{-}3)$$

normalized to the atomic density ρ.

In a neutron diffraction experiment, the quantity measured as a function of angle is the total scattering cross section, which consists of two components: (i) neutrons that scatter coherently (that is, where phase is conserved and the signal of which contains structural information) and (ii) incoherently scattered neutrons that result in a background signal. The scattering amplitude is then determined by the concentration, atomic arrangement, and neutron scattering lengths of the atoms involved. Since different isotopes have different neutron scattering lengths, it is possible to simplify the analysis of the neutron data simply by isotopic exchange experiments and by taking first and second order difference spectra to separate out the partial pair distribution functions. This is clearly set out by Bowron et al. [1] for a mixture of *t*-butanol/water and illustrates how isotopic substitution neutron scattering experiments can assist in distinguishing between intermolecular and intramolecular distributions within a sample, both of which would otherwise contribute to a measured diffraction pattern in a complex and often difficult to interpret combination.

4.1.2
Formation of Deuteriated Samples

In general, isotopic exchange is both expensive and difficult. In the case of many room-temperature ionic liquids, however, the manufacture of deuterated ionic liquids is relatively easily achievable. For example, the general synthesis of 1-alkyl-3-methylimidazolium salts is shown in Scheme 4.1-1 [2]. This methodology allows maximum flexibility in the deuteration on the imidazolium cation; that is, it can be either ring or side chain deuteration or both.

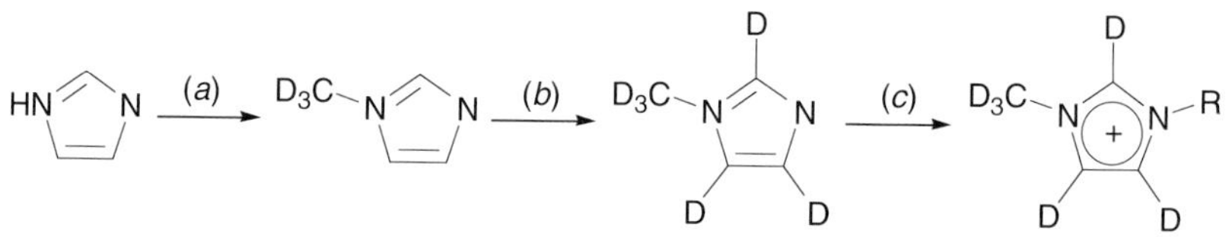

Scheme 4.1-1: Reaction scheme showing a method for deuteration of 1-alkyl-3-methylimidazolium salts: (*a*) CD_3OD, $RuCl_3/(n\text{-}BuO)_3P$, (*b*) D_2O, 10 % Pd/C, and (*c*) RX (CD_3Cl, C_2D_5I).

4.1.3
Neutron Sources

The following sources and instruments dominate studies in the area of liquids and amorphous materials. Although there are a number of sources available, each is optimized for a particular class of experiment. The sources can be split into two types: pulsed neutron sources and reactor sources

4.1.3.1 Pulsed (spallation) neutron sources

One example of a pulsed neutron source is to be found at ISIS, at the Rutherford Appleton Laboratory, UK. This source has the highest flux of any pulsed source in the world at present, and is therefore one of the most suitable for isotopic substitution work, as this class of experiment tends to be flux-limited. At ISIS, two stations are particularly well set up for the examination of liquids.

SANDALS station (Small Angle Neutron Diffractometer for Amorphous and Liquid Samples) This station is optimized for making measurements on liquids and glasses that contain light elements such as H, Li, B, C, N, and O. This set-up relies on collection of data at small scattering angles and the use of high-energy neutrons. This combination of characteristics has the effect of reducing the corrections necessary for the inelastic scattering that otherwise dominates the measured signal and hence complicates the extraction of structural signal information. This instrument is singularly specifically optimized for H-D isotopic substitution experiments.

GEM (GEneral Materials diffractometer) GEM is designed with extremely stable detectors, covering a very large solid angle, and is optimized for collection of data at a very high rate. It is a hybrid instrument that can perform both medium/high-resolution powder diffraction studies on crystalline systems and very accurate total scattering measurements for liquids and glasses. Because of the high stability of the detectors and data acquisition electronics, it is suitable for isotopic substitution work on systems containing elements with only small differences in the isotope neutron scattering lengths, such as ^{12}C and ^{13}C.

ISIS is only one pulsed source available for the study of liquids. Both the USA and Japan have facilities similar to SANDALS and GEM for studying liquids, but with slightly lower neutron intensity: in the forms of the IPNS (Intense Pulsed Neutron Source) at the Argonne National Lab. on the instrument GLAD, and the KEK Neutron Scattering Facility (KENS) on the instrument Hit II, respectively.

4.1.3.2 Reactor sources

The Institute Laue–Langevin (ILL, Grenoble, France) has arguably the premier neutron scattering instrument for total scattering studies of liquid and amorphous materials, in instrument D4C. The neutrons are provided by a reactor source that is very stable and delivers a very high flux. This makes the ILL ideal for isotopic substitution work for elements with atomic numbers greater than that of oxygen. For light elements, the inelasticity corrections are a major problem, as the instrument collects data over a large angular range and with relatively low-energy neutrons. Typically the wavelength of the neutrons used on D4 is 0.7 Å, which is quite long when compared with a source such as ISIS where wavelengths ranging from 0.05 Å to 5.0 Å can be used on an instrument like SANDALS.

Other reactor sources with instruments like D4C but with much lower flux, and hence longer data collection times, are the Laboratoire Leon Brillouin (LLB, Saclay, France), on the instrument 7C2, and the NFL (Studsvik, Sweden), on the instrument SLAD

The following website provides links to all the neutron sites in the world: http://www.isis.rl.ac.uk/neutronSites/.

4.1.4 Neutron Cells for Liquid Samples

As in any scattering experiment, the ideal sample holder is one that does not contribute to the signal observed. In neutron scattering experiments, the typical cells used are either vanadium, which scatters neutrons almost completely incoherently (that is, with almost no structural components in the measured signal), or a null scattering alloy of TiZr alloy. Vanadium cells react with water and are not ideal for studies of hydrated systems, whereas TiZr is more chemically inert; TiZr cells have been used, for example, to study supercritical water and alkali metals in liquid ammonia. In addition, TiZr cells capable of performing measurements at high pressure have been constructed. Figure 4.1-1 shows typical sample cells made from Vanadium and TiZr alloy.

Cells used for high-temperature measurements in furnaces often consist of silica sample tubes, supported by thin vanadium sleeves. The key to the analysis is whether it is possible to have a container that scatters in a sufficiently predictable way, so that its background contribution can be subtracted. With the current neutron flux available from both pulsed and reactor sources, sample volumes of

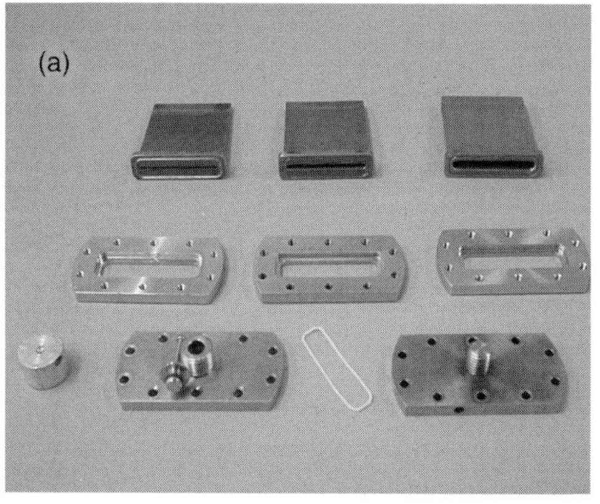

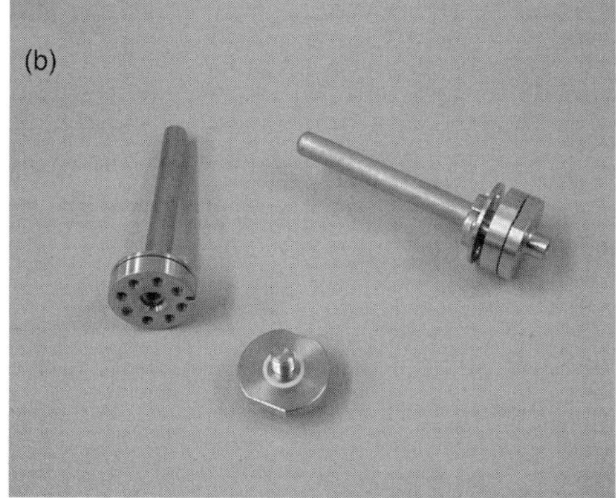

Figure 4.1-1: Liquid sample cells made from (a) TiZr alloy and (b) vanadium.

between 1 and 5 cm^3 are required. Obviously, with increasing flux at the new neutron sources being built in the USA and Japan, the sample sizes will decrease.

4.1.5 Examples

Neutron diffraction has been used extensively to study a range of ionic liquid systems; however, many of these investigations have focussed on high-temperature materials such as NaCl, studied by Enderby and co-workers [3]. A number of liquid systems with relatively low melting points have been reported, and this section summarizes some of the findings of these studies. Many of the salts studied melt above 100 °C, and so are not room-temperature ionic liquids, but the same principles apply to the study of these materials as to the lower melting point salts.

4.1.5.1 Binary mixtures

A number of investigations have focussed on alkali haloaluminates: that is, $(MX)_y(AlX_3)_{1-y}$ mixtures in which M is an alkali metal and X is a halogen (Cl or Br). Blander et al. [4] have used neutron diffraction combined with quantum chemical calculations to investigate the salts formed from KBr and KCl where $y = 0.25$ and 0.33. The authors showed that, for both bromide and chloride salts, $[Al_2X_7]^-$ was the dominant species present, as expected, in full agreement with other spectroscopic techniques such as Raman and infra-red [5]. In the case of chloride, however, as the acidity of the melt increased (that is, as y decreased), although the proportion of $[Al_3X_{10}]^-$ ions did increase, the change was unexpectedly smaller than that predicted by the stoichiometry. A closer relationship with the stoichiometry was found for bromide. The neutron scattering data showed a strong correlation between the liquid structure and that found in, for example, the crystal structure of $K[Al_2Br_7]$. In both the liquid and crystal, the angle of the Al–Br–Al bridge within the $[Al_2Br_7]^-$ anion is found to be approximately 109 ° and both also show that the neighboring structural units pack parallel to each other.

The structures of binary mixtures of $AlCl_3$ with NaCl and LiCl were studied from pure $AlCl_3$ to a 1:1 mixture by Badyal et al. [6]. In the pure $AlCl_3$ liquid, the neutron data indicates that the long held view that isolated Al_2Cl_6 dimers make up the structure may not be the true scenario. The structure is reported to be a sparse liquid network made up of polymeric species containing corner-shared tetrahedra. On addition of the alkali halides, the presence of Al–Cl–Al linkages gradually decreases in proportion to the concentration of the halide added. This coincides with the formation of $[AlCl_4]^-$ species. The neutron scattering also shows that the long-range order within the liquid decreases as the binary salt mixture is formed, consistently with the gradual breakdown of the polymeric aluminium trichloride structure. Of significance in the 1:1 binary mixture is the high level of charge ordering in the system. For example, in the case of LiCl, features at $r = 6.65$, 9.85, and 12.9 Å in the radial distribution function are clearly evident but do not correspond to distances in either of the pure components and therefore are probably associated with spacings between $[AlCl_4]^-$ units.

Ion pairs	$b6_{Li}b_i$	$b\,7_{Li}b_i$
Li-S	0.004	–0.005
Li-C	0.009	–0.012
Li-N	0.013	–0.017
Li-Al	0.005	–0.006
Li-Cl	0.041	–0.054

Table 4.1-1: Comparison of the neutron scattering cross-sections for ^{6}Li and ^{7}Li with all the other atoms present in a binary mixture of LiSCN with $AlCl_3$.

Binary mixtures of LiSCN with $AlCl_3$ have also been studied [7]. In this case, the 1:1 mixture is liquid at ambient temperature and therefore provides a system analogous to room-temperature ionic liquids. Lee et al. used Li isotope substitution to enable the correlations between Li–X to be isolated. The weighting factors between ^{6}Li and ^{7}Li are positive and negative, respectively, and can be used to distinguish features in the partial radial distribution function. Table 4.1-1 compares the combined neutron scattering cross-sections $b_{Li}b_i$ for both isotopes with each other atom i in the liquid. Figure 4.1-2 shows the equivalent total correlation functions for the 1:1 mixture. Clearly the amplitude of the ^{7}Li systems shows negative and reduced features compared with the ^{6}Li sample, and these must therefore be associated with Li–X features.

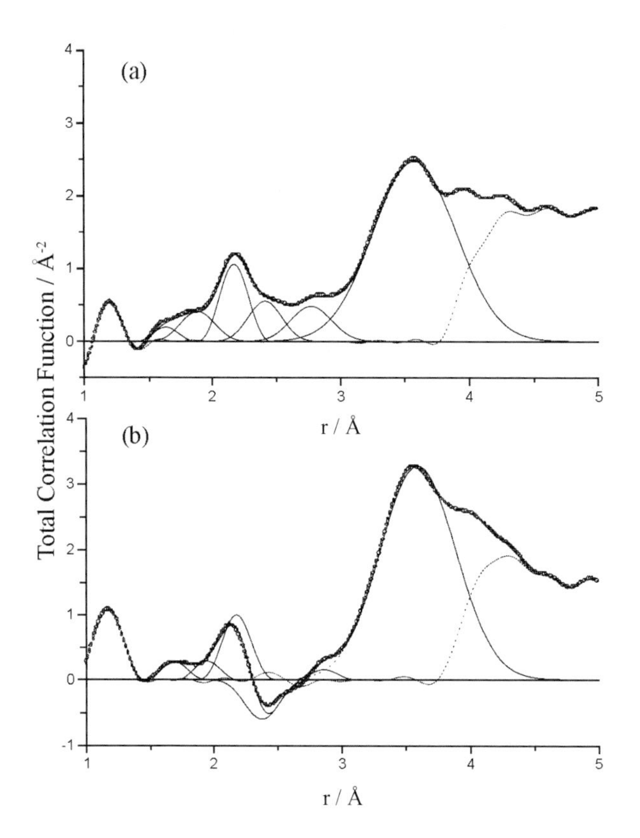

Figure 4.1-2: Total correlation functions for (a) ^{6}LiSCN/$AlCl_3$ and (b) ^{7}LiSCN/$AlCl_3$. The bold lines are the experimental neutron data (•), the fit (–), the Gaussian functions for each of the atomic pairs used to fit the data (–) and the deviation (······) used. Reproduced from reference 7 with permission.

From the data, the liquid is shown to have tetrahedrally coordinated aluminium with three chlorines and the isocyanate group attached. The neutron data clearly shows nitrogen, as opposed to sulfur, coordination to the aluminium center, forming an $AlCl_3NCS^-$ species, which is consistent with a hard base/hard acid interaction as compared with the softer sulfur donation. It was also possible to show that a tetrahedral chloride environment is present around the lithium.

Takahashi et al. have studied the structures of $AlCl_3$/[EMIM]Cl mixtures over a range of concentrations from 46 to 67 mol% $AlCl_3$ [8]. Below 50 mol% $AlCl_3$, the neutron data could be simulated simply by using the isolated ions: that is, $[EMIM]^+$ and $[AlCl_4]^-$. Above 50 mol% $AlCl_3$, $[Al_2Cl_7]^-$ is also known to exist, and at 67 mol% $AlCl_3$ becomes the major anion present. Unlike that of $[AlCl_4]^-$, the geometry of $[Al_2Cl_7]^-$ is changed substantially in the liquid compared with the isolated ions, implying a direct interaction between the imidazolium cation and the anionic species. This is manifested as a decrease in the torsion angle around the central Al–Cl–Al axis, from 57.5° to 26.2°, and hence as a decrease in the Cl–Cl distance across the anion.

Mixtures of HCl and [EMIM]Cl have also been studied [9, 10]. By analysis of the first order differences by hydrogen/deuterium substitution both on the imidazolium ring and the HCl, two intramolecular peaks were observed. These indicated the presence of $[HCl_2]^-$ as an asymmetric species, which, coupled with analysis of the second order differences, allowed the structure in Figure 4.1-3 to be proposed.

4.1.5.2 **Simple salts**

Bowron et al. [11] have performed neutron diffraction experiments on 1,3-dimethylimidazolium chloride ([MMIM]Cl) in order to model the imidazolium room-temperature ionic liquids. The total structure factors, $F(Q)$, for five 1,3-dimethylimidazolium chloride melts – fully protiated, fully deuterated, a 1:1 fully deuterated/fully protiated mixture, ring deuterated only, and side chain deuterated only – were measured. Figure 4.1-4 shows the probability distribution of chloride around a central imidazolium cation as determined by modeling of the neutron data.

As well as charge-ordering in the system, out to two chloride shells, the specific local structure shows strong interactions between the chloride and the ring hydrogens, as well as some interaction between the methyl groups of adjacent imidazolium cations. This is consistent with the crystal structure and implies that the molecular packing and interactions in the first two or three coordination shells are similar in both the crystal and the liquid.

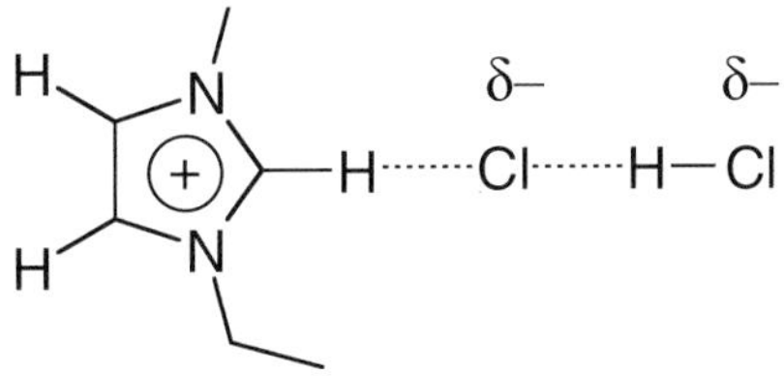

Figure 4.1-3: Proposed structure of the asymmetric $[HCl_2]^-$ ion bound to the $[EMIM]^+$ cation in a binary mixture of HCl/[EMIM]Cl. The figure has been redrawn from reference 10 with permission.

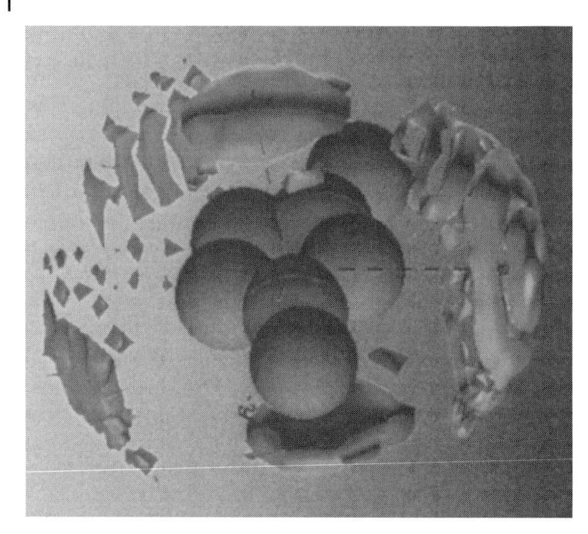

Figure 4.1-4: Chloride probability distribution around a central imidazolium cation as determined by the EPSR model of the neutron data from liquid [MMIM]Cl at 150 °C.

In all these examples, the importance of good simulation and modeling cannot be stressed enough. A variety of methods have been used in this field to simulate the data in the cases studies described above. Blander et al. [4], for example, used a semi-empirical molecular orbital method, MNDO, to calculate the geometries of the free haloaluminate ions and used these as a basis for the modeling of the data by the RPSU model [12]. Badyal et al. [6] used reverse Monte Carlo simulations, whereas Bowron et al. [11] simulated the neutron data from [MMIM]Cl with the Empirical Potential Structure Refinement (EPSR) model [13].

4.1.6 X-ray Diffraction

X-ray diffraction has been used for the study both of simple molten salts and of binary mixtures thereof, as well as for liquid crystalline materials. The scattering process is similar to that described above for neutron diffraction, with the exception that the scattering of the photons arises from the electron density and not the nuclei. The X-ray scattering factor therefore increases with atomic number and the scattering pattern is dominated by the heavy atoms in the sample. Unlike in neutron diffraction, hydrogen (for example) scatters very weakly and its position cannot be determined with any great accuracy.

In contrast with the study of the structure of the molten salts, full analysis of the scattering profile is not generally performed for liquid crystalline materials. In the latter, only the Bragg features are analyzed (that is, for a wavelength λ, incident on the sample at an angle θ to its surface normal, the position of the diffraction peaks are determined by Bragg's law, $n\lambda = 2d\sin\theta$). From the angle of diffraction, the periodicity length, d, may be determined.

In both cases, laboratory X-ray sources may be used and the X-ray measurements taken in θ–2θ geometry. For weakly scattering systems synchrotron radiation is helpful.

4.1.6.1 Cells for liquid samples

Sample cells include Lindemann/capillary tubes (normally < 1 mm in diameter) and aluminium holders. In the latter, thin aluminium windows sandwich the sample in a cylindrical aluminium sample holder. The diffraction from the aluminium is observed in this case, and may be used as a calibration standard. For low-temperature materials, the aluminium window can be replaced by the polymer Kapton. Beryllium may also be used [14]. Sample volumes of between 50 and 100 μL are typically required.

4.1.6.2 Examples

Molten salts and binary mixtures X-ray diffraction has been performed by Takahashi et al. [15] on 1:1 binary mixtures of $AlCl_3$ with LiCl and NaCl. In agreement with the neutron data obtained by Badyal et al. [6] and discussed above, the liquid has a degree of charge ordering, with sets of four $[AlCl_4]^-$ units surrounding a central $[AlCl_4]^-$ unit at distances of 6.75 Å for LiCl and 6.98 Å for NaCl. Similarly, Igarashi et al. [16] have studied a molten LiF/NaF/KF eutectic mixture. For the ion-pairs Li–F, Na–F, and K–F, the nearest-neighbor coordination and distances were almost identical to those found in the individual melts of the component salts.

Binary mixtures that melt close to room temperature, namely $AlCl_3$/*N*-butylpyridinium chloride mixtures, have also been investigated. Takahashi et al. [17] have also shown that for the 1:1 composition, $[AlCl_4]^-$ predominates with a tetrahedral environment. At a ratio of 2:1, $[Al_2Cl_7]^-$ becomes the main species. At high temperature (above 150 °C), some decomposition to $[AlCl_4]^-$ and Al_2Cl_6 was observed.

Liquid crystals A wide range of ionic liquids form liquid-crystalline phases. This is normally achieved by increasing the amphiphilic character of the cation through substitution with longer, linear alkyl groups. The salts have relatively low melting points, close to room temperature when the alkyl chain length (C_n) is small ($n < 10$), and display liquid crystal mesomorphism when $n > 12$. This section describes some of the results of studies in which X-ray diffraction has been used to examine the mesophase and liquid phase. There are also many examples of materials which form liquid-crystalline phases that have been studied by techniques such as NMR, DSC, single-crystal X-ray diffraction, and so on that have not been included (see, for example, [18–21].

Metal-containing systems

Many of the systems studied are based on $[MCl_4]^{2-}$ anion. Neve et al. have extensively studied the formation of liquid-crystalline phases of *N*-alkylpyridinium salts with alkyl chain lengths of n = 12–18 with tetrahalometalate anions based upon Pd(II) [22] and Cu(II) [23]. In general, the liquid-crystalline phases exhibit lamellar-

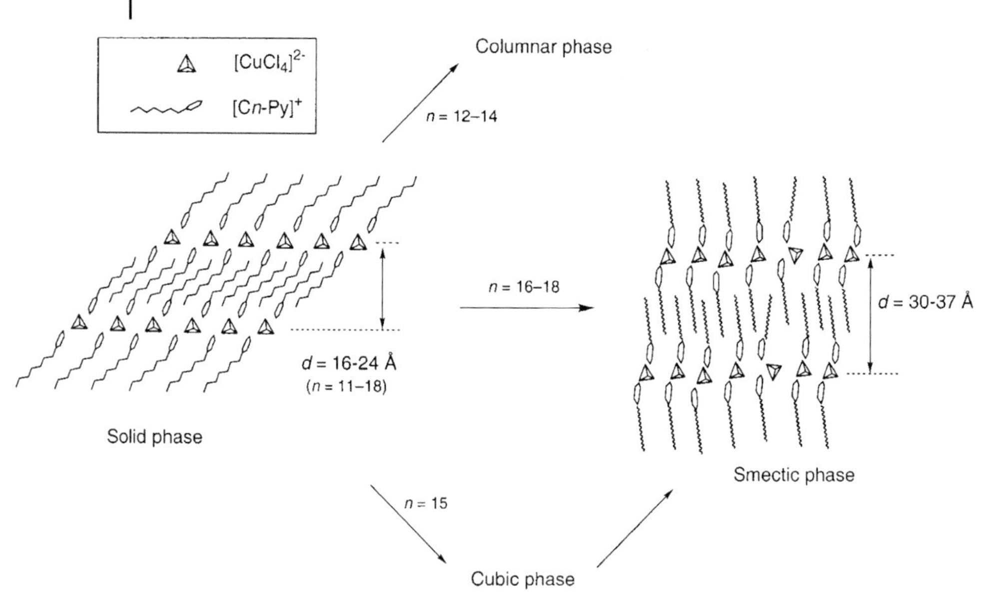

Figure 4.1-5: Schematic showing the changes in structure of *N*-alkylpyridinium tetrachlorocuprate salts with varying alkyl chain lengths. Reproduced from reference 23 with permission.

like structures based upon the smectic A structure. For n = 16, 18, in the cases of $[PdCl_4]^{2-}$ and $[PdBr_4]^{2-}$, this is preceded by an ordered smectic E phase. Cuprate-based pyridinium ionic liquids exhibit a range of structures depending on the alkyl chain length. For C_{12}–C_{18}, each solid-state structure has a layered periodicity. On melting, however, C_{12}–C_{14} exhibit a columnar phase whereas C_{16}–C_{18} simply form a smectic A phase. For n = 15, the solid melts into a cubic phase before transforming into the smectic A phase as seen for longer alkyl chains. Figure 4.1-5 illustrates the changes observed in the latter case.

Similar lamellar structures are formed for 1-alkyl-3-methylimidazolium cations with $[PdCl_4]^{2-}$ when $n > 12$. As with the pyridinium systems, mesomorphic liquid crystal structures based on the smectic A structure are formed [24].

Martin [25] has also shown that ammonium salts display similar behavior. $[Cetyltrimethylammonium]_2[ZnCl_4]$, for example, first melts to an S_C-type liquid crystal at 70 °C and then to an S_A-type mesophase at 160 °C. The broad diffraction features observed in the liquid-crystalline phases are similar to those seen in the original crystal phase and show the retention on melting of some of the order originating from the initial crystal, as shown in Figure 4.1-6.

Needham et al. [26] also used X-ray diffraction to show that, in the case of Mn(II)-, Cd(II)-, and Cu(II)-based C_{12}- and C_{14}-ammonium tetrachlorometalate salts, two mechanistic pathways were present on melting to the mesophase. Each pathway was shown to have a minor and major structural transformation. The minor change was thought to be a torsional distortion of the alkyl chains and the major change the melting of the chains, forming a disordered layer. The order in which the structural changes occur was found to be dependent on the metal and on the alkyl chain length.

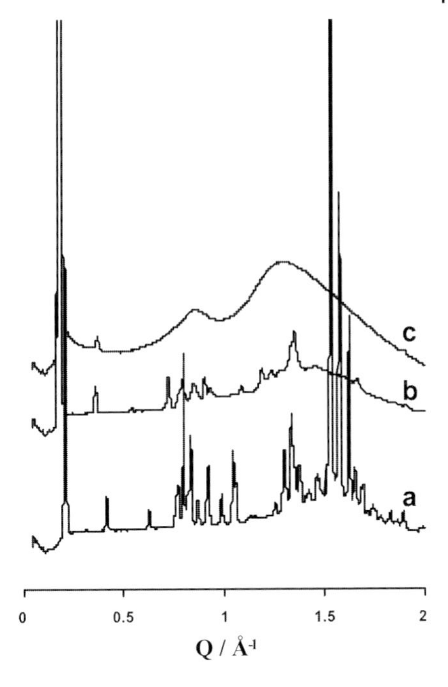

Figure 4.1-6: Small-angle X-ray diffraction data from [cetyltrimethylammonium]$_2$[$ZnCl_4$] at (a) room temperature (solid), (b) 90 °C (S_C phase) and (c) 200 °C (S_A phase). Reproduced from reference 25 with permission.

Non-metal-containing systems

Non-metal-containing salts have been studied extensively. Bradley et al. [27] examined a range of 1-alkyl-3-methylimidazolium-based salts containing chloride, bromide, trifluoromethanesulfonate ([OTf]$^-$), bis(trifluoromethanesulfonyl)imide, and [BF_4]$^-$ anions. In the mesophase, the X-ray data of these salts were consistent with smectic A phases with interlayer spacings of between 22–61 Å, increasing uniformly with increasing alkyl chain length *n*. For a given cation, the mesophase interlayer spacing decreases in the order $Cl^- > Br^- >$ [BF_4]$^- >$ [OTf]$^-$, with the bis(trifluoromethanesulfonyl)imide salts not exhibiting any mesophase structure. The anion dependence of the mesophase interlayer spacing is largest for the anions with greatest ability to form a three-dimensional hydrogen-bonding lattice. On melting to the isotropic liquid, a broad peak is observed in the X-ray scattering data for each salt, as shown in Figure 4.1-7. This peak indicates that some short-range associative structural ordering is still retained even within the isotropic liquid phase.

Similarly, *N*-alkylammonium [28] and alkylphosphonium [29] salts form lamellar phases with smectic bilayer structures. In both cases, X-ray scattering also showed the isotropic liquid not to be completely disordered and still displaying similar features to the mesophase. Buscio et al. [28] showed that in *N*-alkylammonium chlorides the feature was not only much broader than that observed in the mesophase but increased in width with decreasing chain length.

Other examples include ditholium salts, shown in Figure 4.1-8 [30]. The scattering data show that a range of mesophase behavior is present, dependent – as with the metal-containing systems – on alkyl chain length.

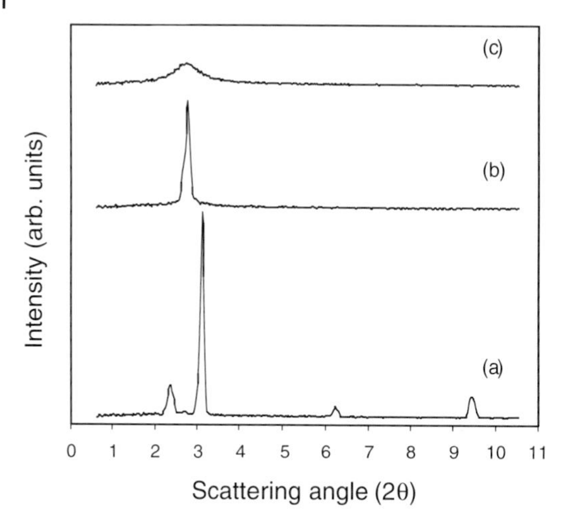

Figure 4.1-7: Small-angle X-ray diffraction data from $[C_{16}MIM][OTf]$ at (a) 50 °C, (b) 70 °C, and (c) 90 °C, in the crystal, SmA_2, and isotropic phases, respectively, on cooling. Reproduced from reference 27 with permission.

For $n = 12$, for example, two transitions within the liquid crystalline region are observed: from a nematic columnar phase (N_{col}) to a hexagonal columnar lattice (D_h), and then finally to a rectangular lattice (D_r). X-ray diffraction data for benzimidazolium salts have also been reported [31], and indicated a switch within the liquid crystalline region from a lamellar β phase to the α phase, which in some examples is not shown by differential scanning calorimetry. Using X-ray diffraction, Bruce and co-workers have propose a new structural model for *N*-alkylpyridinium alkylsulphates [32]. In these liquid-crystalline materials, the *d* spacings obtained are less than the molecular length, but are not associated with tilting of the alkyl chains. The new proposed model shows microdomains of interdigitated and non-interdigitated molecules.

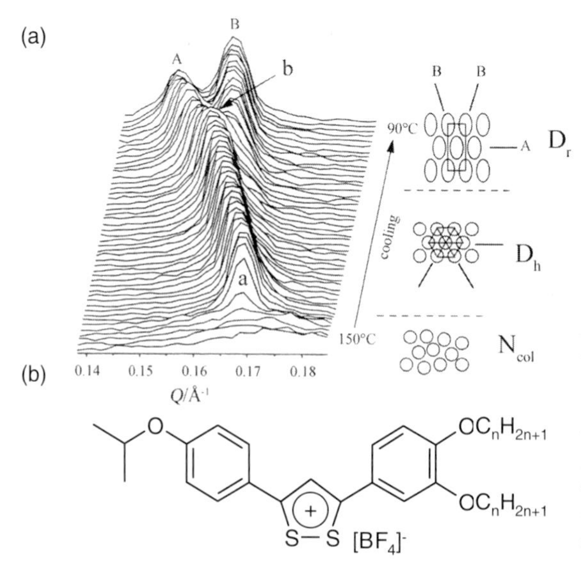

Figure 4.1-8: (a) Small-angle X-ray diffraction data relating to the ditholium salt shown in (b) for $n = 12$. **A** and **B** correspond to the rectangular lattice vectors shown in D_r, and **a** and **b** correspond to the N_{col} to D_h and D_h to D_r phase transitions. Reproduced from reference 30 with permission.

Other liquid-crystalline materials that have been investigated by X-ray scattering include single- and double-chained pyridinium [33] and *N*-substituted 4-(5-alkyl-1,3-dioxan-2-yl)pyridinium salts [34]. In the former case, diffraction analysis allowed an explanation for the differences in mono- and di-substituted salts to be proposed.

In general, X-ray data are used in conjunction with other techniques to obtain as full a picture as possible. For liquid-crystalline materials, differential scanning calorimetry (DSC) and polarizing optical microscopy (POM) are conventionally used.

4.1.7
Extended X-ray Absorption Fine-structure Spectroscopy

Extended X-ray absorption fine-structure (EXAFS) spectroscopy measures X-ray absorption as a function of energy and allows local arrangements of atoms to be elucidated. The absorption results from the excitation of a core electron in an atom. Conventional EXAFS is usually associated with hard X-rays (that is, >3–4 keV), in order to allow measurements to be made outside a vacuum, and requires synchrotron radiation to provide the intensity at the energies involved. At these energies, the core electrons ejected correspond to the 1s (K-edge), 2s (L_I-edge), $2p_{1/2}$(L_{II}-edge), and $2p_{3/2}$(L_{III}-edge) states. As the photon energy is increased past the absorption edge, an oscillatory structure is found, described as the X-ray fine structure. The X-ray fine structure starts at approximately 30 eV past the edge and extends to a range of 1000 eV.

EXAFS is observed as a modulating change in the absorption coefficient caused by the ejected electron wave back-scattering from the surrounding atoms, resulting in interference between ejected and back-scattered waves. It is defined as:

$$\chi(k) = \frac{\mu(k) - \mu_0(k)}{\Delta\mu_0} \qquad (4.1\text{-}4)$$

where $\chi(k)$ is the EXAFS as a function of the wavenumber of the photoelectron k, $\mu(k)$ is the measured absorption above the absorption edge, $\mu_0(k)$ is the absorption spectrum without the EXAFS oscillations (that is, the background), and $\Delta\mu_0$ is a normalization factor.

The wavenumber is defined at a photon energy E above the absorption edge energy E_0, with respect to the mass of the electron m_e.

$$k = \sqrt{\frac{2m_e}{h^2}(E - E_0)} \qquad (4.1\text{-}5)$$

The EXAFS is related to the wavenumber by:

$$\chi(k) \approx \sum_i \frac{N_i f_i(k)}{kr_i^2} \, e^{-2\sigma_i^2 k^2} e^{-2r_i/\lambda} \sin[2kr_i + \alpha_i(k)] \qquad (4.1\text{-}6)$$

Where, $\chi(k)$ is the sum over N_i back-scattering atoms i, where f_i is the scattering amplitude term characteristic of the atom, σ_i is the Debye–Waller factor associated with the vibration of the atoms, r_i is the distance from the absorbing atom, λ is the mean free path of the photoelectron, and α_i is the phase shift of the spherical wave as it scatters from the back-scattering atoms. By taking the Fourier transform of the amplitude of the fine structure (that is, $\chi(k)$), a real-space radial distribution function of the back-scattering atoms around the absorbing atom is produced.

On analysis of the EXAFS data, the local environment around a given absorbing atom – that is, the type, number and distance of the back-scattering atoms – can be obtained. It should be noted that it is not necessary for the surrounding atoms to be formally bonded to the absorbing atom. Typically the distance has an uncertainty of ± 1 % within a radius of approximately 6 Å; however, the error in the coordination number is strongly dependent on the system studied and can be high. In this regard, comparison with standard materials and the use of EXAFS in conjunction with other techniques to ensure a realistic interpretation of the data is vital.

Since the fine structure observed is only associated with the particular absorption edge being studied, and the energy of the absorption edge is dependent on the element and its oxidation state, EXAFS examines the local structure around one particular element, and in some cases, an element in a given oxidation state. A fuller picture can therefore be obtained by studying more than one absorbing element in the sample.

4.1.7.1 Experimental

Measuring EXAFS spectra In general, transmission EXAFS can be used, provided that the concentration of the element to be investigated is sufficiently high. The sample is placed between two ionization chambers, the signals of which are proportional to the incident intensity I_0 and the transmitted intensity through the sample I_t. The transmission of the sample is dependent on the thickness of the sample x and on the absorption coefficient, μ, in a Beer–Lambert relationship:

$$I_t = I_0 e^{-\mu x} \tag{4.1-7}$$

For good spectra to be obtained, the difference between the $ln(I_t/I_0)$ before and after the absorption edge, the edge jump, should be between 0.1 and 1. This may be calculated from the mass absorption coefficient of a sample:

$$\left(\frac{\mu}{\rho}\right)_{sample} = \sum_i w_i \left(\frac{\mu}{\rho}\right)_i \tag{4.1-8}$$

where ρ is the sample density and the mass-weighted average of the mass absorption coefficients of each element in the sample, using weight fractions w_i.

If the edge jump is too large, the sample should be diluted or the path-length decreased. If the edge jump is too small, then addition of more sample is one possibility, although this is dependent on the matrix in which the sample is studied.

For low atomic weight matrices such as carbon-based materials, the path-length can be increased without the transmission of the X-rays being adversely affected. In matrices containing high atomic weight elements, such as chlorine, increasing path-length will result in a larger edge jump, but it will also decrease the overall transmission of the X-rays. For such samples, fluorescence EXAFS may be performed. In this geometry, the emitted X-rays are measured. Optimally, the sample is placed at 45° to the incident X-rays and the X-ray fluorescence is detected at 90° to the direction of the exciting X-rays by use of, for example, a solid-state detector. The X-ray florescence is proportional to the X-rays absorbed by the sample and therefore can be used to measure the EXAFS oscillations. In general, this technique has a poorer signal-to-noise ratio than transmission EXAFS and there are problems with self-absorption effects, requiring dilute or thin sample sizes.

Self-absorption occurs when the path-length is too large [35] and the X-rays emitted have a significant probability of being absorbed by the remainder of the sample before being detected. This has the consequence of reducing the amplitude of the EXAFS oscillations and producing erroneous results. As the sample becomes more dilute this probability decreases. All the atoms in the sample determine the amount of self-absorption: hence the need for thin samples.

Liquid set-ups There are two major methods by which liquid samples are studied; these are shown in Figure 4.1-9.

These consist either of supporting the liquid in an inert, low atomic weight matrix such as graphite or boron nitride, or of sandwiching thin films between low atomic weight plates. The choice of the matrix material used is a balance between its chemical inertness towards the liquid being studied whilst being thermally stable and its being transparent to the X-rays at the absorption energy. The latter becomes less problematic as the energy of the absorption edge increases. Figure 4.1-10 shows an experimental cell which has been used to measure the EXAFS of ionic liquid samples [36].

Analysis A number of commercial software packages to model EXAFS data are available, including the FEFF program developed by Rehr and co-workers [37], GNXAS, developed by Filipponi et al. [38], and EXCURV, developed by Binsted [39]. These analysis packages fit the data to curve wave theory and describe multiple scattering as well as single scattering events. Before analysis, the pre-edge and a smooth

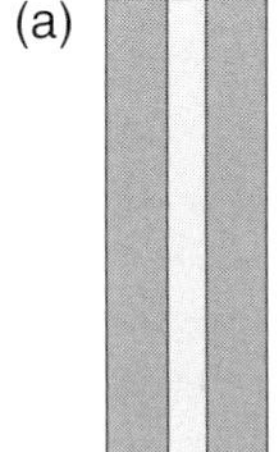

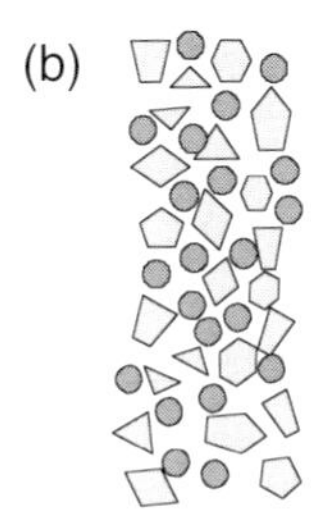

Figure 4.1-9: Schematic of the sample preparation methods used to study liquid EXAFS: (a) thin liquid film sandwich between low atomic weight plates, and (b) the liquid (circles) dispersed in a low atomic number matrix (polyhedrons). The figure has been redrawn from reference 40 with permission.

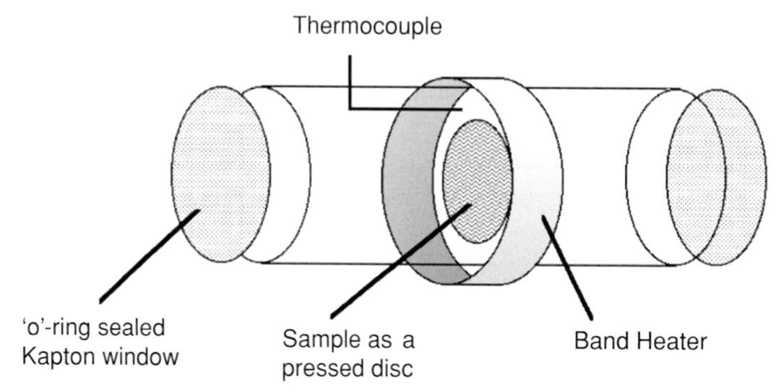

Figure 4.1-10: Schematic of transmission EXAFS cell. Reproduced from reference 36 with permission.

post-edge background function $\mu_0(k)$ is subtracted from the measured spectra. This is one of the most important procedures and can, if performed poorly, result in loss of amplitude of the EXAFS oscillations or unphysical peaks in the Fourier transform.

Two review articles treating the theory more rigorously and summarizing much of the data on general liquid systems have recently been published [40, 41].

4.1.7.2 Examples

Ionic liquid structure To date, EXAFS has only been used to examine the structure of high-temperature molten salts in detail.

Di Cicco and co-workers [42, 43] have examined the structure of molten CuBr with Cu and Br K-edge EXAFS. From the EXAFS data, the Cu–Br bond length distribution was found to be shorter than that derived from neutron data and theoretical models, indicating a more covalent character than previously thought. Similar EXAFS studies on KBr and RbBr are in good agreement with theory, showing high levels of ionicity [43].

Zn and Rb K-edge EXAFS have also been used to examine the melting of Rb_2ZnCl_4 in comparison with the liquid structure of $ZnCl_2$ and RbCl [44]. In molten $ZnCl_2$, the zinc is found to be tetrahedrally coordinated, with the tetrahedra linked by corner-sharing chlorines in a weak extended network. In RbCl, significant disorder is evident in the chloride shell around the rubidium and indicates significant movement of the Rb^+ and Cl^- in the molten state. In the crystal structure of Rb_2ZnCl_4, the chlorine coordination number around the Rb is between 8 and 9, whilst the Zn is found in isolated $ZnCl_4$ units. In the molten state, the EXAFS also indicates isolated $ZnCl_4$ units with a chlorine coordination of 7.6 around the Rb. This may be compared with a chlorine coordination of 4.8 in liquid RbCl. The EXAFS clearly shows that the solid and liquid structures of Rb_2ZnCl_4 are similar and that the melt does not rearrange into a simple combination of the component parts.

In the studies described above, the samples were supported in low atomic weight matrices, melted in situ, and measured in transmission mode. Similarly, second generation ionic liquids have been studied. Carmichael et al. [45] showed that it was possible to support and melt $[EMIM]_2[NiCl_4]$ and $[C_{14}MIM]_2[NiCl_4]$ in inert matrices such as boron nitride, graphite, and lithium fluoride without the EXAFS being affected by the sample matrix used. In these samples, the Ni K-edge EXAFS was investigated between room temperature and 131 °C. Even in LiF, where halide exchange was possible, little difference was found in the Ni coordination on melting.

Species dissolved in ionic liquids A number of systems have been investigated in both chloroaluminate and second generation ionic liquids

Dent et al. [46] studied the dissolution of $[EMIM][MCl_4]$ in $[EMIM]Cl/AlCl_3$ binary mixtures, for M = Mn, Co, and Ni, at $AlCl_3$ mole fractions of 0.35 and 0.60 using the M K-edges. Because of problems associated with the high concentration of chloride it was not possible to perform transmission experiments, and so fluorescence measurements were used. In this case, self-absorption problems were overcome by use of a thin film of liquid pressed between two sheets of polythene sealed in a glove-box. The coordination of Ni, Co, and Mn was found to change from $[MCl_4]^{2-}$ to $[M(AlCl_3)_4]^-$ as the mole fraction of $AlCl_3$ increased. Figure 4.1-11 shows the EXAFS and *pseudo*-radial distribution functions for M = Co in both the acidic and basic chloroaluminate ionic liquids.

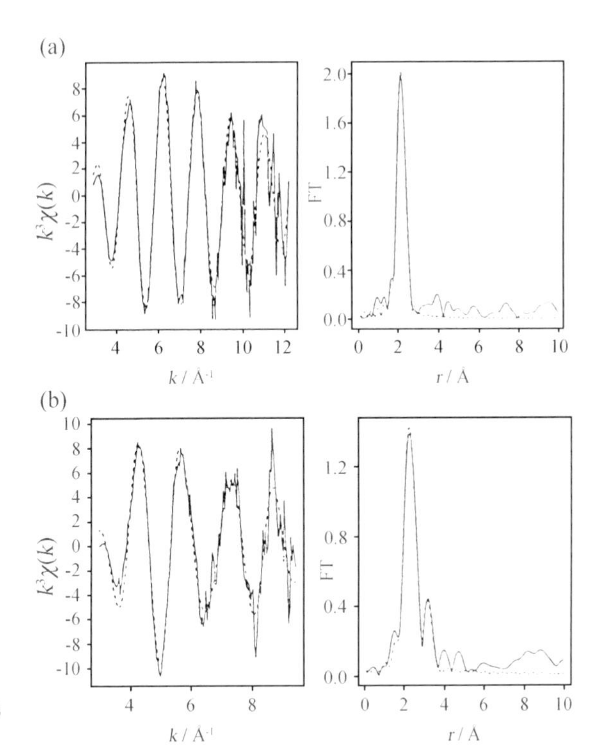

Figure 4.1-11: The EXAFS data and *pseudo*-radial distribution functions of Co(II) in (a) basic and (b) acidic chloroaluminate ionic liquid. Reproduced from reference 46 with permission.

Aluminium coordination was only observed in acidic mixtures (that is, at $AlCl_3$ mole fractions greater than 0.5). The latter was surprising given that at 0.60 $AlCl_3$, almost half the anion species are in the form $[Al_2Cl_7]^-$, yet no coordinating $[Al_2Cl_7]^-$ was observed.

Dent et al. [47] also investigated the V K-edge EXAFS for the dissolution of $[EMIM][VOCl_4]$ and $[NEt_4][VO_2Cl_2]$ in basic $[EMIM]Cl/AlCl_3$ and compared the data with those of solid samples. In both cases the dissolved and the solid samples showed similar EXAFS and no coordination of the chloroaluminate species to, for example, the vanadyl oxygen was found.

Thanks to the decrease in the average atomic weight of the medium compared with chloroaluminate systems, second generation ionic liquids may be studied in transmission. Carmichael et al. [45] have shown that solutions of $[EMIM]_2[NiCl_4]$ in $[BMIM][PF_6]$ may be studied by supporting the liquid between two boron nitride discs. The resulting Ni K-edge EXAFS showed a local structure similar to that of the molten $[EMIM]_2[NiCl_4]$ described above.

Baston et al. [48] studied samples of ionic liquid after the anodization of uranium metal in [EMIM]Cl, using the U L_{III}-edge EXAFS to establish both the oxidation state and the speciation of uranium in the ionic liquid. This was part of an ongoing study to replace high-temperature melts, such as LiCl/KCl [49], with ionic liquids. Although it was expected that, when anodized, the uranium would be in the +3 oxidation state, electrochemistry showed that the uranium was actually in a mixture of oxidation states. The EXAFS of the solution showed an edge jump at 17166.6 eV, indicating a mixture of uranium(IV) and uranium(VI). The EXAFS data and *pseudo*-radial distribution functions for the anodized uranium in [EMIM]Cl are shown in Figure 4.1-12.

Two peaks, corresponding to a 1:1 mixture of $[UCl_6]^{2-}$ and $[UO_2Cl_4]^{2-}$, were fitted, in agreement with the position of the edge. Oxidation to uranium(VI) was surpris-

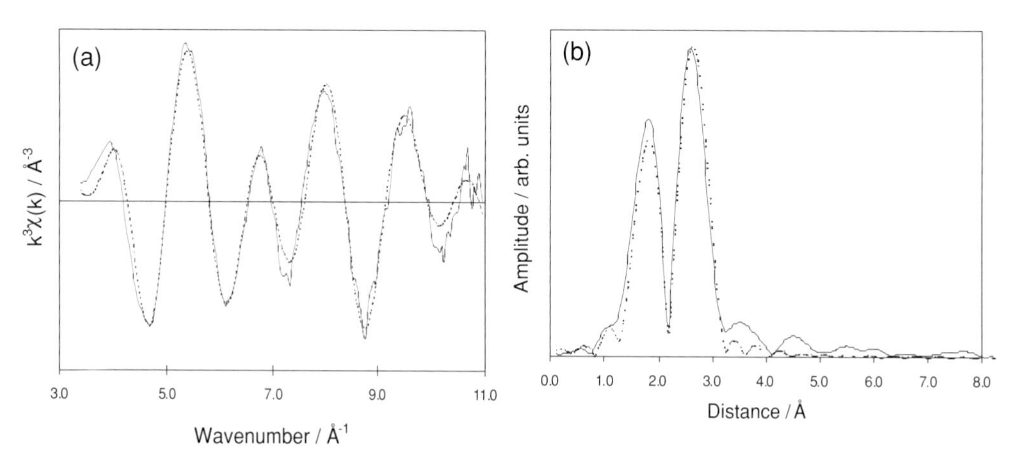

Figure 4.1-12: The experimental (solid line) and fitted (dashed line) U L(III)-edge (a) EXAFS data and (b) *pseudo*-radial distribution function after anodization of uranium in [EMIM]Cl. The figure has been redrawn from reference 48 with permission.

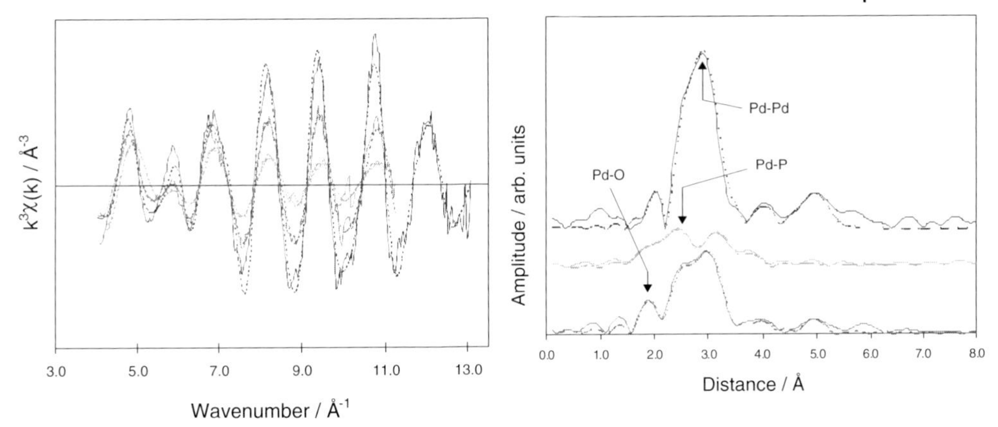

Figure 4.1-13: Comparison of the experimental (solid line) and fitted (dashed line) (a) EXAFS and (b) *pseudo*-radial distribution functions from palladium ethanoate in $[BMIM][PF_6]$ without (—) and with (—) triphenylphosphine at 80 °C and in the presence of triphenylphosphine and reagents at 50 °C for 20 min (—). Reproduced from reference 50 with permission.

ing in this system and may have arisen from the presence of water due to the highly hygroscopic nature of [EMIM]Cl.

In none of the above cases has a reaction been performed whilst taking the EXAFS data. Hamill et al. [50] have investigated catalysis of the Heck reaction by palladium salts and complexes in room-temperature ionic liquids. On dissolution of palladium ethanoate in $[BMIM]^+$ and *N*-butylpyridinium ($[BP]^+$) hexafluorophosphate and tetrafluoroborate ionic liquids, and triethyl-hexyl ammonium bis(trifluoromethanesulfonyl)imide, a gradual change from ethanoate coordination to the formation of palladium metal was observed in the Pd K-edge EXAFS, as shown in Figure 4.1-13.

In pyridinium chloride ionic liquids and in 1,2-dimethyl-3-hexylimidazolium chloride ([HMMIM]Cl) , where the C(2) position is protected by a methyl group, only $[PdCl_4]^{2-}$ was observed, whereas in [HMIM]Cl, the EXAFS showed the formation of a bis-carbene complex. In the presence of triphenylphosphine, Pd–P coordination was observed in all ionic liquids except where the carbene complex was formed. During the Heck reaction, the formation of palladium was found to be quicker than in the absence of reagents. Overall, the EXAFS showed the presence of small palladium clusters of approximately 1 nm diameter formed in solution.

4.1.8 X-ray Reflectivity

Reflectometry is a useful probe with which to investigate the structure of multilayers both in self-supporting films and adsorbed on surfaces [51]. Specular X-ray reflectivity probes the electron density contrast perpendicular to the film. The X-rays irradiate the substrate at a small angle ($<5\ ^{\circ}$) to the plane of the sample, are reflected, and are detected at an equal angle. If a thin film is present on the surface of the substrate, the X-rays may be reflected from the top and the bottom of the film, which gives rise to interference and an oscillatory pattern with changing angle of incidence, known as Kiessig fringes. The pattern obtained is a function of the difference in electron density and roughness at each interface present; rough films give rise to a reduction in the amplitude of the oscillation observed. Analysis of this variation gives information principally about the interfaces, but may also be used to investigate chain layering, in metal soaps, for example [52].

4.1.8.1 Experimental set-up

X-ray reflectivity experiments for thin films of liquids and so on are commonly performed on silicon single-crystal wafers, the X-rays being reflected off the surface of the wafer [53]. To enable good adhesion, the wafers have to be cleaned (in concentrated nitric acid with subsequent UV-O_3 treatment, for example) to remove any trace organics. Deposition of the films can then be performed by spin coating from a solution of the salt in a volatile organic solvent. In general, the spin-coated films are too rough to give good reflectivity spectra and the films need to be pre-annealed. X-ray reflectivity measurements may be performed with a laboratory X-ray source as well as with synchrotron radiation. Figure 4.1-14 shows a typical cell used for reflectivity measurements.

4.1.8.2 Examples

Carmichael et al. [54] have used this technique to compare the structures of thin films of $[C_{18}MIM][PF_6]$, $[C_{18}MIM][BF_4]$, $[BMIM]_2[PdCl_4]$, $[C_{12}MIM]_2[PdCl_4]$, and $[C_{12}MIM][PF_6]$ to bulk solutions as studied by small-angle X-ray scattering. Bragg features were clearly visible for all the salts studied; but in most cases the additional Kiessig fringes were not observed. Figure 4.1-15 shows an example of data collected on a thin film of $[C_{18}MIM][PF_6]$.

The Bragg peaks indicated an ordered local structure within the sample film, and the interlayer spacings were reproduced compared with the bulk samples, with only

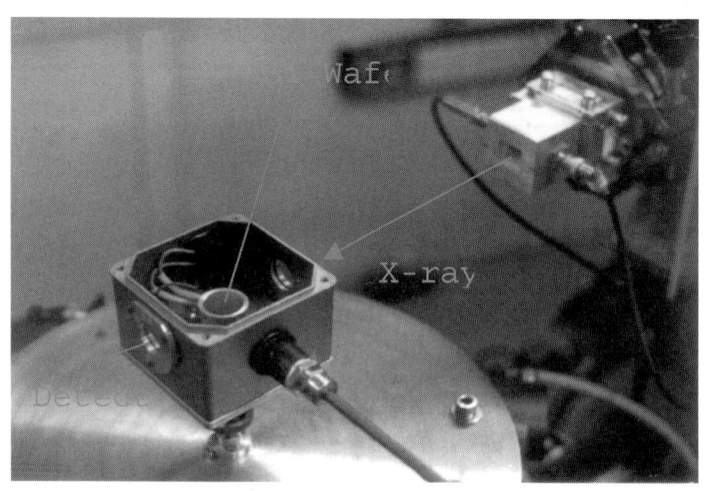

Figure 4.1-14: A typical cell used for X-ray reflectivity measurements.

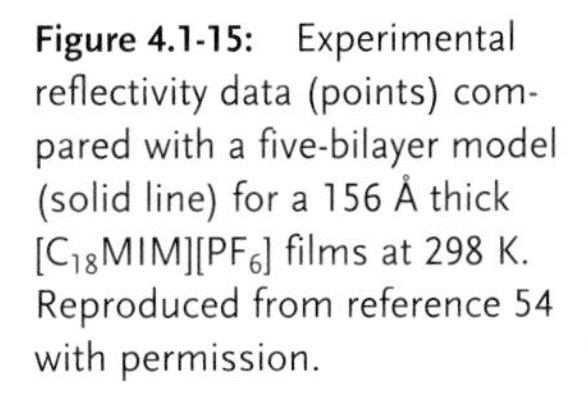
Figure 4.1-15: Experimental reflectivity data (points) compared with a five-bilayer model (solid line) for a 156 Å thick $[C_{18}MIM][PF_6]$ films at 298 K. Reproduced from reference 54 with permission.

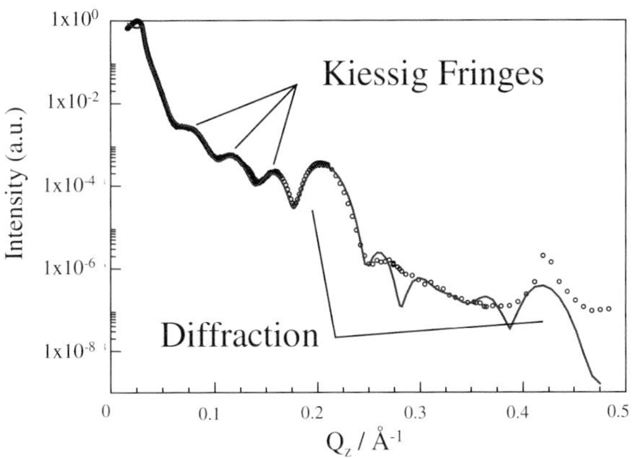

minor shifts in layer spacing. The small changes in layer spacing would be expected, since the thin film structure is not constrained by long-range order effects and hence adopts a slightly different, lower-energy form. The similarity between the bulk samples and the thin film was further demonstrated in the case of $[C_{18}MIM][PF_6]$ by modeling of the Kiessig fringes. This model was comprised of layers of associated 1-ethyl-3-methylimidazolium cation head-groups and hexafluorophosphate anions, denoted as the charged region, separated by hydrocarbon chains. Reasonable fits were only obtained with the charged region at the salt–silicon and the salt–air interfaces.

Although this technique has not been used extensively, it does allow structures of adsorbed layers on solid substrates to be studied. Liquid reflectivity may also be performed with a similar set-up, which relies on a liquid–liquid interface acting as the reflective surface and measures the reflectivity of a thin supported liquid film. This technique has recently been used to investigate water–alkane interfaces [55] and is potentially useful in understanding the interaction of ionic liquids with molecular solvents in which they are immiscible.

4.1.9
Direct Recoil Spectrometry (DRS)

The surface structures of ionic liquids have been studied by direct recoil spectrometry. In this experiment, a pulsed beam of 2–3 keV inert gas ions is scattered from a liquid surface, and the energies and intensities of the scattered and sputtered (recoiled) ions are measured as a function of the incident angle, α, of the ions. Figure 4.1-16 shows a scheme of the process for both the scattered and sputtered ions.

The incident ions cause recoil in the surface atoms. In studies of ionic liquids, only direct recoil – that is, motion in the forward direction – was measured. Watson and co-workers [56, 57] used time-of-flight analysis with a pulsed ion beam to measure the kinetic energies of the scattered and sputtered ions and therefore determine the masses of the recoiled surface atoms. By relating the measured intensities of the

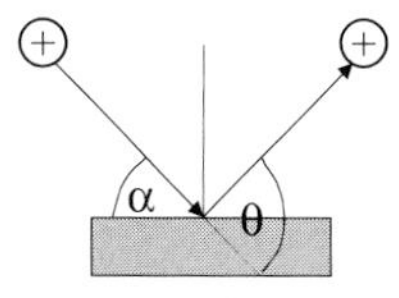

Scattered ions

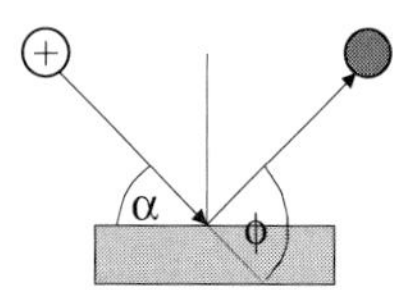

Sputtered(recoiled) ions

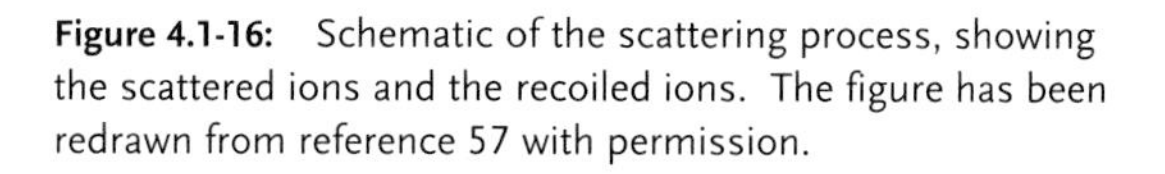

Figure 4.1-16: Schematic of the scattering process, showing the scattered ions and the recoiled ions. The figure has been redrawn from reference 57 with permission.

sputtered atoms to the scattering cross section, the surface concentration may be found. The variation of intensity with incident angle also allowed the orientation of the atoms on the surface to be elucidated. The scattered and sputtered ions are detected at angles of θ and ϕ, respectively, measured with respect to the incident ion beam. For all experiments so far reported on ionic liquids, θ and ϕ have been equal.

4.1.9.1 **Experimental set-up**

Direct recoil spectrometry requires high and ultra-high vacuum conditions for the transport of ions to the sample and to the detector. In this regard, the use of ionic liquids, with their corresponding low vapor pressures, is ideal. To prevent contamination of the surface and any surface charging effects, Watson and co-workers used a rotating stainless steel wheel partially submerged in a reservoir holding the liquid sample, to create a fresh liquid surface continually. Before analysis, the liquid film passed by a blade, leaving a fresh surface approximately 0.1–0.2 mm thick. Figure 4.1-17 shows the typical sample set-up [58].

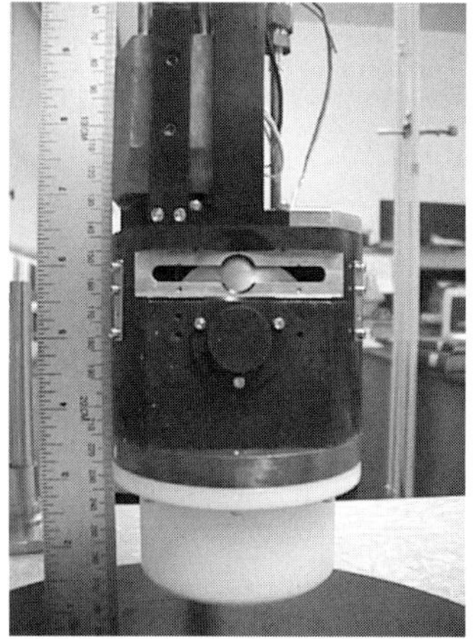

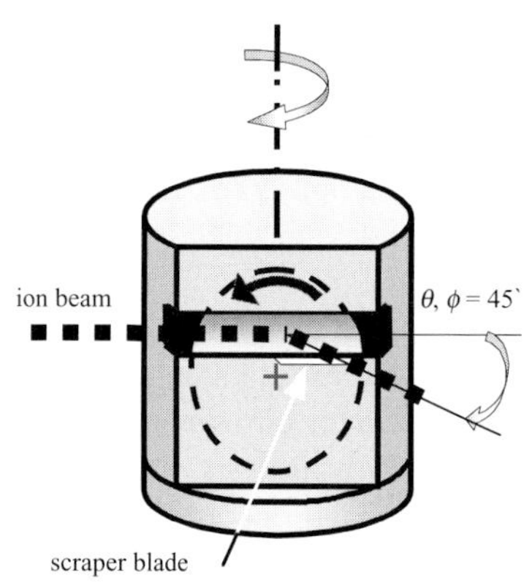

Figure 4.1-17: The experimental set-up used to generate thin films of ionic liquid for analysis by direct recoil spectrometry. Reproduced from reference 58 with permission.

4.1.9.2 Examples

A number of ionic liquids – namely [OMIM][PF_6], [BF_4]$^-$, Br^-, Cl^-; [BMIM][PF_6], [BF_4]$^-$, and [C_{12}MIM][BF_4] – have been studied by DRS. The scattering profile as a function of the incident angle for [OMIM][PF_6] is shown in Figure 4.1-18.

The charged species were in all cases found to concentrate at the surface of the liquid under vacuum conditions. Little surface separation of the anions and cations was observed. For the [PF_6]$^-$ and [BF_4]$^-$ ions, the cation ring was found to prefer a perpendicular orientation to the surface, with the nitrogen atoms closest to the surface. An increase in the alkyl chain length caused the cation to rotate so that the alkyl chain moved into the bulk liquid, away from the surface, forcing the methyl group closer to the surface. For halide ionic liquids, the data were less clear and the cation could be fitted to a number of orientations.

4.1.10 Conclusions

A wide range of structural techniques may be utilized for the study of ionic liquids and dissolved species. Overall, in both high-temperature and low-temperature ionic

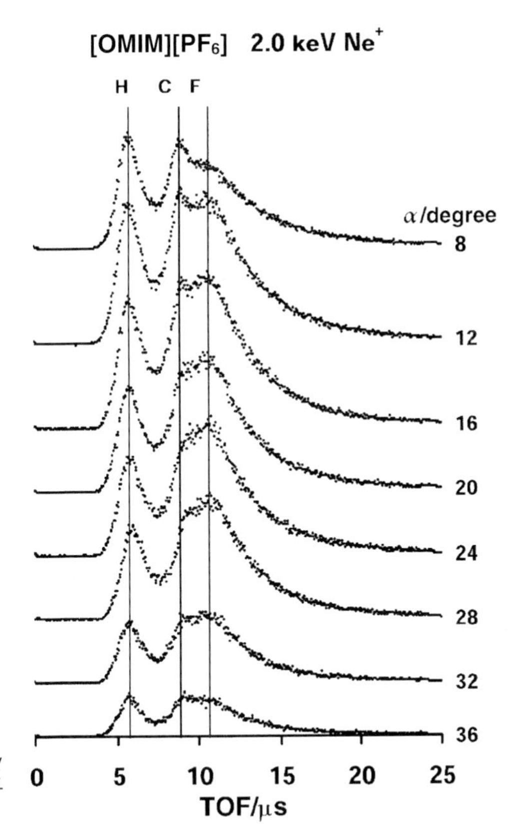

Figure 4.1-18: Ion intensity as a function of incident angle measured using time of flight direct recoil spectrometry on [OMIM][PF_6]. Reproduced from reference 57 with permission.

liquids, as well as for mixtures, a close correlation between the solid structure and liquid is found. In many cases, significant intermediate order is observed, for example in the form of charge ordering.

Acknowledgements

CH would like to thank Daniel Bowron (CLRC, Rutherford Appleton Laboratory), Fred Mosselmanns (CLRC, Daresbury Laboratory), Will Pitner (The Queen's University of Belfast), Nick Terrill (CLRC, Daresbury Laboratory), and Philip Watson (University of Oregon) for contributing figures, as well as for useful discussions and comments in the preparation of this manuscript.

References

1 D.T. Bowron, J.L. Finney, A.K. Soper, *J. Phys. Chem.* **1998**, B *102*, 3551–3563.
2 C. Hardacre, J.D. Holbrey, S.E.J. McMath, *J. Chem. Soc. Chem. Commun.* **2001**, 367–368.
3 F.G. Edwards, J.E. Enderby, R.A. Howe, D.I. Page, *J. Phys. C* **1975**, *8*, 3483–3490; S. Biggin, J.E. Enderby, *J. Phys. C* **1982**, *15*, L305–309.
4 M. Blander, E. Bierwagen, K.G. Calkins, L.A. Curtiss, D.L. Price, M.-L. Saboungi, *J. Chem. Phys.* **1992**, *97*, 2733–2741.
5 S.J. Cyvin, P. Klaeboe, E. Rytter, H.A. Øye, *J. Chem. Phys.* **1970**, *52*, 2776; J. Hvistendahl, P. Klaeboe, E. Rytter, H.A. Øye, *Inorg. Chem.* **1984**, *23*, 706–715.
6 Y.S. Badyal, D.A. Allen, R.A. Howe, *J. Phys.: Condens. Matter* **1994**, *6*, 10,193–10,220.
7 Y.-C. Lee, D.L. Price, L.A. Curtiss, M.A. Ratner, D.F. Shriver, *J. Chem. Phys.* **2001**, *114*, 4591–4594.
8 S. Takahashi, K. Suzuya, S. Kohara, N. Koura, L.A. Curtiss, M.-L. Saboungi, *Z. Phys. Chem.* **1999**, *209*, 209–221.
9 P.C. Truelove, D. Haworth, R.T. Carlin, A.K. Soper, A.J.G. Ellison, D.L. Price, *Proc. 9th Int. Symp. Molten Salts, San Fransisco* **1994**, *3*, 50–57, (Pennington, NJ: Electrochem. Soc.).
10 F.R. Trouw, D.L. Price, *Ann. Rev. Phys. Chem.* **1999**, *50*, 571–601.
11 D.T. Bowron, C. Hardacre, J.D. Holbrey, S.E.J. McMath, A.K. Soper, *J. Chem. Phys. Submitted.*
12 P.A. Egelstaff, D.I. Page, J.G. Powles, *Mol. Phys.* **1971**, *20*, 881 and *Mol. Phys.* **1971**, *22*, 994.
13 A.K. Soper, *Chem. Phys.* **1996**, *202*, 295–306; A.K. Soper, *Chem. Phys.* **2000**, *258*, 121–137.
14 F. Vaslow, A.H. Narten, *J. Chem. Phys.* **1973**, *59*, 4949–4954.
15 S. Takahashi, K. Maruoka, N. Koura, H. Ohno, *J. Chem. Phys.* **1986**, *84*, 408–415; S. Takahashi, T.N. Muneta, N. Koura, H. Ohno, *J. Chem. Soc. Faraday Trans. II* **1985**, *81*, 1107–1115.
16 K. Igarashi, Y. Okamoto, J. Mochinaga, H. Ohno, *J. Chem. Soc. Faraday Trans. I* **1988**, *84*, 4407–4415.
17 S. Takahashi, N. Koura, M. Murase, H. Ohno, *J. Chem. Soc. Faraday Trans. II* **1986**, *82*, 49–60.
18 C.J. Bowlas, D.W. Bruce, K.R. Seddon, *J. Chem. Soc. Chem. Commun.* **1996**, 1625–1626.
19 R. Kind, S. Plesko, H. Arend, R. Blinc, B. Zeks, J. Seliger, B. Lozar, J. Slak, A. Levstik, C. Filipic, V. Zagar, G. Lahajnar, F. Milia, G. Chapuis, *J. Chem. Phys.* **1979**, *71*, 2118–2130.

20 C.M. Gordon, J.D. Holbrey, A. Kennedy, K.R. Seddon, *J. Mater. Chem.* **1998**, *8*, 2627–2636. J.D. Holbrey, K.R. Seddon, *J. Chem. Soc., Dalton Trans.* **1999**, 2133–2139.

21 P.B. Hitchcock, K.R. Seddon, T. Welton, *J. Chem. Soc., Dalton Trans.* **1993**, 2639–2643.

22 F. Neve, A. Crispini, S. Armentano, O. Francescangeli, *Chem. Mater.* **1998**, *10*, 1904–1913.

23 F. Neve, O. Francescangeli, A. Crispini, J. Charmant, *Chem. Mater.* **2001**, *13*, 2032–2041.

24 C. Hardacre, J.D. Holbrey, P.B. McCormac, S.E.J. McMath, M. Nieuwenhuyzen, K.R. Seddon, *J. Mater. Chem.* **2001**, *11*, 346–350.

25 J.D. Martin, R. D. Rogers and K. R. Seddon (Eds) *ACS Symposium Series 818, ACS Washington DC*, **2002**, p. 413–427.

26 G.F. Needham, R.D. Willett, H.F. Franzen, *J. Phys. Chem.* **1984**, *88*, 674–680.

27 A.E. Bradley, C. Hardacre, J. D. Holbrey, S. Johnston, S. E. J. McMath, M. Nieuwenhuyzen, *Chem. Mater,* **2002**, *14*, 629–635.

28 V. Busico, P. Corradini, M. Vacatello, *J. Phys. Chem.* **1982**, *86*, 1033–1034. V. Busico, P. Cernicchlaro, P. Corradini, M. Vacatello, *J. Phys. Chem.* **1983**, *87*, 1631–1635.

29 D.J. Abdallah, A. Robertson, H.-F. Hsu, R.G. Weiss, G., *J. Am. Chem. Soc.* **2000**, 122, 3053–3062.

30 F. Artzner, M. Veber, M. Clerc, A.-M. Levelut, *Liq. Cryst.* **1997**, *23*, 27–33.

31 K.M. Lee, C.K. Lee, I.J.B. Lin, *J. Chem. Soc., Chem. Commun.* **1997**, 899–900.

32 C. Cruz, B. Heinrich, A.C. Ribeiro, D.W. Bruce, D. Guillon, *Liq. Cryst.* **2000** *27*, 1625–1631; D.W. Bruce, S. Estdale, D. Guillon, B. Heinrich, *Liq. Cryst.* **1995**, *19*, 301–305.

33 E.J.R. Sudhölter, J.B.F.N. Engberts, W.H. de Jeu, *J. Phys. Chem.* **1982**, *86*, 1908–1913.

34 Y. Haramoto, S. Ujiie, M. Nanasawa, *Liq. Cryst.* **1996**, *21*, 923–925, Y. Haramoto, M. Nanasawa, S. Ujiie, *Liq. Cryst.* **2001**, *28*, 557–560.

35 J. Jaklevic, J.A. Kirby, M.P. Klein, A.S. Robertson, G.S. Brown, P. Eisenberger, *Solid State Commun.* **1977**, *23*, 1679.

36 J.C. Mikkelsen, J.B. Boyce, R. Allen, *Rev. Sci. Instr.* **1980**, *51*, 388–389.

37 A.L. Ankudinov, B. Ravel, J.J. Rehr, S.D. Conradson, *Phys. Rev. B* **1998**, *58*, 7565–7576. J.J. Rehr, R.C. Albers, *Phys. Rev. B* **1990**, *41*, 8139–8149.

38 A. Filipponi, A. Di Cicco, C.R. Natoli, *Phys. Rev. B* **1995**, *52*, 15,122–15,134.

39 N. Binsted, EXCURV98: CCLRC Daresbury Laboratory computer program, **1998**.

40 A. Filipponi, *J. Phys.: Condensed Matter* **2001**, *13*, R23–60.

41 J.J. Rehr, R.C. Albers, *Rev. Mod. Phys.* **2000**, *72*, 621–654.

42 A. Di Cicco, M. Minicucci, A. Filipponi, *Phys. Rev. Lett.* **1997**, 78, 460–463; M. Minicucci, A. Di Cicco, *Phys. Rev. B* **1997**, *56*, 11,456–11,464.

43 A. Di Cicco, *J. Phys.: Condensed Matter* **1996**, *8*, 9341–9345.

44 L. Hefeng, L. Kunquan, W. Zhonghua, D. Jun, *J. Phys.: Condensed Matter* **1994**, *6*, 3629–3640.

45 A.J. Carmichael, C. Hardacre, J.D. Holbrey, M. Nieuwenhuyzen, K.R. Seddon, *Anal. Chem.* **1999**, *71*, 4572–4574.

46 A.J. Dent, K.R. Seddon, T. Welton, *J. Chem. Soc. Chem. Commun.* **1990**, 315–316.

47 A.J. Dent, A. Lees, R.J. Lewis, T. Welton, *J. Chem. Soc. Dalton Trans.* **1996**, 2787–2792.

48 G.M.N. Baston, A.E. Bradley, T. Gorman, I. Hamblett, C. Hardacre, J.E. Hatter, M.J.F. Healy, B. Hodgson, R. Lewin, K.V. Lovell, G.W.A. Newton, M. Nieuwenhuyzen, W.R. Pitner, D.W. Rooney, D. Sanders, K.R. Seddon, H.E. Simms, R.C. Thied, R. D. Rogers and K. R. Seddon (Eds) *ACS Symposium Series 818, ACS Washington DC*, **2002**, p. 162–177. *Industrial Applications of Ionic Liquids.*

49 J.J. Laidler, J.E. Battles, W.E. Miller, J.P. Ackerman, E.L. Carls, *Progress in Nuclear Energy* **1997**, *31*, 131.

50 N.A. Hamill, C. Hardacre, S.E.J. McMath, *Green Chemistry*, **2002**, *4*, 143–146.
51 X.-L. Zhou, S.-H. Chen, *Phys. Rep.* **1995**, *257*, 223–348.
52 U. Englisch, F. Peñacorada, L. Brehmer, U. Pietsch, *Langmuir* **1999**, *15*, 1833–1841.
53 M.F. Toney, C.M. Mate, K.A. Leach, D. Pocker, *J. Coll. Int. Sci.* **2000**, *225*, 219–226.
54 A.J. Carmichael, C. Hardacre, J.D. Holbrey, M. Nieuwenhuyzen, K.R. Seddon, *Mol. Phys.* **2001**, *99*, 795–800.
55 D.M. Mitrinovic, Z. Zhang, S.M. Williams, Z. Huang, M.L. Schlossman, *J. Phys. Chem.* **1999**, *103*, 1779–1782.
56 T.J. Gannon, G. Law, P.R. Watson, A.J. Carmichael, K.R. Seddon, *Langmuir* **1999**, *15*, 8429–8434; G. Law, P.R. Watson, *Chem. Phys. Lett.* **2001**, *345*, 1–4.
57 G. Law, P.R. Watson, A.J. Carmichael, K.R. Seddon, *Phys. Chem. Chem. Phys.* **2001**, *3*, 2879–2885.
58 M Tassotto, PhD Thesis, Dept. of Physics, Oregon State University, **2000**.

4.2
Quantum Mechanical Methods for Structure Elucidation

W. Robert Carper, Zhizhong Meng, and Andreas Dölle

4.2.1
Introduction

The description of electronic distribution and molecular structure requires quantum mechanics, for which there is no substitute. Solution of the time-independent Schrödinger equation, $\mathbf{H}\psi = E\psi$, is a prerequisite for the description of the electronic distribution within a molecule or ion. In modern computational chemistry, there are numerous approaches that lend themselves to a reasonable description of ionic liquids. An outline of these approaches is given in Scheme 4.2-1 [1]:

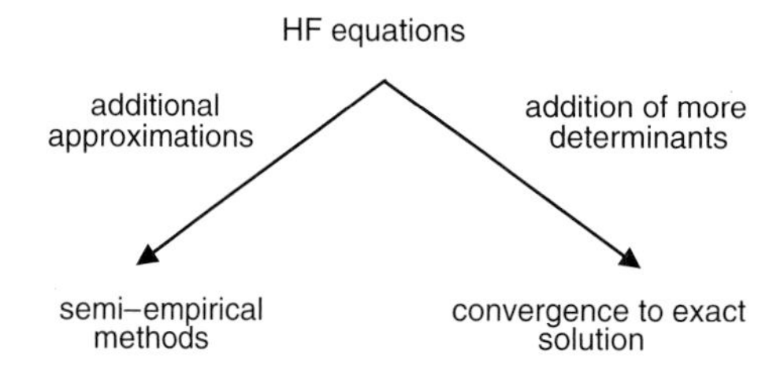

Scheme 4.2-1: Approaches used to describe ionic liquids in computational chemistry.

4.2.2
Choice of Quantum Mechanical Methods

The choices of quantum mechanical method typically include the semi-empirical methods AM1, PM3, and MNDO/d [2–4]. These three methods (and some of their variations) are those most commonly used in the current literature. Of these semi-empirical methods, only MNDO/d includes the effects of d-orbitals. Some of the problems associated with these semiempirical methods include:

(a) rotational barriers for bonds with partial double bond character are too low,
(b) non-bonding interactions, such as in van der Waals complexes or hydrogen bonds, are poorly reproduced with AM1 and
(c) nitrogen-containing groups often give pyramidal structures with PM3, when this is incorrect.

Despite these inconsistencies, the semi-empirical methods produce bond angles, bond lengths and heats of formation that are in reasonable agreement with experimental results. A new version, PM5, will soon be available and is four times more accurate than AM1 or PM3. The advantage of PM5 over the other semi-empirical methods is that d-orbitals are being introduced [5].

The ab initio methods used by most investigators include Hartree–Fock (HF) and Density Functional Theory (DFT) [6, 7]. An ab initio method typically uses one of many basis sets for the solution of a particular problem. These basis sets are discussed in considerable detail in references [1] and [8]. DFT is based on the proof that the ground state electronic energy is determined completely by the electron density [9]. Thus, there is a direct relationship between electron density and the energy of a system. DFT calculations are extremely popular, as they provide reliable molecular structures and are considerably faster than HF methods where correlation corrections (MP2) are included. Although intermolecular interactions in ion-pairs are dominated by dispersion interactions, DFT (B3LYP) theory lacks this term [10–14]. However, DFT theory is quite successful in representing molecular structure, which is usually a primary concern.

The investigator's choice of method (semi-empirical or ab initio) hinges on a number of factors, one of which is simple practicality concerning both time and expense. Semi-empirical methods usually give reasonable molecular structures and thermodynamic values at a fraction of the cost of ab initio calculations. Furthermore, molecular structures calculated by semi-empirical methods are the starting point for more complex ab initio calculations.

The advantages of ab initio calculations are considerable, but they come at a high cost. One of the many factors that affect the cost of computation is the choice of basis set. Often (not always!) one discovers that accuracy of physical parameters (and molecular structure) increases with the size of the basis set. Unfortunately, the formal scaling of HF methods is approximately N^4 (N^3 or less for semiempirical methods), where N is the number of basis functions. Hence, one quickly discovers that the solution of a problem requiring a few hours with a low-order basis set (3-21G(*)) may end up taking several days with a higher-order basis set such as 6-31G(d,p) or 6-31G(dp,p) [1, 8]. Fortunately, many investigators now have access to high-speed parallel processor computers that can handle such large calculations.

4.2.3
Ion-pair Models and Possible Corrections

Typically, the ionic liquid is best considered as an ion-pair (or ion, on an individual basis). The main forces of attraction between the ions are the electrostatic forces and dispersion (van der Waals) forces. At intermediate distances, there is a slight attraction between the electron clouds of the ions. This is due to the phenomenon known as electron correlation, which is important in Hartree–Fock calculations. This correction is not necessary in completely parameterized semi-empirical methods.

The size of the ion-pair may dictate the method of calculation, although the increasing speed of computers, coupled with improved programming, encourage many to begin at the ab initio level of calculation. There are two types of corrections that should be considered with ab initio calculations. If one is working with neutral molecules, then BSSE (basis set superposition error) correction [15, 16] is necessary if one is accurately to determine values of hydrogen bonds in molecular complexes. With ion-pairs, however, the 4 to 8 kJ correction factor with BSSE is minimal compared with the correlation energy correction typically introduced with the use of Møller–Plesset perturbation theory [17]. The use of MP corrections at the 2nd level (MP2) is a computer-intense correction and limits many calculations to very small molecular systems. This fact, coupled with the speed of density functional (DFT) calculations, has encouraged many investigators to use DFT when computer time is either expensive or in short supply [1, 8, 16]. The main advantage of DFT ab initio calculations is that the resulting molecular structure is usually accurate, although energies may be in doubt.

4.2.4 Ab Initio Structures of Ionic Liquids

Figure 4.2-1 shows the calculated ab initio molecular structure of the ionic liquid [BMIM][PF_6] (1-butyl-3-methylimidazolium hexafluorophosphate).

The basis set is 6-31G(d,p), and electron correlation at the MP2 level is included. A similar structure is obtained with the AM1 and PM3 semi-empirical methods. Density functional theory at the B3LYP/6-31G(dp,p) level also produced the same structure for this ion-pair. The only observed differences between the semi-empirical and the ab initio structures were slightly shorter hydrogen bonds (PM3 and AM1) between F1, F2, and F5 and the C2-H (H18) on the imidazolium ring.

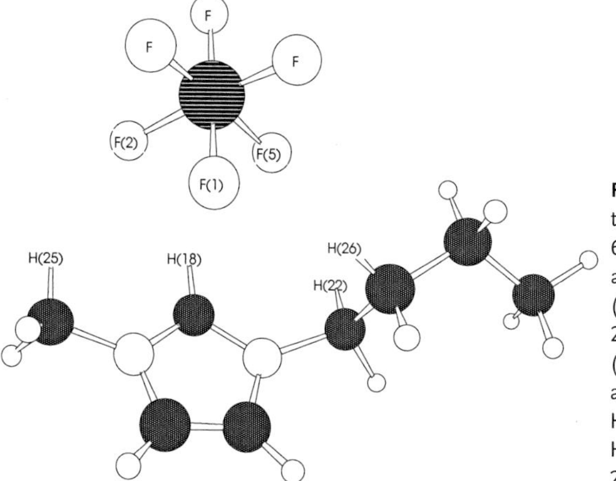

Figure 4.2-1: Molecular structure of [BMIM][PF_6] (MP2/6-31G(d,p)). C-H–F distances are: 2.319 Å (H25-F2), 2.165 Å (H18-F2), 2.655 Å (H18-F1), 2.173 Å (H18-F5), 2.408 Å (H22-F5), 2.467 Å (H26-F1), and 2.671 Å (H26-F5). All H–F distances are less than the H–F van der Waals distance of 2.67 Å (see ref. [18]).

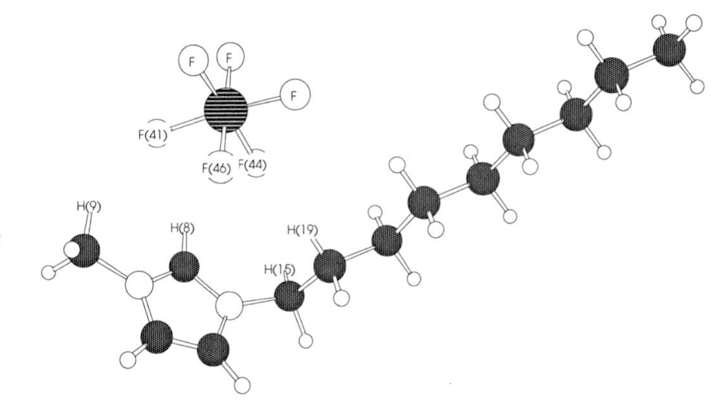

Figure 4.2-2: Molecular structure of [NMIM][PF_6] (MP2/6-31G(d,p)). C-H–F distances are: 2.355 Å (H9-F1), 2.198 Å (H8-F41), 2.636 Å (H8-F46), 2.164 Å (H8-F44), 2.439 Å (H15-F44), 2.403 Å (H19-F46), and 2.582 Å (H19-F44). All H–F distances are less than the H–F van der Waals distance of 2.67 Å (see ref. [18]).

Another ionic liquid, containing a nonyl-rather than a butyl-side chain, is shown in Figure 4.2-2. There is little difference between the basic structures of these two ion-pairs (Figures 4.2-1 and 4.2-2) with respect to the non-bonded interactions (hydrogen bonds) occurring between the F atoms on the anion and the C-H moieties on the imidazolium cation.

4.2.5 DFT Structure of 1-Methyl-3-nonylimidazolium Hexafluorophosphate

Figure 4.2-3 contains the DFT (B3LYP) structure of [NMIM][PF_6] obtained with a 6-31G(d,p) basis set. Here one observes C-H–F hydrogen bonds shorter than those obtained from the MP2/6-31G(d,p) calculation shown in Figure 4.2-2.

Note that DFT structures are as reliable as or more reliable than HF structures obtained with similar or less complex basis sets.

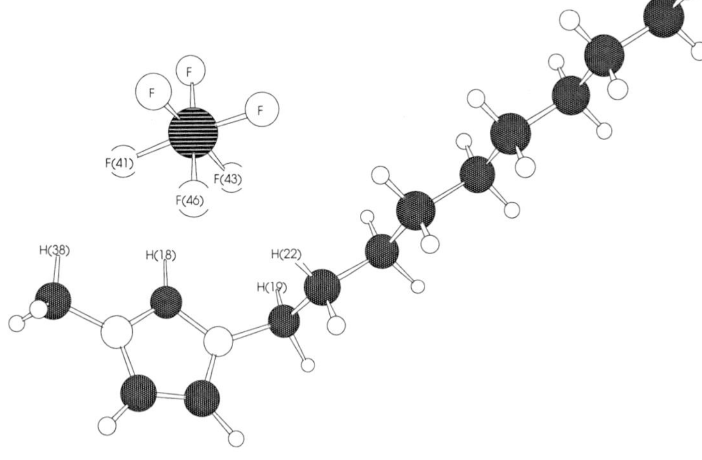

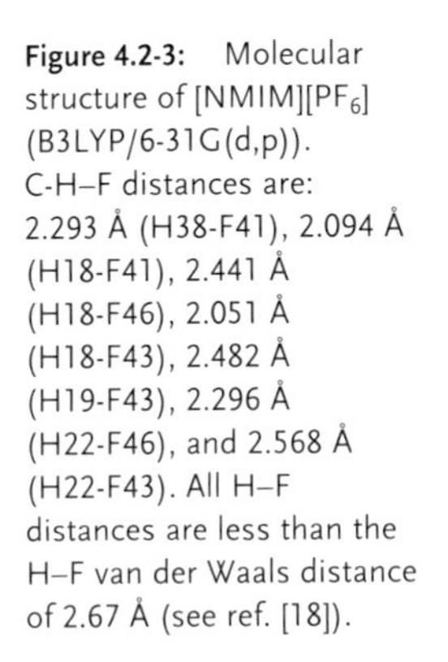

Figure 4.2-3: Molecular structure of [NMIM][PF_6] (B3LYP/6-31G(d,p)). C-H–F distances are: 2.293 Å (H38-F41), 2.094 Å (H18-F41), 2.441 Å (H18-F46), 2.051 Å (H18-F43), 2.482 Å (H19-F43), 2.296 Å (H22-F46), and 2.568 Å (H22-F43). All H–F distances are less than the H–F van der Waals distance of 2.67 Å (see ref. [18]).

4.2.6
Additional Information Obtained from Semi-empirical and Ab Initio Calculations

In addition to the obvious structural information, vibrational spectra can also be obtained from both semi-empirical and ab initio calculations. Computer-generated IR and Raman spectra from ab initio calculations have already proved useful in the analysis of chloroaluminate ionic liquids [19]. Other useful information derived from quantum mechanical calculations include ^{1}H and ^{13}C chemical shifts, quadrupole coupling constants, thermochemical properties, electron densities, bond energies, ionization potentials and electron affinities. As semiempirical and ab initio methods are improved over time, it is likely that investigators will come to consider theoretical calculations to be a routine procedure.

References

1 F. Jensen, *Introduction to Computational Chemistry*, John Wiley & Sons, **1999**, pp. 53–97.
2 M. J. S. Dewar, E. G. Zoebisch, E. F. Healy, J. J. P. Stewart, *J. Am. Chem. Soc.* **1985**, *107*, 3902.
3 J. J. P. Stewart, *J. Comput. Chem.* **1989**, *10*, 209.
4 W. Thiel, A. A. Voityuk, *J. Phys. Chem.* **1996**, *100*, 616.
5 J. J. P. Stewart, personal communication.
6 A. D. Becke, *J. Chem. Phys.* **1992**, *97*, 9173.
7 C. Lee, W. Yang, R. G. Parr, *Phys. Rev. B* **1988**, *37*, 785.
8 D. Young, *Computational Chemistry*, John Wiley & Sons, **2001**, pp. 78–91.
9 P. Hohenberg, W. Kohn, *Phys. Rev. B* **1964**, *136*, 864.
10 J. Nagy, D. F. Weaver, V. H. Smith. Jr., *Mol. Phys.* **1995**, *85*, 1179.
11 E. J. Meijer, M. Sprik, *J. Chem. Phys.* **1996**, *105*, 8684.
12 Y. Andersson, D. C. Langreth, B. I. Lundqvist, *Phys. Rev. Lett.* **1996**, *76*, 102.
13 W. Kohn, Y. Meir, D. E. Makarov, *Phys. Rev. Lett.* **1998**, *80*, 4153.
14 R. L. Rowley, T. Pakkanen, *J. Chem. Phys.* **1999**, *100*, 3368.
15 S. F. Boys, F. Bernardi, *Mol. Phys.* **1970**, *19*, 553.
16 F. B. Van Duijneveldt, *Molecular Interactions* (S. Scheiner ed.), John Wiley & Sons, **1997**, pp. 81–104.
17 C. Møller, M. S. Plesset, *Phys. Rev.* **1934**, *46*, 618.
18 A. Bondi, *J. Phys. Chem.* **1964**, *68*, 441.
19 G. J. Mains, E. A. Nantsis, W. R. Carper, *J. Phys. Chem. A* **2001**, *105*, 4371.

4.3
Molecular Dynamics Simulation Studies

Christof G. Hanke and Ruth M. Lynden-Bell

4.3.1
Performing Simulations

So far, there have been few published simulation studies of room-temperature ionic liquids, although a number of groups have started programs in this area. Simulations of molecular liquids have been common for thirty years and have proven important in clarifying our understanding of molecular motion, local structure and thermodynamics of neat liquids, solutions and more complex systems at the molecular level [1–4]. There have also been many simulations of molten salts with atomic ions [5]. Room-temperature ionic liquids have polyatomic ions and so combine properties of both molecular liquids and simple molten salts.

Atomistic simulations can be carried out at various levels of sophistication and the method of choice is a balance between computational cost and accuracy. The three main types of simulation are classical simulations, fully quantum simulations and hybrid methods. In classical simulations the molecules interact according to a force-field, which must be defined by the user. In quantum simulations the forces on the nuclei are calculated from quantum mechanical electronic energy at each step, which is found by solving approximations to the Schrödinger equation. In hybrid methods, part of the system is treated by quantum mechanics and the rest classically. For simulations of liquids one needs long runs to explore the many possible configurations corresponding to the liquid state. One also needs fairly large system sizes to remove the effects of periodic boundaries. Thus, while a crystalline solid can be simulated by the use of a few unit cells for a few picoseconds, a liquid needs ten to one hundred times as large a system and needs to be simulated for ten to one hundred times as long. This means that classical simulations are the most likely to be useful. The main limitation is that chemical bond formation or breaking cannot be described.

There are many molecular dynamics programs available for simulations, and the book by Allen and Tildesley [6] provides a very helpful introduction for anyone who wishes to perform simulations. The key points are:

- to use a reasonable potential,
- to treat the long-range electrostatics by an accurate method such as the Ewald summation,
- to use a large enough system (say 200 formula units or more) and
- to simulate for a sufficiently long time to sample a sufficient range of configurations typical of the liquid.

In a classical simulation a force-field has to be provided. Experience with molecular liquids shows that surprisingly good results can be obtained with intermolecular potentials based on site–site short-range interactions and a number of charged sites on each molecule, and such models are used for the simulation of systems ranging from simple liquids to biomolecules [1–4]. The short-range interactions are repulsive at short distances, so that the distribution of sites determines the molecular shape in the model. A good description of the electrostatic interactions between different molecules is very important. It is also important to treat the long-range part of the electrostatics carefully. This is best done by the Ewald summation method [6]. Price et al. [7] have developed a model for methyl- and ethylimidazolium ions with charges on the atomic sites. The charges were taken from a distributed multipole analysis of a good quantum chemical calculation of isolated ions with a reasonable-sized basis set with correlations included at the MP2 level. Figure 4.3-1 shows the molecule and the contours of the electrostatic field due to the charges on the sites.

One can see that, while at large distances the contours approach the circular shape expected for an ion, there are considerable distortions near the molecule. The charge is distributed over the ring atoms, the ring protons, and the side chains, particularly on the methyl and methylene groups adjacent to the ring. The charge on the nitrogen atom is negative, while the other charges are all positive. This reflects the electronegativity of the nitrogen atom and is likely to be an important factor in determining the local structure in the liquid. This model was tested by comparison of the predicted and experimental crystal structures.

While this model provides a good description of the electrostatic potential around an isolated ion it does not include the effects of polarization due to the surrounding ions. One might anticipate that a molecule with an aromatic ring would be easily polarizable, and this lack of polarizability is a major shortcoming. The computational cost and problems of parametrizing a polarizable model do not seem worthwhile at this stage of the project. Some justification for this simplification can be taken from recent simulations of triazoles in our group [8]. Triazoles are neutral

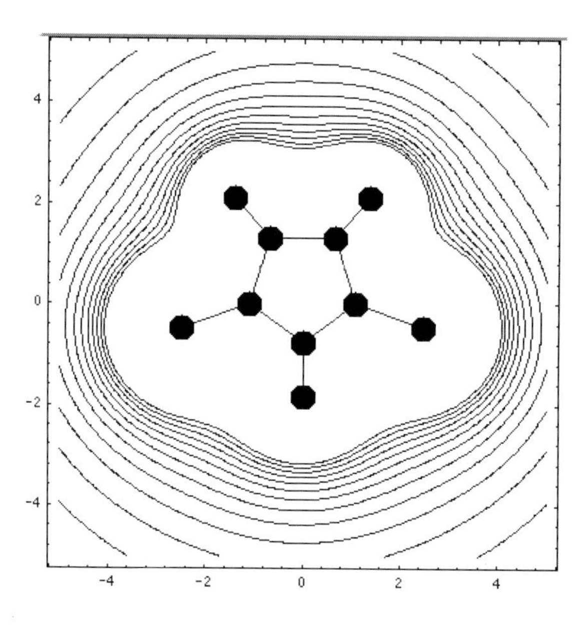

Figure 4.3-1: Contours of the electrostatic potential around the dimethylimidazolium ion.

molecules with five-membered rings containing nitrogen atoms, which have similarities to the imidazolium ions. Our recent work on simulations of a hybrid model of quantum triazoles dissolved in classical liquid water shows surprisingly small charge fluctuations (amplitude about 0.05 e or less). There is, however, a net polarization of the molecule in aqueous solution compared to in the gas phase. It may be useful in the future to try this type of calculation with imidazolium ions in the ionic liquid, although there are technical problems arising from the fact that the ions are charged. At this stage, simulations are being carried out with the fixed-charge model, which should describe the basic physics of the liquid although, given the comments above, we would not expect quantitative agreement. A further approximation frequently used in simulations of molecular liquids is to replace the methyl and methylene groups by single sites (united atoms). This saves between 35 % and 50 % of the computational effort for dimethylimidazolium salts.

4.3.2
What can we Learn?

The simulation gives a sequence of configurations: that is, instantaneous positions and velocities of all the atoms in the system. In a molecular dynamics simulation these are a sequence in time, while Monte Carlo simulations give a sequence generated by random moves. These sequences can be analyzed to give structural information, average energies and pressures, and dynamics. Some of this analysis is normally carried out during the simulation (average energies, for example), while other analyses can be carried out later. The problem is to reduce the data to a manageable and comprehensible form. Structural information for liquids is often presented as radial distribution functions $g_{AB}(r)$. These functions show the ratio of the probability density for finding an atom of type A at distance r from an atom of type B relative to the average density of A atoms. Thus, regions where g is greater than unity

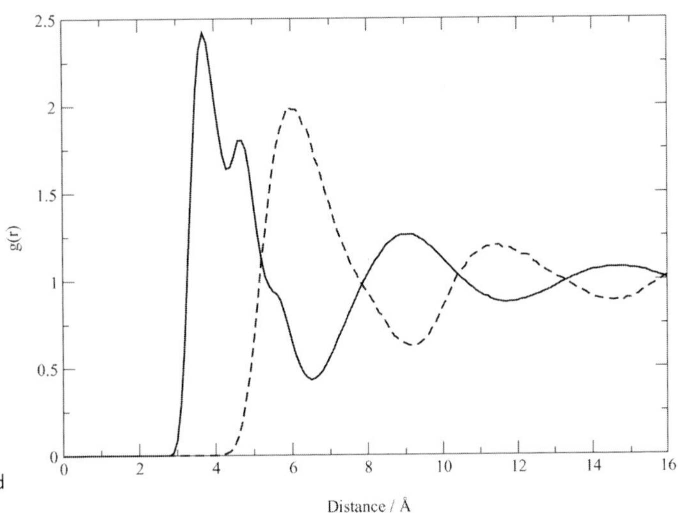

Figure 4.3-2: Radial distribution function for dimethylimidazolium and chloride ions relative to chloride. Full line: cation–anion; dashed line: anion–anion.

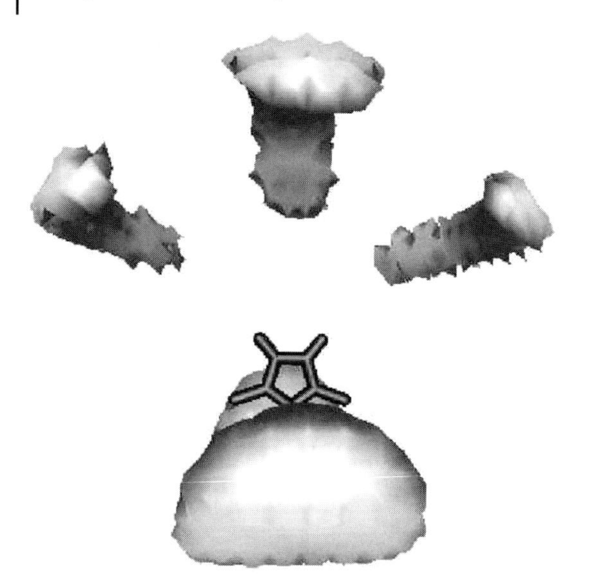

Figure 4.3-3: Three-dimensional distribution function of chloride ions relative to dimethylimidazolium ions.

have an enhanced probability of finding atoms of type B, while regions with g less than unity have a reduced probability. Figure 4.3-2 shows $g(r)$ for chloride ions, relative to the center of a dimethylimidazolium ion (or vice versa) and chloride ions relative to a chloride ion. The successive peaks and troughs are out of phase, showing the charge oscillations, which are quite long-ranged.

These are typical of ionic liquids and are familiar in simulations and theories of molten salts. The indications of structure in the first peak show that the local packing is complex. There are 5 to 6 nearest neighbors contributing to this peak. More details can be seen in Figure 4.3-3, which shows a contour surface of the three-dimensional probability distribution of chloride ions seen from above the plane of the molecular ion. The shaded regions are places at which there is a high probability of finding the chloride ions relative to any imidazolium ion.

Dynamic information such as reorientational correlation functions and diffusion constants for the ions can readily be obtained. Collective properties such as viscosity can also be calculated in principle, but it is difficult to obtain accurate results in reasonable simulation times. Single-particle properties such as diffusion constants can be determined more easily from simulations. Figure 4.3-4 shows the mean square displacements of cations and anions in dimethylimidazolium chloride at 400 K. The rapid rise at short times is due to rattling of the ions in the cages of neighbors. The amplitude of this motion is about 0.5 Å. After a few picoseconds the mean square displacement in all three directions is a linear function of time and the slope of this portion of the curve gives the diffusion constant. These diffusion constants are about a factor of 10 lower than those in normal molecular liquids at room temperature.

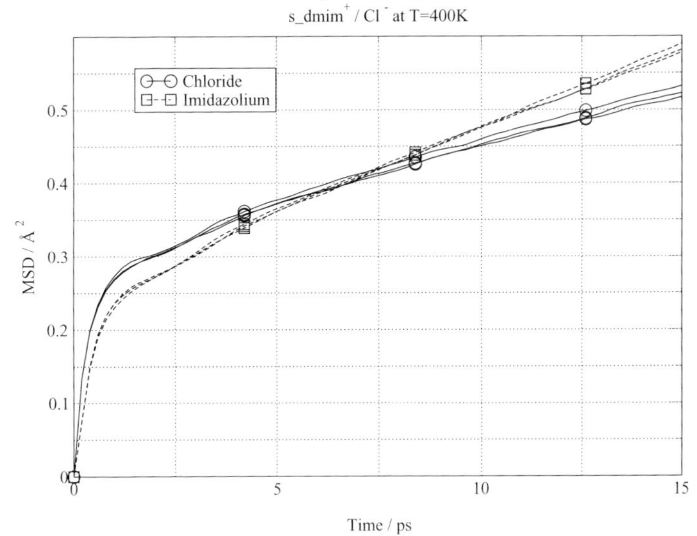

Figure 4.3-4: Mean square displacements of cations (full lines) and anions(dashed lines) in *x*, *y*, and *z* directions as a function of time.

Thermodynamic information can also be obtained from simulations. Currently we are measuring the differences in chemical potential of various small molecules in dimethylimidazolium chloride. This involves gradually transforming one molecule into another and is a computationally intensive process. One preliminary result is that the difference in chemical potential of propane and dimethyl ether is about 17.5 kJ/mol. These molecules are similar in size, but differ in their polarity. Not surprisingly, the polar ether is stabilized relative to the non-polar propane in the presence of the ionic liquid. One can also investigate the local arrangement of the ions around the solute and the contribution of different parts of the interaction to the energy. Thus, while both molecules have a favorable Lennard–Jones interaction with the cation, the main electrostatic interaction is that between the chloride ion and the ether molecule.

References

1 J. E. Shea, C. L. Brooks, *Annu. Rev. Phys. Chem.* **2001**, *52*, 499.

2 P. La Rocca, P. C. Biggin, D. P. Tieleman, M. S. P. Sansom, *Biochim. Biophys. Acta (Biomembranes)* **1999**, *1462*, 185.

3 P. C. Biggin, M. S. P. Sansom, *Biophys. Chem.* **1999**, *76*, 161.

4 P. A. Bopp, A. Kohlmeyer, E. Spohr, *Electrochim. Acta* **1998**, *43*, 2911.

5 P. A. Madden, M. Wilson, *J. Phys., Condens. Matter* **2000**, *12*, A95.

6 M. P. Allen and D. J. Tildesley, *Computer Simulation of Liquids*, Oxford University Press, Oxford **1987**.

7 C. G. Hanke, S. L. Price, R. M. Lynden-Bell, *Mol. Phys.* **2001**, *99*, 801.

8 S. Murdock, G. Sexton, R. M. Lynden-Bell, in preparation.

4.4
Translational Diffusion

Joachim Richter, Axel Leuchter, and Günter Palmer

4.4.1
Main Aspects and Terms of Translational Diffusion

Looking at translational diffusion in liquid systems, at least two elementary categories have to be taken into consideration: self-diffusion and mutual diffusion [1, 2].

In a liquid that is in thermodynamic equilibrium and which contains only one chemical species,[a] the particles are in translational motion due to thermal agitation. The term for this motion, which can be characterized as a random walk of the particles, is *self-diffusion*. It can be quantified by observing the molecular displacements of the single particles. The self-diffusion coefficient D_s is introduced by the Einstein relationship

$$D_s = \lim_{t\to\infty} \frac{1}{6t} \left\langle \left| \vec{r}(t) - \vec{r}(0) \right|^2 \right\rangle \tag{4.4-1}$$

where $\vec{r}(t)$ and $\vec{r}(0)$ denote the locations of a particle at time t and 0, respectively. The brackets indicate that the ensemble average is used.

However, self-diffusion is not limited to one-component systems. As illustrated in Figure 4.4-1, the random walk of particles of each component in any composition of a multicomponent mixture can be observed.

If a liquid system containing at least two components is not in thermodynamic equilibrium due to concentration inhomogenities, transport of matter occurs. This process is called *mutual diffusion*. Other synonyms are chemical diffusion, interdiffusion, transport diffusion, and, in the case of systems with two components, binary diffusion.

The description of mass transfer requires a separation of the contributions of *convection* and mutual diffusion. While convection means macroscopic motion of complete volume elements, mutual diffusion denotes the macroscopically perceptible relative motion of the individual particles due to concentration gradients. Hence, when measuring mutual diffusion coefficients, one has to avoid convection in the system or, at least has to take it into consideration.

Mutual diffusion is usually described by Fick's first law, written here for a system with two components and one-dimensional diffusion in the z-direction:

$$\bar{J}_i = -D_i \frac{\partial c_i}{\partial z} \; (i = 1, 2). \tag{4.4-2}$$

a Components are those substances, the amounts of which can be changed independently from others, while chemical species mean any particles in the sense of chemistry (atoms, molecules, radicals, ions, electrons) which appear at all in the system [3].

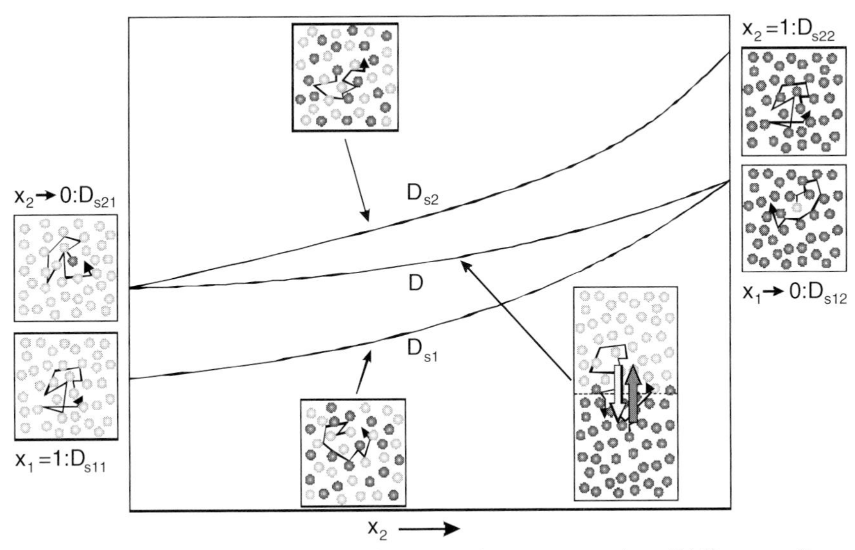

Figure 4.4-1: Self-diffusion and mutual diffusion in a binary mixture. The self-diffusion coefficients are denoted with D_{s1} and D_{s2}, the mutual diffusion coefficient with D. The self-diffusion coefficients of the pure liquids D_{s11} and D_{s22}, respectively, are marked at $x_1 = 1$ and $x_2 = 1$. Extrapolations $x_1 \rightarrow 0$ and $x_2 \rightarrow 0$ give the self-diffusion coefficients D_{s12} and D_{s21}.

Equation 4.4-2 describes the flux density $\vec{J}_i$ (in mol m^{-2} s^{-1}) of component i through a reference plane, caused by the concentration gradient $\partial c_i / \partial z$ (in mol m^{-4}). The factor D_i (in m^2 s^{-1}) is called the diffusion coefficient.

Most mutual diffusion experiments use Fick's second law, which permits the determination of D_i from measurements of the concentration distribution as a function of position and time:

$$\frac{\partial c_i}{\partial t} = D_i \frac{\partial^2 c_i}{\partial z^2} \tag{4.4-3}$$

Solutions for this second-order differential equation are known for a number of initial and boundary conditions [4].

In a system with two components, one finds experimentally the same values for D_1 and D_2 because $\vec{J}_1$ is not independent from $\vec{J}_2$. It follows that the system can be described with only one mutual diffusion coefficient $D = D_1 = D_2$.

In the case of systems containing ionic liquids, components and chemical species have to be differentiated. The methanol/[BMIM][PF_6] system, for example, consists of two components (methanol and [BMIM][PF_6]) but – on the assumption that [BMIM][PF_6] is completely dissociated – three chemical species (methanol, $[BMIM]^+$ and $[PF_6]^-$). If [BMIM][PF_6] is not completely dissociated, one has a fourth species, the undissociated [BMIM][PF_6]. From this it follows that the diffusive transport can be described with three and four flux equations, respectively. The fluxes of $[BMIM]^+$

and $[PF_6]^-$ are not independent, however, because of electroneutrality in each volume of the system. Furthermore, the flux of $[BMIM][PF_6]$ is not independent of the flux of the ions because of the dissociation equilibrium. Thus, the number of independent fluxes is reduced to one, and the system can be described with only one mutual diffusion coefficient. In addition, one has four self-diffusion coefficients – D_s(methanol), $D_s([BMIM]^+)$, $D_s([PF_6]^-)$, and $D_s([BMIM][PF_6])$ – so that five diffusion coefficients are necessary to describe the system completely.

4.4.2
Use of Translational Diffusion Coefficients

Following the general trend of looking for a molecular description of the properties of matter, self-diffusion in liquids has become a key quantity for interpretation and modeling of transport in liquids [5]. Self-diffusion coefficients can be combined with other data, such as viscosities, electrical conductivities, densities, etc., in order to evaluate and improve solvodynamic models such as the Stokes–Einstein type [6–9]. From temperature-dependent measurements, activation energies can be calculated by the Arrhenius or the Vogel–Tamman–Fulcher equation (VTF), in order to evaluate models that treat the diffusion process similarly to diffusion in the solid state with jump or hole models [1, 2, 7].

From the molecular point of view, the self-diffusion coefficient is more important than the mutual diffusion coefficient, because the different self-diffusion coefficients give a more detailed description of the single chemical species than the mutual diffusion coefficient, which characterizes the system with only one coefficient. Owing to its cooperative nature, a theoretical description of mutual diffusion is expected to be more complex than one of self-diffusion [5]. Besides that, self-diffusion measurements are determinable in pure ionic liquids, while mutual diffusion measurements require mixtures of liquids.

From the applications point of view, mutual diffusion is far more important than self-diffusion, because the transport of matter plays a major role in many physical and chemical processes, such as crystallization, distillation or extraction. Knowledge of mutual diffusion coefficients is hence valuable for modeling and scaling-up of these processes.

The need to predict mutual diffusion coefficients from self-diffusion coefficients often arises, and many efforts have been made to understand and predict mutual diffusion data, through approaches such as, for example, the following extension of the Darken equation [5]:

$$D = \left(x_2 D_{21} + x_1 D_{12}\right)\ \Gamma, \text{ with } \Gamma = \frac{\mathrm{d}\ \ln a_1}{\mathrm{d}\ \ln x_1} = \frac{\mathrm{d}\ \ln a_2}{\mathrm{d}\ \ln x_2} \tag{4.4-4}$$

where α_i is the activity of component i. Γ is denoted as the thermodynamic factor.

Systems that are near to ideality can be described satisfactorily with Equation 4.4-4, but the equation does not work very well in systems that are far from thermodynamic ideality, even if the self-diffusion coefficients and activities are known. Since systems with ionic liquids show strong intermolecular forces, there is a need to find better predictions of the mutual diffusion coefficients from self-diffusion coefficients.

Since the prediction of mutual diffusion coefficients from self-diffusion coefficients is not accurate enough to be used for modeling of chemical processes, complete data sets of mutual and self-diffusion coefficients are necessary and valuable.

4.4.3
Experimental Methods

Nowadays, self-diffusion coefficients are almost exclusively measured by NMR methods, through the use of methods such as the 90-δ-180-δ-echo technique (Stejskal and Tanner sequence) [10–12]. The pulse-echo sequence, illustrated in Figure 4.4-2, can be divided into two periods of time τ. After a 90° radio-frequency (RF) pulse the macroscopic magnetization is rotated from the z-axis into the x-y-plane. A gradient pulse of duration δ and magnitude g is applied, so that the spins dephase. After a time τ, a 180° RF pulse reverses the spin precession. A second gradient pulse of equal duration δ and magnitude g follows to tag the spins in the same way. If the spins have not changed their position in the sample, the effects of the two applied gradient pulses compensate each other, and all spins refocus. If the spins have moved due to self-diffusion, the effects of the gradient pulses do not compensate and the echo-amplitude is reduced. The decrease of the amplitude A with the applied gradient is proportional to the movement of the spins and is used to calculate the self-diffusion coefficient.

Popular methods for mutual diffusion measurements in fluid systems are the Taylor dispersion method and interferometric methods, such as Digital Image Holography [13, 14].

With digital image holography it is possible to measure mutual diffusion coefficients in systems that are fairly transparent to laser light and the components of which have a significant difference in their refractive indexes. The main idea of this method is to initiate a diffusion process by creating a so-called step-profile between two mixtures of a binary system with slightly different concentrations. The change

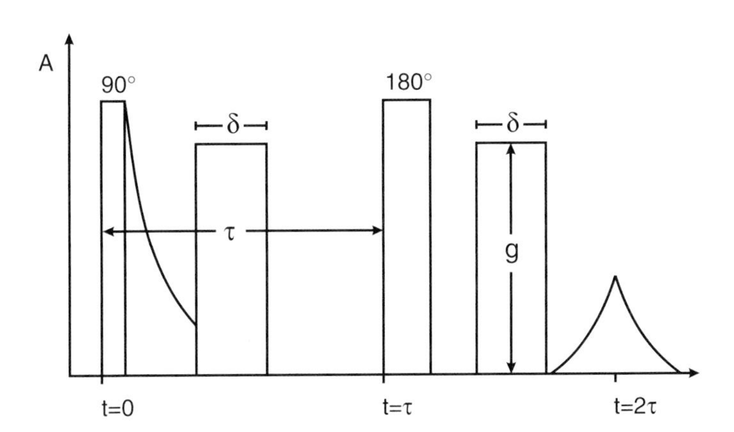

Figure 4.4-2: Pulse-echo sequence in an NMR experiment for the measurement of self-diffusion coefficients.

in the step-profile is associated with the change of the optical phase profile, which can be scanned by a coherent laser beam passing perpendicular to the diffusion axis z through the diffusion cell. The state of the diffusion cell at a certain time is stored as a hologram on a CCD camera. The hologram is processed with holograms taken at different times to produce interference patterns, which indicate the change in the diffusion cell with time. By use of Fick´s second law, the diffusion coefficient can be calculated from a single interference pattern. Mutual diffusion coefficients are accessible over the whole composition range of binary mixtures [15].

With electrochemical methods such as chronoamperometry, cyclovoltammetry (CV), or conductivity measurements, the diffusion coefficients of charged chemical species can be estimated in highly dilute solutions [16, 17].

4.4.4
Results for Ionic Liquids

Typical values of self-diffusion coefficients and mutual diffusion coefficients in aqueous solutions and in molten salt systems such as $(K,Ag)NO_3$ are of the order of 10^{-9} m^2s^{-1}, and the coefficients do not usually vary by more than a factor of 10 over the whole composition range [1, 2, 15]. From measurements in pure ionic liquids we have learned that their self-diffusion coefficients are only of the order of 10^{-11} m^2s^{-1}. From this point of view it is interesting to investigate systems of "ordinary" and ionic liquids. Figure 4.4-3 shows the results of first measurements in the methanol/[BMIM][PF_6] system, which can be seen as a prototype for a system in which an organic and an ionic liquid are mixed.

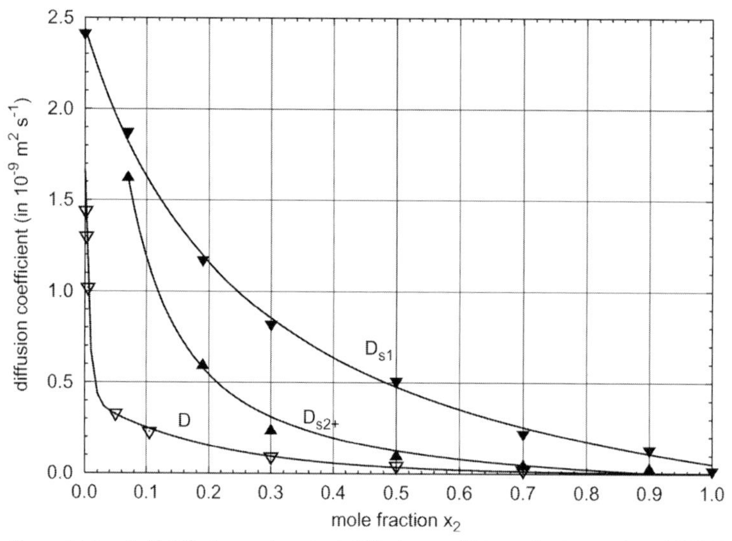

Figure 4.4-3: Self-diffusion and mutual diffusion coefficients in the methanol/[BMIM][PF_6] system. x_2: mole fraction of [BMIM][PF_6], D: mutual diffusion coefficient, D_{s1}: self-diffusion coefficient of methanol, D_{s2+}: self-diffusion coefficient of $[BMIM]^+$.

Self-diffusion coefficients were measured with the NMR spin-echo method and mutual diffusion coefficients by digital image holography. As can be seen from Figure 4.4-3, the diffusion coefficients show the whole bandwidth of diffusion coefficient values, from 10^{-9} m^2s^{-1} on the methanol-rich side, down to 10^{-11} m^2s^{-1} on the [BMIM][PF_6]-rich side. The concentration dependence of the diffusion coefficients on the methanol-rich side is extreme, and shows that special care and attention should be paid in the dimensioning of chemical processes with ionic liquids.

Since this is just the beginning of investigations into the diffusion behavior and intermolecular forces in ionic liquid systems, further experimental work needs to be done both with pure ionic liquids and with systems of mixtures of ionic and organic liquids.

References

1 H.J.V. Tyrrell and K.R. Harris, *Diffusion in Liquids*, Butterworths, London, **1984**.

2 E.L. Cussler, *Diffusion - Mass Transfer in Fluid Systems*, Cambridge University Press, Cambridge, **1984**.

3 R. Haase, *Thermodynamics of irreversible processes*, Dover Publications, Mineola (NY), **1990**.

4 J. Crank, *The Mathematics of Diffusion*, 2nd ed., Clarendon Press, Oxford, **1975**.

5 H. Weingärtner, in: *Diffusion in Condensed Matter* (J. Kärger, P. Heitjans, and R. Haberlandt eds.), Vieweg, Wiesbaden, **1998**.

6 W.R. Carper, G.J. Mains, B.J. Piersma, S.L. Mansfield, and C.K. Larive, *J. Phys. Chem.* **1996**, *100*, 4724.

7 N. Weiden, B. Wittekopf, and K.G. Weil, *Ber. Bunsenges. Phys. Chem.* **1990**, *94*, 353.

8 J.-F. Huang, P.-Y. Chen, I.W. Sun, and S.P. Wang, *Inorganica Chimica Acta* **2001**, *320*, 7.

9 C.K. Larive, M. Lin, B.J. Piersma, and W.R. Carper, *J. Phys. Chem.* **1995**, *99*, 12,409.

10 W.S. Price, *Concepts Magn. Reson.* **1997**, *9*, 299.

11 W.S. Price, *Concepts Magn. Reson.* **1998**, *10*, 197.

12 U. Matenaar, J. Richter, and M.D. Zeidler, *J. Magn. Reson. A* **1996**, *122*, 72.

13 E. Marquardt and J. Richter, *Opt. Eng.* **1998**, *37*, 1514.

14 E. Marquardt, N. Großer, and J. Richter, *Opt. Eng.* **1997**, *36*, 2857.

15 A. Leuchter and J. Richter, *High Temp. Material Processes* **1998**, *2*, 521.

16 C.L. Hussey, I.-W. Sun, S.K.D. Strubinger, and P.A. Barnard, *J. Electrochem. Soc.* **1990**, *137*, 2515.

17 R.A. Osteryoung and M. Lipsztajn, *J. Electrochem. Soc.* **1985**, *132*, 1126.

4.5 Molecular Reorientational Dynamics

Andreas Dölle and W. Robert Carper

4.5.1 Introduction

Models for description of liquids should provide us with an understanding of the dynamic behavior of the molecules, and thus of the routes of chemical reactions in the liquids. While it is often relatively easy to describe the molecular structure and dynamics of the gaseous or the solid state, this is not true for the liquid state. Molecules in liquids can perform vibrations, rotations, and translations. A successful model often used for the description of molecular rotational processes in liquids is the rotational diffusion model, in which it is assumed that the molecules rotate by small angular steps about the molecular rotation axes. One quantity to describe the rotational speed of molecules is the reorientational correlation time τ, which is a measure for the average time elapsed when a molecule has rotated through an angle of the order of 1 radian, or approximately 60°. It is indirectly proportional to the velocity of rotational motion.

4.5.2 Experimental Methods

A particularly important and convenient experimental method with which to obtain information on the reorientational dynamics of molecules is the measurement of longitudinal or spin–lattice relaxation times T_1 of peaks in nuclear magnetic resonance (NMR) spectra [1, 2]. These relaxation times describe how quickly a nuclear spin system reaches thermal equilibrium after disturbance of the system. Longitudinal relaxation is the relaxation process for the magnetization along the z axis, being parallel to the static magnetic field used in NMR spectroscopy. During this relaxation process, energy is exchanged between the spin system and its environment- the lattice. The measurement of ^{13}C relaxation data [3] has great advantages for the study of the reorientational behavior of organic molecules; only one signal is usually obtained for each carbon atom in the molecule, so that the mobility or flexibility of different molecular segments can be studied. Spin diffusion processes, dipolar ^{13}C–^{13}C interactions, and – for ^{13}C nuclei with directly bonded protons – intermolecular interactions can be neglected. The dipolar ^{13}C spin–lattice relaxation rates $1/T_1^{DD}$, which are related to the velocity of the molecular rotational motions (see below), are obtained by measurement of ^{13}C spin–lattice relaxation rates $1/T_1$ and the nuclear Overhauser enhancement (NOE) factors η of the corresponding carbon atoms:

$$\frac{1}{T_1^{DD}} = \frac{\eta}{1.988}\,\frac{1}{T_1}. \qquad (4.5\text{-}1)$$

A simple, but accurate way to determine spin–lattice relaxation rates is the inversion–recovery method [4]. In this experiment, the magnetization is inverted by a 180° radio frequency pulse and relaxes back to thermal equilibrium during a variable delay. The extent to which relaxation is gained by the spin system is observed after a 90° pulse, which converts the longitudinal magnetization into detectable transversal magnetization. The relaxation times for the different peaks in the NMR spectrum can be obtained by means of a routine for determination of the spin–lattice relaxation time, which is usually implemented in the spectrometer software. When the inversion–recovery pulse sequence is applied under ^{1}H broadband decoupling conditions, only one signal is observed for each ^{13}C nucleus and the relaxation is governed by only one time constant $1/T_1$. The NOE factors are obtained by comparing signal intensities I_{dec} from ^{1}H broadband decoupled ^{13}C NMR spectra with those from inverse gated decoupled spectra I_{igdec} with the relationship

$$\eta = \frac{I_{dec}}{I_{igdec}} - 1. \tag{4.5-2}$$

4.5.3 Theoretical Background

Usually, nuclear relaxation data for the study of reorientational motions of molecules and molecular segments are obtained for non-viscous liquids in the extreme narrowing region where the product of the resonance frequency and the reorientational correlation time is much less than unity [1, 3, 5]. The dipolar ^{13}C spin–lattice relaxation rate of ^{13}C nucleus i is then directly proportional to the reorientational correlation time τ_i

$$\left(\frac{1}{T_1^{DD}}\right)_{ij} = n_H (2\pi D_{ij})^2 \tau_p \tag{4.5-3}$$

with the dipolar coupling constant

$$D_{ij} = \frac{\mu_0}{4\pi} \gamma_C \gamma_H \frac{\hbar}{2\pi} r_{ij}^{-3}, \tag{4.5-4}$$

where μ_0 is the magnetic permeability of the vacuum, γ_C and γ_H are the magnetogyric ratios of the ^{13}C and ^{1}H nuclei, respectively, $\hbar = h/2\pi$, with the Planck constant $\hbar$, and r_{ij} is the length of the internuclear vector between ^{13}C nucleus i and interacting proton j. For the relaxation of ^{13}C nuclei with n_H directly bonded protons, only interaction with these protons has to be taken into account.

Ionic liquids, however, are often quite viscous, and the measurements are thus beyond the extreme narrowing region. The relaxation rates hence become frequency-dependent. Under these conditions, the equation for the spin–lattice relaxation rate becomes more complex:

$$\left(\frac{1}{T_1^{DD}}\right)_{ij} = \frac{1}{20} (2\pi D_{ij})^2 [J_i(\omega_C - \omega_H) + 3J_i(\omega_C) + 6J_i(\omega_C + \omega_H)]. \tag{4.5-5}$$

Here, the J_i terms are the spectral densities with the resonance frequencies ω of the ^{13}C and ^{1}H nuclei, respectively. It is now necessary to find an appropriate spectral density to describe the reorientational motions properly (cf. [6, 7]). The simplest spectral density commonly used for interpretation of NMR relaxation data is the one introduced by Bloembergen, Purcell, and Pound [8].

$$J_{\mathrm{BPP},i}(\omega) = \frac{2\tau_{\mathrm{BPP},i}}{1+(\omega\tau_{\mathrm{BPP},i})^2} \tag{4.5-6}$$

Cole and Davidson's continuous distribution of correlation times [9] has found broad application in the interpretation of relaxation data of viscous liquids and glassy solids. The corresponding spectral density is:

$$J_{\mathrm{CD},i}\left(\omega,\tau_{\mathrm{CD},i},\beta_i\right) = \frac{2}{\omega}\frac{\sin\left(\beta_i \arctan\left(\omega\tau_{\mathrm{CD},i}\right)\right)}{\left(1+\left(\omega\tau_{\mathrm{CD},i}\right)^2\right)^{\beta_i/2}}. \tag{4.5-7}$$

Another way to describe deviations from the simple BPP spectral density is the so-called model-free approach of Lipari and Szabo [10]. This takes account of the reduction of the spectral density usually observed in NMR relaxation experiments. Although the model-free approach was first applied mainly to the interpretation of relaxation data of macromolecules, it is now also used for fast internal dynamics of small and middle-sized molecules. For very fast internal motions the spectral density is given by:

$$J_{\mathrm{LS},i}(\omega) = S_i^2 J_i, \tag{4.5-8}$$

which simply means a reduction of the BPP or CD spectral density J_i by the generalized order parameter S^2.

The resonance frequencies of the nuclei are given by the accessible magnetic field strengths through the resonance condition. Since the magnets used for NMR spectroscopy usually have fixed field strengths, the correlation times (that is, the rotational dynamics) have to be varied to leave the extreme narrowing regime. One way to vary the correlation times, and thus the spectral densities and relaxation data, is to change the temperature. The temperature dependence of the correlation times is often given by an Arrhenius equation:

$$\tau_i = \tau_{\mathrm{A},i}\exp(E_{\mathrm{A},i}/RT), \tag{4.5-9}$$

with the gas constant R and the activation energy E_A, interpreted below as a fit parameter representing a measure of the hindrance of the corresponding reorientational process.

4.5.4
Results for Ionic Liquids

The measurement of correlation times in molten salts and ionic liquids has recently been reviewed [11] (for more recent references refer to Carper et al. [12]). We have measured the ^{13}C spin–lattice relaxation rates $1/T_1$ and nuclear Overhauser factors η in temperature ranges in and outside the extreme narrowing region for the neat ionic liquid [BMIM][PF_6], in order to observe the temperature dependence of the spectral density. Subsequently, the models for the description of the reorientational dynamics introduced in the theoretical section (Section 4.5.3) were fitted to the experimental relaxation data. The ^{13}C nuclei of the aliphatic chains can be assumed to relax only through the dipolar mechanism. This is in contrast to the aromatic ^{13}C nuclei, which can also relax to some extent through the chemical-shift anisotropy mechanism. The latter mechanism has to be taken into account to fit the models to the experimental relaxation data (cf. [1] or [3] for more details). Preliminary results are shown in Figures 4.5-1 and 4.5-2, together with the curves for the fitted functions.

Table 4.5-1 gives values for the fit parameters and the reorientational correlation times calculated from the dipolar relaxation rates.

The largest correlation times, and thus the slowest reorientational motion, were shown by the three ^{13}C-^{1}H vectors of the aromatic ring, with values of between approximately 60 and 70 ps at 357 K, values expected for viscous liquids like ionic liquids. The activation energies are also in the typical range for viscous liquids. As can be seen from Table 4.5-1, the best fit was obtained for a combination of the Cole–Davidson with the Lipari–Szabo spectral density, with a distribution parame-

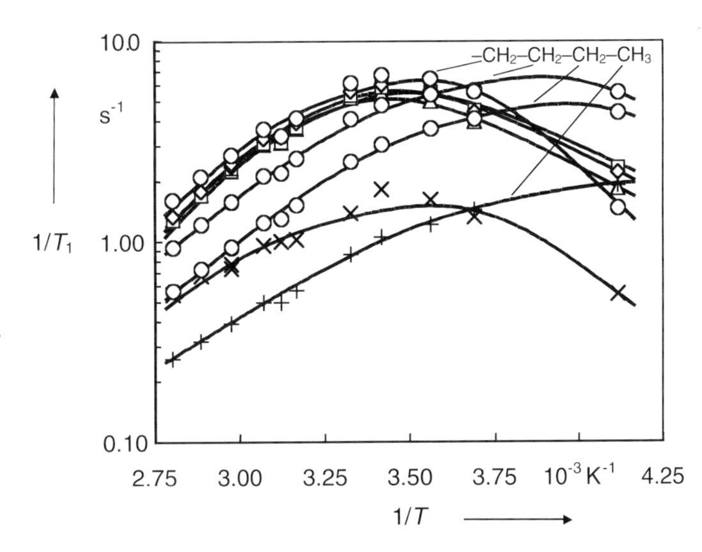

Figure 4.5-1: ^{13}C relaxation rates $1/T_1$ of [BMIM][PF_6] in the neat liquid as a function of reciprocal temperature T (Δ: C2, □ and ◇: C4 and C5, X: CH_3(ring), +: CH_3(butyl group), ○: CH_2, lines: functions calculated with the fitted parameters).

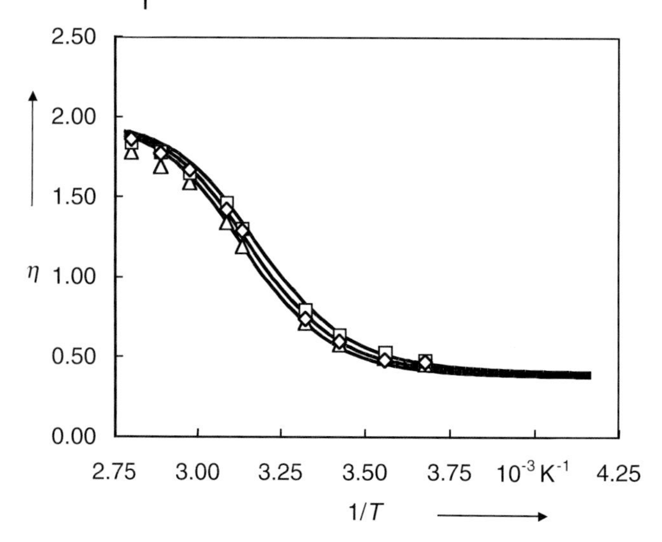

Figure 4.5-2: {^{1}H}-^{13}C NOE factors η for [BMIM][PF_6] in the neat liquid as a function of reciprocal temperature T (Δ: C2, □ and ◇: C4 and C5, lines: functions calculated with the fitted parameters).

ter β of about 0.45. Cole–Davidson spectral densities are often found for glass-forming liquids. The ring can be taken as the rigid part of the molecule without internal flexibility, although a generalized order parameter S^2 of less than unity was observed. The value of approximately 0.70 for S^2 is explained by very fast motions such as librations in the cage of the surrounding molecules and vibrations. The reorientational correlation times of the aliphatic ^{13}C nuclei are smaller than those of their aromatic ring counterparts, indicating the internal motion of the corresponding molecular segments. The flexibility in the butyl chain increases from the methylene group bound to the rigid and slowly moving imidazolium ring to the methyl group at the end. The other methyl group also exhibits fast motion compared to the rigid aromatic ring. The correlation times of the aliphatic carbons relative to those of the ring and their graduation in the chain are similar to those of alkyl chains in hydrocarbons of comparable size [13]. The experimental ^{13}C spin–lattice relaxation rates for the aliphatic carbons could be fitted by a combination of the Lipari–Szabo with the BPP spectral density. The activation energies and the generalized order parameters decrease from the methylene group bound at the ring to the

Table 4.5-1: Reorientational correlation times τ at 357 K and fit parameters activation energy E_A, Cole–Davidson distribution parameter β, and generalized order parameter S^2.

	C2	C4/C5		CH3 (ring)	CH2 (ring)	**CH2**–CH2–CH2	CH2–**CH3**	CH2–**CH3**
τ_i(357 K) (ps)	63	65	68	1.0	46	26	16	4.7
$E_{A,i}$ (kJ·mol^{-1})	38	37	38	27	32	26	26	20
β_i	0.46	0.43	0.44					
S_i^2	0.65	0.73	0.73	0.059	0.37	0.38	0.28	0.075

methyl group at the end of the chain, this again being an indication of the increasing flexibility. The methyl groups had the smallest S^2 value, approximately one tenth of the value for the rigid part of the molecule, which is the typical value for fast methyl group rotation.

The highly detailed results obtained for the neat ionic liquid [BMIM][PF_6] clearly demonstrate the potential of this method for determination of molecular reorientational dynamics in ionic liquids. Further studies should combine the results for the reorientational dynamics with viscosity data in order to compare experimental correlation times with correlation times calculated from hydrodynamic models (cf. [14]). It should thus be possible to draw conclusions about the intermolecular structure and interactions in ionic liquids and about the molecular basis of specific properties of ionic liquids.

References

1 T. C. Farrar, E. D. Becker, *Pulse and Fourier Transform NMR. Introduction to Theory and Methods*, Academic Press, New York, **1971**.
2 M. L. Martin, J.-J. Delpuech, G. J. Martin, *Practical NMR Spectroscopy*, Heyden, London, **1980**.
3 J. R. Lyerla, G. C. Levy, *Top. Carbon-13 NMR Spectrosc.* **1972**, *1*, 79.
4 R. L. Vold, J. S. Waugh, M. P. Klein, D. E. Phelps, *J. Chem. Phys.* **1968**, *48*, 3831.
5 A. Abragam: *The Principles of Nuclear Magnetism*. Oxford University Press, Oxford **1989**.
6 P. A. Beckmann. *Phys. Rep.* **1988**, *171*, 85.
7 A. Dölle, to be published.
8 N. Bloembergen, E. M. Purcell, R. V. Pound, *Phys. Rev.* **1948**, *73*, 679.
9 D. W. Davidson, R. H. Cole, *J. Chem. Phys.* **1951**, *19*, 1484.
10 G. Lipari, A. Szabo, *J. Am. Chem. Soc.* **1982**, *104*, 4546.
11 W. R. Carper, *Molten Salts*, in *Encyclopedia of Nuclear Magnetic Resonance* (D. M. Grant, R. K. Harris eds.), John Wiley & Sons, New York, **1995**.
12 C. E. Keller, B. J. Piersma, G. J. Mains, W. R. Carper, *Inorg. Chem.* **1994**, *33*, 5601; C. E. Keller, W. R. Carper, *J. Phys. Chem.* **1994**, *98*, 6865; C. E. Keller, B. J. Piersma, G. J. Mains, W. R. Carper, *Inorg. Chim Acta* **1995**, *230*, 185; C. E. Keller, B. J. Piersma, W. R. Carper, *J. Phys. Chem.* **1995**, *99*, 12998; C. E. Keller, W. R. Carper, *Inorg. Chim. Acta*, **1995**, *238*, 115; C. K. Larive, M. Lin, B. J. Piersma, W. R. Carper, *J. Phys. Chem.* **1995**, *99*, 12409; W. R. Carper, G. J. Mains, B. J. Piersma, S. L. Mansfield, C. K. Larive, *J. Phys. Chem.* **1996**, *100*, 4724; C. K. Larive, M. Lin, B. S. Kinnear, B. J. Piersma, C. E. Keller, W. R. Carper, *J. Phys. Chem. B* **1998**, *102*, 1717.
13 P. Gruhlke, A. Dölle, *J. Chem. Soc., Perkin Trans. 2* **1998**, 2159.
14 R. P. Klüner, A. Dölle, *J. Phys. Chem. A* **1997**, *101*, 1657.

5
Organic Synthesis

Martyn Earle, Alain Forestièr, Hélène Olivier-Bourbigou, and Peter Wasserscheid

5.1
Stoichiometric Organic Reactions and Acid-Catalyzed Reactions in Ionic Liquids

Martyn Earle

The field of reaction chemistry in ionic liquids was initially confined to the use of chloroaluminate(III) ionic liquids. With the development of "neutral" ionic liquids in the mid-1990s, the range of reactions that can be performed has expanded rapidly. In this chapter, reactions in both chloroaluminate(III) ionic liquids and in similar Lewis acidic media are described. In addition, stoichiometric reactions, mostly in neutral ionic liquids, are discussed. Review articles by several authors are available, including: Welton [1] (reaction chemistry in ionic liquids), Holbrey [2] (properties and phase behavior), Earle [3] (reaction chemistry in ionic liquids), Pagni [4] (reaction chemistry in molten salts), Rooney [5] (physical properties of ionic liquids), Seddon [6, 7] (chloroaluminate(III) ionic liquids and industrial applications), Wasserscheid [8] (catalysis in ionic liquids), Dupont [9] (catalysis in ionic liquids) and Sheldon [10] (catalysis in ionic liquids).

Ionic liquids have been described as "designer solvents" [11]. Properties such as solubility, density, refractive index, and viscosity can be adjusted to suit requirements simply by making changes to the structure of either the anion, or the cation, or both [12, 13]. This degree of control can be of substantial benefit when carrying out solvent extractions or product separations, as the relative solubilities of the ionic and extraction phases can be adjusted to assist with the separation [14]. Also, separation of the products can be achieved by other means such as, distillation (usually under vacuum), steam distillation, and supercritical fluid extraction (CO_2).

To many chemists it may seem daunting to perform reactions in ionic liquids, and the range of ionic liquids or potential ionic liquids available is very large. However, many scientists have found that performing reactions in ionic liquids is straightforward and practical when compared with similar reactions in conventional organic solvents. This is particularly the case when considering reactions normally carried out in noxious and difficult to remove solvents such as dipolar aprotic solvents like dimethyl sulfoxide.

With the growing interest in ionic liquids, reactions were initially performed in various chloroaluminate(III) ionic liquids. Their strong solvating ability was an advantage, but their sensitivity to moisture and strong interactions with certain commonly occurring functional groups limited the scope of reactions in these media. With the discovery of water-stable "neutral" ionic liquids, the range and scope of reactions that can be performed has grown to include most classes of reactions covered in organic chemistry textbooks [15], and the vast majority of reactions in ionic liquids are now carried out in these water-stable variants.

5.1.1 Stoichiometric Organic Reactions

Stoichiometric – or, more simply, non-catalytic – reactions are an important and rapidly expanding area of research in ionic liquids. This section deals with reactions that consume the ionic liquid (or molten salt) or use the ionic liquid as a solvent.

5.1.1.1 Molten salts as reagents

Molten salts have been used for many years, in the form of reagents such as fused KOH, pyridinium chloride, and tetrabutylammonium fluoride (TBAF) [4]. One of the earliest molten salts to be used in synthesis is KOH, with examples dating from 1840 [16]. One common use is in the reaction of fused KOH with arenesulfonic acids to produce phenols. Although KOH has a high melting point (410 °C), impurities such as traces of water or carbonates bring the melting point down. An example is given in Scheme 5.1-1 [17, 18].

A number of examples of the use of molten pyridinium chloride (mp 144 °C) in chemical synthesis are known, dating back to the 1940's. Pyridinium chloride can act both as an acid and as a nucleophilic source of chloride. These properties are exploited in the dealkylation reactions of aromatic ethers [4]. An example involving the reaction of 2-methoxynaphthalene is given in Scheme 5.1-2 [16, 18], and a mechanistic explanation in Scheme 5.1-3 [18].

Pyridinium chloride ([PyH]Cl) has also been used in a number of cyclization reactions of aryl ethers (Scheme 5.1-4) [4, 18]. Presumably the reaction initially proceeds by dealkylation of the methyl ether groups to produce the corresponding phenol. The mechanism of the cyclization is not well understood, but Pagni and Smith have suggested that it proceeds by nucleophilic attack of an Ar-OH or Ar-O^- group on the second aromatic ring (in a protonated form) [4].

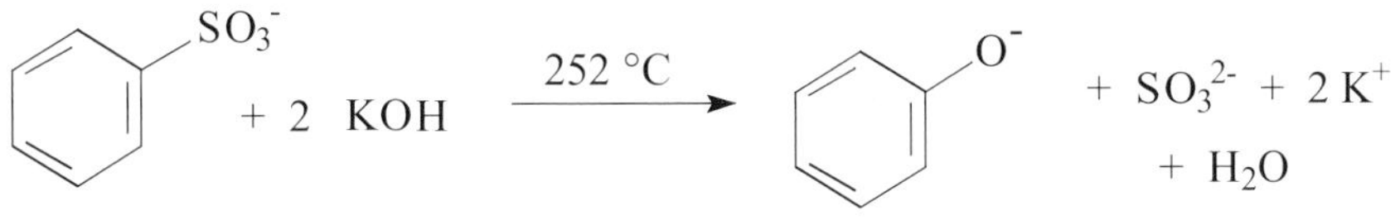

SO_3^- + 2 KOH —252 °C→ O^- + SO_3^{2-} + 2 K^+ + H_2O

Scheme 5.1-1: The reaction of benzenesulfonates with fused KOH.

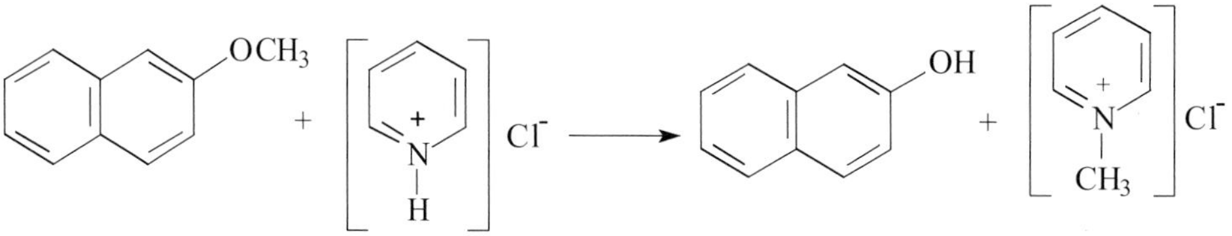

Scheme 5.1-2: The demethylation of 2-methoxynaphthalene to 2-naphthol with pyridinium chloride.

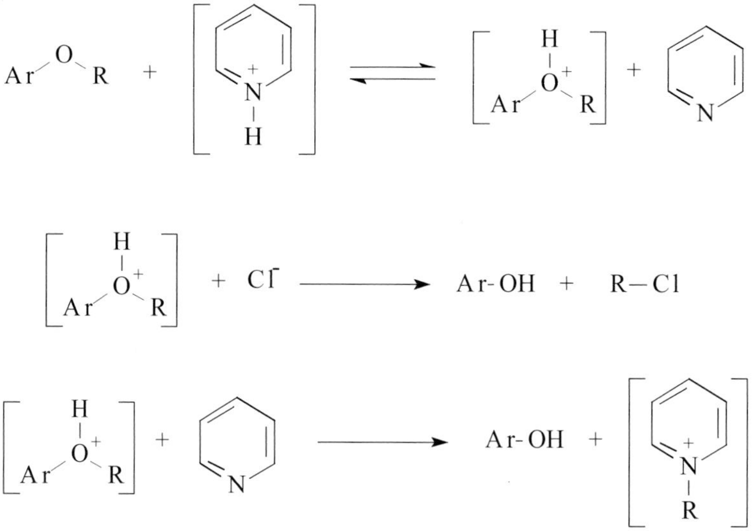

Scheme 5.1-3: A mechanism for the dealkylation of aryl ethers with pyridinium chloride.

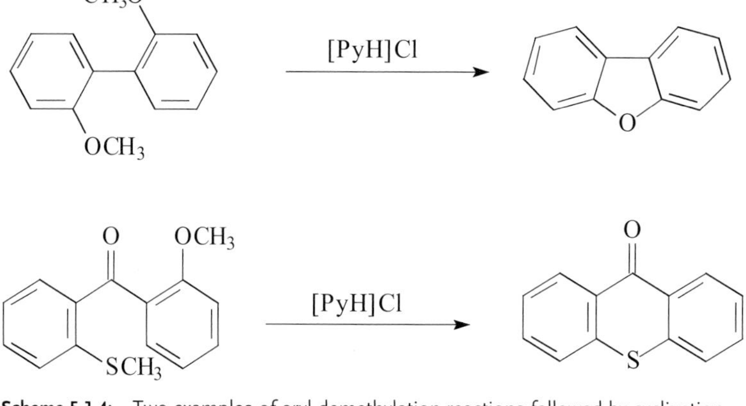

Scheme 5.1-4: Two examples of aryl demethylation reactions followed by cyclization.

Tetrabutylammonium fluoride (TBAF) is usually used in the form of the trihydrate or as a solution in tetrahydrofuran (THF). The pure form is difficult to isolate, owing to decomposition to HF, tributylamine, and but-1-ene [18, 19] on dehydration. It has been used for a variety of reactions, including as a catalyst for various reactions with silicon compounds [20, 21]. One of its main uses is in the cleavage of silyl ether protecting groups [22].

TBAF has been used as a source of fluoride ions in a number of substitution reactions studied by Cox et al. [23]. Alkyl and acyl halides react with TBAF to give the corresponding alkyl or acyl fluoride in good yield. In the reaction between (*R*)-2-tosyloctane and TBAF, the product was (*S*)-2-fluorooctane, confirming an S_N2 mechanism for the reaction (Scheme 5.1-5) [18, 23].

TBAF has also been used in the preparation of various fluorocarbenes. This involved the photolysis of phenyl- or phenoxyfluorodiazirine, which was in turn synthesized from the reaction between TBAF and phenyl or phenoxy halodiazirine, as shown in Scheme 5.1-6 [24, 25].

5.1.1.2 Reactions in chloroaluminate(III) and related ionic liquids*

Reactions in chloroaluminate(III) salts and other related binary salts often proceed smoothly to give products. However, it should be noted that these salts are water-sensitive and must be handled under dry conditions. They react with water to give hydrated aluminium(III) ionic species and HCl. When a reactant or product contains a heteroatomic functional group, such as a ketone, a strong ketone/aluminium(III) chloride adduct is formed. In these cases, this adduct can be difficult to separate from the ionic liquid at the end of a reaction. The isolation of the product often

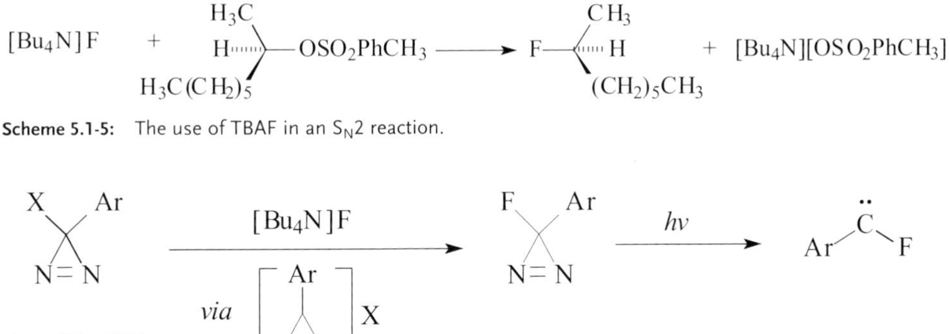

$[Bu_4N]F$ + H_3C, H, $H_3C(CH_2)_5$ C–OSO_2PhCH_3 → F–C(CH_3, H, $(CH_2)_5CH_3$) + $[Bu_4N][OSO_2PhCH_3]$

Scheme 5.1-5: The use of TBAF in an S_N2 reaction.

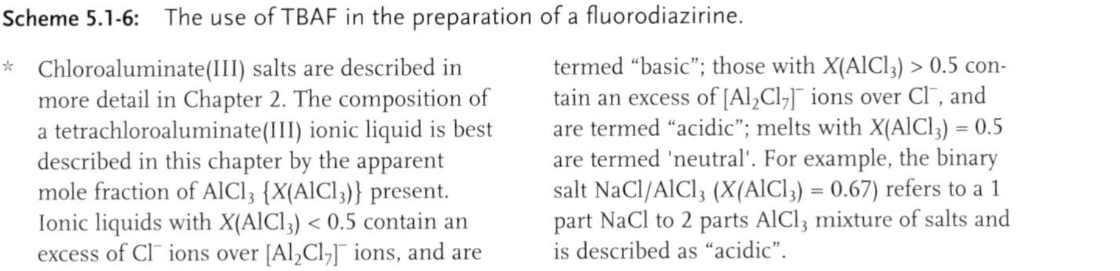

X, Ar diazirine (N=N) —$[Bu_4N]F$→ F, Ar diazirine (N=N) —*hv*→ Ar–C̈–F

via [Ar diazirinium (N=N)] X

Ar = Ph, OPh
X = Cl, Br

Scheme 5.1-6: The use of TBAF in the preparation of a fluorodiazirine.

* Chloroaluminate(III) salts are described in more detail in Chapter 2. The composition of a tetrachloroaluminate(III) ionic liquid is best described in this chapter by the apparent mole fraction of $AlCl_3$ $\{X(AlCl_3)\}$ present. Ionic liquids with $X(AlCl_3) < 0.5$ contain an excess of Cl^- ions over $[Al_2Cl_7]^-$ ions, and are termed "basic"; those with $X(AlCl_3) > 0.5$ contain an excess of $[Al_2Cl_7]^-$ ions over Cl^-, and are termed "acidic"; melts with $X(AlCl_3) = 0.5$ are termed 'neutral'. For example, the binary salt $NaCl/AlCl_3$ $(X(AlCl_3) = 0.67)$ refers to a 1 part NaCl to 2 parts $AlCl_3$ mixture of salts and is described as "acidic".

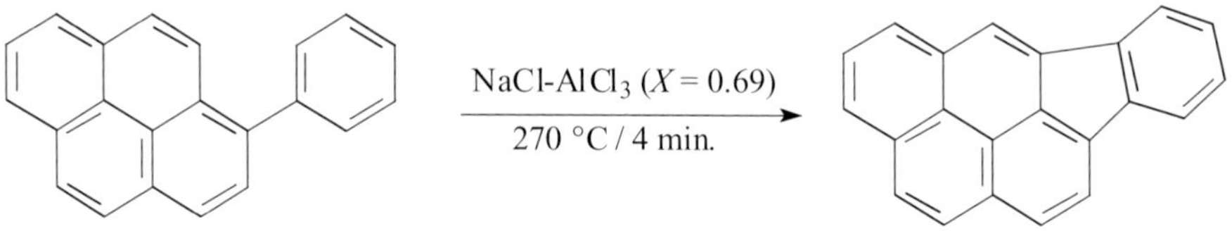

NaCl-AlCl$_3$ (X = 0.69)
270 °C / 4 min.

Scheme 5.1-7: The Scholl reaction of 1-phenylpyrene.

involves destruction of the ionic liquid with water. For products that do not have polar electron-donating functional groups, isolation of the products is straightforward and the ionic liquid can be reused.

One of the first reactions to be carried out in a molten salt (albeit at 270 °C) was the Scholl reaction. This involves the inter- or intramolecular coupling of two aromatic rings. A example of this reaction, in which 1-phenylpyrene was cyclized to indeno[1,2,3-cd]pyrene [26] is given in Scheme 5.1-7. A more elaborate version of the Scholl reaction is shown in Scheme 5.1-8 and involves bicyclization of an aromatic cumulene [27].

Wynberg et al. found that the yields in the cyclization of helicines could be improved from 10 % in an aluminium(III) chloride solution in benzene system to 95 % in a NaCl/AlCl$_3$ (X(AlCl$_3$) = 0.69) molten salt [28]. An example is given in Scheme 5.1-9.

The Scholl reaction involves an overall oxidation of the coupled aromatic rings, yet there is no obvious oxidizing agent. This poses the question of what happens to the two hydrogen atoms that are produced in this reaction. It has been suggested that oxygen (air) may act as the oxidant, but this currently lacks confirmation [18].

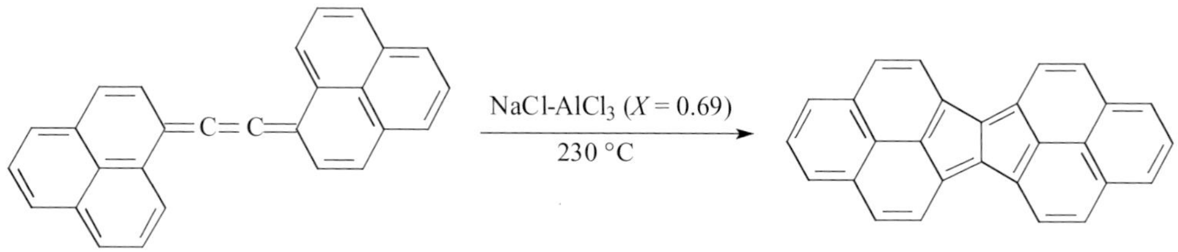

=C=C=
NaCl-AlCl$_3$ (X = 0.69)
230 °C

Scheme 5.1-8: The cyclisation of an aromatic cumulene in a molten salt.

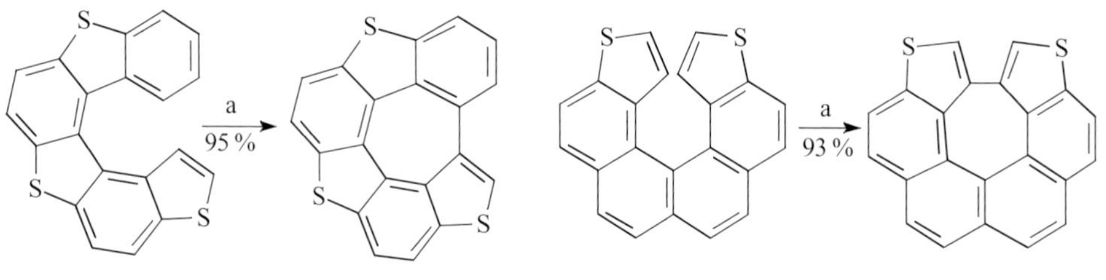

S S S
a
95 %
S S S
S S
a
93 %
S S

Scheme 5.1-9: The Scholl reactions of two helicines. (a = NaCl/AlCl$_3$ (X(AlCl$_3$) = 0.69) at 140 °C).

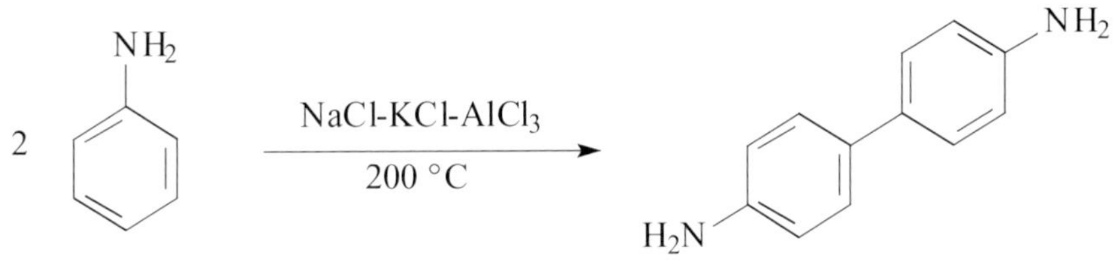

$$2\ \text{aniline} \xrightarrow[200\ ^\circ C]{NaCl\text{-}KCl\text{-}AlCl_3} \text{benzidine}$$

Scheme 5.1-10: The dimerization of aniline to benzidine in a molten salt.

The molten salt $NaCl/KCl/AlCl_3$ (20:20:60) was used in the dimerization of aniline to form benzidine (Scheme 5.1-10) [29].

Buchanan and co-workers studied the behavior of various aromatic compounds in antimony(III) molten salts [30]. These salts can act both as mild Lewis acids and allow redox reactions to take place. The Lewis acidity of the melt can be tuned by controlling the concentration of $[SbCl_2]^+$. Basic melts are formed by addition of a few mol % of a chloride donor such as KCl, whereas acidic melts are formed by addition of chloride acceptors such as $AlCl_3$ (Scheme 5.1-11).

Examples of reactions that have been carried out in these antimony(III) ionic liquids include the cyclizations of 1,2-bis-(9-anthryl)-ethane (Scheme 5.1-12) and 1,2-bis-(1-naphthyl)-ethane (Scheme 5.1-13). A more detailed review of antimony(III) chloride molten salt chemistry has been published by Pagni [4].

Polycyclic aromatic hydrocarbons dissolve in chloroaluminate(III) ionic liquids to give brightly colored solutions (due to the protonated aromatic compound [31]). The

$$SbCl_3 \rightleftharpoons [SbCl_2]^+ + Cl^-$$

$$SbCl_3 + AlCl_3 \longrightarrow [SbCl_2]^+ + [AlCl_4]^- \quad \text{acidic}$$

Scheme 5.1-11: The effect of the addition of aluminium(III) chloride to antimony(III) chloride.

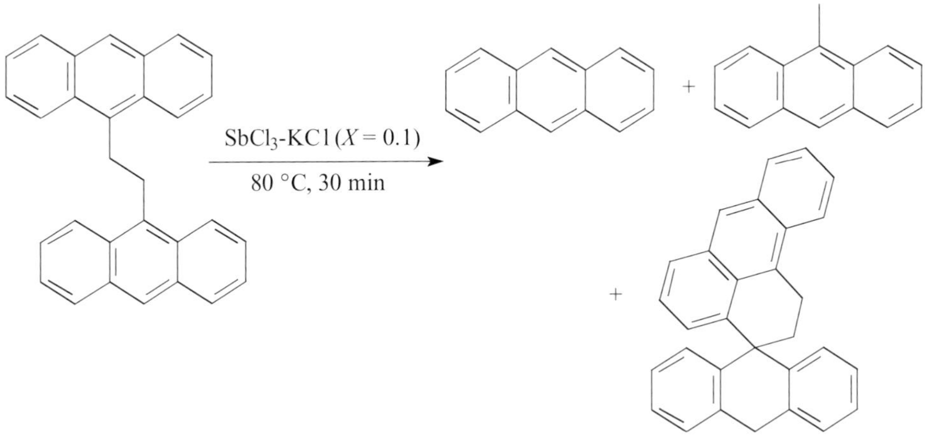

$$\text{1,2-bis-(9-anthryl)-ethane} \xrightarrow[80\ ^\circ C,\ 30\ min]{SbCl_3\text{-}KCl\ (X = 0.1)} \text{products}$$

Scheme 5.1-12: The cyclisation of 1,2-bis-(9-anthryl)-ethane in antimony(III) ionic liquids.

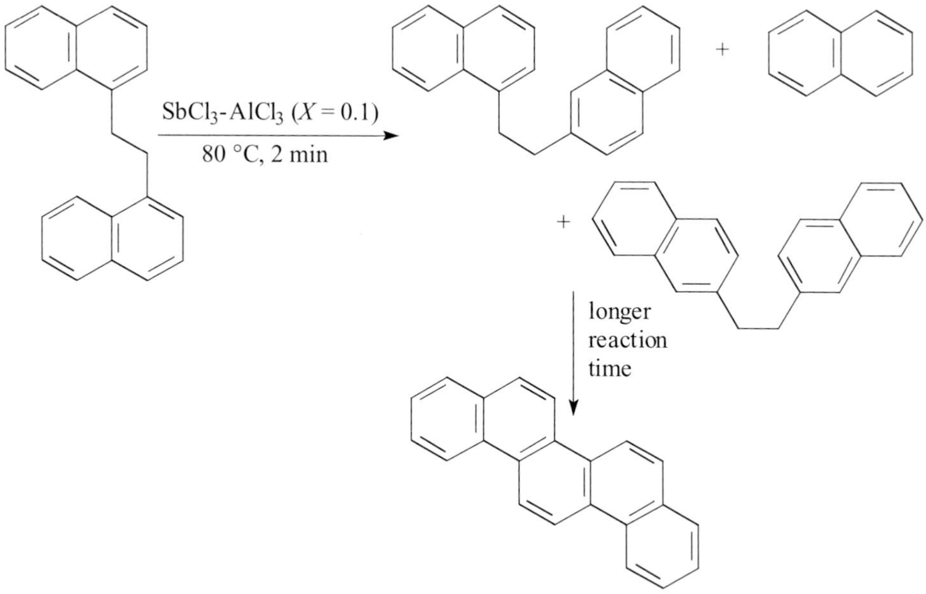

Scheme 5.1-13: Reactions of bisnaphthylethane in antimony(III) ionic liquids.

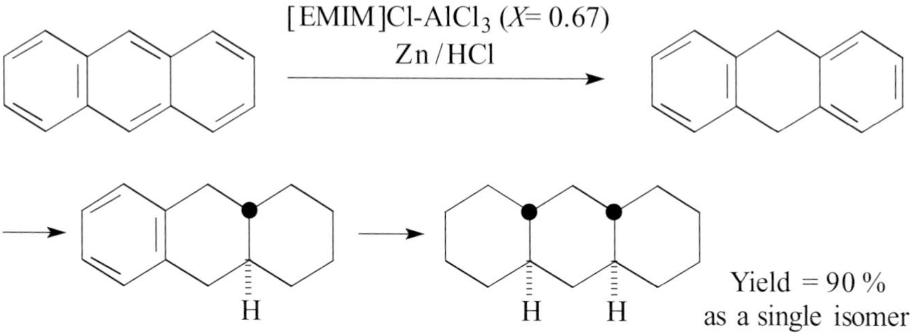

Scheme 5.1-14: The reduction of anthracene to perhydroanthracene.

addition of a reducing agent (such as an electropositive metal and a proton source) results in the selective hydrogenation of the aromatic compound. For example, pyrene and anthracene can be reduced to perhydropyrene and perhydroanthracene at ambient temperatures and pressures (Scheme 5.1-14). Interestingly, only the thermodynamically most stable isomer of the product is obtained [32]. This contrasts with catalytic hydrogenation reactions, which require high temperatures and pressures and expensive platinum oxide catalysts and give rise to isomeric mixtures of products.

Singer and co-workers have shown that benzoyl chloride reacts with ethers to give alkyl benzoates [33] in chloroaluminate(III) ionic liquids. This reaction results in

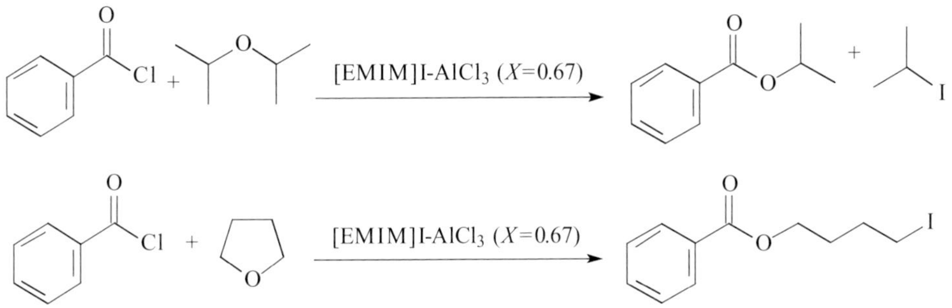

[EMIM]I-$AlCl_3$ (X=0.67)

[EMIM]I-$AlCl_3$ (X=0.67)

Scheme 5.1-15: The acylative cleavage of ethers in an ionic liquid.

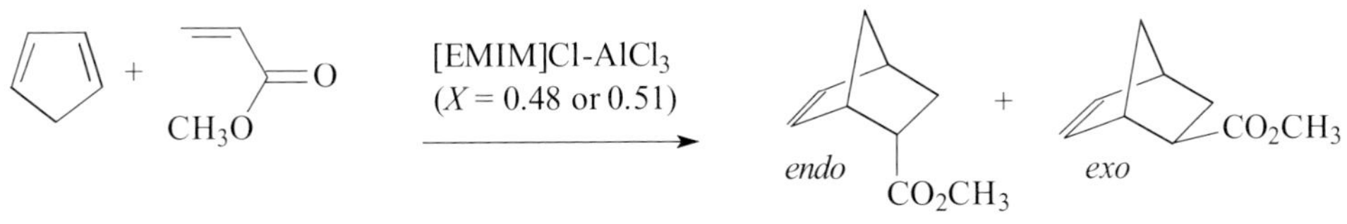

[EMIM]Cl-$AlCl_3$ (X = 0.48 or 0.51)

CH_3O *endo* CO_2CH_3 *exo* CO_2CH_3

Scheme 5.1-16: The Diels-Alder reaction in a chloroaluminate(III) ionic liquid.

the acylative cleavage of ethers, and a number of reactions with cyclic and acyclic ethers have been investigated in the ionic liquid [EMIM]I/$AlCl_3$ ($X(AlCl_3)$ = 0.67). Two examples are shown in Scheme 5.1-15.

Esterification reactions can be catalyzed by the ionic liquid 1-butylpyridinium chloride-aluminium chloride ([BP]Cl/$AlCl_3$ ($X(AlCl_3)$ = 0.33) [34, 35]. Deng and co-workers found that higher yields were obtained than in similar reactions with a sulfuric acid catalyst.

Lee has used chloroaluminate(III) ionic liquids in the Diels–Alder reaction [36]. The *endo*:*exo* ratio rose from 5.25 to 19 on changing the composition of the ionic liquid from $X(AlCl_3)$ = 0.48 to $X(AlCl_3)$ = 0.51 (Scheme 5.1-16). The reaction works well, giving up to 95 % yield, but the moisture-sensitivity of these systems is a major disadvantage, the products being recovered by quenching the ionic liquid in water.

5.1.1.3 Reactions in neutral ionic liquids

Chloroaluminate(III) ionic liquids are excellent media in many processes, but suffer from several disadvantages, such as their moisture-sensitivity and the difficulties in separation of products containing heteroatoms. Furthermore, these ionic liquids often have to be quenched (usually in water) at the end of a chemical reaction, and are lost in the form of acidic aqueous waste. Research is, therefore, shifting to the investigation of ionic liquids that are more stable to water. This allows for straightforward product separation and ease of handling. In particular, a number of ionic liquids have been found to be hydrophobic (immiscible with water), but readily dissolve many organic molecules (with the exception of alkanes, some ethers, and alkylated aromatic compounds such as toluene). An example of this is the ionic liquid [BMIM][PF_6] [37], which forms triphasic solutions with alkanes and water [38]. This multiphasic behavior has important implications for clean synthesis and is analogous to the use of fluorous phases in some chemical processes [39]. For example, a reaction can be performed in the ionic liquid, the products separated by distillation or steam stripping, and a by-product extracted with water or an organic solvent.

Diels–Alder reactions Neutral ionic liquids have been found to be excellent solvents for the Diels–Alder reaction. The first example of a Diels–Alder reaction in an ionic liquid was the reaction of methyl acrylate with cyclopentadiene in $[EtNH_3][NO_3]$ [40], in which significant rate enhancement was observed. Howarth et al. investigated the role of chiral imidazolium chloride and trifluoroacetate salts (dissolved in dichloromethane) in the Diels–Alder reactions between cyclopentadiene and either crotonaldehyde or methacroline [41]. It should be noted that this paper describes one of the first examples of a chiral cationic ionic liquid being used in synthesis (Scheme 5.1-17). The enantioselectivity was found to be < 5 % in this reaction for both the *endo* (10 %) and the *exo* (90 %) isomers.

A study of the Diels–Alder reaction was carried out by Earle et al. [42]. The rates and selectivities of reactions between ethyl acrylate (EA) and cyclopentadiene (CP) in water, 5 M lithium perchlorate in diethyl ether (5 M LPDE), and $[BMIM][PF_6]$ were compared. The reactions in the ionic liquid $[BMIM][PF_6]$ were marginally faster than in water, but both were slower than in 5 M LPDE [42, 43] (see Table 5.1-1 and Scheme 5.1-18). It should be noted that these three reactions give up to 98 % yields if left for 24 hours. The *endo*:*exo* selectivity in $[BMIM][PF_6]$ was similar to that in 5 M LPDE, and considerably greater than that in water (Table 5.1-1).

In the reaction between isoprene (IP) and methyl vinyl ketone (MVK), the selectivities between the two isomers produced in this reaction can be improved from 4:1 to 20:1 by the addition of a mild Lewis acid such as zinc(II) iodide (5 mol %) to the ionic liquid $[BMIM][PF_6]$ (Scheme 5.1-18). One of the key benefits of this is that the

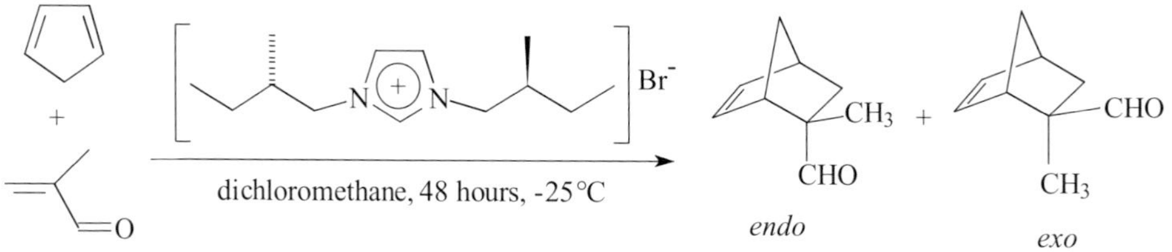

Br^-

dichloromethane, 48 hours, -25°C

CH_3 + CHO

CHO CH_3

endo *exo*

Scheme 5.1-17: Use of a chiral ionic liquid in a Diels-Alder reaction.

Table 5.1-1: Diels-Alder reactions in various solvents.

Solvent	*Diene*	*Dienophile*	*Product*	*Time*	*Yield*	*a:b ratio*
$[BMIM][PF_6]$	CP	EA	**1a** + **1b**	1	36	8.0
5 M LPDE	CP	EA	**1a** + **1b**	1	61	8.0
Water	CP	EA	**1a** + **1b**	1	30	3.5
$[BMIM][PF_6]$[a]	IP	MVK	**2a** + **2b**	6	98	20
$[BMIM][PF_6]$	IP	MVK	**2a** + **2b**	18	11	4

[a] 5 mol % ZnI_2 added, IP = isoprene.

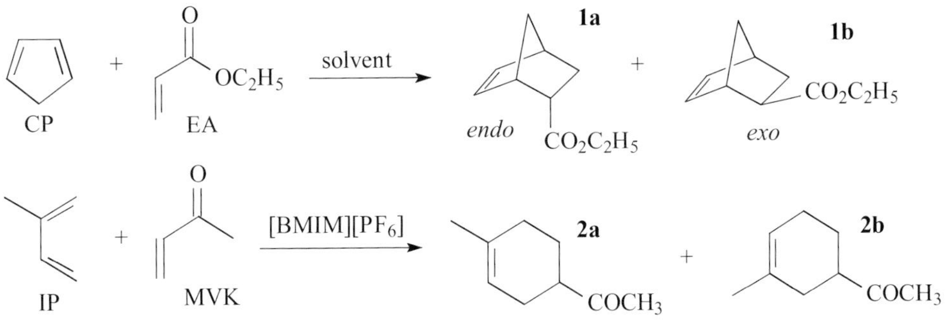

Scheme 5.1-18: The Diels-Alder reaction in different solvents (results are given in Table 5.1-1).

ionic liquid and catalyst can be recycled and reused after solvent extraction or direct distillation of the product from the ionic liquid. The reaction was also carried out in the chiral ionic liquid [BMIM][lactate] (Figure 5.1-1). This was found to give the fastest reaction rates of all the ionic liquids tested, and also the lowest *endo:exo* selectivity. The products of the Diels–Alder reaction were found to be racemic and no chiral induction was observed [42].

A similar study performed by Welton and co-workers studied the rate and selectivities of the Diels–Alder reaction between cyclopentadiene and methyl acrylate in a number of neutral ionic liquids [44]. It was found that *endo:exo* ratios decreased slightly as the reaction proceeded, and were dependent on reagent concentration and ionic liquid type. Subsequently, they went on to demonstrate that the ionic liquids controlled the *endo:exo* ratios through a hydrogen bond (Lewis acid) interaction with the electron-withdrawing group of the dienophile.

The use of molten salts based on phosphonium tosylates has also been reported for Diels–Alder reactions [45]. These salts have higher melting points than most ionic liquids in common use, and so the reactions were performed in a sealed tube. The authors claim very high selectivities in the reactions between isoprene and MVK or methyl acrylate. A new class of room-temperature ionic liquids based on phosphonium salts has been described, and has also been used for a number of Diels–Alder reactions [5]. Kitazume and Zulfiqar have investigated the aza-Diels–Alder reaction in 1-ethyl-1,8-diazabicyclo[5,4,0]undec-7-enium trifluoromethanesulfonate [EDBU][OTf] [46] (Figure 5.1-2). This reaction involved the scandium(III) trifluoromethanesulfonate-catalyzed reaction between an imine (usually generated in situ from an aldehyde and an amine) and a diene. An example of this reaction is given in Scheme 5.1-19. The yields in this reaction were high (80–99 %) and it was found that the ionic liquid could be recycled and reused.

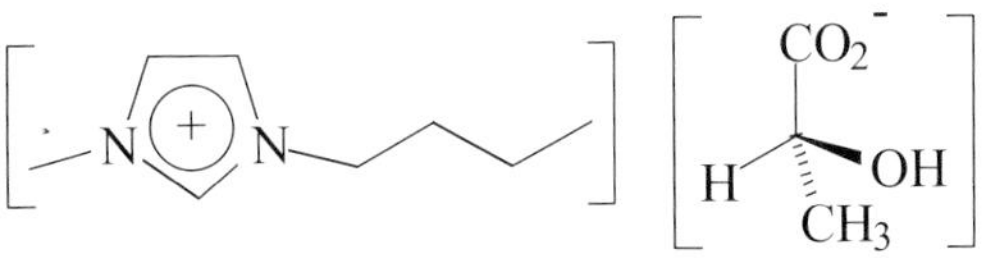

Figure 5.1-1: An example of a chiral ionic liquid used in the Diels–Alder reaction.

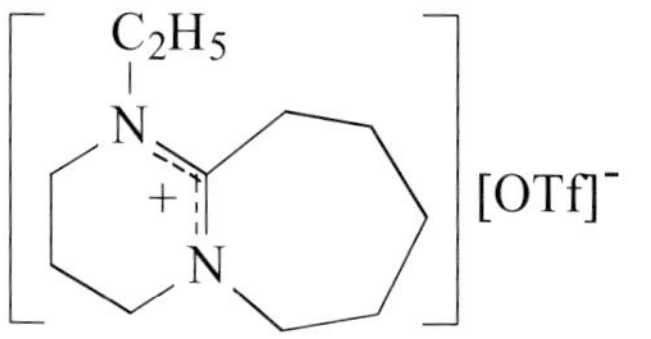

Figure 5.1-2: The structure of 1-ethyl-1,8-diazabicyclo[5,4,0]undec-7-enium trifluoromethanesulfonate [EDBU][OTf].

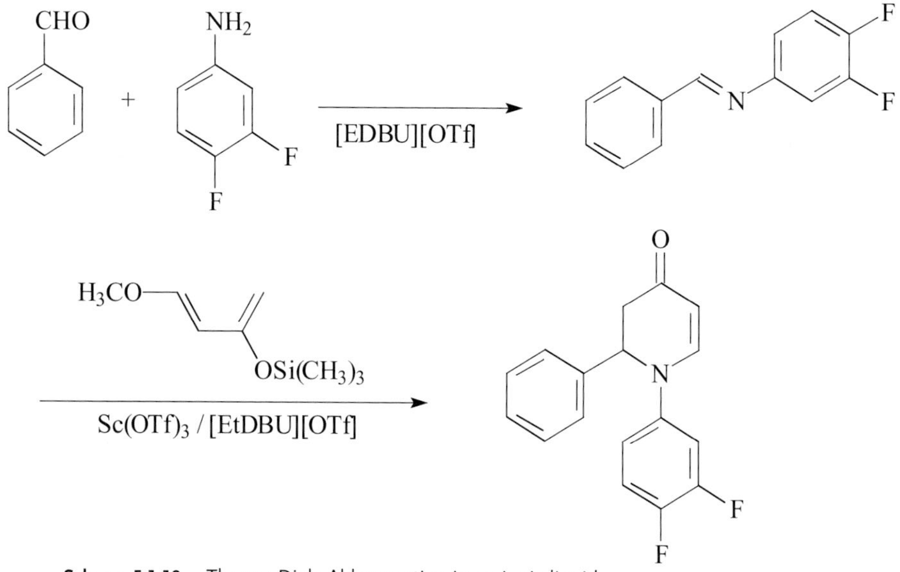

Scheme 5.1-19: The aza-Diels-Alder reaction in an ionic liquid.

Nucleophilic displacement reactions One of the most common reactions in organic synthesis is the nucleophilic displacement reaction. The first attempt at a nucleophilic substitution reaction in a molten salt was carried out by Ford and co-workers [47, 48, 49]. Here, the rates of reaction between halide ion (in the form of its triethylammonium salt) and methyl tosylate in the molten salt triethylhexylammonium triethylhexylborate were studied (Scheme 5.1-20) and compared with similar reactions in dimethylformamide (DMF) and methanol. The reaction rates in the molten salt appeared to be intermediate in rate between methanol and DMF (a dipolar aprotic solvent known to accelerate S_N2 substitution reactions).

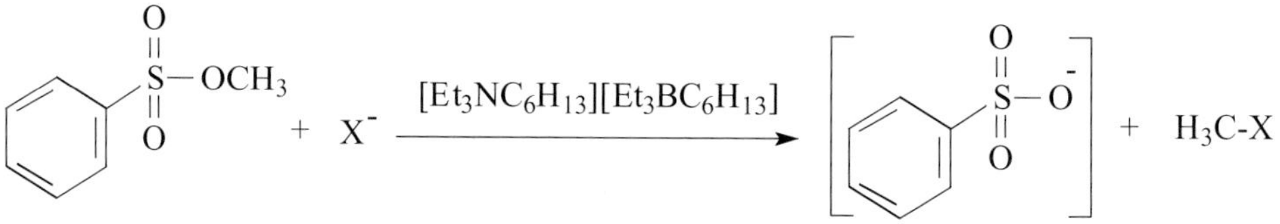

Scheme 5.1-20: The reaction between halide and methyl tosylate in triethylhexylammonium triethylhexylborate.

The alkylation of sodium 2-naphthoxide with benzyl bromide in tetrabutylammonium and tetrabutylphosphonium halide salts was investigated by Brunet and Badri [50] (Scheme 5.1-21). The yields in this reaction were quantitative, and alkylation occurred predominantly on the oxygen atom of the naphthoxide ion (typically 93–97 %). The rate of the reaction was slower in the chloride salts, due to the benzyl bromide reacting with chloride ion to give the less reactive benzyl chloride.

Indole and 2-naphthol undergo alkylation on the nitrogen and oxygen atoms, respectively (Scheme 5.1-22), when treated with an alkyl halide and base (usually NaOH or KOH) in [BMIM][PF_6] [51].

These reactions occur with similar rates to those carried out in dipolar aprotic solvents such as DMF or DMSO. An advantage of using the room-temperature ionic liquid for this reaction is that the lower reaction temperatures result in higher selectivities for substitution on the oxygen or nitrogen atoms. The by-product (sodium or potassium halide) of the reaction can be extracted with water and the ionic liquid recycled.

A quantitative study of the nucleophilic displacement reaction of benzoyl chloride with cyanide ion in [BMIM][PF_6] was investigated by Eckert and co-workers [52]. The separation of the product, 1-phenylacetonitrile, from the ionic liquid was achieved by distillation or by extraction with supercritical CO_2. The 1-phenylacetonitrile was then treated with KOH in [BMIM][PF_6] to generate an anion, which reacted with 1,4-dibromobutane to give 1-cyano-1-phenylcyclopentane (Scheme 5.1-23). This was in turn extracted from the ionic liquid with supercritical CO_2. These

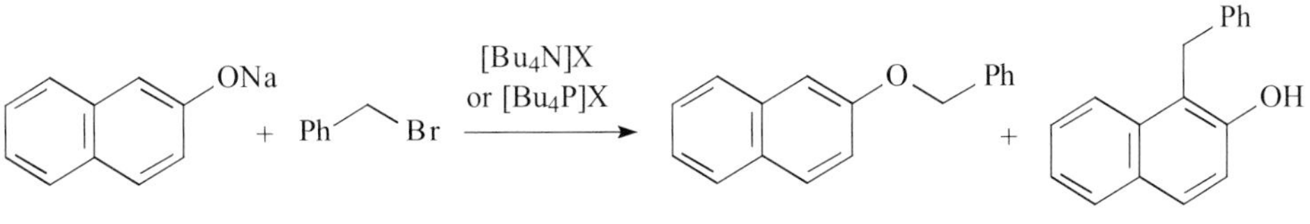

Scheme 5.1-21: The benzylation of sodium 2-naphthoxide with benzyl bromide in ammonium or phosphonium halide salts (X = Cl, Br).

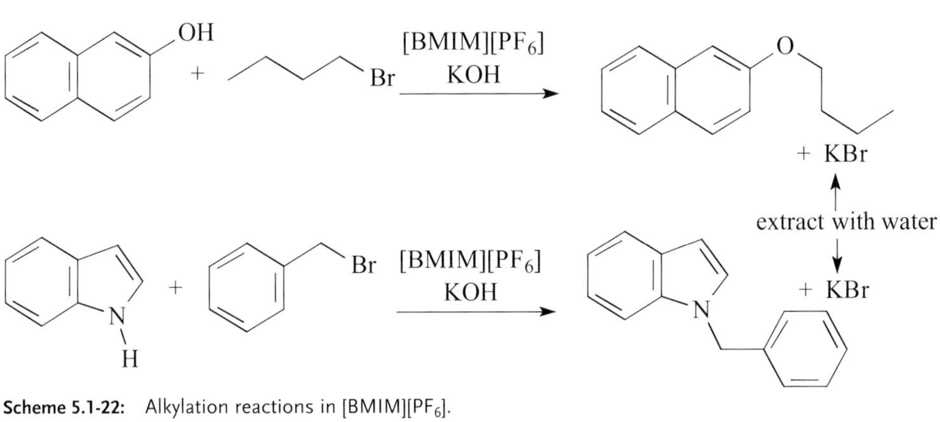

Scheme 5.1-22: Alkylation reactions in [BMIM][PF_6].

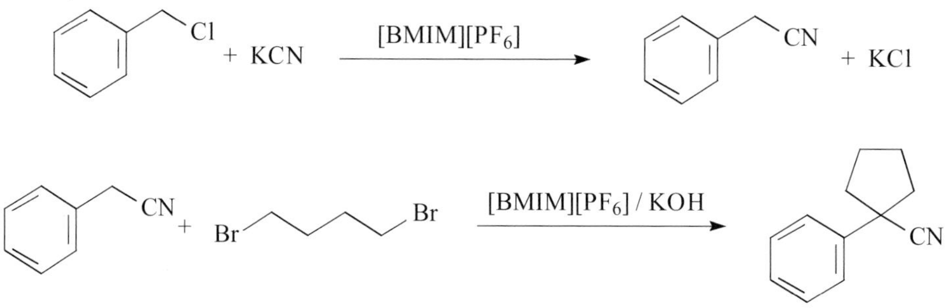

Scheme 5.1-23. The reaction of cyanide with benzyl chloride to produce 1-phenylacetonitrile, and subsequent treatment with 1,4-dibromobutane.

reactions resulted in a build-up of KCl or KBr in the ionic liquid, which was removed by washing the ionic liquid with water.

As a demonstration of the complete synthesis of a pharmaceutical in an ionic liquid, Pravadoline was selected, as the synthesis combines a Friedel–Crafts reaction and a nucleophilic displacement reaction (Scheme 5.1-24) [53]. The alkylation of 2-methylindole with 1-(*N*-morpholino)-2-chloroethane occurs readily in [BMIM][PF_6] and [BMMIM][PF_6] (BMMIM = 1-butyl-2,3-dimethylimidazolium), in 95–99 % yields, with potassium hydroxide as the base. The Friedel–Crafts acylation step in [BMIM][PF_6] at 150 °C occurs in 95 % yield and requires no catalyst.

Reactions involving organometallic reagents in neutral ionic liquid The addition of organometallic reagents to carbonyl compounds is an important reaction in organic chemistry, the Grignard reaction being one example of this. Procedures that

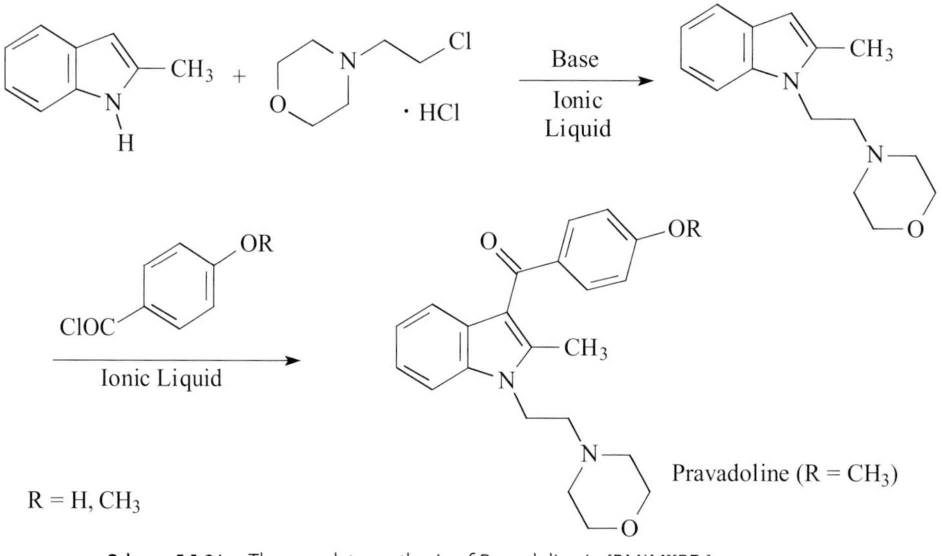

Scheme 5.1-24: The complete synthesis of Pravadoline in [BMIM][PF_6].

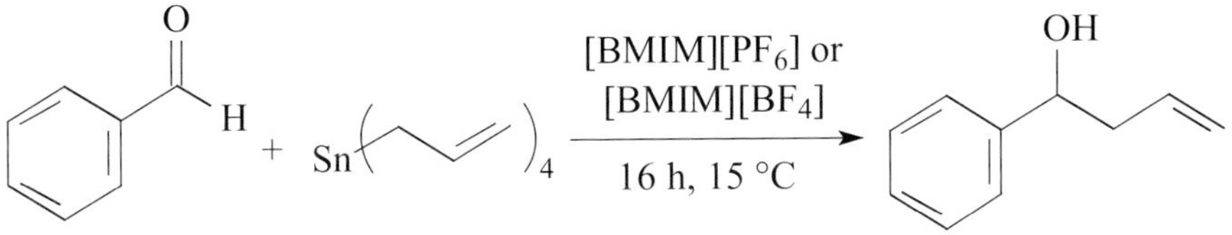

[BMIM][PF_6] or [BMIM][BF_4]

16 h, 15 °C

Scheme 5.1-25: Allylation of aldehydes in [BMIM][PF_6] or [BMIM][BF_4].

achieve similar results in ionic liquids are hence desirable. Gordon and McClusky [54] have reported the formation of homoallylic alcohols though the addition of allyl stannanes to aldehydes in the ionic liquids [BMIM][BF_4] and [BMIM][PF_6] (Scheme 5.1-25). It was found that the ionic liquid could be recycled and reused over several reaction cycles.

Kitazume and Kasai [55] have investigated the Reformatsky reaction in three ionic liquids. This reaction involves treatment of an α-bromo ester with zinc to give an α-zinc bromide ester, which in turn reacts with an aldehyde to give an addition product. An example is given in Scheme 5.1-26. Moderate to good yields (45–95 %) were obtained in ionic liquids such as [EDBU][OTf] for the reactions between ethyl bromoacetate or ethyl bromodifluoroacetate and benzaldehyde [55].

Reactions between aldehydes and alkynes to give propargyl alcohols are also described in Kitazume and Kasai's paper [55]. Here, various aldehydes such as benzaldehyde or 4-fluorobenzaldehyde were treated with alkynes such as phenylethyne or pent-1-yne in three ionic liquids: [EDBU][OTf], [BMIM][PF_6], and [BMIM][BF_4] (Scheme 5.1-27). A base (DBU) and $Zn(OTf)_2$ were required for the reaction to be effective; the yields were in the 50–70 % range. The best ionic liquid for this reaction depended on the individual reaction.

McCluskey et al. have also used [BMIM][BF_4] as a solvent for the allylation of aldehydes and Weinreb amides [56]. Similar diastereoselectivities and similar or slightly lower yields were obtained in this ionic liquid, compared with reactions carried

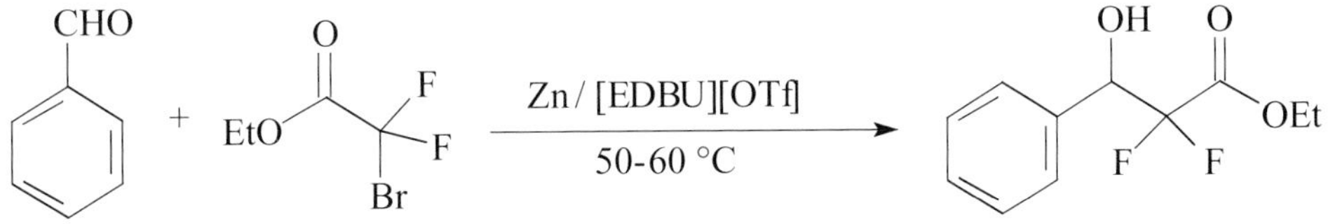

Zn / [EDBU][OTf]

50-60 °C

Scheme 5.1-26: The Reformatsky reaction in ionic liquids.

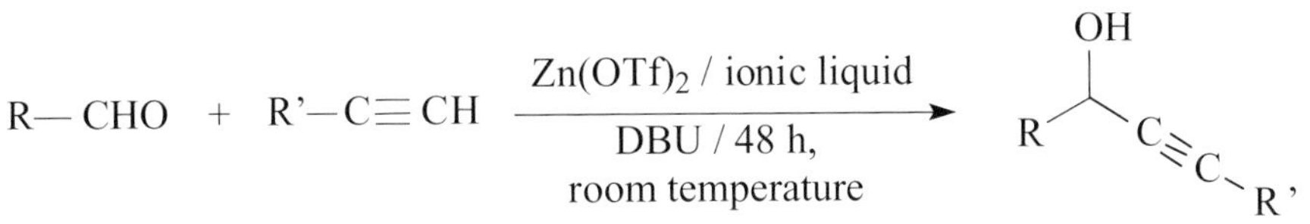

R— CHO + R'—C≡CH

$Zn(OTf)_2$ / ionic liquid

DBU / 48 h, room temperature

Scheme 5.1-27: The zinc triflate-catalyzed coupling of alkynes with aldehydes to give propargyl alcohols in an ionic liquid.

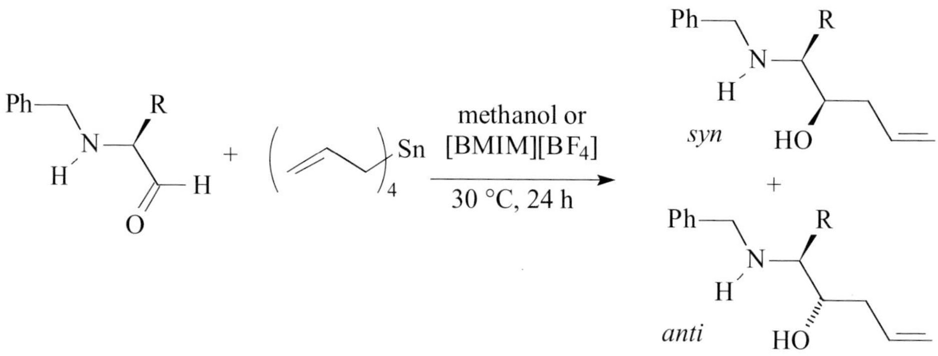

Scheme 5.1-28: The reaction between tetraallylstannane and an aldehyde in methanol or [BMIM][BF_4].

Table 5.1-2: The yields and selectivities for the reaction shown in Scheme 5.1-28.

R	*syn- : anti-*	*% Yield [BMIM][BF$_4$]*	*% Yield methanol*	*d.e. (%)*
CH_3	82:18	72	87	64
$CH(CH_3)_2$	93:7	70	74	86
$PhCH_2$	93:7	73	82	86

out in methanol (Scheme 5.1-28). The lower yield assigned to the reaction in the ionic liquid (see Table 5.1-2) is thought to be due to difficulty in extracting the product from the ionic liquid.

Ionic liquids such as [BMIM][BF_4] and [EMIM][PF_6] have been used in the trialkylborane reduction of aldehydes to alcohols (Scheme 5.1-29) [57]. In the reduction of benzaldehyde with tributylborane, similar yields (90–96 %) were obtained for the ionic liquids [EMIM][BF_4], [EMIM][PF_6], [BMIM][BF_4], and [BMIM][PF_6]. The effect of electron-releasing and electron-withdrawing groups on the aromatic aldehyde were investigated. In general, electron-withdrawing groups such as halogen give near quantitative yields, but electron-releasing groups such as methoxy reduced the reaction rate and yield.

Miscellaneous reactions in neutral ionic liquids Kitazume et al. have also investigated the use of [EDBU][OTf] as a medium in the formation of heterocyclic compounds [58]. Compounds such as 2-hydroxymethylaniline readily condense with

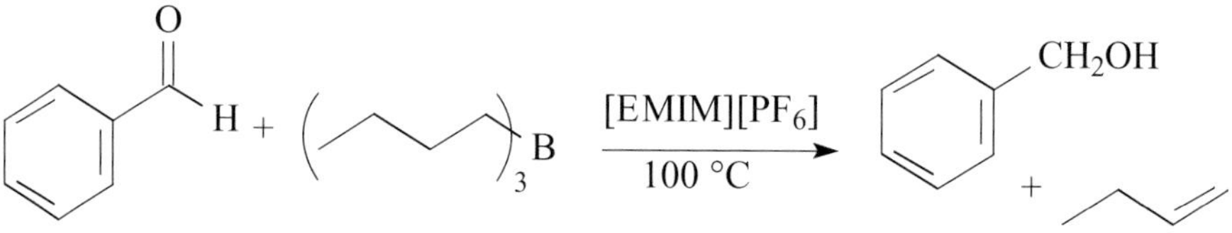

Scheme 5.1-29: The reduction of benzaldehyde in [EMIM][PF_6].

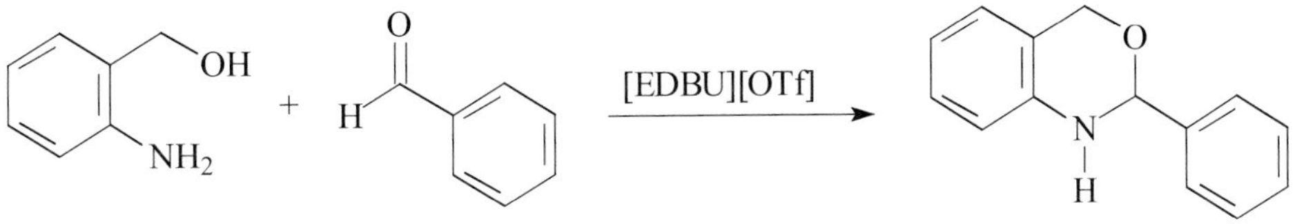

Scheme 5.1-30: The formation of 2-phenylbenzoxazine in [EDBU][OTf].

benzaldehyde to give the corresponding benzoxazine (Scheme 5.1-30). The product of the reaction is readily extracted with solvents such as diethyl ether, and the ionic liquid can be recycled and reused.

Beckmann rearrangements of several ketoximes were performed in room-temperature ionic liquids based on 1,3-dialkylimidazolium or alkylpyridinium salts containing phosphorus compounds (such as PCl_5) by Deng and Peng [59] (Scheme 5.1-31, BP = 1-butylpyridinium). Turnover numbers of up to 6.6 were observed, but the authors did not mention whether the ionic liquid could be reused.

The first examples of Horner–Wadsworth–Emmons reactions have been reported by Kitazume and Tanaka [60]. Here the ionic liquid [EDBU][OTf] was used in the synthesis of α-fluoro-α,β-unsaturated esters (Scheme 5.1-32). It was found that when K_2CO_3 was used as a base, the *E* isomer was the major product and that when DBU was used as a base, the *Z* isomer was the major product. The reaction was also performed in [EMIM][BF_4] and [EMIM][PF_6], but gave lower yields than with [EDBU][OTf] [60].

Davis and co-workers have carried out the first examples of the Knoevenagel condensation and Robinson annulation reactions [61] in the ionic liquid [HMIM][PF_6] (HMIM = 1-hexyl-3-methylimidazolium) (Scheme 5.1-33). The Knoevenagel condensation involved the treatment of propane-1,3-dinitrile with a base (glycine) to generate an anion. This anion added to benzaldehyde and, after loss of a water molecule, gave 1,1-dicyano-2-phenylethene. The product was separated from the ionic liquid by extraction with toluene.

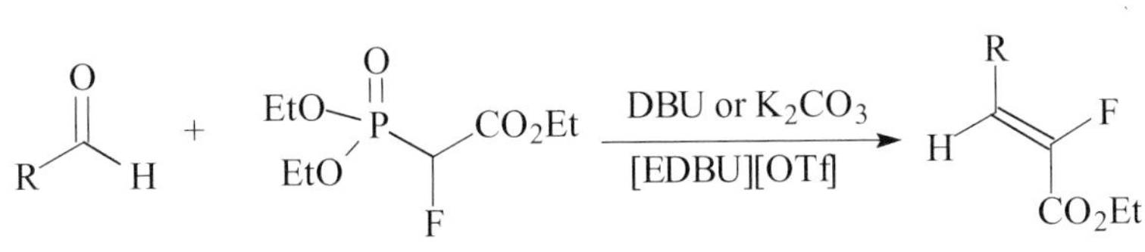

Scheme 5.1-31: The Beckmann rearrangement in ionic liquids.

Scheme 5.1-32: The Horner-Wadsworth-Emmons reaction in an ionic liquid.

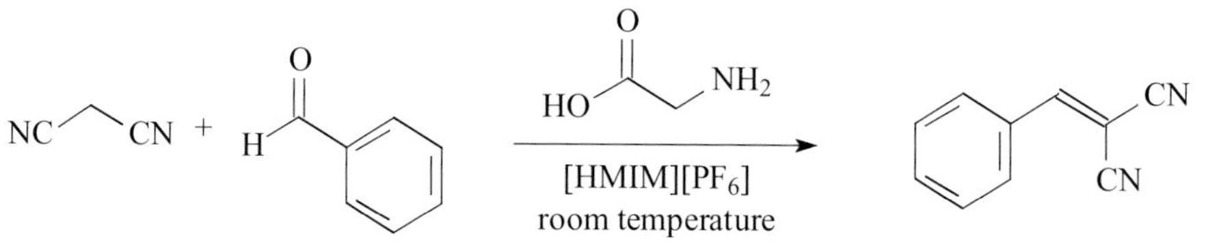

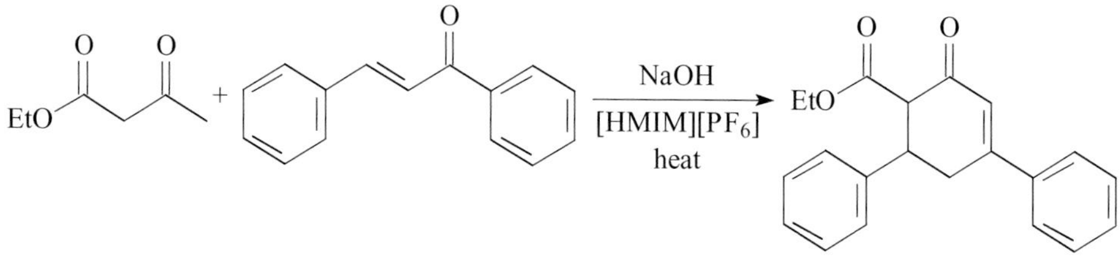

Scheme 5.1-33: The Knoevenagel condensation and the Robinson annulation in $[HMIM][PF_6]$.

The Robinson annulation of ethyl acetoacetate and *trans*-chalcone proceeded smoothly to give 6-ethoxycarbonyl-3,5-diphenyl-2-cyclohexenone in 48 % yield. The product was separated from the ionic liquid by solvent extraction with toluene. In both these reactions, the ionic liquid $[HMIM][PF_6]$ was recycled and reused with no reduction in the product yield.

Deng and Peng have found that certain ionic liquids catalyze the Biginelli reaction [62]. Usually, this reaction is catalyzed by Lewis acids such as $InCl_3$, $[Fe(H_2O)_6]Cl_3$, or $BF_3.O(C_2H_5)_2$, or by acid catalysts such as Nafion-H. The reaction was found to give yields in the 77–99 % range in the ionic liquids $[BMIM][PF_6]$ or $[BMIM][BF_4]$ for the examples in Scheme 5.1-34. The reaction fails if there is no ionic liquid present or in the presence of tetrabutylammonium chloride.

Singer and Scammells have investigated the γ-MnO_2 oxidation of codeine methyl ether (CME) to thebaine in the ionic liquid $[BMIM][BF_4]$ [63]. The ionic liquid was used in different ways and with mixed results (Scheme 5.1-35). For example, the oxidation of CME in the ionic liquid gave 38 % yield after 120 hours. A similar reaction under biphasic conditions (with diethyl ether) gave a 36 % yield of thebaine. This reaction gave a 25 % yield of thebaine when carried out in tetrahydrofuran

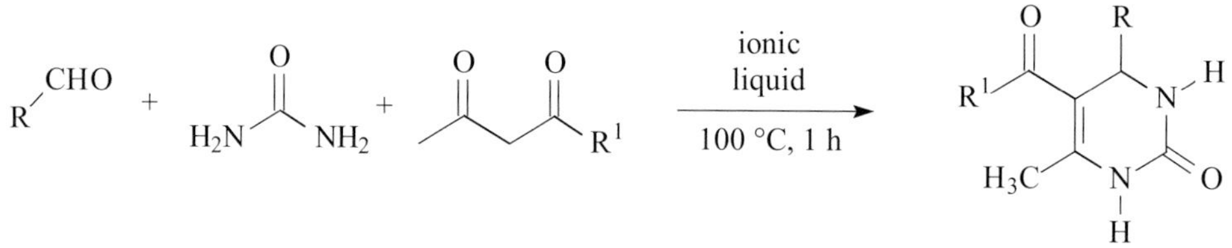

Scheme 5.1-34: The Biginelli reaction in an ionic liquid. R = C_6H_5, 4-(H_3CO)-C_6H_4, 4-Cl-C_6H_4, 4-(O_2N)-C_6H_4, C_5H_{11}. R^1 = OC_2H_5, CH_3.

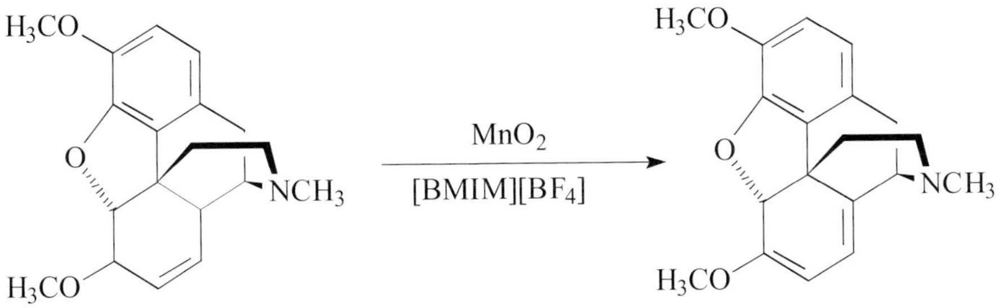

Scheme 5.1-35: The oxidation of CME to thebaine in [BMIM][BF_4].

(THF). The authors found that the yield could be increased to 95 % by sonication of the reaction vessel, for the reaction in THF. The ionic liquid was then used to extract the manganese by-products and impurities from an ethyl acetate solution of the product [63].

A novel use of the salt [BMIM][PF_6] is to enhance microwave absorption and hence accelerate the rate of a reaction. Ley found that [BMIM][PF_6] enhanced the rate of the microwave-promoted thionation of amides by a polymer-supported thionating agent [64].

Hardacre et al. have developed a procedure for the synthesis of deuterated imidazoles and imidazolium salts [65]. The procedure involves the platinum- or palladium-catalyzed deuterium exchange of 1-methyl-d^3-imidazole with D_2O to give 1-methylimidazole-d^6, followed by treatment with a deuterated alkyl halide.

5.1.2 Acid-Catalyzed Reactions

5.1.2.1 Electrophilic substitutions and additions

This section deals with Brønsted acid and Lewis acid catalyzed reactions, excluding Friedel–Crafts reactions, but including reactions such as nitrations, halogenations, and Claisen rearrangements. Friedel–Crafts reactions are discussed in the subsequent Sections 5.1.2.2 and 5.1.2.3.

The first example of an electrophilic nitration in an ionic liquid was performed by Wilkes and co-workers [66]. A number of aromatic compounds were nitrated with KNO_3 dissolved in chloroaluminate(III) ionic liquids. A number of nitration reactions have also been carried out by Laali et al. [67]. The reactions of nitrates, preformed nitronium salts, and alkyl nitrates with aromatic compounds have been performed in a wide range of ionic liquids. Reactions between toluene and [NO_2][BF_4], for example, have been performed with varying degrees of success in [EMIM]Cl, [EMIM][$AlCl_4$], [EMIM][Al_2Cl_7], [EMIM][BF_4], [EMIM][PF_6], and [EMIM][OTf]. Of these, the reaction in [EMIM][BF_4] (Scheme 5.1-36) gave the best yield (71 %, *o*:*p* ratio = 1.17:1), but only after the imidazolium ring had undergone nitration (Figure 5.1-3).

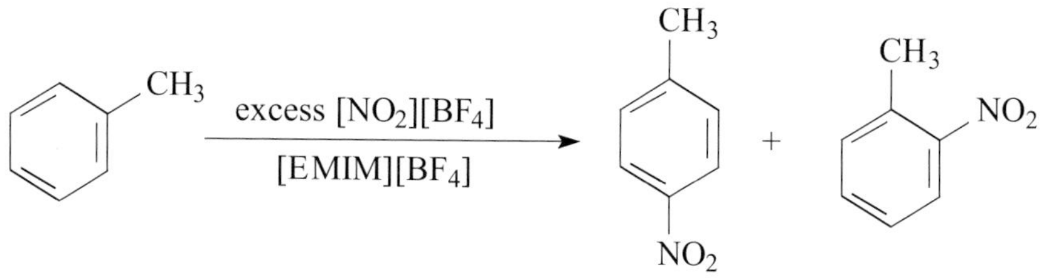

CH_3 — excess $[NO_2][BF_4]$ / $[EMIM][BF_4]$ → CH_3, NO_2 + CH_3, NO_2

Scheme 5.1-36: The nitration of toluene with $[NO_2][BF_4]$ in $[EMIM][BF_4]$.

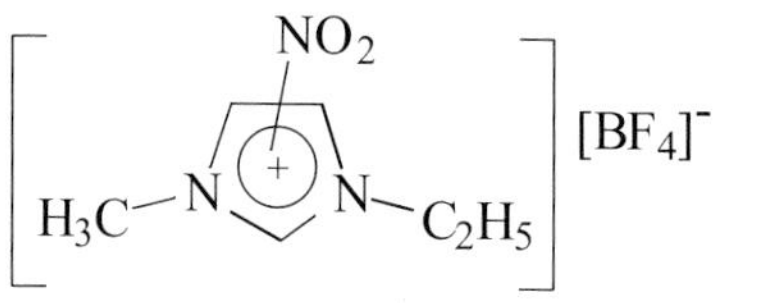

NO_2 H_3C–N N–C_2H_5 $[BF_4]^-$

Figure 5.1-3: The nitroimidazolium ionic liquid.

Other methods of nitration that Laali investigated were with isoamyl nitrate in combination with a Brønsted or Lewis acid in several ionic liquids, with [EMIM][OTf] giving the best yields (69 %, 1.0:1.0 *o*:*p* ratio). In the ionic liquid $[HNEt(^iPr)_2]$ $[CF_3CO_2]$ (m.p. = 92–93 °C), toluene was nitrated with a mixture of $[NH_4][NO_3]$ and trifluoroacetic acid (TFAH) (Scheme 5.1-37). This gave ammonium trifluoroacetate $[NH_4][TFA]$ as a by-product, which could be removed from the reaction vessel by distillation (sublimation).

Wilkes and co-workers have investigated the chlorination of benzene in both acidic and basic chloroaluminate(III) ionic liquids [66]. In the acidic ionic liquid $[EMIM]Cl/AlCl_3$ ($X(AlCl_3) > 0.5$), the chlorination reaction initially gave chlorobenzene, which in turn reacted with a second molecule of chlorine to give dichlorobenzenes. In the basic ionic liquid, the reaction was more complex. In addition to the

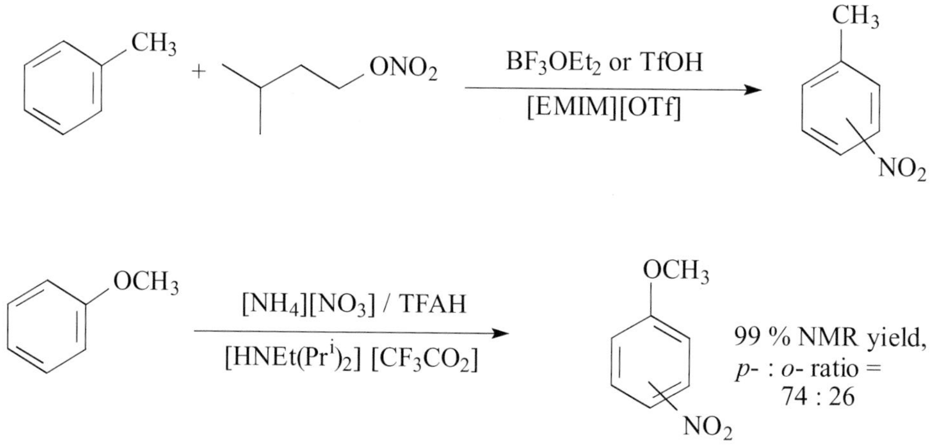

CH_3 + ONO_2 — BF_3OEt_2 or TfOH / [EMIM][OTf] → CH_3, NO_2

OCH_3 — $[NH_4][NO_3]$ / TFAH / $[HNEt(Pr^i)_2]$ $[CF_3CO_2]$ → OCH_3, NO_2

99 % NMR yield,
p- : *o*- ratio =
74 : 26

Scheme 5.1-37: Aromatic nitration reactions in ionic liquids.

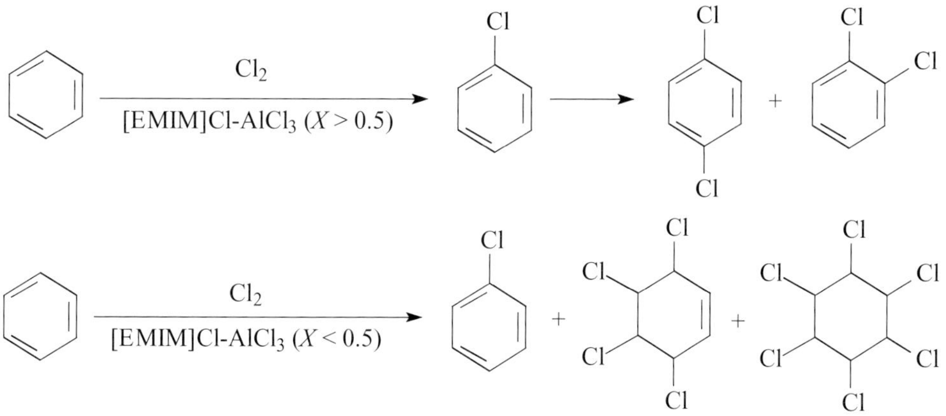

Scheme 5.1-38: The chlorination of benzene in acidic and basic chloroaluminate ionic liquids.

formation of chlorobenzene, addition products of chlorine and benzene were observed. These addition products included various isomers of tetrachlorocyclohexene and hexachlorocyclohexane (Scheme 5.1-38).

Another common reaction is the chlorination of alkenes to give 1,2-dihaloalkanes. Patell et al. reported that the addition of chlorine to ethene in acidic chloroaluminate(III) ionic liquids gave 1,2-dichloroethane [68]. Under these conditions, the imidazole ring of imidazolium ionic liquid is chlorinated. Initially, the chlorination occurs at the 4- and 5-positions of the imidazole ring, and is followed by much slower chlorination at the 2-position. This does not affect the outcome of the alkene chlorination reaction and it was found that the chlorinated imidazolium ionic liquids are excellent catalysts for the reaction (Scheme 5.1-39).

In an attempt to study the behavior and chemistry of coal in ionic liquids, 1,2-diphenylethane was chosen as a model compound and its reaction in acidic pyridinium chloroaluminate(III) melts ([PyH]Cl/$AlCl_3$ was investigated [69]. At 40 °C, 1,2-diphenylethane undergoes a series of alkylation and dealkylation reactions to give a mixture of products. Some of the products are shown in Scheme 5.1-40. Newman also investigated the reactions of 1,2-diphenylethane with acylating agents such as acetyl chloride or acetic anhydride in the pyridinium ionic liquid [70] and with alcohols such as isopropanol [71].

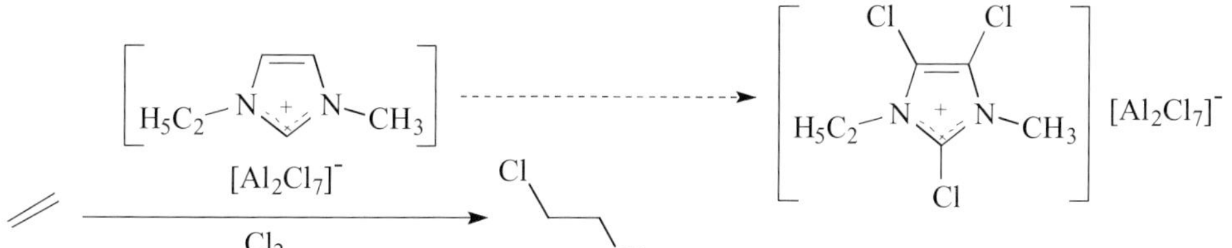

Scheme 5.1-39: The chlorination of ethene to give 1,2-dichloroethane.

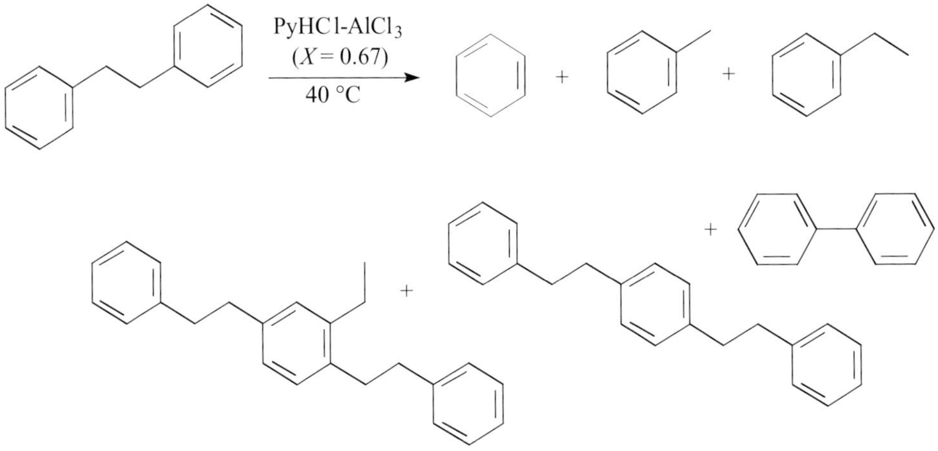

PyHCl-$AlCl_3$
(X = 0.67)
40 °C

Scheme 5.1-40: The reaction of 1,2-diphenylethane with PyHCl/$AlCl_3$ (X($AlCl_3$) = 0.67).

Kitazume and Zulfiqar have investigated the Claisen rearrangement of several aromatic allyl ethers in ionic liquids, catalyzed by scandium(III) trifluoromethanesulfonate [72]. The reaction initially gave the 2-allylphenol but this reacted further to give 2-methyl-2,3-dihydrobenzo[*b*]furan (Scheme 5.1-41). The yields in this reaction were highly dependant on the ionic liquid chosen, with [EDBU][OTf] giving the best yields (e.g., 91 % for R = 6-CH_3). Reactions in [BMIM][BF_4] and [BMIM][PF_6] gave low yields (9–12 %).

In order to confirm that 2-allylphenol was an intermediate in the reaction, the authors subjected 2-allylphenol to the same reaction conditions and found that it rearranged to give 2-methyl-2,3-dihydrobenzo[*b*]furan. On treatment of 2-methyl-2-propenyl phenyl ether (Scheme 5.41) under similar conditions, 2,3-diisopropylbenzo[*b*]furan was isolated in 15 % yield. A mechanistic scheme is given by the authors (Scheme 5.1-42). It involves the Claisen rearrangement of 2-methyl-2-propenyl phenyl ether to 2-(2-methyl-2-propenyl)-phenol, followed by a transalkylation of a 2-methylpropenyl group to the phenyl OH group. This undergoes further rearrangements and cyclization to give the 2,3-diisopropylbenzo[*b*]furan [72].

Lee et al. have investigated the Lewis acid-catalyzed three-component synthesis of α-amino phosphonates [73]. This was carried out in the ionic liquids [BMIM][PF_6],

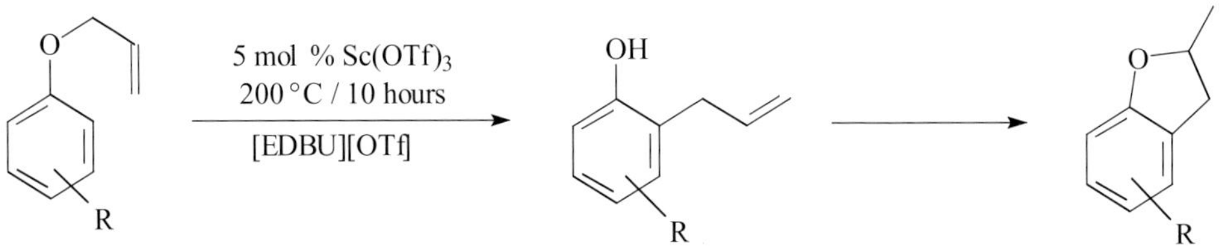

O
5 mol % Sc(OTf)$_3$
200 °C / 10 hours
[EDBU][OTf]
OH
O
R
R
R

Scheme 5.1-41: Claisen rearrangements of several phenyl allyl ethers (R = H, 4-CH_3, 6-CH_3).

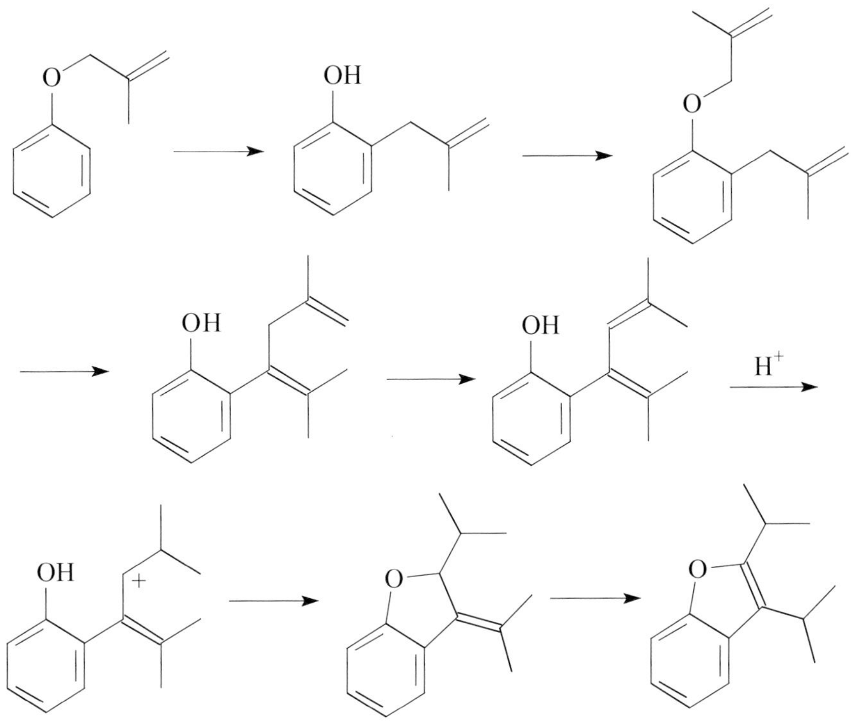

Scheme 5.1-42: Proposed mechanism for the formation of 2,3-diisopropylbenzo[*b*]furan.

[BMIM][OTf], [BMIM][BF_4], and [BMIM][SbF_6], and the results were compared with a similar reaction carried out in dichloromethane (Scheme 5.1-43).

Lee found that the reaction gave good yields (70–99 %) in the ionic liquids [BMIM][PF_6], [BMIM][OTf], and [BMIM][SbF_6] with Lewis acids such as $Yb(OTf)_3$, $Sc(OTf)_3$, $Dy(OTf)_3$, $Sm(OTf)_3$, and $InCl_3$. The reaction was also performed in [BMIM][PF_6] or dichloromethane with $Sm(OTf)_3$ as the catalyst. The ionic liquid reaction gave a yield of 99 %, compared with 70 % for the reaction in dichloromethane [73].

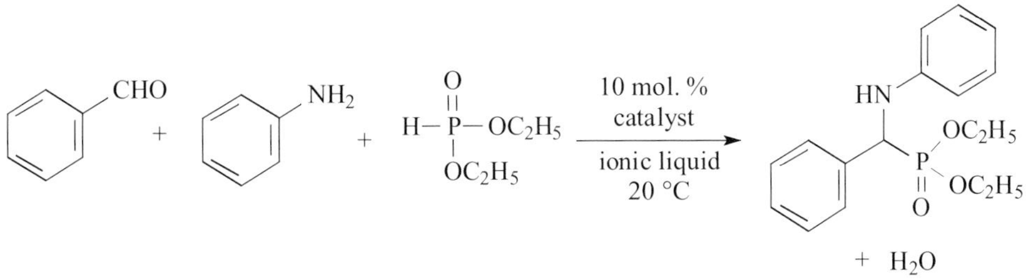

Scheme 5.1-43: Three-component reaction of benzaldehyde, aniline, and diethyl phosphonate in ionic liquids, catalyzed by lanthanide triflates and indium(III) chloride.

5.1.2.2 Friedel–Crafts alkylation reactions

Friedel–Crafts reactions have been studied in detail by Olah [74, 75]. These reactions result in the formation of carbon–carbon bonds and are catalyzed by strong Brønsted or Lewis acids.

The Friedel–Crafts alkylation reaction usually involves the interaction of an alkylation agent such as an alkyl halide, alcohol, or alkene with an aromatic compound, to form an alkylated aromatic compound (Scheme 5.1-44).

It should be noted that Scheme 5.1-44 shows idealized Friedel–Crafts alkylation reactions. In practice, there are a number of problems associated with the reaction. These include polyalkylation reactions, since the products of a Friedel–Crafts alkylation reaction are often more reactive than the starting material. Also, isomerization and rearrangement reactions can occur, and can result in a large number of products [74, 75]. The mechanism of Friedel–Crafts reactions is not straightforward, and it is possible to propose two or more different mechanisms for a given reaction. Examples of the typical processes occurring in a Friedel–Crafts alkylation reaction are given in Scheme 5.1-45 for the reaction between 1-chloropropane and benzene.

The chemical behavior of Franklin acidic chloroaluminate(III) ionic liquids (where $X(AlCl_3) > 0.50$) [6] is that of a powerful Lewis acid. As might be expected, it catalyzes reactions that are conventionally catalyzed by aluminium(III) chloride, without suffering the disadvantage of the low solubility of aluminium(III) chloride in many solvents.

The first examples of alkylation reactions in molten salts were reported in the 1950's. Baddeley and Williamson performed a number of intramolecular cyclization reactions [76] (Scheme 5.1-46), carried out in mixtures of sodium chloride and aluminium chloride. The reactions were run at below the melting point of the pure salt, and it is presumed that the mixture of reagents acts to lower the melting point.

Baddeley also investigated the cyclization of alkenes in the $NaCl/AlCl_3$ molten salt. An example is given in Scheme 5.1-47 [77].

Mendelson et al. [78] also investigated a number of cyclization reactions. One of these involved the cyclodehydration of *N*-benzylethanolamine chloride in a molten salt derived from $AlCl_3$ and NH_4Cl ($X(AlCl_3) = 0.73$). This gave rise to the corresponding tetrahydroisoquinoline in 41–80 % yield, as shown in Scheme 5.1-48.

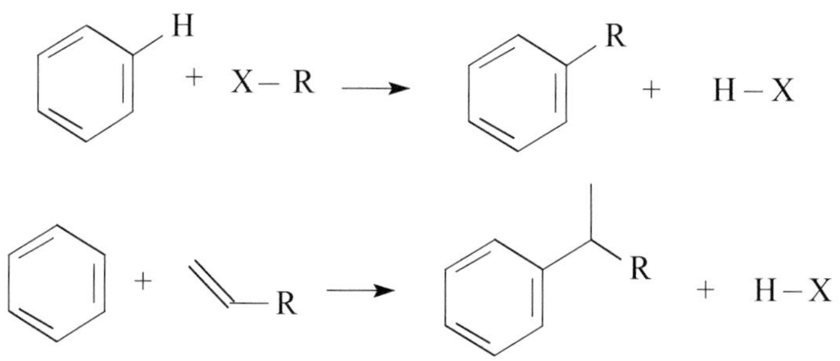

H + X–R ⟶ (Ph–R) + H–X

(benzene) + (alkene)–R ⟶ (Ph–CH(CH3)–CH2R) + H–X

Scheme 5.1-44: The Friedel-Crafts alkylation reaction (R = alkyl, X = leaving group).

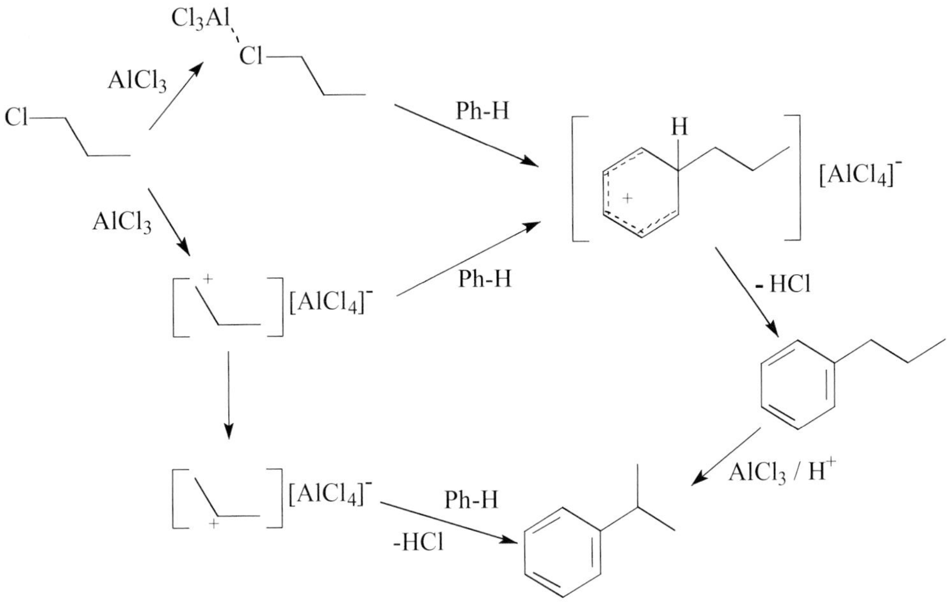

Cl_3Al
Cl
$AlCl_3$
Cl
Ph-H
H
$[AlCl_4]^-$
$AlCl_3$
+
$[AlCl_4]^-$
Ph-H
- HCl
$[AlCl_4]^-$
Ph-H
$AlCl_3$ / H^+
-HCl

Scheme 5.1-45: The reaction between 1-chloropropane and benzene under Friedel-Crafts conditions.

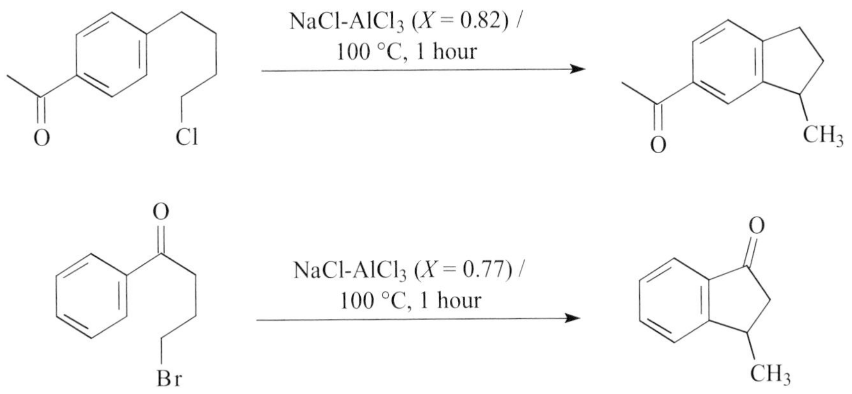

O
Cl
NaCl-$AlCl_3$ ($X = 0.82$) /
100 °C, 1 hour
O
CH_3

O
NaCl-$AlCl_3$ ($X = 0.77$) /
100 °C, 1 hour
Br
O
CH_3

Scheme 5.1-46: The intramolecular cyclization of alkyl chlorides and bromides.

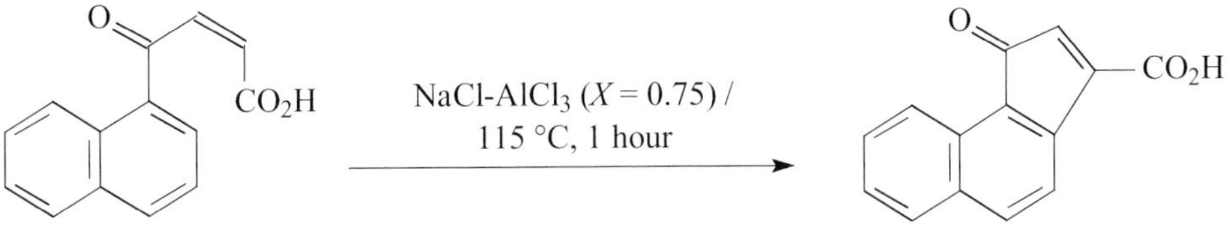

O
CO_2H
NaCl-$AlCl_3$ ($X = 0.75$) /
115 °C, 1 hour
O
CO_2H

Scheme 5.1-47: The intramolecular cyclization of an alkene.

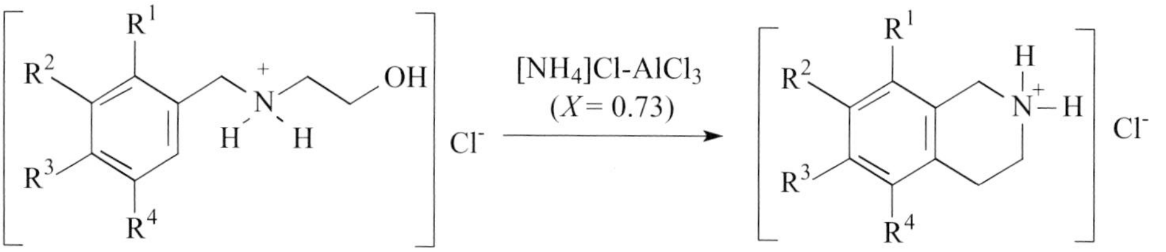

$$[NH_4]Cl\text{-}AlCl_3 \; (X = 0.73)$$

Scheme 5.1-48: The cyclodehydration of *N*-benzylethanolamine chloride.

Boon et al. investigated the reactions of benzene and toluene in room-temperature ionic liquids based on $[EMIM]Cl/AlCl_3$ mixtures [79]. The reactions of various alkyl chlorides with benzene in the ionic liquid $[EMIM]Cl/AlCl_3$ ($X(AlCl_3) = 0.60$ or 0.67) were carried out, and the product distributions are given in Table 5.1-3 and Scheme 5.1-49. The methylation of benzene with methyl chloride proceeds to give predominantly dimethylbenzene (xylenes) and tetramethylbenzene, with about 10 % hexamethylbenzene. In the propylation of benzene with 1-chloropropane, not only does polyalkylation occur, but there is a considerable degree of isomerization of the *n*-propyl group to the isopropyl isomer (Scheme 5.1-49). In the butylation, complete isomerization of the butyl side chain occurs, to give only *sec*-butyl benzenes.

Piersma and Merchant have studied the alkylation of benzene with various chloropentanes in $[EMIM]Cl/AlCl_3$ ($X(AlCl_3) = 0.55$) [80]. Treatment of 1-chloropentane with benzene gave a mixture of products, with only a 1 % yield of the unisomerized *n*-pentylbenzene. The major products of the reaction had all undergone isomerization (Scheme 5.1-50).

Details of two related patents for the alkylation of aromatic compounds with chloroaluminate(III) ionic or chlorogallate(III) ionic liquid catalysts have become available. The first, by Seddon and co-workers [81], describes the reaction between ethene and benzene to give ethylbenzene (Scheme 5.1-51). This is carried out in an

Table 5.1-3: The products from the reactions between alkyl chlorides and benzene in $[EMIM]Cl/AlCl_3$ ($X(AlCl_3) = 0.60$ or 0.67).

R-Cl	*X*	*R-Cl : C_6H_6: IL*	*Mono-*	*Di-*	*Tri-*	*Tetra-*	*Penta-*	*Hexa-*
Methyl[a]	0.67	xs : 1 : 1	1.5	58.5	1.5	26.8	1.4	10.2
Ethyl[a]	0.67	xs : 1 : 1	11.5	10.8	33.4	24.4		1.5
n-Propyl[b]	0.60	1.25 : 1.25 : 1	24.8	19.9	55.3			
n-Butyl[b,c]	0.60	1.33 : 1.33 : 1	25.0	26.3	48.7			
Cyclohexyl	0.60	10 : 10 : 1	35.0	30.0	34.4			
Benzyl[d]	0.60	0.78 : 1.17 : 1	50.0	34.5	15.6			

[a] At reflux temperature of alkyl halide. [b] Room temperature in dry-box. [c] Only *sec*-butyl products formed. [d] Tar formed, only a small amount of alkylated product isolated.

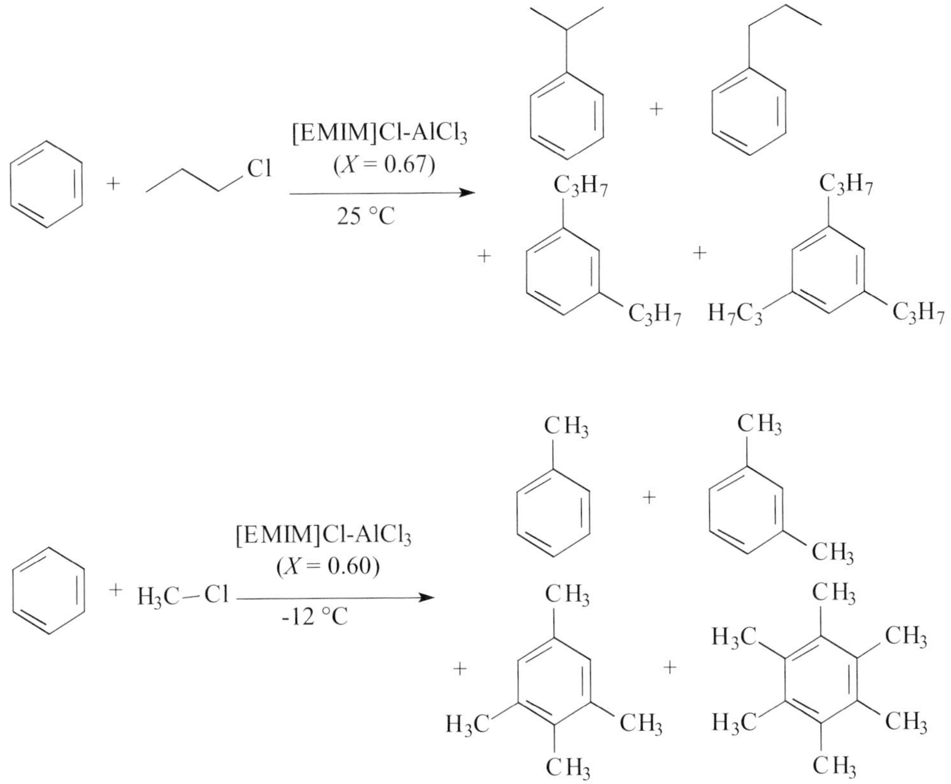

[EMIM]Cl-$AlCl_3$ ($X = 0.67$), 25 °C

C_3H_7 C_3H_7 C_3H_7 H_7C_3 C_3H_7

[EMIM]Cl-$AlCl_3$ ($X = 0.60$), -12 °C

H_3C-Cl CH_3 CH_3 CH_3 CH_3 CH_3 H_3C CH_3 CH_3 H_3C CH_3 H_3C CH_3 CH_3

Scheme 5.1-49: The alkylation of benzene with methyl chloride or *n*-propyl chloride in an ionic liquid.

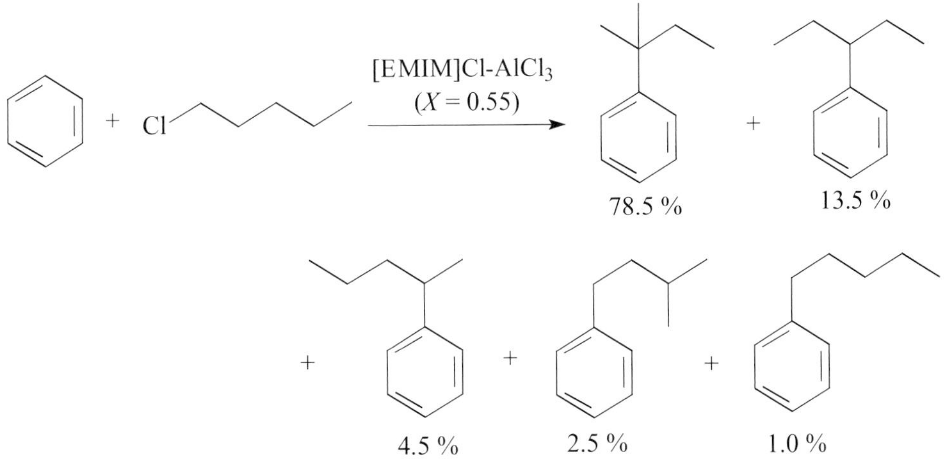

[EMIM]Cl-$AlCl_3$ ($X = 0.55$)

Cl 78.5 % 13.5 % 4.5 % 2.5 % 1.0 %

Scheme 5.1-50: The reaction between 1-chloropentane and benzene in [EMIM]Cl/$AlCl_3$ ($X(AlCl_3) = 0.55$).

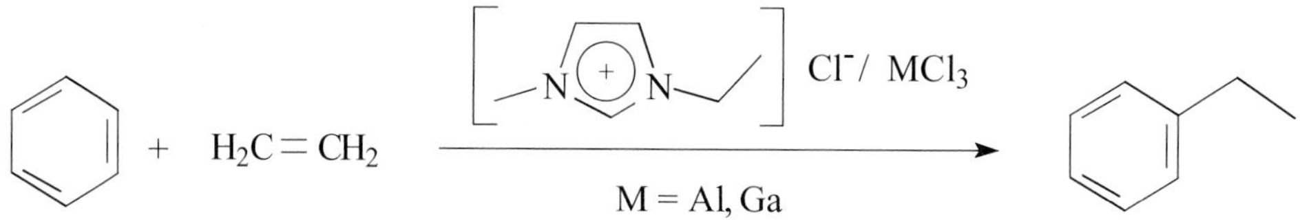

$$\text{benzene} + H_2C{=}CH_2 \xrightarrow[\text{M = Al, Ga}]{[\text{EMIM}]Cl^- / MCl_3} \text{ethylbenzene}$$

Scheme 5.1-51: The alkylation of aromatic compounds in chloroaluminate(III) or chlorogallate(III) ionic liquids.

acidic ionic liquid based on an imidazolium cation, and is claimed for ammonium, phosphonium, and pyridinium cations. The anion exemplified in the patent is a chloroaluminate(III) and a claim is made for chlorogallate(III) anions and various mixtures of anions.

The second patent, by Wasserscheid and co-workers [82], also describes the reaction of benzene with ethene in ionic liquids, but exemplifies a different ionic liquid suitable for this reaction (Scheme 5.1-52).

The production of linear alkyl benzenes (LABs) is carried out on a large scale for the production of surfactants. The reaction involves the reaction between benzene and a long-chain alkene such as dodec-1-ene and often gives a mixture of isomers. Greco et al. have used a chloroaluminate(III) ionic liquid as a catalyst in the preparation of LABs [83] (Scheme 5.1-53).

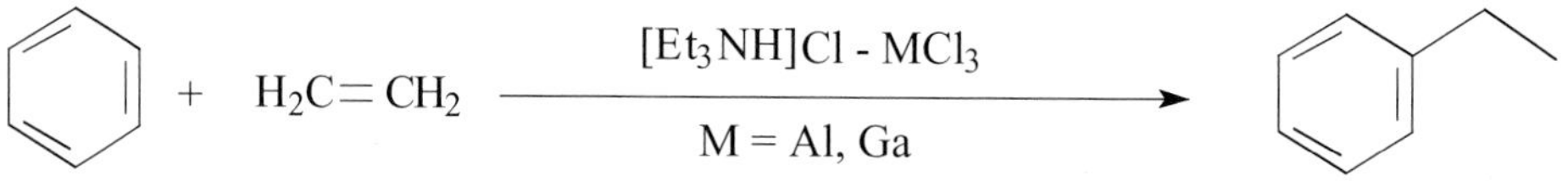

$$\text{benzene} + H_2C{=}CH_2 \xrightarrow[\text{M = Al, Ga}]{[Et_3NH]Cl - MCl_3} \text{ethylbenzene}$$

Scheme 5.1-52: The reaction between benzene and ethene in a triethylammonium ionic liquid.

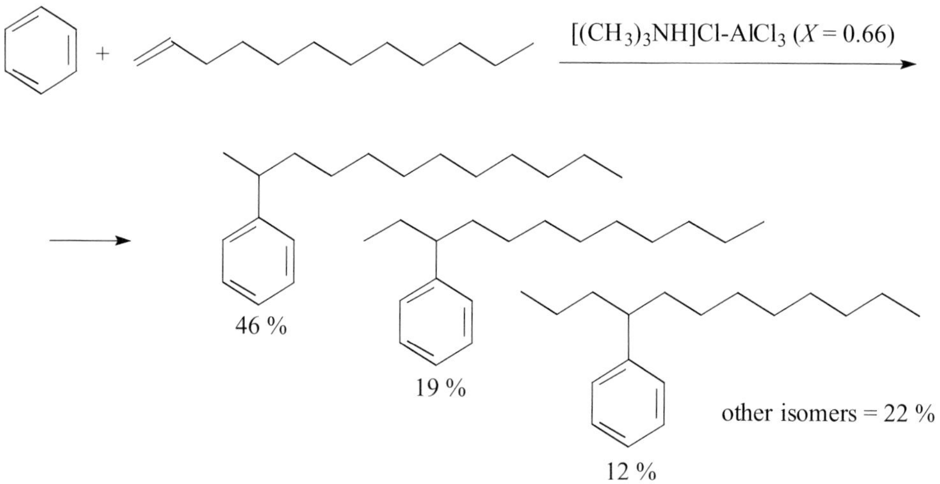

$$\text{benzene} + \text{dodec-1-ene} \xrightarrow{[(CH_3)_3NH]Cl\text{-}AlCl_3\ (X = 0.66)}$$

46 %

19 %

12 %

other isomers = 22 %

Scheme 5.1-53: The reaction between dodec-1-ene and benzene with an ionic liquid as a catalyst.

Keim and co-workers have carried out various alkylation reactions of aromatic compounds in ionic liquids substantially free of Lewis acidity [84]. An example is the reaction between benzene and decene in [BMIM][HSO_4], which was used together with sulfuric acid as the catalyst (Scheme 5.1-54). These authors have also claimed that these acid-ionic liquids systems can be used for esterification reactions.

The methodology of a Lewis acid dissolved in an ionic liquid has been used for Friedel–Crafts alkylation reactions. Song [85] has reported that scandium(III) triflate in [BMIM][PF_6] acts as an alkylation catalyst in the reaction between benzene and hex-1-ene (Scheme 5.1-55).

The ionic liquids that were found to give the expected hexylbenzenes were [BMIM][PF_6], 1-pentyl-3-methylimidazolium hexafluorophosphate, [HMIM][PF_6], [EMIM][SbF_6], and [BMIM][SbF_6]. The reaction did not succeed in the corresponding tetrafluoroborate or trifluoromethanesulfonate ionic liquids. For the successful reactions, conversions of 99 % of the hexene into products occurred, with 93–96 % of the products being the monoalkylated product (Scheme 5.1-55). The authors noted that the successful reactions all took place in the hydrophobic ionic liquids. It should be noted that the $[PF_6]^-$ and $[SbF_6]^-$ ions are less stable to hydrolysis reactions, resulting in the formation of HF, than the $[BF_4]^-$ or $[OTf]^-$ ions, and so the possibility of these reactions being catalyzed by traces of HF cannot be excluded [86].

The alkylation of a number of aromatic compounds through the use of a chloroaluminate(III) ionic liquid on a solid support has been investigated by Hölderich and co-workers [87, 88]. Here the alkylation of aromatic compounds such as benzene, toluene, naphthalene, and phenol with dodecene was performed using the ionic liquid [BMIM]Cl/$AlCl_3$ supported on silica, alumina, and zirconia. With benzene, monoalkylated dodecylbenzenes were obtained (Scheme 5.1-56).

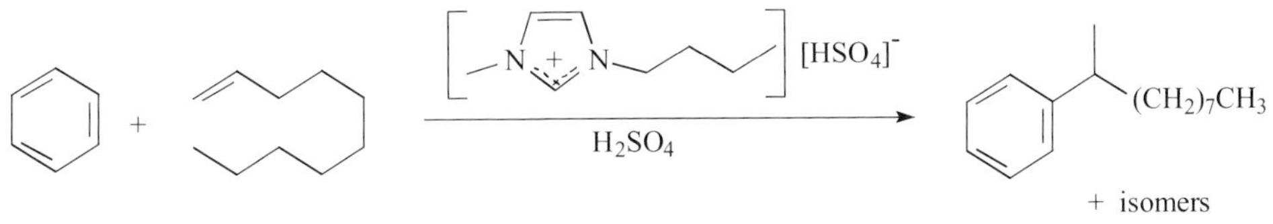

[HSO_4]$^-$

H_2SO_4

$(CH_2)_7CH_3$

+ isomers

Scheme 5.1-54: The sulfuric acid-catalyzed alkylation of benzene in a hydrogensulfate ionic liquid.

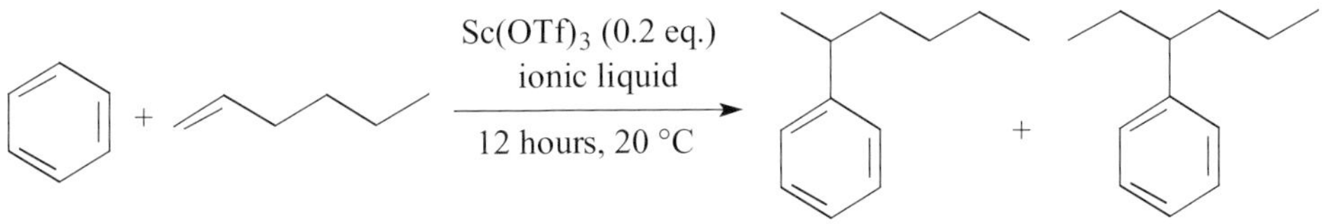

$Sc(OTf)_3$ (0.2 eq.)
ionic liquid
12 hours, 20 °C

Scheme 5.1-55: The alkylation of benzene with hex-1-ene, catalyzed by scandium(III) triflate in ionic liquids.

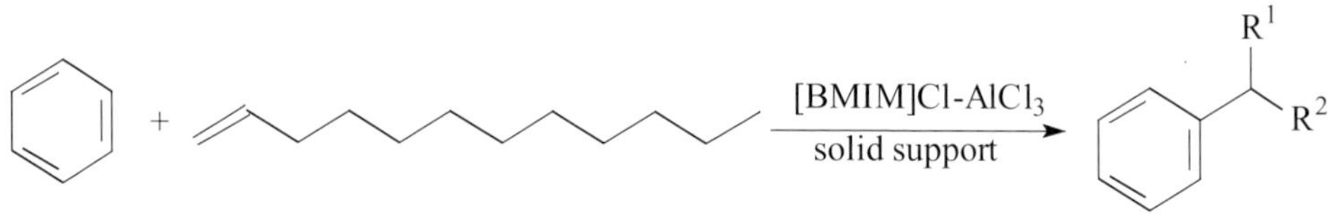

$$\text{Benzene} + \text{dodecene} \xrightarrow[\text{solid support}]{[BMIM]Cl\text{-}AlCl_3} \text{PhCH}(R^1)(R^2)$$

$R^1 = CH_3$, $R^2 = (CH_2)_9CH_3$ 2-phenyldodecane
$R^1 = CH_2CH_3$, $R^2 = (CH_2)_8CH_3$ 3-phenyldodecane
$R^1 = (CH_2)_2CH_3$, $R^2 = (CH_2)_7CH_3$ 4-phenyldodecane
$R^1 = (CH_2)_3CH_3$, $R^2 = (CH_2)_6CH_3$ 5-phenyldodecane
$R^1 = (CH_2)_4CH_3$, $R^2 = (CH_2)_5CH_3$ 6-phenyldodecane

Scheme 5.1-56: The alkylation of benzene with dodecene with an ionic liquid on a solid support.

The product distribution in the reaction of benzene with dodecene was determined for a number of catalysts (Table 5.1-4). As can be seen, the reaction with the zeolite H-Beta gave predominantly the 2-phenyldodecane, whereas the reaction in the pure ionic liquid gave a mixture of isomers, with selectivity similar to that of aluminium chloride. The two supported ionic liquid reactions (H-Beta / IL and T 350 / IL) again gave product distributions similar to aluminium(III) chloride (T350 is a silica support made by Degussa).

Raston has reported an acid-catalyzed Friedel–Crafts reaction [89] in which compounds such as 3,4-dimethoxyphenylmethanol were cyclized to cyclotriveratrylene (Scheme 5.1-57). The reactions were carried out in tributylhexylammonium bis(trifluoromethanesulfonyl)amide $[NBu_3(C_6H_{13})][(CF_3SO_2)_2N]$ with phosphoric or *p*-toluenesulfonic acid catalysts. The product was isolated by dissolving the ionic liquid/catalyst in methanol and filtering off the cyclotriveratrylene product as white crystals. Evaporation of the methanol allowed the ionic liquid and catalyst to be regenerated.

Table 5.1-4: The product distribution dependency on the catalyst used for the reaction of benzene with dodecene. IL = $[BMIM]Cl/AlCl_3$ ($X(AlCl_3) = 0.6$), Temperature = 80 °C, with 6 mol% catalyst and benzene to dodecene ratio = 10:1.

Catalyst	*2-Phenyl dodecane*	*3-Phenyl dodecane*	*4-Phenyl dodecane*	*5-Phenyl dodecane*	*6-Phenyl dodecane*
$AlCl_3$	46.4	19.4	12.7	12.1	9.5
IL ($X(AlCl_3) = 0.6$)	36.7	19.0	15.0	15.5	13.8
T 350 / IL	42.9	22.8	13.0	11.8	9.4
H-Beta	75.7	19.0	3.8	1.1	0.4
H-Beta / IL	43.9	21.2	12.4	12.0	10.5

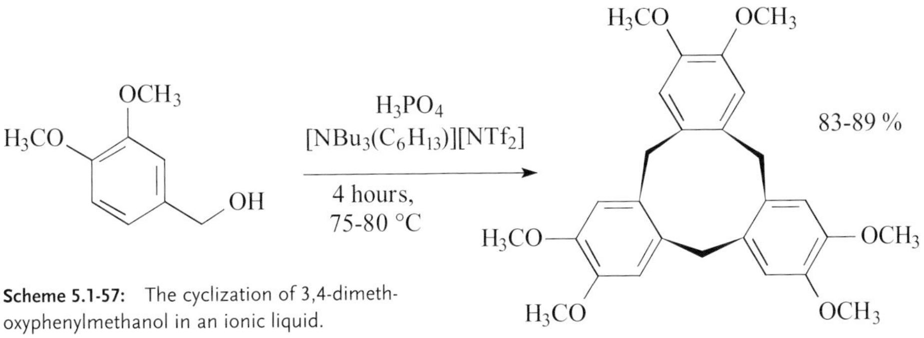

H_3PO_4
$[NBu_3(C_6H_{13})][NTf_2]$
4 hours, 75-80 °C
83-89 %

Scheme 5.1-57: The cyclization of 3,4-dimethoxyphenylmethanol in an ionic liquid.

5.1.2.3 Friedel–Crafts acylation reactions

Friedel–Crafts acylation reactions usually involve the interaction of an aromatic compound with an acyl halide or anhydride in the presence of a catalyst, to form a carbon–carbon bond [74, 75]. As the product of an acylation reaction is less reactive than its starting material, monoacylation usually occurs. The "catalyst" in the reaction is not a true catalyst, as it is often (but not always) required in stoichiometric quantities. For Friedel–Crafts acylation reactions in chloroaluminate(III) ionic liquids or molten salts, the ketone product of an acylation reaction forms a strong complex with the ionic liquid, and separation of the product from the ionic liquid can be extremely difficult. The products are usually isolated by quenching the ionic liquid in water. Current research is moving towards finding genuine catalysts for this reaction, some of which are described in this section.

The first example of a Friedel–Crafts acylation reaction in a molten salt was carried out by Raudnitz and Laube [90]. It involved the reaction between phthalic anhydride and hydroquinone at 200 °C in $NaCl/AlCl_3$ ($X(AlCl_3) = 0.69$) (Scheme 5.1-58).

Scholl and co-workers [91] performed the acylation of 1-benzoylpyrene with 4-methylbenzoyl chloride in a $NaCl/AlCl_3$ ($X(AlCl_3) = 0.69$) molten salt (110–120 °C). This gave 1-benzoyl-6-(4-methylbenzoyl)-pyrene as the major product (Scheme 5.1-59).

Bruce et al. carried out the cyclization of 4-phenylbutyric acid to tetralone in $NaCl/AlCl_3$ ($X(AlCl_3) = 0.68$) at 180–200 °C [92]. The reaction between valerolactone and hydroquinone to give 3-methyl-4,7-dihydroxyindanone was also performed by Bruce, using the same ionic liquid and reaction conditions. These are shown in Scheme 5.1-60.

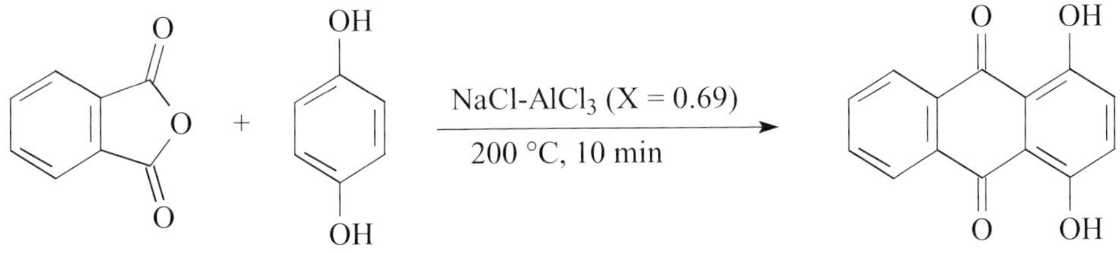

$NaCl\text{-}AlCl_3$ (X = 0.69)
200 °C, 10 min

Scheme 5.1-58: The reaction between phthalic anhydride and hydroquinone in $NaCl/AlCl_3$ ($X(AlCl_3) = 0.69$).

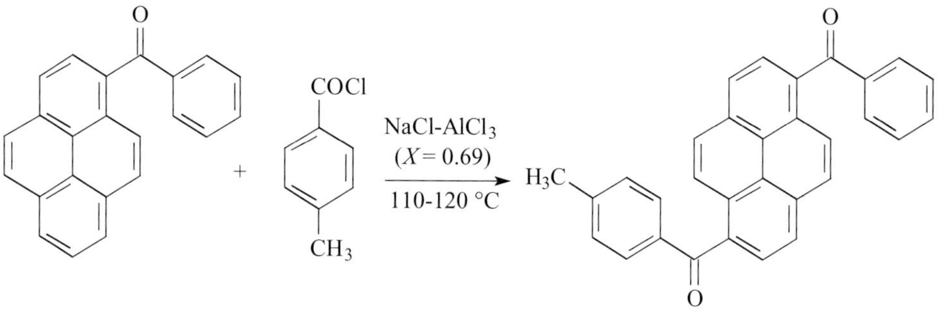

Scheme 5.1-59: The acylation of 1-benzoylpyrene in $NaCl/AlCl_3$ ($X(AlCl_3) = 0.69$).

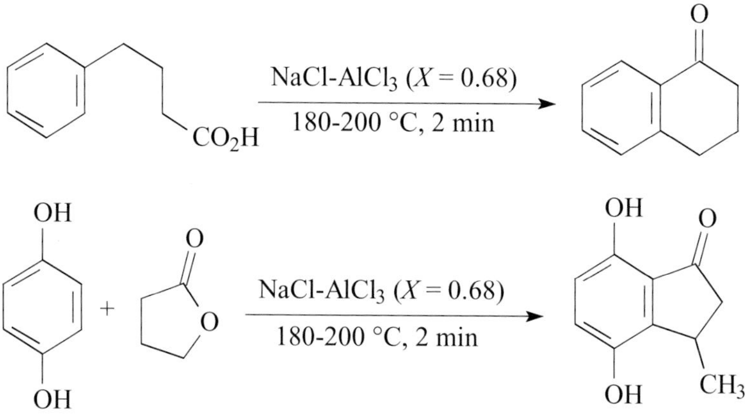

Scheme 5.1-60: The use of $NaCl/AlCl_3$ ($X(AlCl_3) = 0.68$) in the formation of cyclic ketones.

The Fries rearrangement can be viewed as a type of Friedel–Crafts acylation reaction. Two examples of this reaction are given in Scheme 5.1-61. The first is the rearrangement of 4,4'-diacetoxybiphenyl to 4,4'-dihydroxy-3,3'-diacetoxybiphenyl in a $NaCl/AlCl_3$ ($X(AlCl_3) = 0.69$) molten salt [93]. The second example is the rearrangement of phenyl 3-chloropropionate to 2'-hydroxy-3-chloropropiophenone, followed by cyclization to an indanone [94].

One of the problems with the $NaCl/AlCl_3$ molten salts is their high melting points and corresponding high reaction temperatures. The high reaction temperatures tend to cause side reactions and decomposition of the products of the reaction. Hence, a number of reactions have been carried out under milder conditions, in room-temperature ionic liquids. The first example of a Friedel–Crafts acylation in such an ionic liquid was performed by Wilkes and co-workers [66, 79] (Scheme 5.1-62). The rate of the acetylation reaction was found to be dependent on the concentration of the $[Al_2Cl_7]^-$ ion, suggesting that this ion was acting as the Lewis acid in the reaction. Wilkes went on to provide evidence that the acylating agent is the acetylium ion $[H_3CCO]^+$ [66, 79].

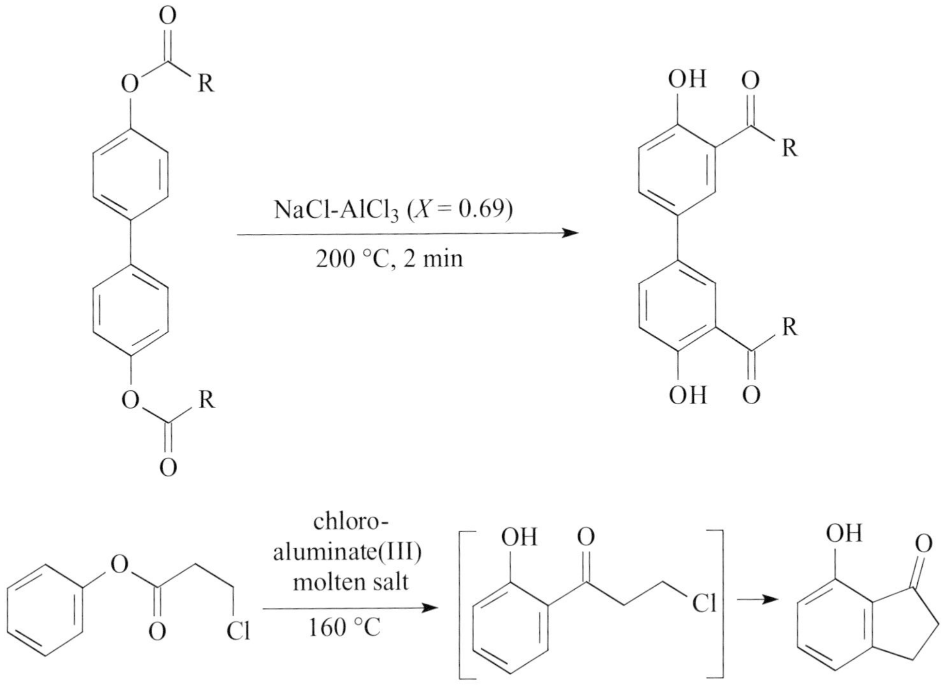

NaCl-AlCl$_3$ ($X = 0.69$)
200 °C, 2 min

chloro-aluminate(III) molten salt
160 °C

Scheme 5.1-61: The Fries rearrangement in chloroaluminate(III) molten salts.

[EMIM]-AlCl$_3$ ($X = 0.67$)

Scheme 5.1-62: The acetylation of benzene in a room-temperature ionic liquid.

A number of commercially important fragrance molecules have been synthesized by Friedel–Crafts acylation reactions in these ionic liquids. Traseolide® (5-acetyl-1,1,2,6-tetramethyl-3-isopropylindane) (Scheme 5.1-63) has been made in high yield in the ionic liquid [EMIM]Cl/AlCl$_3$ (X(AlCl$_3$) = 0.67) [95].

For the acylation of naphthalene, the ionic liquid gives the highest reported selectivity for the 1-position [95]. The acetylation of anthracene at 0 °C was found to be a reversible reaction. The initial product of the reaction between acetyl chloride (1.1 equivalents) and anthracene is 9-acetylanthracene, formed in 70 % yield in less than 5 minutes. The 9-acetylanthracene was then found to undergo diacetylation reactions, giving the 1,5- and 1,8-diacetylanthracenes and anthracene after 24 hours (Scheme 5.1-64).

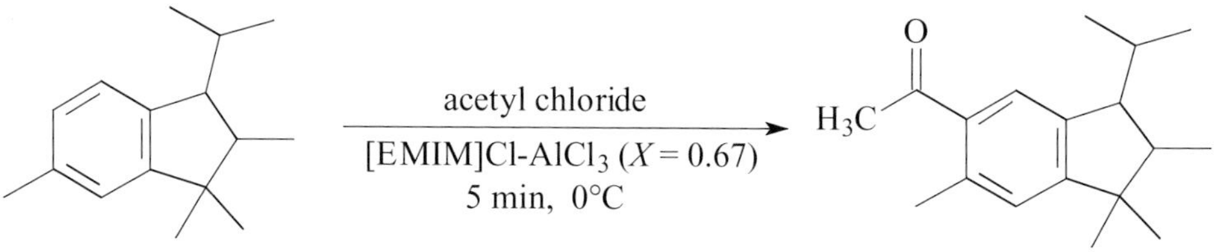

acetyl chloride
$[EMIM]Cl\text{-}AlCl_3$ ($X = 0.67$)
5 min, 0°C

Scheme 5.1-63: The acetylation of 1,1,2,6-tetramethyl-3-isopropylindane in $[EMIM]Cl/AlCl_3$ ($X(AlCl_3) = 0.67$).

H_3CCOCl
$[EMIM]Cl\text{-}(AlCl_3)$ ($X = 0.67$)
fast
$[H_3CCO]^+$ slow
$[H_3CCO]^+$
slow
$[H_3CCO]^+$
major isomer
minor isomer

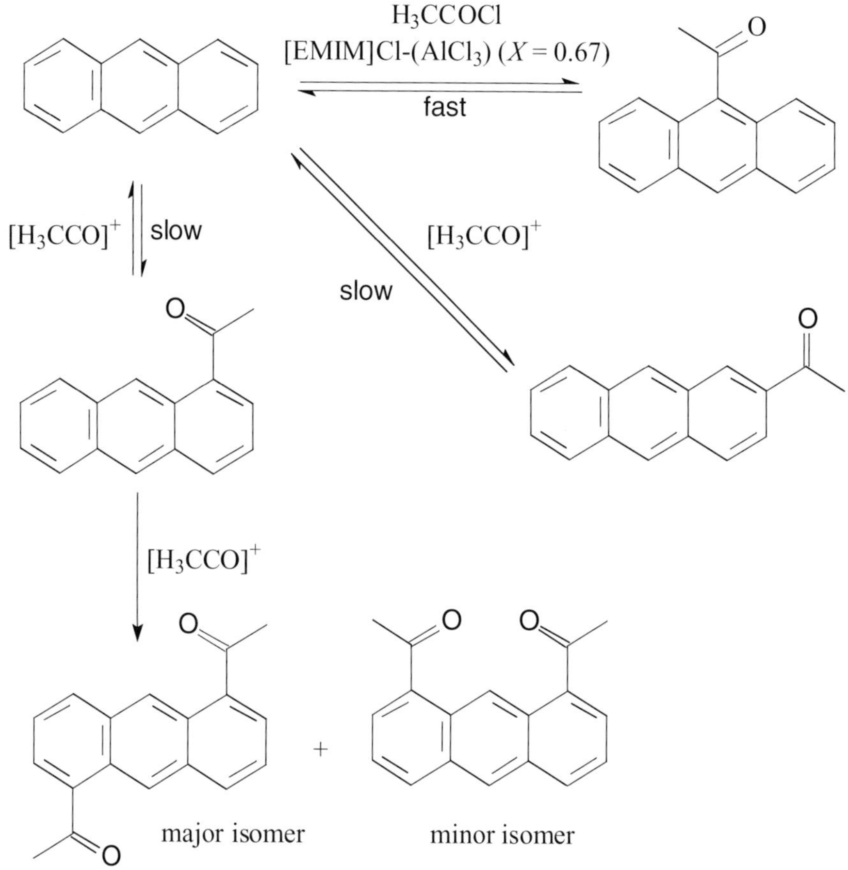

Scheme 5.1-64: The acetylation of anthracene in $[EMIM]Cl/AlCl_3$ ($X(AlCl_3) = 0.67$).

This was confirmed by taking a sample of 9-acetylanthracene and allowing it to isomerize in the ionic liquid. This gave a mixture of anthracene, 1,5-diacetylanthracene and 1,8-diacetylanthracene. It should be noted that a proton source was needed for this reaction to occur, implying an acid-catalyzed mechanism (Scheme 5.1-65) [95].

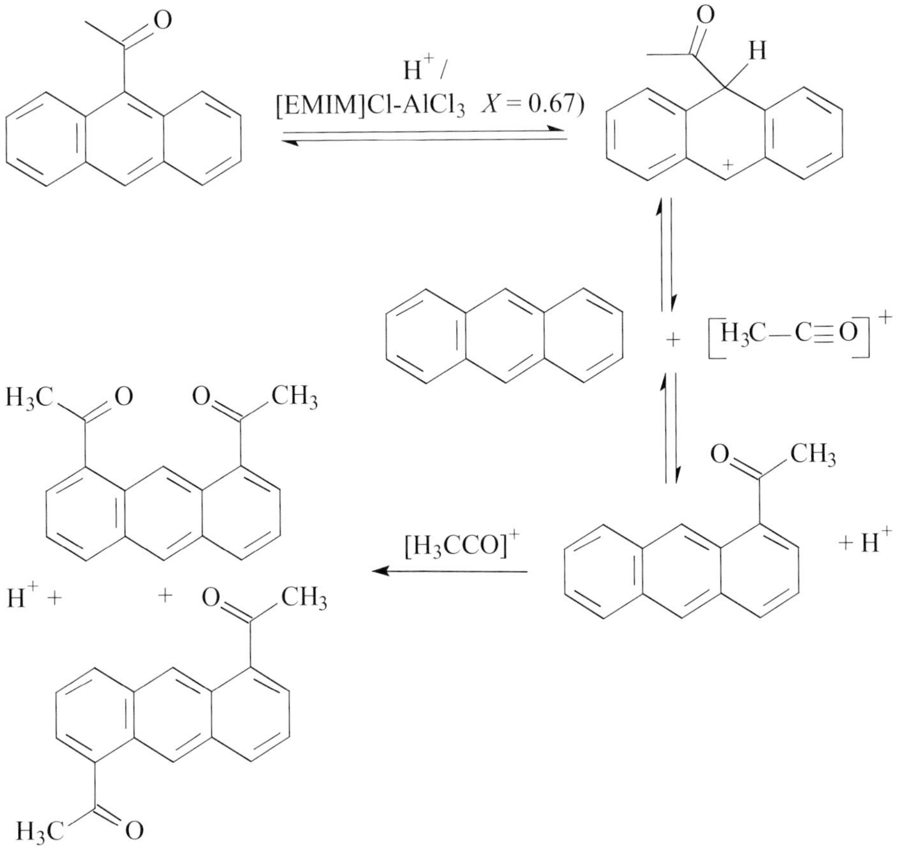

Scheme 5.1-65: Proposed mechanism for the isomerization of 9-acetylanthracene in $[EMIM]Cl/AlCl_3$ ($X(AlCl_3) = 0.67$).

The Friedel–Crafts acylation reaction has also been performed in iron(III) chloride ionic liquids, by Seddon and co-workers [96]. An example is the acetylation of benzene (Scheme 5.1-66). Ionic liquids of the type $[EMIM]Cl/FeCl_3$ ($0.50 < X(FeCl_3) < 0.62$) are good acylation catalysts, with the added benefit that the ketone product of the reaction can be separated from the ionic liquid by solvent extraction, provided that $X(FeCl_3)$ is in the range 0.51–0.55.

The ability of iron(III) chloride genuinely to catalyze Friedel–Crafts acylation reactions has also been recognized by Hölderich and co-workers [97]. By immobilizing the ionic liquid $[BMIM]Cl/FeCl_3$ on a solid support, Hölderich was able to acetylate mesitylene, anisole, and *m*-xylene with acetyl chloride in excellent yield. The performance of the iron-based ionic liquid was then compared with that of the corresponding chlorostannate(II) and chloroaluminate(III) ionic liquids. The results are given in Scheme 5.1-67 and Table 5.1-5. As can be seen, the iron catalyst gave superior results to the aluminium- or tin-based catalysts. The reactions were also carried out in the gas phase at between 200 and 300 °C. The acetylation reac-

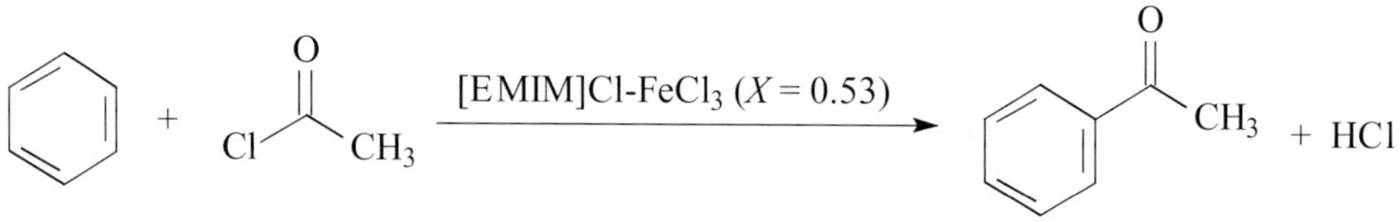

Scheme 5.1-66: The acetylation of benzene in an iron(IIII) chloride-based ionic liquid.

tion was complicated by two side reactions. In the reaction between acetyl chloride and *m*-xylene, for example, the decomposition of acetyl chloride to ketene and the formation of 1-(1-chlorovinyl)-2,4-dimethylbenzene were also found to occur [97].

Rebeiro and Khadilkar have investigated the reactions between trichloroalkanes and aromatic compounds. For example, the benzoylation of aromatic compounds in ionic liquids was performed with benzotrichloride, giving ketones on aqueous workup [98].

5.1.2.4 Cracking and isomerization reactions

Cracking and isomerization reactions occur readily in acidic chloroaluminate(III) ionic liquids. A remarkable example of this is the reaction of poly(ethene), which is converted into a mixture of gaseous alkanes of formula (C_nH_{2n+2}, where n = 3–5) and cyclic alkanes with a hydrogen to carbon ratio of less than two (Figure 5.1-4, Scheme 5.1-68) [99].

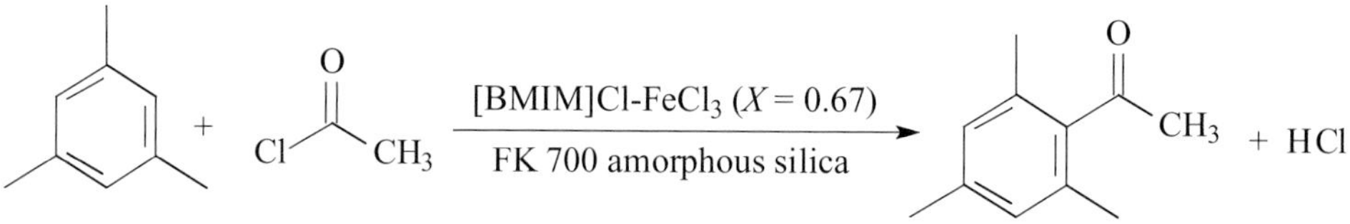

Scheme 5.1-67: The acetylation of aromatics with supported ionic liquids (FK 700 is a type of amorphous silica made by Degussa).

Table 5.1-5: The acylation of aromatics in batch reactions at 100 °C, for 1 hour. Ratio of aromatic compound to acetylating agent = 5:1, mes. = mesitylene.

Ionic liquid	*Reaction*	*Molar ratio IL : Ar-H*	*% Conversion*	*% Selectivity*
$[BMIM]Cl/AlCl_3$	mes. + AcCl	1:205	68.1	98
$[BMIM]Cl/AlCl_3$	anisole + Ac_2O	1:45	8.3	96
$[BMIM]Cl/AlCl_3$	*m*-xylene + AcCl	1:205	3.5	96
$[BMIM]Cl/FeCl_3$	mes. + AcCl	1:205	94.7	95
$[BMIM]Cl/FeCl_3$	anisole + Ac_2O	1:45	100	98
$[BMIM]Cl/FeCl_3$	*m*-xylene + AcCl	1:205	33.8	79
$[BMIM]Cl/SnCl_2$	anisole + Ac_2O	1:45	19.7	94
$[BMIM]Cl/SnCl_2$	*m*-xylene + AcCl	1:205	3.6	95

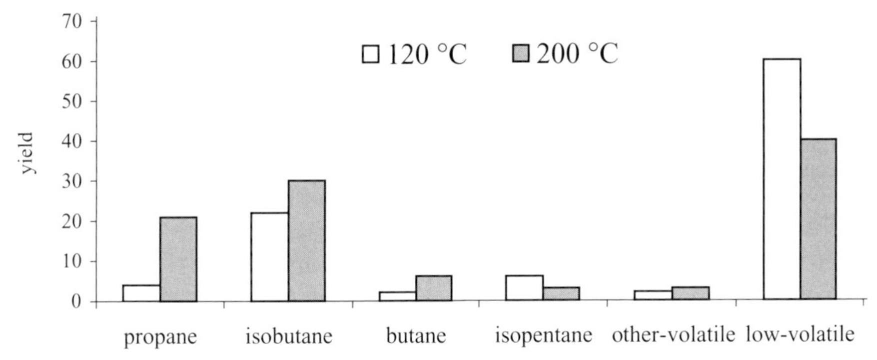

Figure 5.1-4: The products from the ionic liquid cracking of high-density polyethylene at 120 °C and at 200 °C.

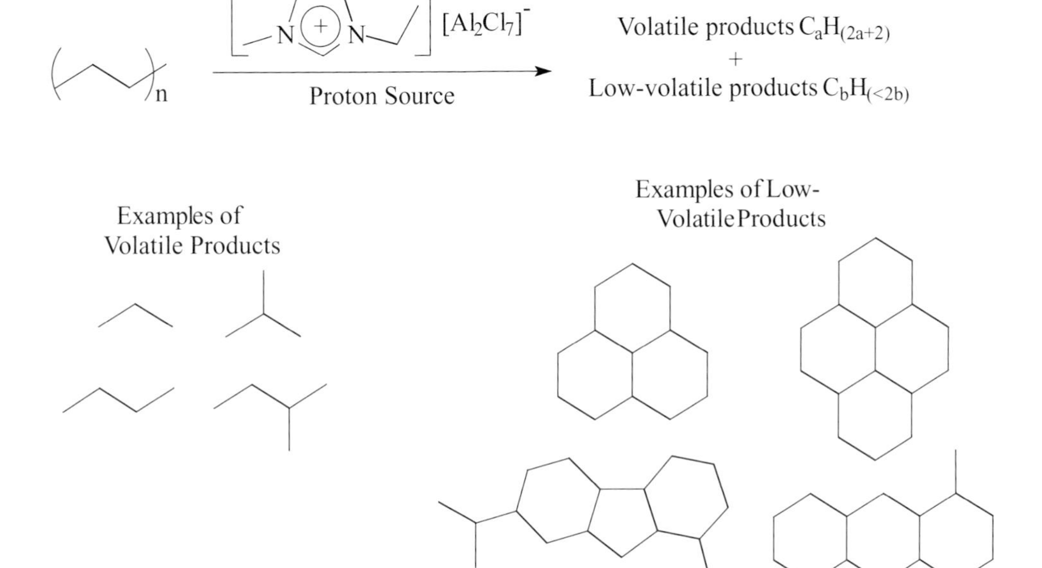

Scheme 5.1-68: The cracking of poly(ethene) in an ionic liquid.

The distribution of the products obtained from this reaction depends upon the reaction temperature (Figure 5.1-4) and differs from those of other poly(ethene) recycling reactions in that aromatics and alkenes are not formed in significant concentrations. Another significant difference is that this ionic liquid reaction occurs at temperatures as low as 90 °C, whereas conventional catalytic reactions require much higher temperatures, typically 300–1000 °C [100]. A patent filed for the Secretary of State for Defence (UK) has reported a similar cracking reaction for lower molecular weight hydrocarbons in chloroaluminate(III) ionic liquids [101]. An

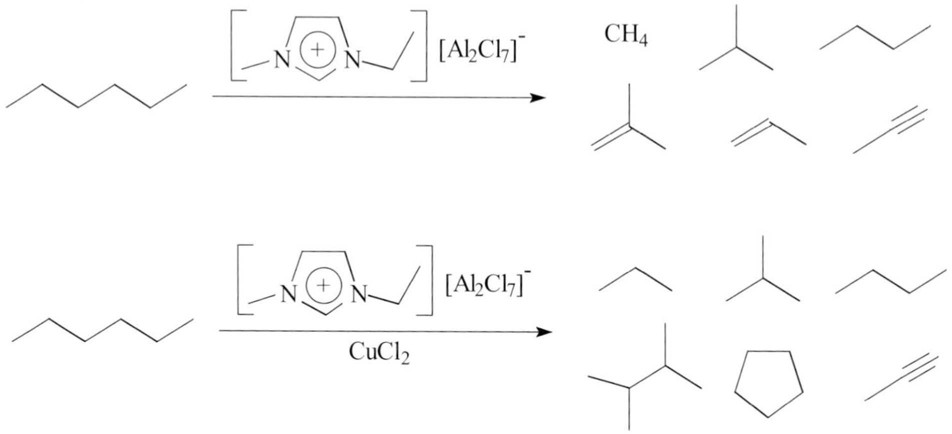

$[Al_2Cl_7]^-$

CH_4

$[Al_2Cl_7]^-$

$CuCl_2$

Scheme 5.1-69: The cracking of hexane in [EMIM]Cl/AlCl$_3$ ($X(AlCl_3) = 0.67$) with and without added copper(II) chloride.

example is the cracking of hexane to products such as propene and isobutene (Scheme 5.1-69). The reaction was also performed with added copper(II) chloride, which gave a significantly different product distribution.

A similar reaction occurs with fatty acids (such as stearic acid) or methyl stearate, which undergo isomerization, cracking, dimerization, and oligomerization reactions. This has been used to convert solid stearic acid into the more valuable liquid isostearic acid [102] (Scheme 5.1-70). The isomerization and dimerization of oleic acid and methyl oleate have also been found to occur in chloroaluminate(III) ionic liquids [103].

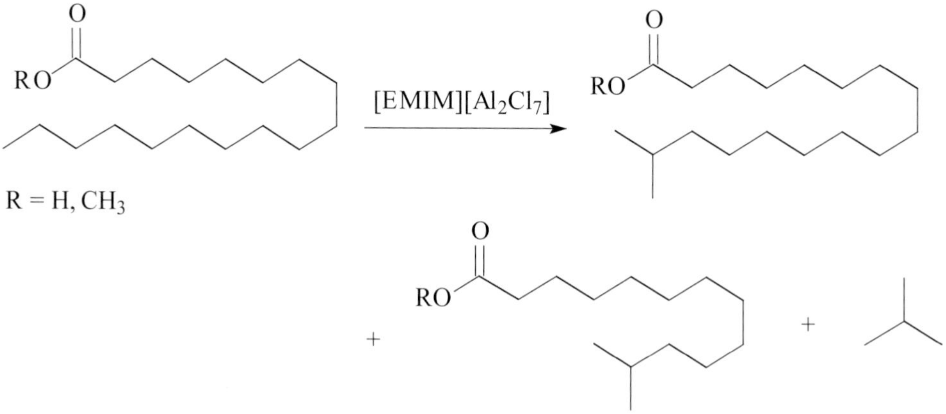

O RO

R = H, CH_3

$[EMIM][Al_2Cl_7]$

O RO

\+ O RO +

Scheme 5.1-70: Cracking and isomerization of fatty acids and fatty acid methyl esters in chloroaluminate(III) ionic liquids.

References

1 T. Welton, *Chem. Rev.* **1999**, *99*, 2071–2083.
2 J. D. Holbrey. K. R. Seddon, *Clean Prod. Proc.* **1999**, *1*, 223–236.
3 M. J. Earle, K. R. Seddon, *Pure Appl. Chem.* **2000**, *72*, 1391–1398.
4 R. M. Pagni, in *Advances in Molten Salt Chemistry*, Vol. 6 (G. Mamantov, J. Braunstein eds.), Elsevier, Oxford, **1987**, 211–346.
5 D. W. Rooney, K. R. Seddon, in *The Handbook of Solvents* (G. Wypych ed.), ChemTech Publishing, New York, **2001**, 1459–1484.
6 K. R. Seddon, *J. Chem. Tech. Biotech.* **1997**, *68*, 351–356.
7 K. R. Seddon, *Green Chem*, **2002**, *4*, 147–151.
8 P. Wasserscheid, W. Keim, *Angew. Chem. Int. Ed.* **2000**, *39*, 3773–3789.
9 J. Dupont, C. S. Consorti, J. Spencer, *J. Braz. Chem. Soc.* **2000**, *11*, 337–344.
10 R. A. Sheldon, *Chem. Commun.* **2001**, 2399–2407.
11 M. Fremantle, *Chem. Eng. News* **1998** (30th March), *76*, 32–37.
12 C. M. Gordon, J. D. Holbrey, A. R. Kennedy, K. R. Seddon, *J. Mater. Chem.* **1998**, *8*, 2627–2636.
13 K. R. Seddon, A. Stark, M. J. Torres, *Pure App. Chem.* **2000**, *72*, 2275–2287.
14 A. E. Visser, R. P. Swatloski, R. D. Rogers, *Green Chem.* **2000**, *2*, 1–4.
15 J. March, *Advanced Organic Chemistry*, 4th ed., Wiley, Chichester, **1992**.
16 M. Fieser, L. Fieser, *Reagents for Organic Synthesis*, Vol. 1, Wiley-Interscience, New York, **1967**, 936, and references therein.
17 C. M. Suter, *The Organic Chemistry of Sulfur, Tetravalent Sulfur Compounds*, Wiley-Interscience, New York, **1944**, 420.
18 G. P. Smith, R. M. Pagni in *Molten Salt Chemistry, An introduction to selected Applications* (G. Mamantov, R. Marassi eds.), D. Reidel Publishing Co., Dordrecht, **1987**, 383–416.
19 R. K. Sharma, J. L. Fry, *J. Org. Chem.* **1983**, *48*, 2112–2114.
20 E. Nakamura, M. Shimizu, I. Kuwajima, J. Sakata, K. Yokoyama, R. Noyori, *J. Org. Chem.* **1983**, *48*, 932–945.
21 G. Majetich, A. Casares, D. Chapman, M. Behnke, *J. Org. Chem.* **1986**, *51*, 1745–1753.
22 T. W. Green, P. G. M. Wuts, *Protective Groups in Organic Synthesis*, Wiley, New York, **1991**.
23 D. P. Cox, J. Terpinski, W. Lawrynowicz, *J. Org. Chem.* **1984**, *49*, 3216–3219.
24 R. A. Moss, W. Lawrynowicz, *J. Org. Chem.* **1984**, *49*, 3828–3830.
25 D. P. Cox, R. A. Moss, J. Terpinski, *J. Am. Chem. Soc.* **1983**, *105*, 6513–6514.
26 P. Studt, *Lebigs Ann. Chem.* **1978**, 528–529.
27 K. Nakasuji, K. Yoshida and I. Murata, *J. Am. Chem. Soc.* **1983**, *105*, 5136–5137.
28 M. B. Groen, H. Schadenberg, H. Wynberg, *J. Org. Chem.* **1971**, *36*, 2797–2809.
29 H. Imaizumi, S. Sekiguchi, K. Matsui, *Bull. Chem. Soc. Jpn.* **1977**, *50*, 948–952.
30 A. C. Buchanan III, D. M. Chapman, G. P. Smith, *J. Org. Chem.* **1985**, *50*, 1702–1711.
31 G. P. Smith, A. S. Dworkin, R. M. Pagni, S. P. Zingg, *J. Am. Chem. Soc.* **1989**, *111*, 525–530.
32 C. J. Adams, M. J. Earle, K. R. Seddon, *Chem. Commun.* **1999**, 1043–1044.
33 L. Green, I. Hemeon, R. D. Singer, *Tetrahedron Lett.* **2000**, *41*, 1343–1345.
34 Z. Ma, Y. Q. Deng, F. Shi, Can. Patent CN 1247856, **2000**.
35 Y. Q. Deng, F. Shi, J. J. Beng, K. Qiao, *J. Mol. Cat. A* **2001**, *165*, 33–36.
36 C. W. Lee, *Tetrahedron Lett.* **1999**, *40*, 2461–2462.
37 J. D. Huddleston, H. D. Willauer, R. P. Swatloski, A. E. Visser, R. D. Rogers, *Chem. Commun.* **1998**, 1765–1766.
38 A. J. Carmichael, M. J. Earle, J. D. Holbrey, P. B. McCormac, K. R. Seddon, *Org. Lett.* **1999**, *1*, 997–1000.

39 L. P. Barthel–Rosa, J. A. Gladysz, *Coord. Chem. Rev.* **1999**, *192*, 587–605.
40 D. A. Jaeger, C. E. Trucker, *Tetrahedron Lett.* **1989**, *30*, 1785–1788.
41 J. Howarth, K. Hanlon, D. Fayne, P. B. McCormac, *Tetrahedron. Lett.* **1997**, *38*, 3097–3099.
42 M. J. Earle, P. B. McCormac, K. R. Seddon, *Green Chem.* **1999**, *1*, 23–25.
43 M. J. Earle, in Ioniy Liquids, R. D. Rogers, K. R. Seddon eds., *ACS Symposium Series*, **2002**, *818*, 90–105.
44 T. Fisher, A. Sethi, T. Welton, J. Woolf, *Tetrahedron Lett.* **1999**, *40*, 793–795.
45 P. Ludley, N. Karodia, *Tetrahedron Lett.* **2001**, *42*, 2011–2014.
46 F. Zulfiqar, T. Kitazume, *Green Chem.* **2000**, *2*, 137–139.
47 W. T. Ford, R. J. Hauri, D. J. Hart, *J. Org, Chem.* **1973**, *38*, 3916–3918.
48 W. T. Ford, R. J. Hauri, *J. Am. Chem. Soc.* **1973**, *95*, 7381–7391.
49 W. T. Ford, *J. Org, Chem.* **1973**, *38*, 3614–3615.
50 M. Badri, J.-J. Brunet, *Tetrahedron Lett.* **1992**, *33*, 4435–4438.
51 M. J. Earle, P. B. McCormac, K. R. Seddon, *Chem. Commun.* **1998**, 2245–2246.
52 C. Wheeler, K. N. West, C. L. Liotta, C. A. Eckert, *Chem. Commun.* **2001**, 887–888.
53 M. J. Earle, P. B. McCormac, K. R. Seddon, *Green Chem.* **2000**, *2*, 261–262.
54 C. M. Gordon, A. McClusky, *Chem. Commun.* **1999**, 143–144.
55 T. Kitazume, K. Kasai, *Green Chem.* **2001**, *3*, 30–32.
56 A. McCluskey, J. Garner, D. J. Young, S. Caballero, *Tetrahedron Lett*, **2000**, *41*, 8147–8151.
57 G. W. Kabalka, R. R. Malladi, *Chem. Commun.* **2000**, 2191.
58 T. Kitazume, F. Zulfiqar, G. Tanaka, *Green Chem.*, **2000**, *2*, 133–136.
59 P. P. Peng, Y. Q. Deng, *Tetrahedron Lett.* **2001**, *42*, 403–405.
60 T. Kitazume, G. Tanaka, *J. Fluorine. Chem.* **2000**, *106*, 211–215.
61 D. W. Morrison, D. C. Forbes, J. H. Davis Jr., *Tetrahedron Lett.* **2001**, *42*, 6053–6055.
62 J. Peng, Y. Deng, *Tetrahedron Lett.* **2001**, *42*, 5917–5919.
63 R. D. Singer, P. J. Scammells, *Tetrahedron Lett.* **2001**, 42, 6831–6833.
64 S. V. Ley, A. G. Leach, R. I. Storer, *J. Chem. Soc., Perkin Trans. 1* **2001**, 358–361.
65 C. Hardacre, J. D. Holbrey, S. E. McMath, *Chem. Commun.* **2001**, 367–368.
66 J. A. Boon, S. W. Lander Jr., J. A. Levisky, J. L. Pflug, L. M. Skrznecki-Cooke, J. S. Wilkes, *Proceedings of the Joint International Symposium on Molten Salts*, 6th ed., **1987**, 979–990.
67 K. K. Laali, V. J. Gettwert, *J. Org. Chem.* **2001**, 66, 35–40.
68 Y. Patell, N. Winterton, K. R. Seddon, World Patent WO 0037400, **2000**.
69 D. S. Newman, T. H. Kinstle, G. Thambo, *Proceedings of the Joint International Symposium on Molten Salts*, 6th ed., **1987**, 991–1001.
70 D. S. Newman, T. H. Kinstle, G. Thambo, *J. Electrochem. Soc.* **1987**, *134*, C512.
71 D. S. Newman, R. E. Winans, R. L. McBeth, *J. Electrochem. Soc.* **1984**, *131*, 1079–1083.
72 F. Zulfiqar, T. Kitazume, *Green Chem.* **2000**, *2*, 296–298.
73 S.-gi Lee, J. H. Park, J. Kang, J. K. Lee, *Chem. Commun.* **2001**, 1698–1699.
74 G. A. Olah, *Friedel–Crafts and Related Reactions*, Interscience , New York, **1963**.
75 G. A. Olah, *Friedel Crafts Chemistry*, Wiley–Interscience, New York, **1973**.
76 G. Baddeley, R. Williamson, *J. Chem. Soc.* **1953**, 2120–2123.
77 G. Baddeley, G. Holt, S. M. Makar, M. G. Ivinson, *J. Chem. Soc.* **1952**, 3605–3607.
78 W. L. Mendelson, C. B. Spainhour, S. S. Jones, B. L. Lamb, K. L. Wert, *Tetrahedron Lett.* **1980**, *21*, 1393–1396.
79 J. A. Boon, J. A. Levisky, J. L. Pflug, J. S. Wilkes, *J. Org. Chem*, **1986**, *51*, 480–483.
80 B. J. Piersma, M. Merchant in *Proceedings of the 7th International Symposium on Molten Salts* (C. L. Hussey, S. N. Flengas, J. S. Wilkes, Y. Ito eds.), The Electrochemical Society, Pennington, **1990**, 805–821.

81 P. K. G. Hodgson, M. L. M. Morgan, B. Ellis, A. A. K. Abdul-Sada, M. P. Atkins, K. R. Seddon, *US Patent*, US 5994602, **1999**.
82 P. Wasserscheid, B. Ellis, H. Fabienne, *World Patent*, WO 0041809, **2000**.
83 C. C. Greco, S. Fawzy, S. Lieh-Jiun, *US Patent*, US 5824832, **1998**.
84 W. Keim, W. Korth, P. Wasserscheid, *World Patent*, WO 0016902, **2000**.
85 C. E. Song, W. H. Shim, E. J. Roh, J. H. Choi, *Chem. Commun.* **2000**, 1695–1696.
86 M. J. Earle, unpublished results, **1998**.
87 C. P. DeCastro, E. Sauvage, M. H. Valkenberg, W. F. Hölderich, *World Patent*, WO 0132308, **2001**.
88 C. P. DeCastro, E. Sauvage, M. H. Valkenberg, W. F. Hölderich, *J. Catal.* **2000**, *196*, 86–94.
89 J. L. Scott, D. R. MacFarlan, C. L. Raston, C. M. Teoh, *Green Chem.* **2000**, *2*, 123–126.
90 H. Raudnitz, G. Laube, *Ber.* **1929**, *62*, 509.
91 R. Scholl, K. Meyer, J. Donat, *Chem. Ber.* **1937**, *70*, 2180–2189.
92 D. B. Bruce, A. J. S. Sorrie, R. H. Thomson, *J. Chem. Soc.* **1953**, 2403–2408.
93 G. C. Misra, L. M. Pande, G. C. Joshi, A. K. Misra, *Aust. J. Chem.* **1972**, *25*, 1579–1581.
94 S. Wagatsuma, H. Higuchi, T. Ito, T. Nakano, Y. Naoi, K. Sakai, T. Matsui, Y. Takahashi, A. Nishi, S. Sano, *Org. Prep. Proced. Int.* **1973**, *5*, 65–70.
95 C. J. Adams, M. J. Earle, G. Roberts, K. R. Seddon, *Chem. Commun.* **1998**, 2097–2098.
96 P. N. Davey, M. J. Earle, C. P. Newman, K. R. Seddon, *World Patent* WO 99 19288, **1999**.
97 M. H. Valkenberg, C. deCastro, W. F. Hölderich, *App. Catal. A* **2001**, *215*, 185–190.
98 G. L. Rebeiro, B. M. Khadilkar, *Syn. Commun.* **2000**, *30*, 1605–1608.
99 C. J. Adams, M. J. Earle, K. R. Seddon, *Green Chem.* **2000**, *2*, 21–24.
100 R. W. J. Westerhout, J. A. M. Kuipers, W. P. M. van Swaaij, *Ind. Eng. Chem. Res.* **1998**, *37*, 841–847.
101 P. N. Barnes, K. A. Grant, K. J. Green, N. D. Lever, World Patent WO 0040673, **2000**.
102 C. J. Adams, M. J. Earle, J. Hamill, C. M. Lok, G. Roberts, K. R. Seddon, World Patent WO 9807680, **1998**.
103 C. J. Adams, M. J. Earle, J. Hamill, C. M. Lok, G. Roberts, K. R. Seddon, World Patent WO 9807679, **1998**.

5.2 Transition Metal Catalysis in Ionic Liquids

Peter Wasserscheid

Many transition metal complexes dissolve readily in ionic liquids, which enables their use as solvents for transition metal catalysis. Sufficient solubility for a wide range of catalyst complexes is an obvious, but not trivial, prerequisite for a versatile solvent for homogenous catalysis. Some of the other approaches to the replacement of traditional volatile organic solvents by “greener” alternatives in transition metal catalysis, namely the use of supercritical CO_2 or perfluorinated solvents, very often suffer from low catalyst solubility. This limitation is usually overcome by use of special ligand systems, which have to be synthesized prior to the catalytic reaction.

In the case of ionic liquids, special ligand design is usually not necessary to obtain catalyst complexes dissolved in the ionic liquid in sufficiently high concentrations.

However, it should be mentioned that the dissolution process of a solid, crystalline complex in an (often relatively viscous) ionic liquid can sometimes be slow. This is due to restricted mass transfer and can be speeded up either by increasing the exchange surface (ultrasonic bath) or by reducing the ionic liquid's viscosity. The latter is easily achieved by addition of small amounts of a volatile organic solvent that dissolves both the catalyst complex and the ionic liquid. As soon as the solution is homogeneous, the volatile solvent is then removed in vacuo.

Since no special ligand design is usually required to dissolve transition metal complexes in ionic liquids, the application of ionic ligands can be an extremely useful tool with which to immobilize the catalyst in the ionic medium. In applications in which the ionic catalyst layer is intensively extracted with a non-miscible solvent (i.e., under the conditions of biphasic catalysis or during product recovery by extraction) it is important to ensure that the amount of catalyst washed from the ionic liquid is extremely low. Full immobilization of the (often quite expensive) transition metal catalyst, combined with the possibility of recycling it, is usually a crucial criterion for the large-scale use of homogeneous catalysis (for more details see Section 5.3.5).

The first example of homogeneous transition metal catalysis in an ionic liquid was the platinum-catalyzed hydroformylation of ethene in tetraethylammonium trichlorostannate (mp. 78 °C), described by Parshall in 1972 (Scheme 5.2-1, a)) [1]. In 1987, Knifton reported the ruthenium- and cobalt-catalyzed hydroformylation of internal and terminal alkenes in molten $[Bu_4P]Br$, a salt that falls under the now accepted definition for an ionic liquid (see Scheme 5.2-1, b)) [2]. The first applications of room-temperature ionic liquids in homogeneous transition metal catalysis were described in 1990 by Chauvin et al. and by Wilkes et al.. Wilkes et al. used weekly acidic chloroaluminate melts and studied ethylene polymerization in them with Ziegler–Natta catalysts (Scheme 5.2-1, c)) [3]. Chauvin's group dissolved nickel catalysts in weakly acidic chloroaluminate melts and investigated the resulting ionic catalyst solutions for the dimerization of propene (Scheme 5.2-1, d)) [4].

The potential of ionic liquids as novel media for transition metal catalysis received a substantial boost from the work of Wilkes' group, when in 1992 they described the synthesis of non-chloroaluminate, room-temperature liquid systems with significantly enhanced stability to hydrolysis, such as low-melting tetrafluoroborate melts [5]. In contrast to the chloroaluminate ionic liquids, these "second generation ionic liquids" offer high tolerance to functional groups, which opens up a much larger range of applications, especially for transition metal catalysis. The first successful catalytic reactions in tetrafluoroborate ion-based ionic liquids included the rhodium-catalyzed hydrogenation and hydroformylation of olefins [6]. Nowadays, the tetrafluoroborate and the (published slightly later [7]) hexafluorophosphate ionic liquids are among the "workhorses" for transition metal catalysis in ionic liquids. They – like some other ionic liquids with weakly coordinating anions – combine the properties of relatively polar yet non-coordinating solvents. This special combination makes them extremely suitable solvents for reactions involving electrophilic catalysts [8]. Moreover, these ionic liquids are now widely commercially available [9], so research groups and companies focussing on catalyt-

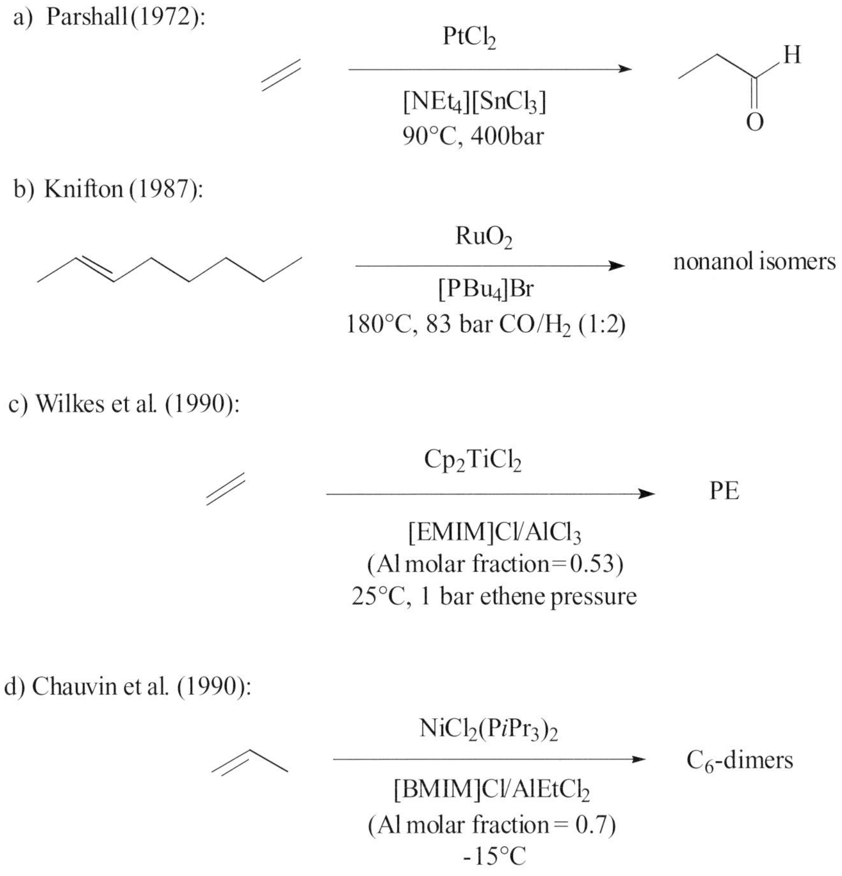

a) Parshall(1972):

$\xrightarrow[\text{[NEt}_4\text{][SnCl}_3\text{]; 90°C, 400bar}]{\text{PtCl}_2}$

b) Knifton (1987):

$\xrightarrow[\text{[PBu}_4\text{]Br; 180°C, 83 bar CO/H}_2\text{ (1:2)}]{\text{RuO}_2}$ nonanol isomers

c) Wilkes et al. (1990):

$\xrightarrow[\text{[EMIM]Cl/AlCl}_3\text{ (Al molar fraction=0.53); 25°C, 1 bar ethene pressure}]{\text{Cp}_2\text{TiCl}_2}$ PE

d) Chauvin et al. (1990):

$\xrightarrow[\text{[BMIM]Cl/AlEtCl}_2\text{ (Al molar fraction= 0.7); -15°C}]{\text{NiCl}_2\text{(P}i\text{Pr}_3\text{)}_2}$ C_6-dimers

Scheme 5.2-1: Early examples of transition metal catalysis in ionic liquids.

ic applications do not necessarily have to go through all of the synthetic work themselves (for the synthesis of ionic liquids, and especially for the quality requirements related to their applications as solvents in homogeneous catalysis, see Chapter 2).

However, a number of limitations are still evident when tetrafluoroborate and hexafluorophosphate ionic liquids are used in homogeneous catalysis. The major aspect is that these anions are still relatively sensitive to hydrolysis. The tendency to anion hydrolysis is of course much less pronounced than that of the chloroaluminate melts, but it still occurs and this has major consequences for their use in transition metal catalysis. For example, the $[PF_6]^-$ anion of 1-butyl-3-methylimidazolium ([BMIM]) hexafluorophosphate was found (in the author's laboratories) to hydrolyze completely after addition of excess water when the sample was kept for 8 h at 100 °C. Gaseous HF and phosphoric acid were formed. Under the same conditions, only small amounts of the tetrafluoroborate ion of $[BMIM][BF_4]$ was converted into HF and boric acid [10]. The hydrolytic formation of HF from the anion of the ionic liquid under the reaction conditions causes the following problems with regard to their use as solvents for transition metal catalysis: a) loss or partial loss of the ionic liquid solvent, b) corrosion problems related to the HF formed, and c) deactivation of the transition metal catalyst through its irreversible complexation by F^- ions. Consequently, the application of tetrafluoroborate and hexafluorophosphate ionic liquids is effectively restricted – at least in technical applications – to those applications in which water-free conditions can be achieved at acceptable costs or – even more so – in which the catalyst or the substrates used are water-sensitive anyway, so that the reaction is traditionally carried out under inert conditions.

In 1996, Grätzel, Bonhôte and co-workers published syntheses and properties of ionic liquids with anions containing CF_3 and other fluorinated alkyl groups [11]. These do not show the same sensitivity towards hydrolysis as $[BF_4]^-$- and $[PF_6]^-$-containing systems. In fact, heating of $[BMIM][(CF_3SO_2)_2N]$ with excess water to 100 °C for 24 h did not reveal any hint of anion hydrolysis [10]. Successful catalytic experiments with these ionic liquid systems have been reported in, for example, the hydrovinylation of styrene catalyzed by a cationic nickel complex in $[EMIM][(CF_3SO_2)_2N]$ [12]. However, despite the very high stabilities of these salts to hydrolysis and a number of other very suitable properties (such as low viscosity, high thermal stability, easy preparation in halogen-free form due to the miscibility gap with water), the high price of $[(CF_3SO_2)_2N]^-$ and of related anions may be a major problem for their practical application in larger quantities (the Li salt is commercially available from both Rhodia and 3M). Moreover, the presence of fluorine in the anion may still be problematic even if hydrolysis is not an issue. In addition to the elevated price of the anion (in itself related to the presence of fluorine), the disposal of spent ionic liquids of this type, by combustion, for example, is more complicated due to the presence of the fluorine.

In this context, the use of ionic liquids with halogen-free anions may become more and more popular. In 1998, Andersen et al. published a paper describing the use of some phosphonium tosylates (all with melting points >70 °C) in the rhodium-catalyzed hydroformylation of 1-hexene [13]. More recently, in our laboratories, we found that ionic liquids with halogen-free anions and with much lower melting points could be synthesized and used as solvents in transition metal catalysis. $[BMIM][n\text{-}C_8H_{17}SO_4]$ (mp = 35 °C), for example, could be used as catalyst solvent in the rhodium-catalyzed hydroformylation of 1-octene [14].

The author anticipates that the further development of transition metal catalysis in ionic liquids will, to a significant extent, be driven by the availability of new ionic liquids with different anion systems. In particular, cheap, halogen-free systems combining weak coordination to electrophilic metal centers and low viscosity with high stability to hydrolysis are highly desirable.

Very recently, Olivier-Bourbigou and Magna [15], Sheldon [16], and Gordon [17] have published three excellent reviews presenting a comprehensive overview of current work in transition metal catalysis involving ionic liquids, with slightly different emphases. All three update previously published reviews on the same topic, by Wasserscheid and Keim [18], Welton [19] and Seddon and Holbrey [20].

Without doubt, this extensive reviewing practice is a clear sign of the busy research activity in the field. However, there is clearly no need at present for another complete list of publications. Consequently, this section will aim to derive general principles from the work published so far, in order to provide a better understanding of the scope and limitations of the actual development of transition metal catalysis in ionic liquids. In this way, the author hopes to encourage scientists working in the field of transition metal catalysis to test and further develop ionic liquids as a "tool box" for their future research.

The section is divided into two major parts. Section 5.2.1 presents general motivation, successful concepts and current strategies, as well as those aspects that still represent limiting factors for transition metal catalysis in ionic liquids. In Section 5.2.2, selected applications are described in more detail, to demonstrate how the strategies explained earlier have already been used to produce new, superior catalytic systems. For the selection of these applications, the maturity of the research, the degree of understanding, the general significance, and the potential to transfer key results to future, new applications have been taken as criteria. The (obviously somewhat subjective) selection includes transition metal-catalyzed hydrogenation, oxidation, hydroformylation, Pd-catalyzed C-C-coupling and dimerization/oligomerization reactions that have been carried out using of ionic liquids.

5.2.1 Why use Ionic Liquids as Solvents for Transition Metal Catalysis?

5.2.1.1 Their nonvolatile nature

Probably the most prominent property of an ionic liquid is its lack of vapor pressure. Transition metal catalysis in ionic liquids can particularly benefit from this on economic, environmental, and safety grounds.

Obviously, the use of a nonvolatile ionic liquid simplifies the distillative workup of volatile products, especially in comparison with the use of low-boiling solvents, where it may save the distillation of the solvent during product isolation. Moreover, common problems related to the formation of azeotropic mixtures of the volatile solvents and the product/by-products formed are avoided by use of a nonvolatile ionic liquid. In the Rh-catalyzed hydroformylation of 3-pentenoic acid methyl ester it was even found that the addition of ionic liquid was able to stabilize the homogeneous catalyst during the thermal stress of product distillation (Figure 5.2-1) [21]. This option may be especially attractive technically, due to the fact that the stabilizing effects could already be observed even with quite small amounts of added ionic liquid.

As in stoichiometric organic reactions, the application of nonvolatile ionic liquids can contribute to the reduction of atmospheric pollution. This is of special relevance for non-continuous reactions, in which complete recovery of a volatile organic solvent is usually difficult to integrate into the process.

As well as this quite obvious environmental aspect, the switch from a volatile, flammable, organic solvent to an ionic liquid may significantly improve the safety of a given process. This will be especially true in oxidation reactions in which air or pure oxygen are used as oxidants; the use of common organic solvents is often restricted due to the potential formation of explosive mixtures between oxygen and

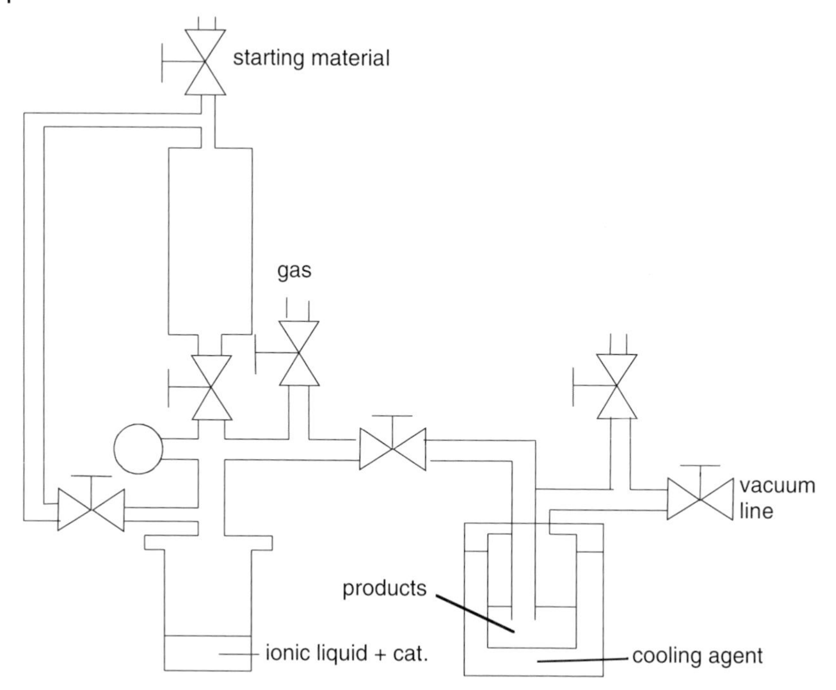

Figure 5.2-1: Stabilization of the active rhodium catalyst by addition of the ionic liquid [BMIM][PF_6] as co-solvent during distillative product isolation - apparatus for distillative product isolation from the ionic catalyst layer.

the volatile organic solvent in the gas phase. Although no example of this issue being addressed by use of an ionic liquid has yet been published in the open literature, there is no doubt that the application of nonvolatile solvents should open up new ways to overcome these problems. It may therefore be anticipated that there will be some future technical applications for ionic liquids in which solely the advantage of their nonvolatile character is used, largely for safety reasons.

5.2.1.2 New opportunities for biphasic catalysis

In comparison to heterogeneous catalyzed reactions, homogeneous catalysis offers several important advantages. The catalyst complex is usually well defined and can be rationally optimized by ligand modification. Every metal center can be active in the reaction. The reaction conditions are usually much milder (T usually < 200 °C), and selectivities are often much higher than with heterogeneous catalysts.

These advantages notwithstanding, the proportion of homogeneous catalyzed reactions in industrial chemistry is still quite low. The main reason for this is the difficulty in separating the homogeneously dissolved catalyst from the products and by-products after the reaction. Since the transition metal complexes used in homogeneous catalysis are usually quite expensive, complete catalyst recovery is crucial in a commercial situation.

Biphasic catalysis in a liquid–liquid system is an ideal approach through which to combine the advantages of both homogeneous and heterogeneous catalysis. The reaction mixture consists of two immiscible solvents. Only one phase contains the catalyst, allowing easy product separation by simple decantation. The catalyst phase can be recycled without any further treatment. However, the right combination of catalyst, catalyst solvent, and product is crucial for the success of biphasic catalysis [22]. The catalyst solvent has to provide excellent solubility for the catalyst complex without competing with the reaction substrate for the free coordination sites at the catalytic center.

Even more attractive is the possibility of optimizing the reaction's activity and selectivity by means of a biphasic reaction mode. This can be achieved by *in situ* extraction of catalyst poisons or reaction intermediates from the catalytic layer. To benefit from this potential, however, even more stringent requirements have to be fulfilled by the catalyst solvent, since this now has to provide a specific, very low solubility for the substances that are to be extracted from the catalyst phase under the reaction conditions. Figure 5.2-2 demonstrates this concept, shown for an oligomerization reaction. The dimer selectivity of the oligomerization of compound A can be significantly enhanced if the reaction is carried out in biphasic mode with a catalyst solvent with a high preferential solubility for A. The produced A–A is readily extracted from the catalyst phase into the product layer, which reduces the chance of the formation of higher oligomers.

From all this, it becomes understandable why the use of traditional solvents (such as water or butanediol) for biphasic catalysis has only been able to fulfil this potential in a few specific examples [23], whereas this type of highly specialized liquid–liquid biphasic operation is an ideal field for the application of ionic liquids, mainly due to their exactly tunable physicochemical properties (see Chapter 3 for more details).

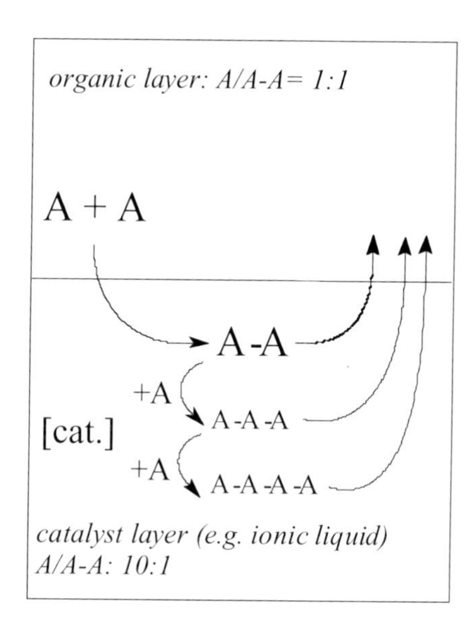

Figure 5.2-2: Enhanced dimer selectivity in the oligomerization of compound A due to a biphasic reaction mode with a catalyst solvent of high preferential solubility for A.

As well as the potential to enhance catalyst lifetime by recycling, or to improve a reaction's selectivity and a catalyst's activity by in situ extraction, biphasic catalysis also represents a very efficient way to reuse the (relatively expensive) ionic liquid itself. Thus – for a commercial application – the ionic liquid may be seen as an investment for the process (in an ideal case) or at least as a "working solution", meaning that only a small amount has to be replaced after a certain time of application. Obviously, multiphasic catalysis is the most promising way to use ionic liquids in catalysis in general and in transition metal catalysis in particular. Therefore, most groups dealing with transition metal catalysis in ionic liquids explore their potential under multiphasic reaction conditions. Consequently, most of the examples given in Section 5.2.4 have been carried out under biphasic catalysis conditions.

Because of the great importance of liquid–liquid biphasic catalysis for ionic liquids, all of Section 5.3 is dedicated to specific aspects relating to this mode of reaction, with special emphasis on practical, technical, and engineering needs. Finally, Section 5.4 summarizes a very interesting recent development for biphasic catalysis with ionic liquids, in the form of the use of ionic liquid/compressed CO_2 biphasic mixtures in transition metal catalysis.

5.2.1.3 Activation of a transition metal catalyst in ionic liquids

Apart from the activation of a biphasic reaction by extraction of catalyst poisons as described above, an ionic liquid solvent can activate homogeneously dissolved transition metal complexes by chemical interaction.

In general, it is possible for a chemical interaction between an ionic liquid solvent and a dissolved transition metal complex to be either activating or deactivating. It is therefore crucial to understand these chemical interactions in order to benefit from this potential and to avoid deactivation. Everything comes back to the rather obvious question of how much the presence of a specific ionic liquid influences the electronic and steric properties of the active catalyst complex and – perhaps even more importantly – to what extent the ionic liquid influences the availability of free coordination sites at the catalytic center for the substrates that are supposed to undergo the catalysis. Clearly, exact knowledge both of the catalytic mechanism in common organic solvents and of the chemical properties of the ionic liquid is very helpful in understanding these different effects.

In the following section, the nature of the chemical interactions between an ionic liquid and a transition metal catalyst is systematically developed according to the role of the ionic liquid in the different systems.

5.2.2 The Role of the Ionic Liquid

Depending on the coordinative properties of the anion and on the degree of the cation's reactivity, the ionic liquid can be regarded as an "innocent" solvent, as a ligand (or ligand precursor), as a co-catalyst, or as the catalyst itself.

5.2.2.1 The ionic liquid as "innocent" solvent

Ionic liquids with weakly coordinating, inert anions (such as $[(CF_3SO_2)_2N]^-$, $[BF_4]^-$, or $[PF_6]^-$ under anhydrous conditions) and inert cations (cations that do not coordinate to the catalyst themselves, nor form species that coordinate to the catalyst under the reaction conditions used) can be looked on as "innocent" solvents in transition metal catalysis. In these cases, the role of the ionic liquid is solely to provide a more or less polar, more or less weakly coordinating medium for the transition metal catalyst, but which additionally offers special solubility for feedstock and products.

However, the chemical inertness of these "innocent" ionic liquids does not necessarily mean that the reactivity of a transition metal catalyst dissolved in the ionic liquid is equal to the reactivity observed in common organic solvents. This becomes understandable from the fact that many organic solvents applied in catalytic reactions do not behave as innocent solvents, but show significant coordination to the catalytic center. The reason why these solvents are nevertheless used in catalysis is that some polar or ionic catalyst complexes are not soluble enough in weakly coordinating organic solvents. For example, many cationic transition metal complexes are known to be excellent oligomerization catalysts [24]. However, their usually poor solubilities in non-polar solvents often require, if organic solvents are used, a compromise between the solvation and the coordination properties of the solvent. In order to achieve sufficient solubility of the metal complex a solvent of higher polarity is required, and this may compete with the substrate for the coordination sites at the catalytic center. Consequently, the use of an inert, weakly coordinating ionic liquid in these cases can result in a clear enhancement of catalytic activity, since some ionic liquids are known to combine high solvation power for polar catalyst complexes (polarity) with weak coordination (nucleophilicity) [25]. It is this combination of properties of the ionic liquids that cannot be attained with water or common organic solvents.

5.2.2.2 Ionic liquid as solvent and co-catalyst

Ionic liquids formed by treatment of a halide salt with a Lewis acid (such as chloroaluminate or chlorostannate melts) generally act both as solvent and as co-catalyst in transition metal catalysis. The reason for this is that the Lewis acidity or basicity, which is always present (at least latently), results in strong interactions with the catalyst complex. In many cases, the Lewis acidity of an ionic liquid is used to convert the neutral catalyst precursor into the corresponding cationic active form. The activation of Cp_2TiCl_2 [26] and $(ligand)_2NiCl_2$ [27] in acidic chloroaluminate melts and the activation of $(PR_3)_2PtCl_2$ in chlorostannate melts [28] are examples of this kind of activation (Eqs. 5.2-1, 5.2-2, and 5.2-3).

$$Cp_2TiCl_2 + [cation][Al_2Cl_7] \rightleftharpoons [Cp_2TiCl][AlCl_4] + [cation][AlCl_4] \quad (5.2\text{-}1)$$

$$(ligand)_2NiCl_2 + [cation][Al_2Cl_7] + [cation][Al_2EtCl_6] \rightleftharpoons$$
$$[(ligand)Ni\text{-}CH_2\text{-}CH_3][AlCl_4] + 2\,[cation][AlCl_4] + AlCl_3\text{-}ligand \quad (5.2\text{-}2)$$

$$(PR_3)_2PtCl_2 + [cation][Sn_2Cl_5] \rightleftharpoons [(PR_3)_2PtCl][SnCl_3] + [cation][SnCl_3] \quad (5.2\text{-}3)$$

In cases in which the ionic liquid is not directly involved in creating the active catalytic species, a co-catalytic interaction between the ionic liquid solvent and the dissolved transition metal complex still often takes place and can result in significant catalyst activation. When a catalyst complex is, for example, dissolved in a slightly acidic ionic liquid, some electron-rich parts of the complex (e.g., lone pairs of electrons in the ligand) will interact with the solvent in a way that will usually result in a lower electron density at the catalytic center (for more details see Section 5.2.3).

If this higher electrophilicity of the catalytic center results in higher catalytic activity (as in oligomerization reactions of most olefins, for example), then there is a very good chance of activating the catalyst system in a slightly acidic ionic liquid. In fact, this is the reason why many Ni-catalyzed oligomerization reactions of propene and butene have been carried out in slightly acidic or buffered chloroaluminate ionic liquids.

This type of co-catalytic influence is well known in heterogeneous catalysis, in which for some reactions an acidic support will activate a metal catalyst more efficiently than a neutral support. In this respect, the acidic ionic liquid can be considered as a liquid acidic support for the transition metal catalysts dissolved in it.

As one would expect, in those cases in which the ionic liquid acts as a co-catalyst, the nature of the ionic liquid becomes very important for the reactivity of the transition metal complex. The opportunity to optimize the ionic medium used, by variation of the halide salt, the Lewis acid, and the ratio of the two components forming the ionic liquid, opens up enormous potential for optimization. However, the choice of these parameters may be restricted by some possible incompatibilities with the feedstock used. Undesired side reactions caused by the Lewis acidity of the ionic liquid or by strong interaction between the Lewis acidic ionic liquid and, for example, some oxygen functionalities in the substrate have to be considered.

5.2.2.3 **Ionic liquid as solvent and ligand/ligand precursor**

Both the cation and the anion of an ionic liquid can act as a ligand or ligand precursor for a transition metal complex dissolved in the ionic liquid.

Anions of the ionic liquid may, to some degree, act as ligands if the catalytic center is cationic, depending on their coordination strength. Indeed, it has been clearly demonstrated that the anion of a cationic transition metal complex is replaced to a large extent by the ionic liquid's anion if they are different [12]. While most ionic liquid anions used in catalysis are chosen so as to interact as weakly as possible with the catalytic center, this situation may change dramatically if the ionic liquid's anion undergoes decomposition reactions. If, for example, the hexafluorophosphate anion of an ionic liquid hydrolyses in contact with water, strongly coordinating fluoride ions are liberated, and will act as strong ligands and catalyst poisons to many transition metal complexes.

With respect to the ionic liquid's cation the situation is quite different, since catalytic reactions with anionic transition metal complexes are not yet very common in ionic liquids. However, an imidazolium moiety as an ionic liquid cation can act as a ligand precursor for the dissolved transition metal. Its transformation into a ligand under the reaction conditions has been observed in three different ways: a) formation of metal carbene complexes by deprotonation of the imidazolium cation, b) formation of metal–carbene complexes by oxidative addition of the imidazolium cation to the metal center, and c) dealkylation of the imidazolium cation and formation of a metal–imidazole complex. These different ways are shown in a general form in Scheme 5.2-2.

The first reaction pathway for the *in situ* formation of a metal–carbene complex in an imidazolium ionic liquid is based on the well known, relatively high acidity of the H atom in the 2-position of the imidazolium ion [29]. This can be removed (by basic ligands of the metal complex, for example) to form a metal–carbene complex (see Scheme 5.2-2, route a)). Xiao and co-workers demonstrated that a Pd imidazolylidene complex was formed when $Pd(OAc)_2$ was heated in the presence of [BMIM]Br [30]. The isolated Pd carbene complex was found to be active and stable in Heck coupling reactions (for more details see Section 5.2.4.4). Welton et al. were later able to characterize an isolated Pd–carbene complex obtained in this way by X-ray spectroscopy [31]. The reaction pathway to the complex is displayed in Scheme 5.2-3.

However, formation of the metal carbene complex was not observed in pure, halide-free [BMIM][BF_4], indicating that the formation of carbene depends on the

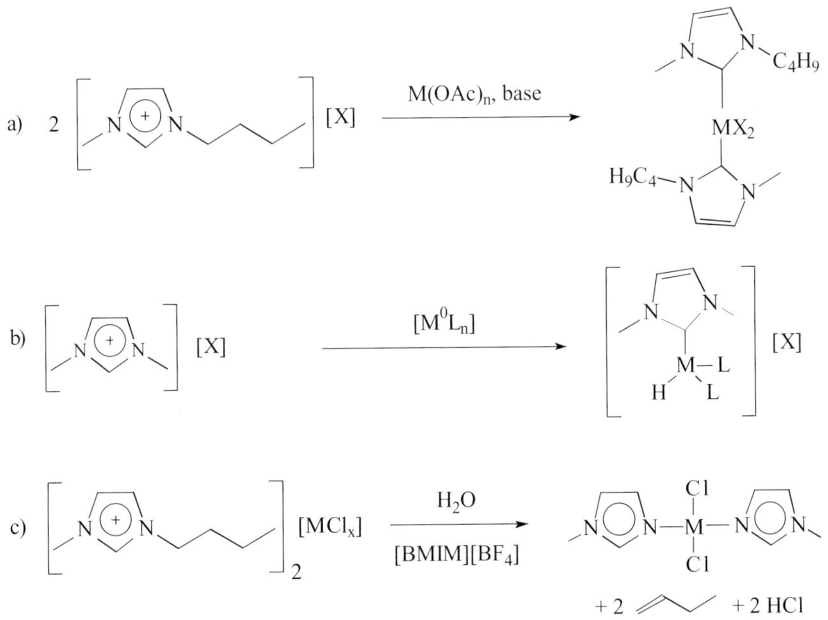

a) 2 [imidazolium] [X] —$M(OAc)_n$, base→ bis(carbene) MX_2 complex (C_4H_9, H_9C_4)

b) [imidazolium] [X] —[M^0L_n]→ [carbene–M(H)(L)–L] [X]

c) [imidazolium]$_2$ [MCl_x] —H_2O, [BMIM][BF_4]→ imidazole–M(Cl_2)–imidazole + 2 butene + 2 HCl

Scheme 5.2-2: Different potential routes for in situ ligand formation from an imidazolium cation of an ionic liquid.

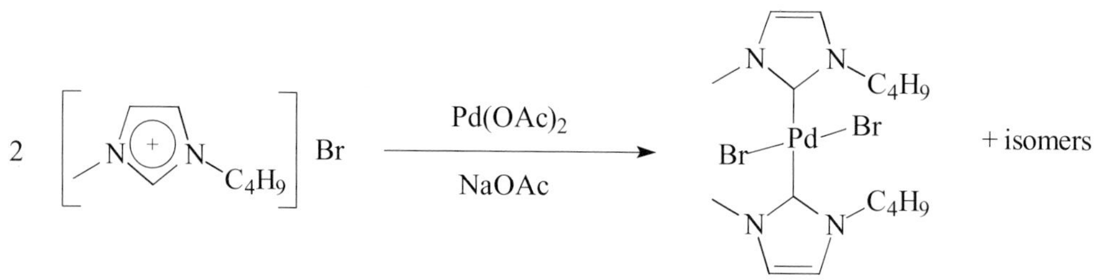

Scheme 5.2-3: Formation of a Pd-carbene complex by deprotonation of the imidazolium cation.

nucleophilicity of the ionic liquid's anion. To avoid the formation of metal carbene complexes by deprotonation of the imidazolium cation under basic conditions, the use of 2-methyl-substituted imidazolium is frequently suggested. However, it should be mentioned here that strong bases can also abstract a proton here, to form methyleneimidazolidene species that may also act as a strong ligand to electrophilic metal centers.

Another means of *in situ* metal–carbene complex formation in an ionic liquid is the direct oxidative addition of the imidazolium cation to a metal center in a low oxidation state (see Scheme 5.2-2, route b)). Cavell and co-workers have observed oxidative addition on heating 1,3-dimethylimidazolium tetrafluoroborate with $Pt(PPh_3)_4$ in refluxing THF [32]. The Pt-carbene complex formed can decompose by reductive elimination. Winterton et al. have also described the formation of a Pt-carbene complex by oxidative addition of the [EMIM] cation to $PtCl_2$ in a basic $[EMIM]Cl/AlCl_3$ system (free Cl^- ions present) under ethylene pressure [33]. The formation of a Pt–carbene complex by oxidative addition of the imidazolium cation is displayed in Scheme 5.2-4.

In the light of these results, it becomes important to question whether a particular catalytic result obtained in a transition metal-catalyzed reaction in an imidazolium ionic liquid is caused by a metal carbene complex formed *in situ*. The following simple experiments can help to verify this in more detail: a) variation of ligands in the catalytic system, b) application of independently prepared, defined metal carbene complexes, and c) investigation of the reaction in pyridinium-based ionic liquids. If the reaction shows significant sensitivity to the use of different ligands, if the application of the independently prepared, defined metal–carbene complex

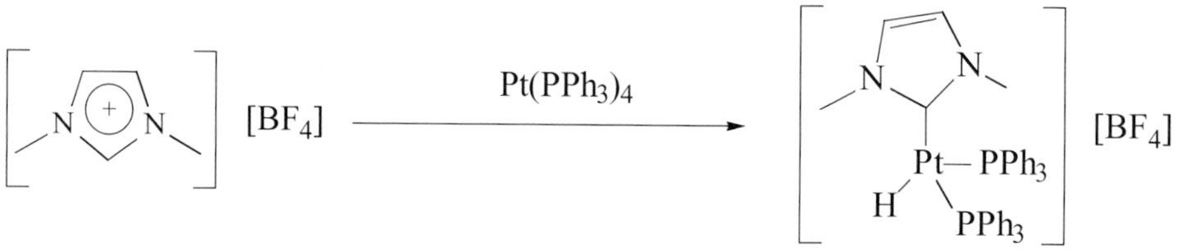

Scheme 5.2-4: Formation of a Pt-carbene complex by oxidative addition of the imidazolium cation.

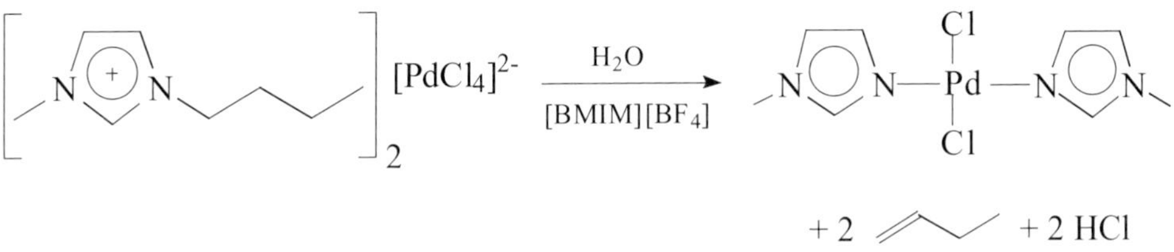

Scheme 5.2-5: Formation of the active Pd-catalyst from $[BMIM]_2$ $PdCl_4$ for the hydrodimerization of 1,3-butadiene.

shows a different reactivity than the catalytic system under investigation, or if the catalytic result in the pyridinium ionic liquid is similar to that in the imidazolium system, then significant influence of a metal carbene complex formed *in situ* is unlikely. Of course, even then, *in situ* formation of a metal carbene complex cannot be totally excluded, but its lifetime may be very short so that significant influence on the catalysis does not take place.

Finally, a third means of ligand formation from an imidazolium cation, described by Dupont and co-workers, should be mentioned here [34]. They investigated the hydrodimerization/telomerization of 1,3-butadiene with palladium(II) compounds in $[BMIM][BF_4]$ and described the activation of the catalyst precursor complex $[BMIM]_2[PdCl_4]$ by a palladium(IV) compound formed by oxidative addition of the imidazolium nitrogen atom and the alkyl group with cleavage of the C–N bond of the $[BMIM]^+$ ion, resulting in bis(methylimidazole) dichloropalladate (Scheme 5.2-5). However, this reaction was only observed in the presence of water.

5.2.2.4 **Ionic liquid as solvent and transition metal catalyst**

Acidic chloroaluminate ionic liquids have already been described as both solvents and catalysts for reactions conventionally catalyzed by $AlCl_3$, such as catalytic Friedel–Crafts alkylation [35] or stoichiometric Friedel–Crafts acylation [36], in Section 5.1. In a very similar manner, Lewis-acidic transition metal complexes can form complex anions by reaction with organic halide salts. Seddon and co-workers, for example, patented a Friedel–Crafts acylation process based on an acidic chloroferrate ionic liquid catalyst [37].

However, ionic liquids acting as transition metal catalysts are not necessarily based on classical Lewis acids. Dyson et al. recently reported the ionic liquid $[BMIM][Co(CO)_4]$ [38]. The system was obtained as an intense blue-green colored liquid by metathesis between [BMIM]Cl and $Na[Co(CO)_4]$. The liquid was used as a catalyst in the debromination of 2-bromoketones to their corresponding ketones.

In general, the incorporation of an active transition metal catalyst into the anion of an ionic liquid appears to be an attractive concept for applications in which a high catalyst concentration is needed.

5.2.3
Methods of Analysis of Transition Metal Catalysts in Ionic Liquids

Many transition metal-catalyzed reactions have already been studied in ionic liquids. In several cases, significant differences in activity and selectivity from their counterparts in conventional organic media have been observed (see Section 5.2.4). However, almost all attempts so far to explain the special reactivity of catalysts in ionic liquids have been based on product analysis. Even if it is correct to argue that a catalyst is more active because it produces more product, this is not the type of explanation that can help in the development of a more general understanding of what happens to a transition metal complex under catalytic conditions in a certain ionic liquid. Clearly, much more spectroscopic and analytical work is needed to provide better understanding of the nature of an active catalytic species in ionic liquids and to explain some of the observed "ionic liquid" effects on a rational, molecular level.

In general, most of the methods used to analyze the chemical nature of the ionic liquid itself, as described in Chapter 4, should also be applicable, in some more sophisticated form, to study the nature of a catalyst dissolved in the ionic liquid. For attempts to apply spectroscopic methods to the analysis of active catalysts in ionic liquids, however, it is important to consider three aspects: a) as with catalysis in conventional media, the lifetime of the catalytically active species will be very short, making it difficult to observe, b) in a realistic catalytic scenario the concentration of the catalyst in the ionic liquid will be very low, and c) the presence and concentration of the substrate will influence the catalyst/ionic liquid interaction. These three concerns alone clearly show that an ionic liquid/substrate/catalyst system is quite complex and may be not easy to study by spectroscopic methods.

One obvious approach involves the application of *in situ* NMR spectroscopy. However, this method often suffers from the relatively low concentration of the catalyst in the ionic liquid. Moreover, ^{1}H and ^{13}C NMR spectroscopic investigations are difficult, since the intense signals of the ionic liquid make clear detection of the dissolved catalyst difficult. Several approaches to overcome the latter problem have been suggested. Hardacre and co-workers have described the synthesis and application of fully deuterated ionic liquids [39]. Alternatively, deuterium can be selectively introduced into the ligand of the transition metal catalyst in order to study the complex dissolved in the ionic liquid by in situ ^{2}H NMR spectroscopy [40]. The latter method has been used to investigate the activation of the square-planar Ni-complex (η-4-cycloocten-1-yl](1,5-diphenyl-2,4-pentanedionato-O,O′)nickel in slightly acidic chloroaluminate ionic liquids. The deuterated analogue of this complex was prepared according to Scheme 5.2-6, by treatment of 1,5-diphenyl-2,4-pentanedione with NaH, followed by hydrolysis with D_2O. The deuterated ligand was dried and treated with dicyclooctadienyl nickel $Ni(COD)_2$.

^{2}HNMR spectra of the deuterated complex obtained in CH_2Cl_2 and in [EMIM]Cl/$AlCl_3$ (1:1.2) are displayed in Figure 5.2-3.

While the deuterated complex shows the expected NMR signals in CH_2Cl_2 (two signals from the complex and one signal from the solvent), the ^{2}H NMR spectrum

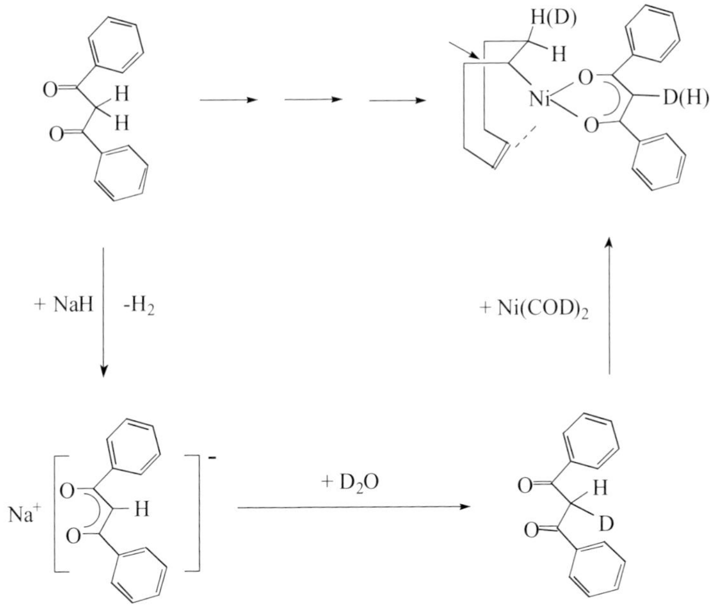

Scheme 5.2-6: Synthesis of a deuterated analogue of the square-planar Ni-complex (η-4-cycloocten-1-yl](1,5-diphenyl-2,4-pentanedionato-O,O')nickel for ^{2}H NMR investigations.

obtained from the complex in the slightly acidic chloroaluminate ionic liquid shows only one signal, indicating that the abstraction of COD is more efficient in the ionic liquid medium. Moreover, the deuterium signal of the acac ligand undergoes a significant downfield shift, suggesting intense electronic interaction between the ligand and the Lewis acidic centers of the melt. These interactions, which should result in an increased electrophilicity of the Ni-center, help to explain the activation of Ni-acac complexes in slightly acidic chloroaluminate ionic liquids.

This example should illustrate that *in situ* NMR spectroscopy can be a powerful tool with which to study catalysts dissolved in ionic liquids, if the signals of the metal complex can be detected in sufficient intensity independently from the signals of the ionic liquid.

If this is not possible for any reason, an alternative way to obtain some insight into interactions between the catalyst complex and the ionic liquid may be to record changes in the ionic liquid during the catalytic process in an indirect manner. This method has been successfully used by the author's group to understand the activation of $(PPh_3)_2PtCl_2$ in chlorostannate ionic liquids in more detail. The change in color from yellow to red during the dissolution of the complex in the ionic liquid was attributed to the abstraction of chloride from the Pt-complex by the acidic $[Sn_2Cl_5]^-$ species of the ionic liquid. It proved possible to support this assumption by recording the Lewis acidity of the chlorostannate ionic liquid by ^{119}Sn NMR before and after the addition of $(PPh_3)_2PtCl_2$ [28]. The results of this investigation corresponded very well to an acid–base reaction of both chloride atoms of the platinum complex with the acidic ionic liquid.

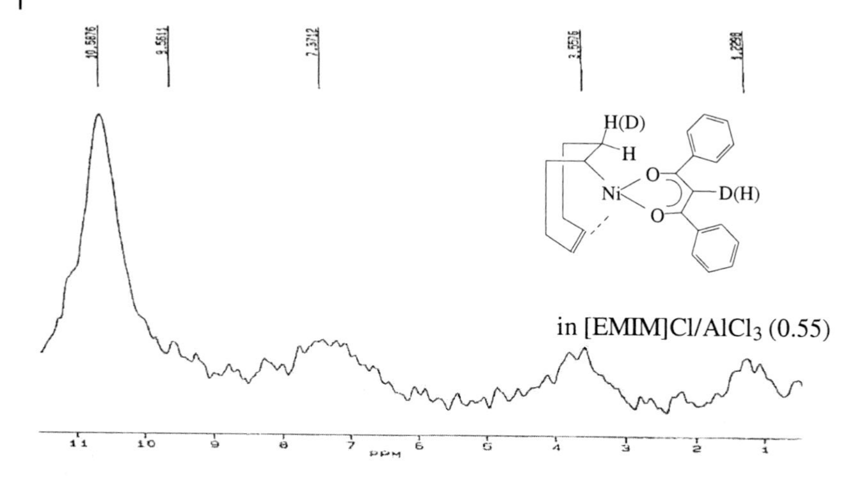

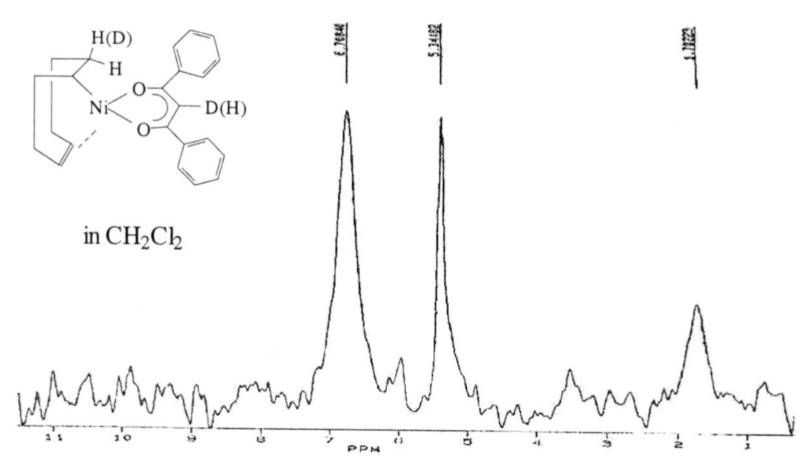

Figure 5.2-3: ^{2}H NMR spectra of the deuterated analogue of the square planar Ni-complex (η-4-cycloocten-1-yl](1,5-diphenyl-2,4-pentanedionato-O,O')nickel recorded [EMIM]Cl/$AlCl_3$ [X($AlCl_3$) = 0.55] and in CH_2Cl_2.

In addition to *in situ* NMR spectroscopy, other methods such as *in situ* IR spectroscopy, EXAFS, and electrochemistry should be very useful for the investigation of active catalytic species in ionic liquids. However, far too little effort has been directed to this end in recent years.

This is surprising in view of the fact that a great deal of effort was made to study transition metal complexes in chloroaluminate ionic liquids in the 1980s and early 1990s (see Section 6.1 for some examples). The investigations at this time generally started with electrochemical studies [41], but also included spectroscopic and complex chemistry experiments [42].

Obviously, with the development of the first catalytic reactions in ionic liquids, the general research focus turned away from basic studies of metal complexes dissolved in ionic liquids. Today there is a clear lack of fundamental understanding of many catalytic processes in ionic liquids on a molecular level. Much more fundamental work is undoubtedly needed and should be encouraged in order to speed up the future development of transition metal catalysis in ionic liquids.

5.2.4 Selected Examples of the Application of Ionic Liquids in Transition Metal Catalysis

5.2.4.1 Hydrogenation

In general, transition metal-catalyzed hydrogenation reactions in ionic liquids are particularly promising. On the one hand, a large number of known, ionic hydrogenation catalysts are available [43]. On the other, the solubility of many alkenes and the availability of hydrogen in many ionic liquids appear to be sufficiently high for good reaction rates to be achieved. In this context it is noteworthy that the availability of hydrogen results not only from its solubility under equilibrium conditions, but also reflects the ease of its transfer from the gas phase into the melt. Since the diffusion of hydrogen into ionic liquids has been found to be relatively fast, the latter contribution is of special importance [44]. Finally, the miscibility gap between the saturated reaction products and the ionic liquid is often large, so that a biphasic procedure is possible in the majority of cases.

The first successful hydrogenation reactions in ionic liquids were studied by the groups of de Souza [45] and Chauvin [46] in 1995. De Souza et al. investigated the Rh-catalyzed hydrogenation of cyclohexene in 1-*n*-butyl-3-methylimidazolium ([BMIM]) tetrafluoroborate. Chauvin et al. dissolved the cationic "Osborn complex" $[Rh(nbd)(PPh_3)_2][PF_6]$ (nbd = norbornadiene) in ionic liquids with weakly coordinating anions (e.g., $[PF_6]^-$, $[BF_4]^-$, and $[SbF_6]^-$) and used the obtained ionic catalyst solutions for the biphasic hydrogenation of 1-pentene as seen in Scheme 5.2-7.

Although the reactants have only limited solubility in the catalyst phase, the rates of hydrogenation in $[BMIM][SbF_6]$ are almost five times faster than for the comparable reaction in acetone. All ionic catalyst solutions tested could be reused repeatedly. The loss of rhodium through leaching into the organic phase lay below the detection limit of 0.02 %. These results are of general importance for the field of

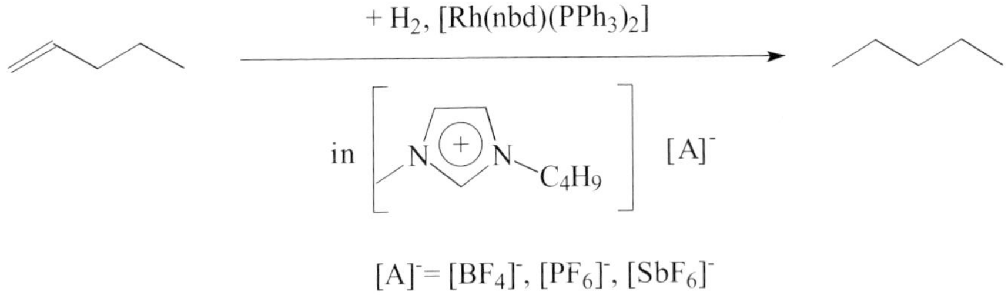

$+ H_2$, $[Rh(nbd)(PPh_3)_2]$

in [1-methyl-3-C_4H_9-imidazolium] $[A]^-$

$[A]^- = [BF_4]^-, [PF_6]^-, [SbF_6]^-$

Scheme 5.2-7: Biphasic hydrogenation of 1-pentene with the cationic "Osborn complex" $[Rh(nbd)(PPh_3)_2][PF_6]$ (nbd = norbornadiene) in ionic liquids with weakly coordinating anions.

biphasic catalysis, since this was the first time that a rhodium catalyst was able to be "immobilized" in a polar solution without the use of specially designed ligands.

Chauvin's group described the selective hydrogenation of cyclohexadiene to cyclohexene through making use of the biphasic reaction system [46]. Since the solubility of cyclohexadiene in [BMIM][SbF_6] is about five times higher than the solubility of cyclohexene in the same ionic liquid, the latter was obtained in 98 % selectivity at 96 % conversion.

Rhodium- and cobalt-catalyzed hydrogenation of butadiene and 1-hexene [47, 48] and the Ru-catalyzed hydrogenation of aromatic compounds [49] and acrylonitrile–butadiene copolymers [50] have also been reported to be successful in ionic liquids.

An example of a stereoselective hydrogenation in ionic liquids was recently successfully demonstrated by Drießen-Hölscher et al. On the basis of investigations into the biphasic water/*n*-heptane system [51], the ruthenium-catalyzed hydrogenation of sorbic acid to *cis*-3-hexenoic acid in the [BMIM][PF_6]/MTBE system was studied [52], as shown in Scheme 5.2-8.

In comparison with polar organic solvents (such as glycol) a more than threefold increase in activity with comparable selectivity for *cis*-3-hexenoic acid was observed in the ionic liquid. This is explained by partial deactivation (through complexation) of the active catalytic center in those polar organic solvents that are able to dissolve the cationic Ru catalyst. In contrast, the ionic liquid [BMIM][PF_6] is known to combine high solvation power for ionic metal complexes with relatively weak coordination strength. In this way, the catalyst can be dissolved in a "more innocent" environment than is the case if polar organic solvents are used. After the biphasic hydrogenation of sorbic acid, the ionic catalyst solution could be recovered by phase separation and reused repeatedly. Other examples of selective hydrogenation of dienes by use of cobalt [47] and palladium [53] catalysts have been reported by Dupont and de Souza.

A number of enantioselective hydrogenation reactions in ionic liquids have also been described. In all cases reported so far, the role of the ionic liquid was mainly to open up a new, facile way to recycle the expensive chiral metal complex used as the hydrogenation catalyst.

Chauvin et al. hydrogenated α-acetamidocinnamic acid to (*S*)-phenylalanine in the presence of a [Rh(cod)(–)-(diop)][PF_6] catalyst in a [BMIM][SbF_6] melt with 64 % *ee* [46].

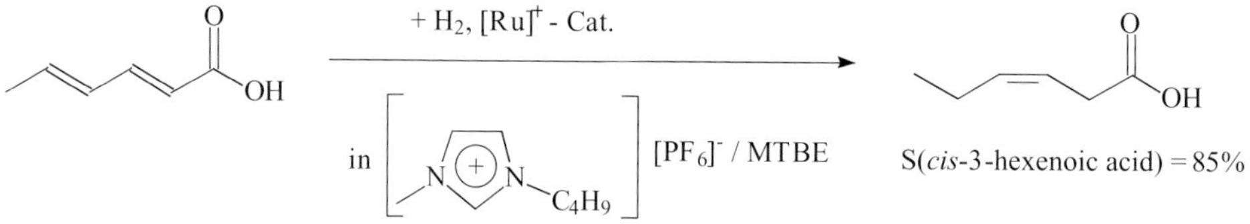

O OH + H_2, $[Ru]^+$ - Cat. in [N(+)N C_4H_9] $[PF_6]^-$ / MTBE O OH S(*cis*-3-hexenoic acid) = 85%

Scheme 5.2-8: Stereoselective hydrogenation of sorbic acid in the [BMIM][PF_6]/MTBE biphasic system.

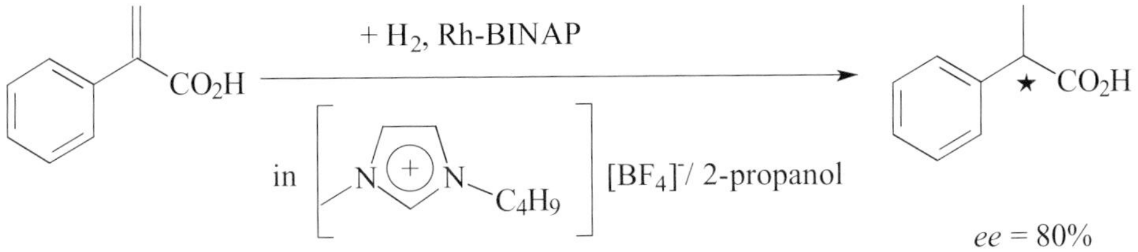

CO_2H $\xrightarrow{+ H_2, \text{ Rh-BINAP}}$ $\star$ CO_2H

in [BMIM] $[BF_4]^-$ / 2-propanol (C_4H_9)

$ee = 80\%$

Scheme 5.2-9: Hydrogenation of 2-phenylacrylic acid to (*S*)-2-phenylpropionic acid with the chiral complex $[RuCl_2(S)\text{-BINAP}]_2NEt_3$ as catalyst in $[BMIM][BF_4]$.

Dupont et al. were able to obtain up to 80 % *ee* in the conversion of 2-phenylacrylic acid into (*S*)-2-phenylpropionic acid with the chiral $[RuCl_2(S)\text{-BINAP}]_2NEt_3$ complex as catalyst in $[BMIM][BF_4]$ melts (Scheme 5.2-9) [54].

Both reactions were carried out under two-phase conditions with the help of an additional organic solvent (such as *i*PrOH). The catalyst could be reused with the same activity and enantioselectivity after decantation of the hydrogenation products. A more recent example, again by de Souza and Dupont, has been reported. They made a detailed study of the asymmetric hydrogenation of α-acetamidocinnamic acid and the kinetic resolution of methyl (±)-3-hydroxy-2-methylenebutanoate with chiral Rh(I) and Ru(II) complexes in $[BMIM][BF_4]$ and $[BMIM][PF_6]$ [55]. The authors described the remarkable effects of the molecular hydrogen concentration in the ionic catalyst layer on the conversion and enantioselectivity of these reactions. The solubility of hydrogen in $[BMIM][BF_4]$ was found to be almost four times higher than in $[BMIM][PF_6]$.

Hydrogenation reactions were among the first transformations to be successfully carried out in reaction systems consisting of an ionic liquid and compressed CO_2 [56, 57]. While the conceptual aspects of this innovative, biphasic reaction mode are covered in more detail in Section 5.4, the specific applications reported by Tumas et al. [56] and Jessop et al. [57] once more demonstrate the great potential of transition metal-catalyzed hydrogenation in ionic liquids. Tumas and co-workers investigated the hydrogenation of olefins in the biphasic system $[BMIM][PF_6]/scCO_2$. After reaction, the ionic catalyst layer could be separated by simple decantation and could be reused up to four times [56].

Jessop and co-workers studied asymmetric hydrogenation reactions with the catalyst complex $Ru(OAc)_2$(tolBINAP) dissolved in $[BMIM][PF_6]$. In both reactions under investigation – the hydrogenation of tiglic acid (Scheme 5.2.10) and the hydrogenation of the precursor of the anti-inflammatory drug ibuprofen (Scheme 5.2.11) – no CO_2 was present during the catalytic transformation. However, $scCO_2$ was used in both cases to extract the reaction products from the reaction mixture when the reaction was complete.

Finally, a special example of transition metal-catalyzed hydrogenation in which the ionic liquid used does not provide a permanent biphasic reaction system should be mentioned. The hydrogenation of 2-butyne-1,4-diol, reported by Dyson et al., made use of an ionic liquid/water system that underwent a reversible two-

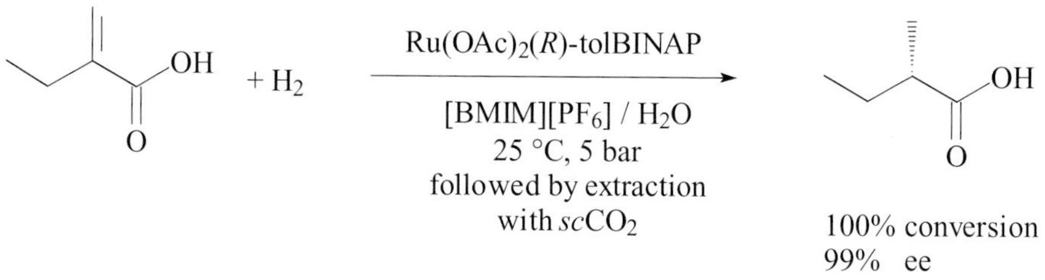

OH + H_2 → Ru(OAc)$_2$(*R*)-tolBINAP; [BMIM][PF$_6$] / H_2O; 25 °C, 5 bar; followed by extraction with *sc*CO_2 → OH

100% conversion
99% ee

Scheme 5.2-10: Ru-catalyzed asymmetric hydrogenation of tiglic acid, followed by product extraction with *sc*CO_2.

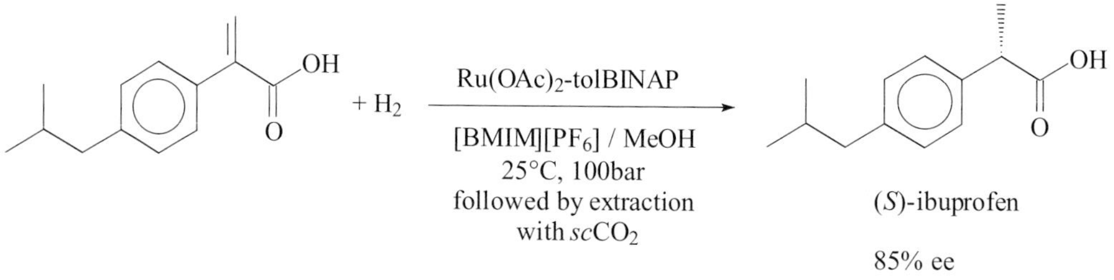

OH + H_2 → Ru(OAc)$_2$-tolBINAP; [BMIM][PF$_6$] / MeOH; 25°C, 100bar; followed by extraction with *sc*CO_2 → OH

(*S*)-ibuprofen

85% ee

Scheme 5.2-11: Ru-catalyzed asymmetric hydrogenation of isobutylatropic acid, followed by extraction of the product ibuprofen with *sc*CO_2.

phase/single-phase transformation upon a temperature switch [58]. At room temperature, the ionic liquid 1-methyl-3-*n*-octyl imidazolium ([OMIM]) tetrafluoroborate containing the cationic Rh catalyst formed a separate layer with water containing the substrate. At 80 °C however, a homogeneous single-phase reaction could be carried out.

Temperature-dependent phase behavior was first applied to separate products from an ionic liquid/catalyst solution by de Souza and Dupont in the telomerization of butadiene and water [34]. This concept is especially attractive if one of the substrates shows limited solubility in the ionic liquid solvent.

5.2.4.2 **Oxidation reactions**

Catalytic oxidation reactions in ionic liquids have been investigated only very recently. This is somewhat surprising in view of the well known oxidation stability of ionic liquids, from electrochemical studies [11], and the great commercial importance of oxidation reactions. Moreover, for oxidation reactions with oxygen, the nonvolatile nature of the ionic liquid is of real advantage for the safety of the reaction. While the application of volatile organic solvents may be restricted by the formation of explosive mixtures in the gas phase, this problem does not arise if a nonvolatile ionic liquid is used as the solvent.

Howarth oxidized various aromatic aldehydes to the corresponding carboxylic acids with Ni(acac)$_2$ dissolved in [BMIM][PF$_6$] as the catalyst and oxygen at atmospheric pressure as the oxidant [59]. However, this reaction cannot be considered a real challenge. Moreover, the catalyst loading used for the described reaction was rather high (3 mol%).

Ley et al. reported oxidation of alcohols catalyzed by an ammonium perruthenate catalyst dissolved in [NEt_4]Br and [EMIM][PF_6] [60]. Oxygen or N-methylmorpholine N-oxide is used as the oxidant and the authors describe easy product recovery by solvent extraction and mention the possibility of reusing the ionic catalyst solution.

The oxidation of alkenes and allylic alcohols with the urea-H_2O_2 adduct (UHP) as oxidant and methyltrioxorhenium (MTO) dissolved in [EMIM][BF_4] as catalyst was described by Abu-Omar et al. [61]. Both MTO and UHP dissolved completely in the ionic liquid. Conversions were found to depend on the reactivity of the olefin and the solubility of the olefinic substrate in the reactive layer. In general, the reaction rates of the epoxidation reaction were found to be comparable to those obtained in classical solvents.

Song and Roh investigated the epoxidation of compounds such as 2,2-dimethylchromene with a chiral Mn^{III}(salen) complex (Jacobsen catalyst) in a mixture of [BMIM][PF_6] and CH_2Cl_2 (1:4 v/v), using NaOCl as the oxidant (Scheme 5.2-12) [62].

The authors describe a clear enhancement of the catalyst activity by the addition of the ionic liquid even if the reaction medium consisted mainly of CH_2Cl_2. In the presence of the ionic liquid, 86 % conversion of 2,2-dimethylchromene was observed after 2 h. Without the ionic liquid the same conversion was obtained only after 6 h. In both cases the enantiomeric excess was as high as 96 %. Moreover, the ionic catalyst solution could be reused several times after product extraction, although the conversion dropped from 83 % to 53 % after five recycles; this was explained, according to the authors, by a slow degradation process of the Mn^{III} complex.

A very exciting way to combine electrochemistry and transition metal catalysis in ionic liquids was reported by Gaillon and Bedioui [63], who investigated the electroassisted activation of molecular oxygen by Jacobsen's epoxidation catalysts dissolved in [BMIM][PF_6] and were able to provide evidence for the formation of the highly reactive oxomanganese(V) intermediate, which was not detectable in organic solvents. This may open new perspectives for clean, electroassisted oxidation reactions with molecular oxygen in ionic liquids.

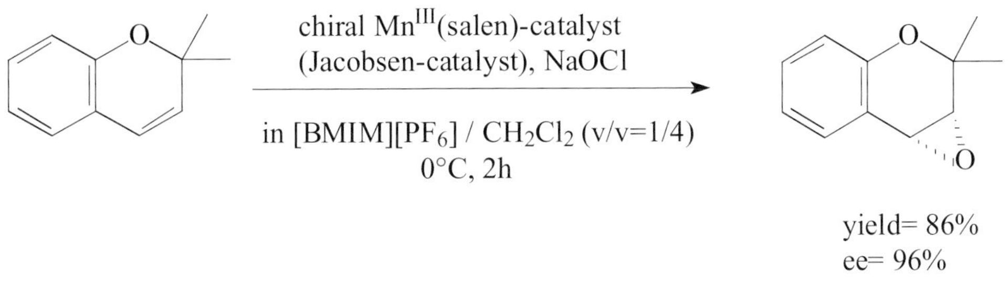

chiral Mn^{III}(salen)-catalyst
(Jacobsen-catalyst), NaOCl

in [BMIM][PF_6] / CH_2Cl_2 (v/v=1/4)
0°C, 2h

yield= 86%
ee= 96%

Scheme 5.2-12: Mn-catalyzed asymmetric epoxidation in a [BMIM][PF_6]/CH_2Cl_2 (v/v = 1/4) solvent mixture.

Finally, it should be mentioned that ionic liquids have successfully been used in classical, stoichiometric oxidation reactions as well. Singer et al., for example, described the application of [BMIM][BF_4] in the oxidation of codeine methyl ether to thebaine [64]. The ionic liquid was used here as a very convenient solvent to extract excess MnO_2 and associated impurities from the reaction mixture.

5.2.4.3 **Hydroformylation**

In hydroformylation, biphasic catalysis is a well established method for effective catalyst separation and recycling. In the case of Rh-catalyzed hydroformylation reactions, this principle is implemented technically in the Ruhrchemie–Rhône–Poulenc process, in which water is used as the catalyst phase [65]. Unfortunately, this process is limited to C2-C5-olefins, due to the low water solubility of higher olefins. Nevertheless, the hydroformylation of many higher olefins is of commercial interest. One example is the hydroformylation of 1-octene for the selective synthesis of linear nonanal. This can be obtained highly selectively by application of special ligand systems around the catalytic center. However, the additional costs associated with the use of these ligands make it even more economically attractive to develop new methods for efficient catalyst separation and recycling. In this context, biphasic catalysis with an ionic liquid as catalyst layer is a highly promising approach.

As early as 1972 Parshall described the platinum-catalyzed hydroformylation of ethene in tetraethylammonium trichlorostannate melts [1]. [NEt_4][$SnCl_3$], the ionic liquid used for these investigations, has a melting point of 78 °C. Recently, platinum-catalyzed hydroformylation in the room-temperature chlorostannate ionic liquid [BMIM]Cl/$SnCl_2$ was studied in the author's group. The hydroformylation of 1-octene was carried out with remarkable *n/iso* selectivities (Scheme 5.2-13) [66].

Despite the limited solubility of 1-octene in the ionic catalyst phase, a remarkable activity of the platinum catalyst was achieved [turnover frequency (TOF) = 126 h^{-1}]. However, the system has to be carefully optimized to avoid significant formation of hydrogenated by-product. Detailed studies to identify the best reaction conditions revealed that, in the chlorostannate ionic liquid [BMIM]Cl/$SnCl_2$ [$X(SnCl_2)$ = 0.55],

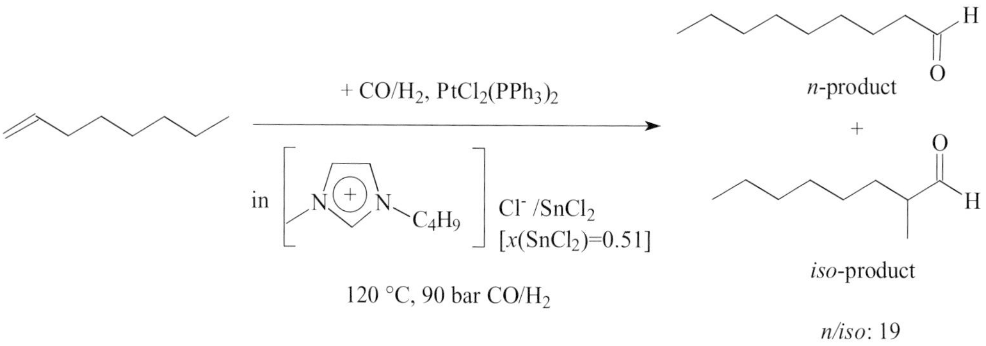

+ CO/H_2, $PtCl_2(PPh_3)_2$

in [BMIM] $Cl^-/SnCl_2$ [$x(SnCl_2)$=0.51]

120 °C, 90 bar CO/H_2

n-product + *iso*-product

n/iso: 19

Scheme 5.2-13: Biphasic, Pt-catalyzed hydroformylation of 1-octene with a slightly acidic [BMIM]Cl/$SnCl_2$ ionic liquid as catalyst layer.

the highest ratio of hydroformylation to hydrogenation was found at high syn-gas pressure and low temperature. At 80 °C and 90 bar CO/H_2-pressure, more than 90 % of all products were *n*-nonanal and *iso*-nonanal, the ratio between these two hydroformylation products being as high as 98.6:1.4 (*n*/*iso* = 72.4) [66].

Moreover, these experiments reveal some unique properties of the chlorostannate ionic liquids. In contrast to other known ionic liquids, the chlorostannate system combine a certain Lewis acidity with high compatibility to functional groups. The first resulted, in the hydroformylation of 1-octene, in the activation of $(PPh_3)_2PtCl_2$ by a Lewis acid–base reaction with the acidic ionic liquid medium. The high compatibility to functional groups was demonstrated by the catalytic reaction in the presence of CO and hydroformylation products.

Ruthenium- and cobalt-catalyzed hydroformylation of internal and terminal alkenes in molten $[PBu_4]Br$ was reported by Knifton as early as in 1987 [2]. The author described a stabilization of the active ruthenium-carbonyl complex by the ionic medium. An increased catalyst lifetime at low synthesis gas pressures and higher temperatures was observed.

The first investigations of rhodium-catalyzed hydroformylation in room-temperature liquid molten salts were published by Chauvin et al. in 1995 [6, 67]. The hydroformylation of 1-pentene with the neutral $Rh(CO)_2(acac)$/triarylphosphine catalyst system was carried out as a biphasic reaction with $[BMIM][PF_6]$ as the ionic liquid. With none of the ligands tested, however, was it possible to combine high activity, complete retention of the catalyst in the ionic liquid, and high selectivity for the desired linear hydroformylation product at that time. The use of PPh_3 resulted in significant leaching of the Rh-catalyst out of the ionic liquid layer. In this case, the catalyst is active in both phases, which makes a clear interpretation of solvent effects on the reactivity difficult. The catalyst leaching could be suppressed by the application of sulfonated triaryl phosphine ligands, but a major decrease in catalytic activity was found with these ligands (TOF = 59 h^{-1} with tppms, compared to 333 h^{-1} with PPh_3). Moreover, all of the ligands used in Chauvin's work showed poor selectivity to the desired linear hydroformylation product (*n*/*iso* ratio between 2 and 4). Obviously, the Rh-catalyzed, biphasic hydroformylation of higher olefins in ionic liquids requires the use of ligand systems specifically designed for this application. These early results thus stimulated research into other immobilizing, ionic ligand systems that would provide good catalyst immobilization without deactivation of the catalyst.

A pioneering ligand system specially designed for use in ionic liquids was described in 2000 by Salzer et al. [68]. Cationic ligands with a cobaltocenium backbone were successfully used in the biphasic, Rh-catalyzed hydroformylation of 1-octene. 1,1'-Bis(diphenylphosphino) cobaltocenium hexafluorophosphate (cdpp) proved to be an especially promising ligand. The compound can be synthesized as shown in Scheme 5.2-14, by mild oxidation of 1,1'-bis(diphenylphosphino)cobaltocene with C_2Cl_6 and anion-exchange with $[NH_4][PF_6]$ in acetone (for detailed ligand synthesis see [68]).

The results obtained in the biphasic hydroformylation of 1-octene are presented in Table 5.2-1. In order to evaluate the properties of the ionic diphosphine ligand

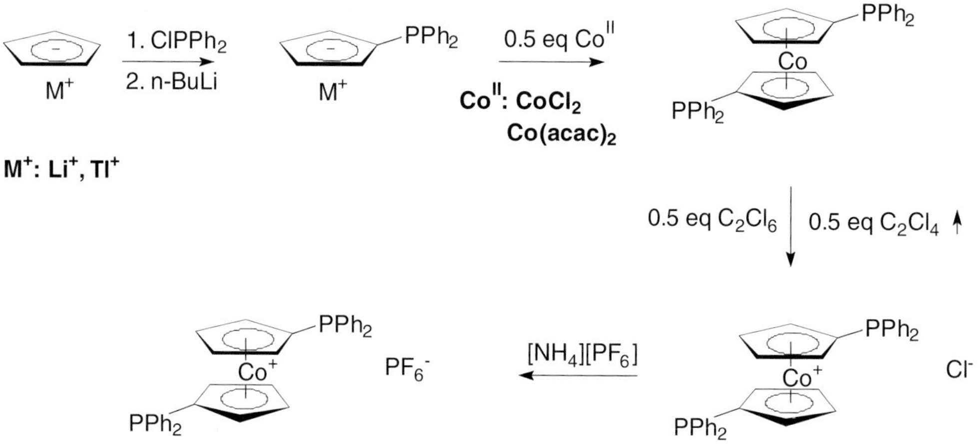

Scheme 5.2-14: Synthesis of 1,1'-bis(diphenylphosphino)cobaltocenium hexafluorophosphate.

with the cobaltocenium backbone, the results with the cdpp ligand are compared with those obtained with PPh_3, two common neutral bidentate ligands, and with Natppts as standard anionic ligand [68].

It is noteworthy that a clear enhancement of selectivity for the linear hydroformylation product is observed only with cdpp (Table 5.2-1, entry e). With all other ligands, the *n/iso* ratios are in the 2 to 4 range. While this is in accordance with known results in the case of PPh_3 (entry a) and dppe (entry c) (in comparison to monophasic hydroformylation [69]) and also with reported results in the case of Natppts (entry b; in comparison to the biphasic hydroformylation of 1-pentene in [BMIM][PF_6] [46]), it is more remarkable for the bidentate metallocene ligand dppf.

Taking into account the high structural similarity of dppf and cdpp, their different influence on the reaction's selectivity has to be attributable to electronic effects. The electron density at the phosphorus atoms is significantly lower in the case of cdpp, due to the electron-withdrawing effect of the formal cobalt(III) central atom

Entry	*Ligand*	*TOF* (h^{-1})	*n/iso*	*S (n-ald)(%)*[a]
a	PPh_3	426	2.6	72
b	tppts	98	2.6	72
c	dppe	35	3.0	75
d	dppf	828	3.8	79
e	cdpp	810	16.2	94

Conditions: ligand/Rh: 2:1, CO/H_2 = 1:1, t = 1 h, T = 100 °C, p = 10 bar, 1-octene/Rh = 1000, 5 mL [BMIM][PF_6]; dppe: bis(diphenylphosphinoethane); dppf: 1,1′-bis(diphenylphosphino)ferrocene; a) S (*n*-ald) = selectivity for *n*-nonanal in the product.

Table 5.2-1: Comparison of different phosphine ligands in the Rh-catalyzed hydroformylation of 1-octene in [BMIM][PF_6].

in the ligand. This interpretation is supported by previous work by Casey et al. [70] and Duwell et al. [71], who described positive effects of ligands with electron-poor phosphorus atoms in selective hydroformylation reactions, which they attribute to their ability to allow back-bonding from the catalytically active metal atom. It has to be pointed out that with the phosphinocobaltocenium ligand cdpp the reaction takes place almost exclusively in the ionic liquid phase (almost clear and colorless organic layer, less than 0.5 % Rh in the organic layer). An easy catalyst separation by decantation was possible. Moreover, it was found that the recovered ionic catalyst solution could be reused at least one more time with the same activity and selectivity as in the original run [68].

Cationic phosphine ligands containing guanidiniumphenyl moieties were originally developed in order to make use of their pronounced solubility in water [72, 73]. They were shown to form active catalytic systems in Pd-mediated C–C coupling reactions between aryl iodides and alkynes (Castro–Stephens–Sonogashira reaction) [72, 74] and Rh-catalyzed hydroformylation of olefins in aqueous two-phase systems [75].

It was recently found that the modification of neutral phosphine ligands with cationic phenylguanidinium groups represents a very powerful tool with which to immobilize Rh-complexes in ionic liquids such as [BMIM][PF_6] [76]. The guanidinium-modified triphenylphosphine ligand was prepared from the corresponding iodide salt by anion-exchange with [NH_4][PF_6] in aqueous solution, as shown in Scheme 5.2-15. The iodide can be prepared as previously described by Stelzer et al. [73].

In contrast to when PPh_3 is used as the ligand, the reaction takes place solely in the ionic liquid layer when the guanidinium-modified triphenylphosphine is applied. In the first catalytic run the hydroformylation activity was found to be somewhat lower than with PPh_3 (probably due to the fact that some of the activity observed with PPh_3 takes place in the organic layer). However, thanks to the excellent immobilization of the Rh-catalyst with the guanidinium-modified ligand [leaching is $<$ 0.07 % per run according to ICP analysis (detection limit)], the cat-

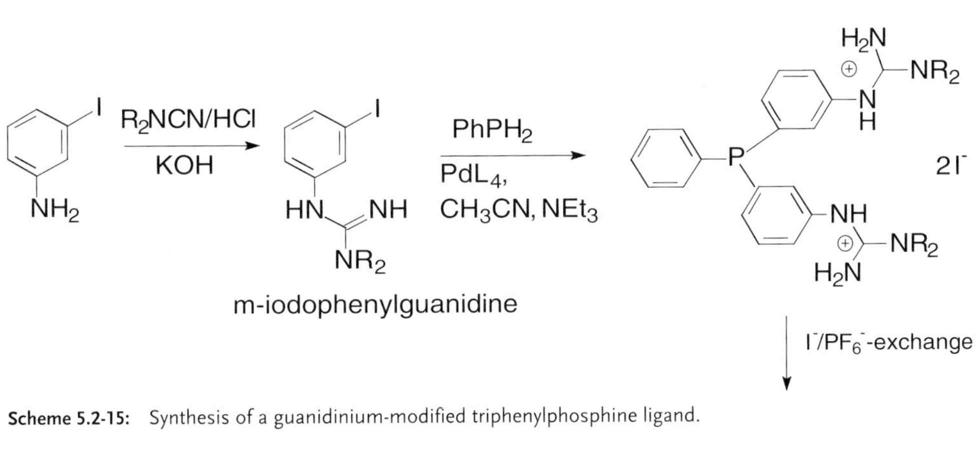

Scheme 5.2-15: Synthesis of a guanidinium-modified triphenylphosphine ligand.

alytic activity does not drop over the first ten recycling runs. For the recycling runs the organic layer was decanted after each run (under normal atmosphere) and the ionic catalyst layer was retained in the autoclave for the next hydroformylation experiment. Even after five recycling runs, the overall catalytic activity obtained with the ionic catalyst solution containing the guanidinium-modified ligand was higher than could be achieved with the simple PPh_3 ligand. With both ligands the *n/iso* ratio of the hydroformylation products was in the expected range of 1.7–2.8.

Further development is directed towards the adaptation of this immobilization concept to a ligand structure that promises better regioselectivity in the hydroformylation reaction. It is well known that diphosphine ligands with large natural P-metal-P bite angles form highly regioselective hydroformylation catalysts [77]. Here, xanthene-type ligands (P-metal-P ~ 110°) developed by van Leeuven's group proved to be especially suitable, allowing, for example, an overall selectivity of 98 % for the desired linear aldehyde in the hydroformylation of 1-octene [78, 79].

While unmodified xanthene ligands (compound a in Figure 5.2-4) show highly preferential solubility in the organic phase in the biphasic 1-octene/[BMIM][PF_6] mixture even at room temperature, the application of the guanidinium-modified xanthene ligand (compound b in Figure 5.2-4) resulted in excellent immobilization of the Rh-catalyst in the ionic liquid.

The guanidinium-modified ligand is synthesized by treatment of the xanthenediphosphine [80] with iodophenylguanidine in a Pd(0)-catalyzed coupling reaction. The ligand was tested in Rh-catalyzed hydroformylation in ten consecutive recycling runs, the results of which are presented in Figure 5.2-5. It is noteworthy that the catalytic activity increases during the first runs, achieving a stable level only after the forth recycling run. This behavior is attributable not only to a certain catalyst pre-forming but also to iodoaromate impurities in the ligand used. These are probably slowly washed out of the catalyst layer during the first catalytic runs.

After ten consecutive runs the overall turnover number reaches up to 3500 mol 1-octene converted per mol Rh-catalyst. In agreement with these recycling experiments, no Rh could be detected in the product layer by AAS or ICP, indicating leaching of less then 0.07 %. In all experiments, very good selectivities for the linear aldehyde were obtained, thus proving that the attachment of the guanidinium moiety onto the xanthene backbone had not influenced its known positive effect on

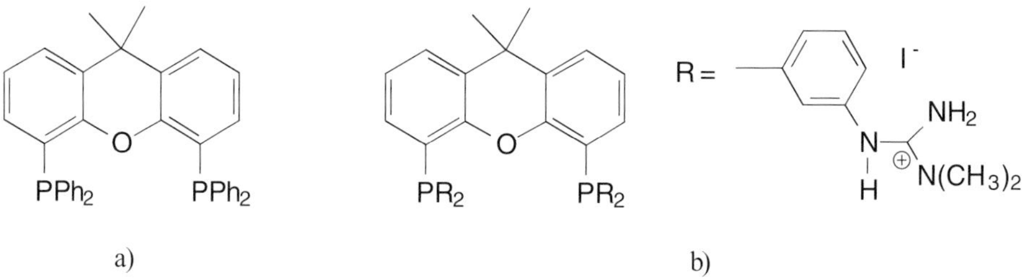

Figure 5.2-4: Unmodified (a) and guanidinium-modified (b) xanthene ligand as used in the biphasic, Rh-catalyzed hydroformylation of 1-octene.

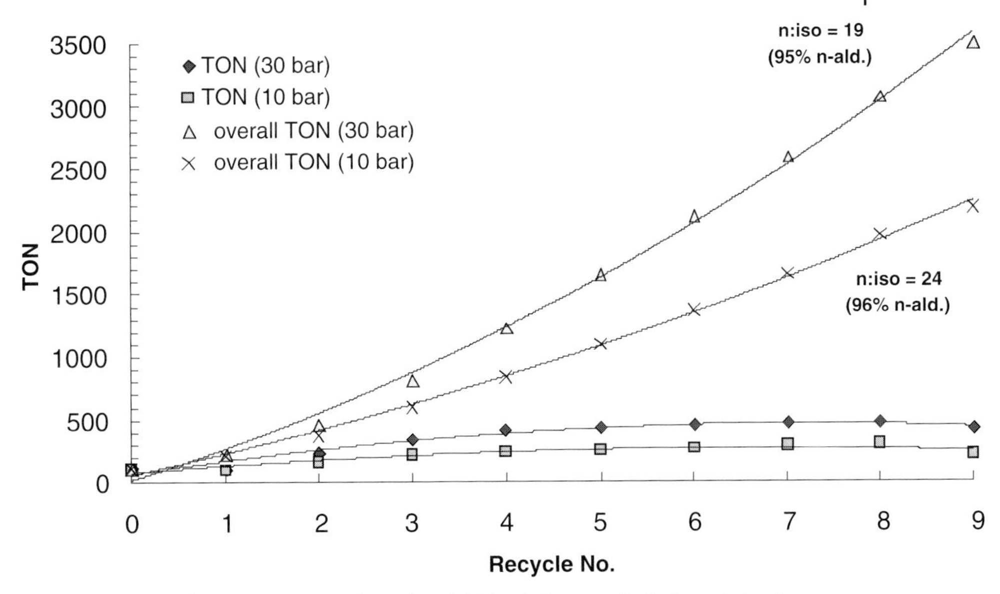

Figure 5.2-5: Recycling experiments - Rh-catalyzed, biphasic 1-octene hydroformylation in [BMIM][PF_6] with a guanidinium-modified diphosphine ligand with xanthene backbone.

the regioselectivity of the reaction. Thus, these results demonstrate that the modification of known phosphine ligands with guanidinium groups is a simple and very efficient method by which to immobilize transition metal complexes fully in ionic liquids.

Alternatively, methods for the immobilization of phosphine ligands by attaching them to ionic groups with high similarity to the ionic liquid's cation have been reported. Both pyridinium-modified phosphine ligands [81] and imidazolium-modified phosphine ligands [82, 83] have been synthesized and applied in Rh-catalyzed hydroformylation (see Figure 5.2-6). While the presence of the ionic group resulted in better immobilization of the Rh-catalyst in the ionic liquid in all cases, no outstanding reactivity or selectivity were observed with these ligands. This is not really surprising, since all these ligands are electronically and sterically closely related to PPh_3.

So far, research in the field of biphasic, Rh-catalyzed hydroformylation in ionic liquids has largely been dominated by attempts to improve the immobilization of the phosphine ligands in the ionic medium. Although the development of such ionic ligands is undoubtedly an important prerequisite for all future research in biphasic hydroformylation using ionic liquids, one should not forget other research activities with slightly different goals.

Olivier-Bourbigou's group, for example, has recently shown that phosphite ligands can be used in Rh-catalyzed hydroformylation in ionic liquids as well as the well known phosphine systems [81]. Since phosphite ligands are usually unstable in aqueous media, this adds (apart from the much better solubility of higher olefins in

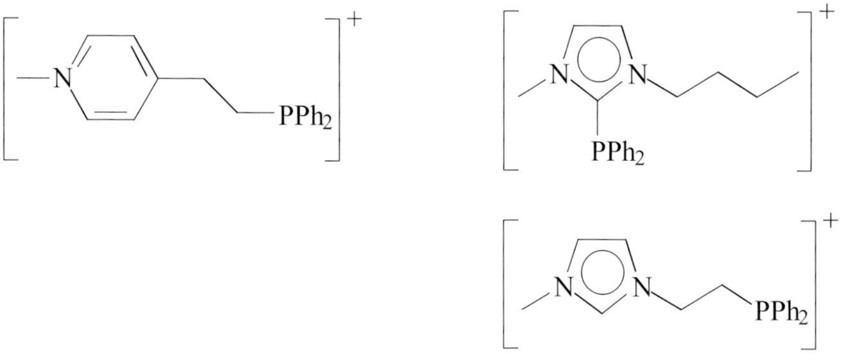

Figure 5.2-6: Cationic diphenylphosphine ligands as used in the biphasic, Rh-catalyzed hydroformylation of 1-octene in, for example, $[BMIM][PF_6]$.

ionic liquids) another important advantage to biphasic hydroformylation using ionic liquids in comparison to the well known biphasic reaction in water.

Another interesting recent development is the continuous, Rh-catalyzed hydroformylation of 1-octene in the unconventional biphasic system $[BMIM][PF_6]/scCO_2$, described by Cole-Hamilton et al. [84]. This specific example is described in more detail, together with other recent work in ionic liquid/$scCO_2$ systems, in Section 5.4.

Finally, research efforts to replace hexafluorophosphate (and other halogen-containing) ionic liquids by some cheap and halogen-free ionic liquids in Rh-catalyzed hydroformylation should be mentioned. The first attempts in this direction were made by Andersen et al. [13]. These authors investigated the hydroformylation of 1-hexene in high-melting phosphonium salts, such as butyltriphenylphosphonium tosylate (mp = 116–117 °C). Obviously, the high melting point of the salts used makes the processing of the reaction difficult, although the authors describe an easy product isolation by pouring the product off from the solid catalyst medium at room temperature.

In the author's group, much lower-melting benzenesulfonate, tosylate, or octylsulfate ionic liquids have recently been obtained in combination with imidazolium ions. These systems have been successfully applied as catalyst media for the biphasic, Rh-catalyzed hydroformylation of 1-octene [14]. The catalyst activities obtained with these systems were in all cases equal to or even higher than those found with the commonly used $[BMIM][PF_6]$. Taking into account the much lower costs of the ionic medium, the better hydrolysis stability, and the wider disposal options relating to, for example, an octylsulfate ionic liquid in comparison to $[BMIM][PF_6]$, there is no real reason to center future hydroformylation research around hexafluorophosphate ionic liquids.

5.2.4.4 Heck, Suzuki, Stille, and Negishi coupling reactions

The Heck reaction and other related transformations for selective C–C couplings are receiving a great deal of attention among synthetic chemists, due to their versatility for fine chemical synthesis. However, these reactions suffer in many cases from the instability of the Pd-catalysts used, resulting in high catalyst consumption and difficult processing.

The use of ionic liquids as reaction media for the palladium-catalyzed Heck reaction was first described by Kaufmann et al., in 1996 [85]. Treatment of bromobenzene with butyl acrylate to provide butyl *trans*-cinnamate succeeded in high yield in molten tetraalkylammonium and tetraalkylphosphonium bromide salts, without addition of phosphine ligands (Scheme 5.2-16).

The authors describe a stabilizing effect of the ionic liquid on the palladium catalyst. In almost all reactions no precipitation of elemental palladium was observed, even at complete conversion of the aromatic halide. The reaction products were isolated by distillation from the nonvolatile ionic liquid.

Extensive studies of the Heck reaction in low-melting salts have been presented by Hermann and Böhm [86]. Their results indicate that the application of ionic solvents offers clear advantages over that of commonly used organic solvents (e.g., DMF), especially for conversions of the commercially interesting chloroarenes. Additional activation and stabilization was observed with almost all catalyst systems tested. Among the ionic solvent systems investigated, molten $[NBu_4]Br$ (mp = 103 °C) proved to be an especially suitable reaction medium. On treatment of bromobenzene with styrene, with diiodo-bis(1,3-dimethylimidazolin-2-ylidene)-palladium(II) as catalyst, the yield of stilbene could be increased from 20 % (DMF) to over 99 % ($[NBu_4]Br$) under otherwise identical conditions. Again, distillative product separation from the nonvolatile ionic catalyst solution was possible. The latter could be reused up to thirteen times without any significant drop in activity. Additional advantages of the new solvent strategy lie in the excellent solubility of all reacting molecules in the ionic solvent and the possibility of using cheap inorganic bases.

This work was followed up by other research groups, using different substrates and other Pd-precursor/ligand combinations in molten $[NBu_4]Br$ for Heck coupling.

Muzart et al. described the coupling of aryl iodides and bromides with allylic alcohols to give the corresponding β-arylated carbonyl compounds [87]. Calò et al

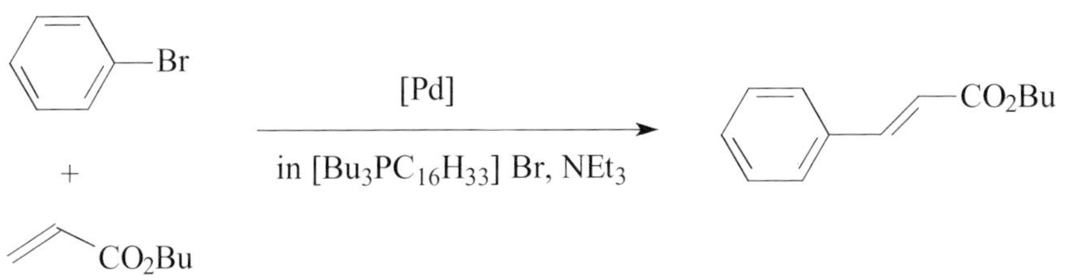

Br

[Pd]

CO_2Bu

+

in $[Bu_3PC_{16}H_{33}]$ Br, NEt_3

CO_2Bu

Scheme 5.2-16: Pd-catalyzed Heck reaction between butyl acrylate and bromobenzene, carried out in a phosphonium bromide salt.

described the Heck coupling of substituted acrylates with bromobenzene in molten $[NBu_4]Br$, catalyzed by Pd-benzothiazole carbene complexes [88]. The same solvent was found to be essential in investigations carried out by Buchmeiser et al. into the Pd-catalyzed Heck coupling of aryl chlorides and the amination of aryl bromides [89].

The use of imidazolium-based ionic liquids in Pd-catalyzed Heck reactions always carries with it the possibility of in situ formation of Pd-carbene complexes (for more details see Section 5.2.2.3). The formation of these under the conditions of the Heck reaction was confirmed by investigations by Xiao et al. [30], who described a significantly enhanced reactivity of the Heck reaction in [BMIM]Br in relation to the same reaction in $[BMIM][BF_4]$ and explained this difference by the fact that formation of Pd-carbene complexes was observed only in the bromide melt.

The regioselective arylation of butyl vinyl ether was carried out by the same group, using $Pd(OAc)_2$ as catalyst precursor and 1,3-bis(diphenylphosphino)-propane (dppp) as the ligand, dissolved in $[BMIM][BF_4]$ (Scheme 5.2-17) [90].

The results in the ionic liquid were compared with those obtained in four conventional organic solvents. Interestingly, the reaction in the ionic liquid proceeded with very high selectivity to give the α-arylated compound, whereas variable mixtures of the α- and β-isomers were obtained in the organic solvents DMF, DMSO, toluene, and acetonitrile. Furthermore, no formation of palladium black was observed in the ionic liquid, while this was always the case with the organic solvents.

Seddon's group described the option of carrying out Heck reactions in ionic liquids that do not completely mix with water. These authors studied different Heck reactions in the triphasic $[BMIM][PF_6]$/water/hexane system [91]. While the $[BMIM]_2[PdCl_4]$ catalyst used remains in the ionic liquid, the products dissolve in the organic layer, with the salt formed as a by-product of the reaction ([H-base]X) being extracted into the aqueous phase.

Finally, some recently published Heck couplings of aryl iodides, including the use of Pd(0) nanoparticles formed in situ [92] and heterogeneous Pd on carbon [93] should be mentioned here.

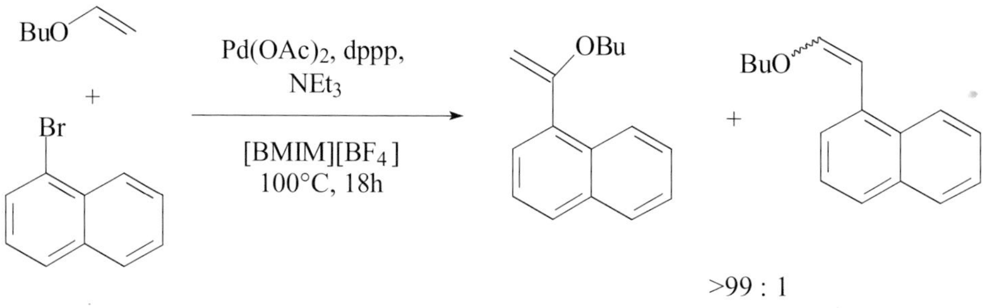

BuO + Br (1-bromonaphthalene) → [$Pd(OAc)_2$, dppp, NEt_3; $[BMIM][BF_4]$, 100°C, 18h] → OBu (α-product) + BuO (β-product)

>99 : 1

Scheme 5.2-17: Pd-catalyzed, regioselective arylation of butyl vinyl ether in a $[BMIM][BF_4]$ ionic liquid.

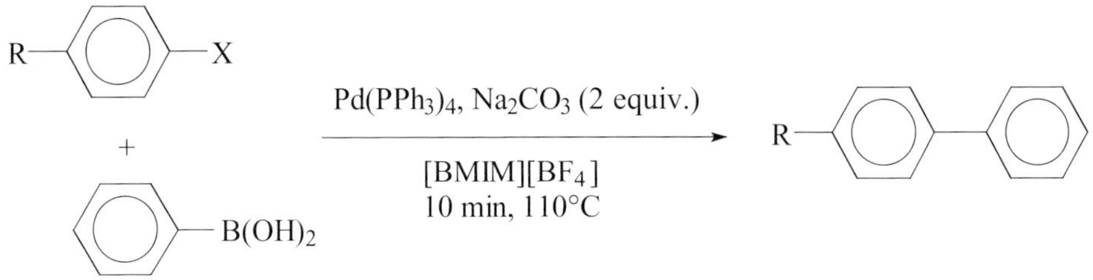

R–C6H4–X + C6H5–B(OH)2 → ($Pd(PPh_3)_4$, Na_2CO_3 (2 equiv.), [BMIM][BF_4], 10 min, 110°C) → R–C6H4–C6H5

Scheme 5.2-18: Pd-catalyzed Suzuki cross-coupling reaction in a [BMIM][BF_4] ionic liquid.

Suzuki cross-coupling reactions using $Pd(PPh_3)_4$ as catalyst in [BMIM][BF_4] have been reported by Welton et al.. (Scheme 5.2-18) [94]. The best results were achieved by pre-heating the aryl halide to 110 °C in the ionic liquid with the Pd-complex. The arylboronic acid and Na_2CO_3 were later added to start the reaction. Several advantages over the reaction as performed under the conventional Suzuki conditions were described. The reaction showed significantly enhanced activity in the ionic liquid (TOF = 455 h^{-1} in [BMIM][BF_4], in comparison to 5 h^{-1} under conventional Suzuki conditions). The formation of the homo-coupling aryl by-product was suppressed. Moreover, the ionic catalyst layer could be reused after extraction of the products with ether and removal of the by-products ($NaHCO_3$ and $NaXB(OH)_2$) with excess water. No deactivation was observed with this procedure over three further reaction cycles.

A number of Stille coupling reactions have been reported by Handy et al. [95]. With $PdCl_2(PhCN)_2/Ph_3As/CuI$ in [BMIM][BF_4], good yields and good catalyst recyclability (up to five times) were reported for the reaction between α-iodenones and vinyl and aryl stannanes (Scheme 5.2-19). However, the reported reaction rates were significantly lower than those obtained in NMP.

Knochel et al. described Pd-catalyzed Negishi cross-coupling reactions between zinc organometallics and aryl iodides in [BMMIM][BF_4]. Scheme 5.2-20 illustrates the reaction for the formation of a 3-substituted cyclohexenone from 3-iodo-2-cyclohexen-1-one [82].

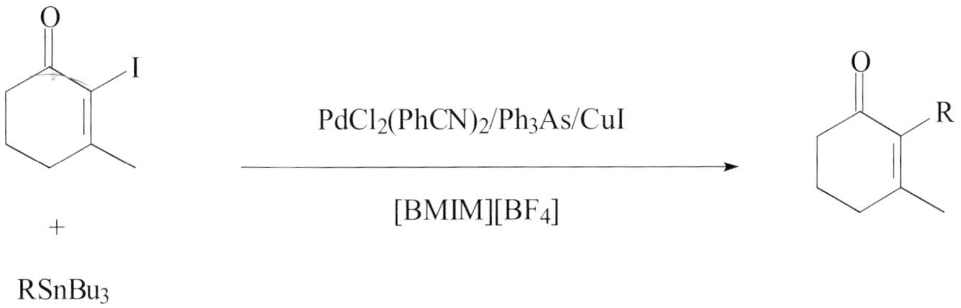

α-iodoenone + $RSnBu_3$ → ($PdCl_2(PhCN)_2/Ph_3As/CuI$, [BMIM][$BF_4$]) → α-R-enone

Scheme 5.2-19: Pd-catalyzed Stille coupling of α-iodoenones with vinyl and aryl stannanes in [BMIM][BF_4].

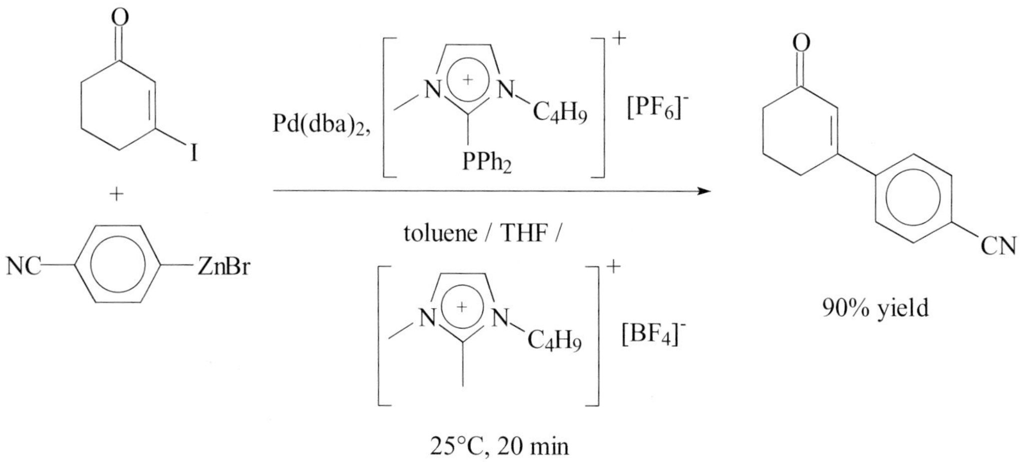

Scheme 5.2-20: Pd-catalyzed cross-coupling of organozinc compounds (Negishi cross-coupling) in [BMMIM][BF_4].

The reaction was carried out in an ionic liquid/toluene biphasic system, which allowed easy product recovery from the catalyst by decantation. However, attempts to recycle the ionic catalyst phase resulted in significant catalyst deactivation after only the third recycle.

5.2.4.5 Dimerization and oligomerization reactions

In dimerization and oligomerization reactions, ionic liquids have already proven to be a highly promising solvent class for the transfer of established catalytic systems into biphasic catalysis.

Biphasic catalysis is not a new concept for oligomerization chemistry. On the contrary, the oligomerization of ethylene was the first commercialized example of a biphasic, catalytic reaction. The process is known under the name "Shell Higher Olefins Process (SHOP)", and the first patents originate from as early as the late 1960's.

While the SHOP uses 1,4-butanediol as the catalyst phase, it turned out in subsequent years of research that many highly attractive catalyst systems for dimerization and oligomerization were not compatible with polar organic solvents or water. This was because the electrophilicity of the metal center is a key characteristic for its catalytic activity in oligomerization. The higher the electrophilicity of the metal center, the higher – usually – is its catalytic activity, but its compatibility with polar organic solvents or water is at the same time lower. Consequently, many cationic transition metal complexes may be excellent oligomerization catalysts [24], but none of these systems could be used in the biphasic reaction mode with water or polar organic solvents as the catalyst phase.

One technically important example of an oligomerization that could not be carried out in a liquid–liquid biphasic mode with polar organic solvents or water is the

Ni-catalyzed dimerization of propene and/or butenes, which was intensively studied in the 1960's [96] and later commercialized as the "Dimersol process" by the Institut Français du Pétrole (IFP). The active catalytic species is formed in situ through the reaction between a Ni(II) source and an alkylaluminium co-catalyst. The reaction takes place in a monophasic reaction mode in an organic solvent or – technically preferred – in the alkene feedstock. After the reaction, the catalyst is destroyed by addition of an aqueous solution of a base and the precipitated Ni salt is filtered off and has to be disposed of. Twenty-five Dimersol units are currently in operation, producing octane booster for gasoline with a total processing capacity of 3.4m tons per year. In view of the significant consumption of nickel and alkylaluminiums associated with the monophasic Dimersol process, it is not surprising that IFP research teams were looking for new solvent approaches to allow a biphasic version of the Dimersol chemistry. Chloroaluminate ionic liquids proved to be highly attractive in this respect.

As early as 1990, Chauvin and his co-workers from IFP published their first results on the biphasic, Ni-catalyzed dimerization of propene in ionic liquids of the $[BMIM]Cl/AlCl_3/AlEtCl_2$ type [4]. In the following years the nickel-catalyzed oligomerization of short-chain alkenes in chloroaluminate melts became one of the most intensively investigated applications of transition metal catalysts in ionic liquids to date.

Because of its significance, some basic principles of the Ni-catalyzed dimerization of propene in chloroaluminate ionic liquids should be presented here. Table 5.2-2 displays some reported examples, selected to explain the most important aspects of oligomerization chemistry in chloroaluminate ionic liquids [97].

The Ni-catalyzed oligomerization of olefins in ionic liquids requires a careful choice of the ionic liquid's acidity. In basic melts (Table 5.2-2, entry (a)), no dimerization activity is observed. Here, the basic chloride ions prevent the formation of free coordination sites on the nickel catalyst. In acidic chloroaluminate melts, an oligomerization reaction takes place even in the absence of a nickel catalyst (entry (b)). However, no dimers are produced, but a mixture of different oligomers is

Table 5.2-2: Selected results from Ni-catalyzed propene dimerization in chloroaluminate ionic liquids.

	Ionic liquid	*Composition of the ionic liquid (molar ratio)*	*Ni-complex*	*Activity (kg g(Ni)$^{-1}$ h^{-1})*	*Product DMB/ M2P/nH* [(a)]
a)	$[BMIM]Cl/AlCl_3$	1/0.8	$NiBr_2L_2$ [(b)]	0	
b)	$[BMIM]Cl/AlCl_3$	1/1.5		[c]	
c)	$[BMIM]Cl/ AlEtCl_2$	1/1.2	$NiCl_2$	2.5	5/74/21
d)	$[BMIM]Cl/ AlEtCl_2$	1/1.2	$NiCl_2(iPr_3P)_2$	2.5	74/24/2
e)	$[BMIM]Cl/ AlCl_3/AlEtCl_2$	1/1.2/0.1	$NiCl_2(iPr_3P)_2$	12.5	83/15/2

T = –15 °C; (a) DMB = dimethylbutenes, M2P = methylpentenes, *n*H = *n*-hexenes,
(b) L = 2-methylallyl; (c) highly viscous oligomers from cationic oligomerization were obtained.

formed by cationic oligomerization. Superacidic protons and the reactivity of the acidic anions $[Al_2Cl_7]^-$ and $[Al_3Cl_{10}]^-$ may account for this reactivity.

The addition of alkylaluminium compounds is known to suppress this undesired cationic oligomerization activity. In the presence of $NiCl_2$ as catalyst precursor, the ionic catalyst solution is formed and shows high activity for the dimerization (entry (c)). Without added phosphine ligands, a product distribution with no particular selectivity is obtained. In the presence of added phosphine ligand, the distribution of regioisomers in the C6-fraction is influenced by the steric and electronic properties of the ligand used in the same way as known from the catalytic system in organic solvents [96] (entry (d)). At longer reaction times, a decrease in the selectivity to highly branched products is observed. It has been postulated that a competing reaction of the basic phosphine ligand with the hard Lewis acid $AlCl_3$ takes place. This assumption is supported by the observation that the addition of a soft competing base such as tetramethylbenzene can prevent the loss in selectivity.

Unfortunately, investigations with ionic liquids containing high amounts of $AlEtCl_2$ showed several limitations, including the reductive effect of the alkylaluminium affecting the temperature stability of the nickel catalyst. At very high alkylaluminium concentrations, precipitation of black metallic nickel was observed even at room temperature.

From these results, the Institut Français du Pétrole (IFP) has developed a biphasic version of its established monophasic "Dimersol process", which is offered for licensing under the name "Difasol process" [98]. The "Difasol process" uses slightly acidic chloroaluminate ionic liquids with small amounts of alkylaluminiums as the solvent for the catalytic nickel center. In comparison to the established "Dimersol process", the new biphasic ionic liquid process drastically reduces the consumption of Ni-catalyst and alkylaluminiums. Additional advantages arise from the good performance obtained with highly diluted feedstocks and the significantly improved dimer selectivity of the "Difasol process" (for more detailed information see Section 5.3).

Closely related catalytic systems have also been used for the selective dimerization of ethene to butenes [99]. Dupont et al. dissolved $[Ni(MeCN)_6][BF_4]_2$ in the slightly acidic $[BMIM]Cl/AlCl_3/AlEtCl_2$ chloroaluminate system (ratio = 1: 1.2: 0.25) and obtained 100 % butenes at –10 °C and 18 bar ethylene pressure (TOF = 1731 h^{-1}). Unfortunately, the more valuable 1-butene was not produced selectively, with a mixture of all linear butene isomers (i.e., 1-butene, *cis*-2-butene, *trans*-2-butene) being obtained.

More recently, biphasic ethylene oligomerization reactions that make use of catalytic metals other than nickel have been described in chloroaluminate ionic liquids. Olivier-Bourbigou et al. dissolved the tungsten complex $[Cl_2W{=}NPh(PMe_3)_3]$ in a slightly acidic $[BMIM]Cl/AlCl_3$ ionic liquid and used this ionic catalyst solution in ethylene oligomerization without addition of a co-catalyst [100]. At 60 °C and 40 bar a product distribution of 81 % butenes, 18 % hexenes, and 1 % higher oligomers was obtained with good activity (TOF = 1280 h^{-1}). However, the selectivity for the more valuable 1-olefins was found to be relatively low (65 %). The selective, chromium-catalyzed trimerization of ethylene to 1-hexene in a biphasic reaction system using alkylchloroaluminate ionic liquids was reported in a patent by SASOL [101].

$[Ni(MeCN)_6][BF_4]_2$ dissolved in the slightly acidic chloroaluminate system $[BMIM]Cl/AlCl_3/AlEtCl_2$ (ratio = 1: 1.2: 0.25) has been used not only for the dimerization of ethene but also – at 10 °C and under atmospheric pressure – for the dimerization of butenes [102]. The reaction showed high activity under these conditions, with a turnover frequency of 6840 h^{-1} and a productivity of 6 kg oligomer per gram Ni per hour. The distribution of the butene dimers obtained (typically 39±1 % dimethylhexenes, 56±2 % monomethylheptenes, and 6±1 % *n*-octenes) was reported to be independent of the addition of phosphine ligands. Moreover, the product mix was independent of feedstock, with both 1-butene and 2-butenes yielding the same dimer distribution, with only 6 % of the linear product. This clearly indicates that the catalytic system used here is not only an active oligomerization catalyst but also highly active for isomerization.

The selective, Ni-catalyzed, biphasic dimerization of 1-butene to linear octenes has been studied in the author's group. A catalytic system well known for its ability to form linear dimers from 1-butene in conventional organic solvents – namely the square-planar Ni-complex (η-4-cycloocten-1-yl](1,1,1,5,5,5,-hexafluoro-2,4-pentanedionato-O,O')nickel [(H-COD)Ni(hfacac)] [103] – was therefore used in chloroaluminate ionic liquids.

For this specific task, ionic liquids containing alkylaluminiums proved unsuitable, due to their strong isomerization activity [102]. Since, mechanistically, only the linkage of two 1-butene molecules can give rise to the formation of linear octenes, isomerization activity in the solvent inhibits the formation of the desired product. Therefore, slightly acidic chloroaluminate melts that would enable selective nickel catalysis without the addition of alkylaluminiums were developed [104]. It was found that an acidic chloroaluminate ionic liquid buffered with small amounts of weak organic bases provided a solvent that allowed a selective, biphasic reaction with [(H-COD)Ni(hfacac)].

The function of the base is to trap any free acidic species, which might initiate cationic side reactions, in the melt. A suitable base has to fulfil a number of requirements. Its basicity has to be in the appropriate range to provide enough reactivity to eliminate all free acidic species in the melt. At the same time, it has to be non-coordinating with respect to the catalytically active Ni center. Another important feature is a very high solubility in the ionic liquid. During the reaction, the base has to remain in the ionic catalyst layer even under intensive extraction of the ionic liquid by the organic layer. Finally, the base has to be inert to the 1-butene feedstock and to the oligomerization products.

The use of pyrrole and N-methylpyrrole was found to be preferable. Through the addition of N-methylpyrrole, all cationic side reactions could be effectively suppressed, and only dimerization products produced by Ni-catalysis were obtained. In this case the dimer selectivity was as high as 98 %. Scheme 5.2-21 shows the catalytic system that allowed the first successful application of [(H-COD)Ni(hfacac)] in the biphasic linear dimerization of 1-butene.

Comparison of the dimerization of 1-butene with [(H-COD)Ni(hfacac)] in chloroaluminate ionic liquids with the identical reaction in toluene is quite instructive. First of all, the reaction in the ionic liquid solvent is biphasic with no detectable

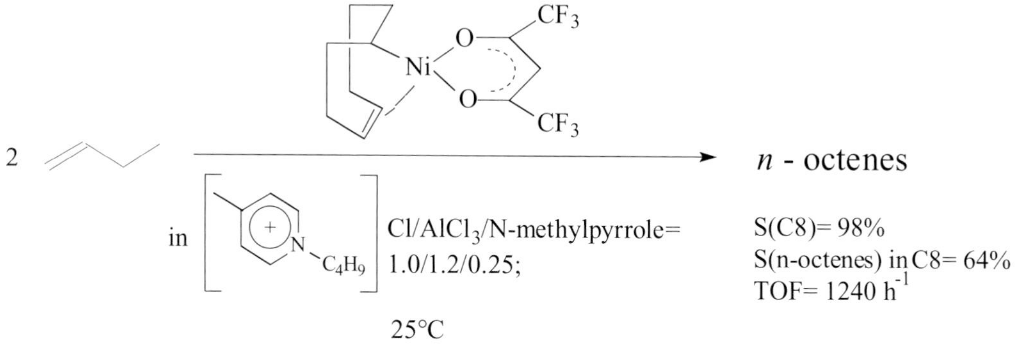

$$2 \text{ butene} \xrightarrow[\text{in } [\text{4-methyl-}N\text{-butylpyridinium}]\text{Cl}/\text{AlCl}_3/N\text{-methylpyrrole}=1.0/1.2/0.25;\ 25^\circ\text{C}]{[(\text{H-COD})\text{Ni(hfacac)}]} n\text{ - octenes}$$

S(C8)= 98%
S(n-octenes) in C8= 64%
TOF= 1240 h^{-1}

Scheme 5.2-21: Ni-catalyzed, biphasic, linear dimerization in a slightly acidic, buffered chloroaluminate ionic liquid.

catalyst leaching, enabling easy catalyst separation and recycling. While [(H-COD)Ni(hfacac)] in toluene requires an activation temperature of 50 °C, the reaction proceeds in the ionic liquid even at −10 °C. This indicates that the catalyst activation, believed to be the formation of the active Ni-hydride complex, proceeds much more efficiently in the chloroaluminate solvent (for more details on mechanistic studies see Section 5.2.3). Furthermore, the product selectivities obtained in both solvents show significantly higher dimer selectivities in the biphasic case. This can be understood by considering the fact that the C8-product is much less soluble than the butene feedstock in the ionic liquid (by about a factor of 4). During the reaction, rapid extraction of the C8-product into the organic layer takes place, thus preventing subsequent C12-formation. The linear selectivity is high in both solvents, although somewhat lower in the ionic liquid solvent.

To produce reliable data on the lifetime and overall activity of the ionic catalyst system, a loop reactor was constructed and the reaction was carried out in continuous mode [105]. Some results of these studies are presented in Section 5.3, together with much more detailed information about the processing of biphasic reactions with an ionic liquid catalyst phase.

Biphasic oligomerization with ionic liquids is not restricted to chloroaluminate systems. Especially in those cases where the – at least – latent acidity or basicity of the chloroaluminate causes problems, neutral ionic liquids with weakly coordinating anions can be used with great success.

As already mentioned above, the Ni-catalyzed oligomerization of ethylene in chloroaluminate ionic liquids was found to be characterized by high oligomerization and high isomerization activity. The latter results in a rapid consecutive transformation of the α-olefins formed into mixtures of far less valuable internal olefins. Higher α-olefins (HAOs) are an important group of industrial chemicals that find a variety of end uses. Depending on their chain length, they are components of plastics (C4-C6 HAOs in copolymerization), plasticizers (C6-C10 HAOs through hydroformylation/hydrogenation, lubricants (C10-C12 HAOs through oligomerization), and surfactants (C12-C16 HAOs through arylation/sulfonation).

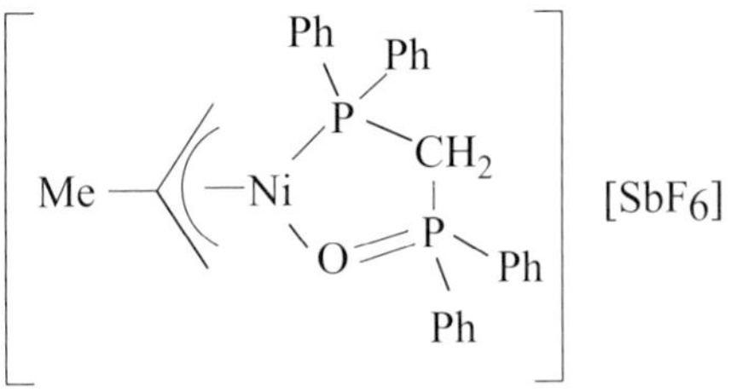

Figure 5.2-7: The cationic Ni-complex [(mall)Ni(dppmo)][SbF_6] as used for the biphasic oligomerization of ethylene to α-olefins in, for example, [BMIM][PF_6].

In addition to the neutral nickel/phosphine complexes used in the Shell Higher Olefins Process (SHOP), cationic Ni-complexes such as [(mall)Ni(dppmo)][SbF_6] (see Figure 5.2-7) have attracted some attention as highly selective and highly active catalysts for ethylene oligomerization to HAOs [106].

However, all attempts to carry out biphasic ethylene oligomerization with this cationic catalyst in traditional organic solvents, such as 1,4-butanediol (as used in the SHOP) resulted in almost complete catalyst deactivation by the solvent. This reflects the much higher electrophilicity of the cationic complex [(mall)Ni(dppmo)][SbF_6] in relation to the neutral Ni-complexes used in the SHOP.

It was recently demonstrated in the author's group that the use of hexafluorophosphate ionic liquids allows, for the first time, selective, biphasic oligomerization of ethylene to 1-olefins with the aid of the cationic Ni-complex [(mall)Ni(dppmo)][SbF_6] (Scheme 5.2-22) [25, 107].

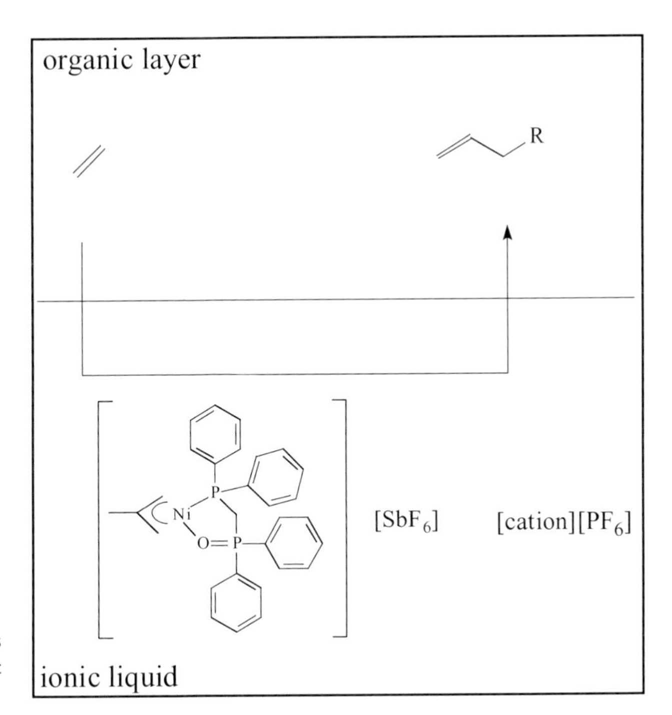

Scheme 5.2-22: Biphasic ethylene oligomerization with cationic Ni-complexes in a [BMIM][PF_6] ionic liquid.

Obviously, the ionic liquid's ability to dissolve the ionic catalyst complex, in combination with low solvent nucleophilicity, opens up the possibility for biphasic processing. Furthermore it was found that the biphasic reaction mode in this specific reaction resulted in improved catalytic activity and selectivity and in enhanced catalyst lifetime.

The higher activity of the catalyst [(mall)Ni(dppmo)][SbF_6] in [BMIM][PF_6] (TOF = 25,425 h^{-1}) relative to the reaction under identical conditions in CH_2Cl_2 (TOF = 7591 h^{-1}) can be explained by the fast extraction of products and side products out of the catalyst layer and into the organic phase. A high concentration of internal olefins (from oligomerization and consecutive isomerization) at the catalyst is known to reduce catalytic activity, due to the formation of fairly stable Ni-olefin complexes.

The selectivity of the ethylene oligomerization reaction is clearly influenced by the biphasic reaction mode. The oligomers were found to be much shorter in the biphasic system, due to restricted ethylene availability at the catalytic center when dissolved in the ionic liquid. This behavior correlates with the ethylene solubility in the different solvents under the reaction conditions. Ethylene solubility in 10 ml CH_2Cl_2 was determined to be 6.51 g at 25 °C/50 bar, in comparison with only 1.1 g ethylene dissolved in [BMIM][PF_6] under identical conditions. Since the rate of ethylene insertion is dependent on the ethylene concentration at the catalyst, but the rate of β-H-elimination is not, it becomes understandable that a low ethylene availability at the catalytic active center would favor the formation of shorter oligomers. In good agreement with this, a shift of the oligomer distribution was observed if the ionic liquid's cation was modified with longer alkyl chains. With increasing alkyl chain length, the obtained oligomer distribution gradually became broader, following the higher ethylene solubility in these ionic liquids. However, all biphasic oligomerization experiments still showed much narrower oligomer distributions than found in the case of the monophasic reaction in CH_2Cl_2 (under identical conditions).

As well as the oligomer distribution, the selectivity for 1-olefins is of great technical relevance. Despite the much higher catalytic activity, this selectivity was even slightly higher in [BMIM][PF_6] than in CH_2Cl_2. The overall 1-hexene selectivity in C6-products is 88.5 % in [BMIM][PF_6], against 85.0 % in CH_2Cl_2. Interestingly, smaller quantities of the internal hexenes (formed by subsequent isomerization of 1-olefins) are obtained in the case of biphasic oligomerization with the ionic liquid solvent. This is explained by the much lower solubility of the higher oligomerization products in the catalyst solvent [BMIM][PF_6]. Since the 1-olefins formed are quickly extracted into the organic layer, consecutive isomerization of these products at the Ni-center is suppressed relative to the monophasic reaction in CH_2Cl_2.

It is noteworthy that the best results could be obtained only with very pure ionic liquids and by use of an optimized reactor set-up. The contents of halide ions and water in the ionic liquid were found to be crucial parameters, since both impurities poisoned the cationic catalyst. Furthermore, the catalytic results proved to be highly dependent on all modifications influencing mass transfer of ethylene into the ionic catalyst layer. A 150 ml autoclave stirred from the top with a special stirrer

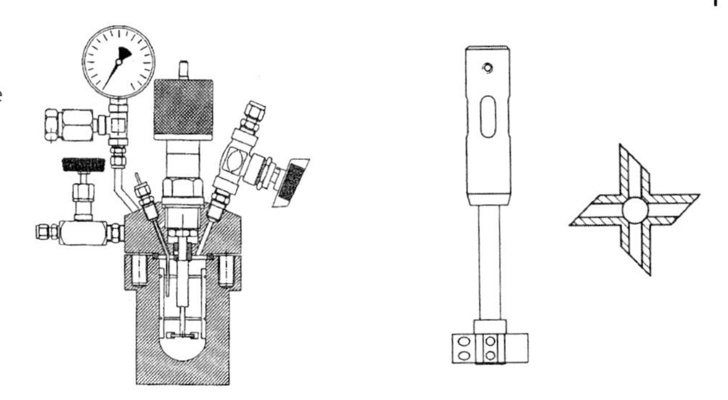

Figure 5.2-8: A 150 ml autoclave with special stirrer design to maximize ethylene intake into an ionic liquid catalyst layer.

designed to maximize ethylene intake into the ionic liquid and also equipped with baffles to improve the liquid–liquid mixing (see Figure 5.2-8) gave far better results than a standard autoclave stirred with a magnetic stirrer bar.

Finally, it was possible to demonstrate that the ionic catalyst solution can, in principle, be recycled. By repetitive use of the ionic catalyst solution, an overall activity of 61,106 mol ethylene converted per mol catalyst could be achieved after two recycle runs.

An example of a biphasic, Ni-catalyzed co-dimerization in ionic liquids with weakly coordinating anions has been described by the author's group in collaboration with Leitner et al. [12]. The hydrovinylation of styrene in the biphasic ionic liquid/compressed CO_2 system with a chiral Ni-catalyst was investigated. Since it was found that this reaction benefits particularly from this unusual biphasic solvent system, more details about this specific application are given in Section 5.4.

Dupont and co-workers studied the Pd-catalyzed dimerization [108] and cyclodimerization [109] of butadiene in non-chloroaluminate ionic liquids. The biphasic dimerization of butadiene is an attractive research goal since the products formed, 1,3,5-octatriene and 1,3,6-octatriene, are sensitive towards undesired polymerization, so that separation by distillation is usually not possible. These octatrienes are of some commercial relevance as intermediates for the synthesis of fragrances, plasticizers, and adhesives. Through the use of $PdCl_2$ with two equivalents of the ligand PPh_3 dissolved in [BMIM][PF_6], [BMIM][BF_4], or [BMIM][CF_3SO_3], it was possible to obtain the octatrienes with 100 % selectivity (after 13 % conversion) (Scheme 5.2-23) [108]. The turnover frequency (TOF) was in the range of 50 mol butadiene converted per mol catalyst per hour, which represents a substantial increase in catalyst activity in comparison to the same reaction under otherwise identical conditions (70 °C, 3 h, butadiene/Pd = 1250) in THF (TOF = 6 h^{-1}).

The cyclodimerization of 1,3-butadiene was carried out in [BMIM][BF_4] and [BMIM][PF_6] with an *in situ* iron catalyst system. The catalyst was prepared by reduction of [$Fe_2(NO)_4Cl_2$] with metallic zinc in the ionic liquid. At 50 °C, the reaction proceeded in [BMIM][BF_4] to give full conversion of 1,3-butadiene, and 4-vinylcyclohexene was formed with 100 % selectivity. The observed catalytic activity corresponded to a turnover frequency of at least 1440 h^{-1} (Scheme 5.2-24).

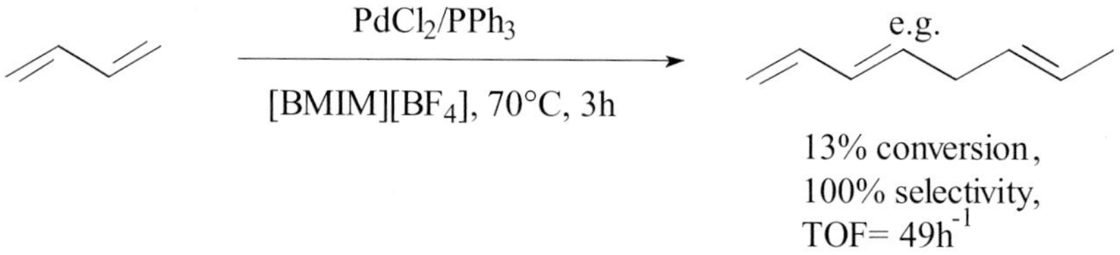

$PdCl_2/PPh_3$

[BMIM][BF_4], 70°C, 3h

e.g.

13% conversion,
100% selectivity,
TOF= $49h^{-1}$

Scheme 5.2-23: Biphasic, Pd-catalyzed dimerization of butadiene in [BMIM][BF_4].

[$Fe_2(NO)_4Cl_2$], Zn

[BMIM][BF_4], 50°C

100% conversion,
100% selectivity,
TOF= $1440h^{-1}$

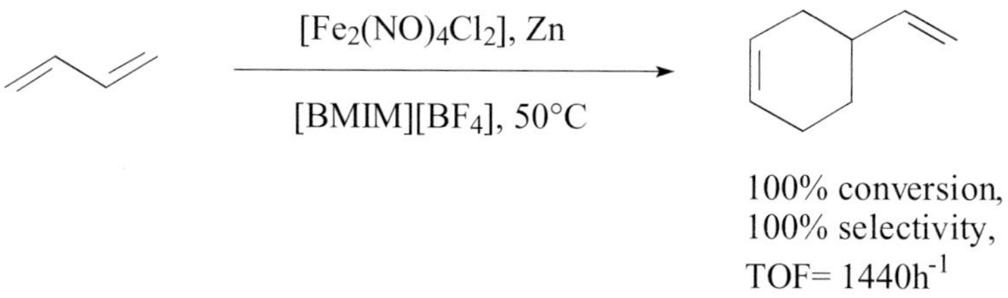

Scheme 5.2-24: Biphasic, Fe-catalyzed cyclotrimerization of butadiene in [BMIM][BF_4].

The authors correlate the observed catalytic activity with the solubility of the 1,3-butadiene feedstock in the ionic liquid, which was found to be twice as high in the tetrafluoroborate ionic liquid as in the corresponding hexafluorophosphate system. It is noteworthy that the same reaction in a monophasic systems with toluene as the solvent was found to be significantly less active (TOF = 240 h^{-1}).

5.2.5
Concluding Remarks

Obviously, there are many good reasons to study ionic liquids as alternative solvents in transition metal-catalyzed reactions. Besides the engineering advantage of their nonvolatile natures, the investigation of new biphasic reactions with an ionic catalyst phase is of special interest. The possibility of adjusting solubility properties by different cation/anion combinations permits systematic optimization of the biphasic reaction (with regard, for example, to product selectivity). Attractive options to improve selectivity in multiphase reactions derive from the preferential solubility of only one reactant in the catalyst solvent or from the *in situ* extraction of reaction intermediates from the catalyst layer. Moreover, the application of an ionic liquid catalyst layer permits a biphasic reaction mode in many cases where this would not be possible with water or polar organic solvents (due to incompatibility with the catalyst or problems with substrate solubility, for example).

In addition to the applications reported in detail above, a number of other transition metal-catalyzed reactions in ionic liquids have been carried out with some success in recent years, illustrating the broad versatility of the methodology. Butadiene telomerization [34], olefin metathesis [110], carbonylation [111], allylic alkylation [112] and substitution [113], and Trost–Tsuji-coupling [114] are other examples of high value for synthetic chemists.

However, research into transition metal catalysis in ionic liquids should not focus only on the question of how to make some specific products more economical or ecological by use of a new solvent and, presumably, a new multiphasic process. Since it bridges the gap between homogeneous and heterogeneous catalysis, in a novel and highly attractive manner, the application of ionic liquids in transition metal catalysis gives access to some much more fundamental and conceptual questions for basic research.

In many respects, transition metal catalysis in ionic liquids is in fact better regarded as heterogeneous catalysis on a liquid support than as conventional homogeneous catalysis in an organic solvent. As in heterogeneous catalysis, support–catalyst interactions are known in ionic liquids and can give rise to catalyst activation. Product separation from an ionic catalyst layer is often easy (at least if the products are not too polar and have low boiling points), as in classical heterogeneous catalysis. However, mass transfer limitation problems (when the chemical kinetics are fast) and some uncertainty concerning the exact microenvironment around the catalytically active center represent common limitations for transition metal catalysis both in ionic liquids and in heterogeneous catalysis.

Of course, the use of a liquid catalyst immobilization phase still produces some very important differences in comparison to classical heterogeneous supports. Obviously, by use of a liquid, ionic catalyst support it is possible to integrate some classical features of traditional homogenous catalysis into this type of "heterogeneous" catalysis. For example, a defined transition metal complex can be introduced and immobilized in an ionic liquid to provide opportunities to optimize the selectivity of a reaction by ligand variation, which is a typical approach in homogeneous catalysis. Reaction conditions in ionic liquid catalysis are still mild, as typically used in homogenous catalysis. Analysis of the active catalyst in an ionic liquid immobilization phase is, in principle, possible by the same methods as developed for homogeneous catalysis, which should enable more rational catalyst design in the future.

In comparison with traditional biphasic catalysis using water, fluorous phases, or polar organic solvents, transition metal catalysis in ionic liquids represents a new and advanced way to combine the specific advantages of homogeneous and heterogeneous catalysis. In many applications, the use of a defined transition metal complex immobilized on a ionic liquid "support" has already shown its unique potential. Many more successful examples – mainly in fine chemical synthesis – can be expected in the future as our knowledge of ionic liquids and their interactions with transition metal complexes increases.

References

1 G. W. Parshall, *J. Am. Chem. Soc.* **1972**, *94*, 8716–8719.

2 J. F. Knifton, *J. Mol. Catal.* **1987**, *43*, 65–78.

3 R. T. Carlin, R. A. Osteryoung, *J. Mol. Catal.* **1990**, *63*, 125–129.

4 Y. Chauvin, B. Gilbert, I. Guibard, *J. Chem. Soc. Chem. Commun.* **1990**, 1715–1716.

5 J. S. Wilkes, M. J. Zaworotko, *J. Chem. Soc. Chem. Commun.* **1992**, 965–967.

6 a) P. A. Z. Suarez, J. E. L. Dullius, S. Einloft, R. F. de Souza, J. Dupont, *Polyhedron* **1996**, *15*, 1217–1219; b) Y. Chauvin, L. Mußmann, H. Olivier, *Angew. Chem.* **1995**, *107*, 2941–2943, *Angew. Chem. Int. Ed. Engl.* **1995**, *34*, 1149–1155.

7 J. Fuller, R. T. Carlin, H. C. de Long, D. Haworth, *J. Chem. Soc. Chem. Commun.* **1994**, 299–300.

8 P. Wasserscheid, C. M. Gordon, C. Hilgers, M. J. Maldoon, I. R. Dunkin, *Chem. Comm.* **2001**, 1186–1187.

9 A list of current commercial suppliers of tetrafluoroborate and hexafluorophosphate salts includes (with no guarantee of completeness): Solvent Innovation GmbH, Köln (www.solvent-innovation.com), Sachem Inc. (www.sacheminc.com), Fluka (www.fluka.com), Acros Organics (www. acros.be), and Wako (www.wako-chem.co.jp).

10 A. Bösmann, P. Wasserscheid, unpublished results.

11 P. Bonhôte, A.-P. Dias, N. Papageorgiou, K. Kalyanasundaram, M. Grätzel, *Inorg. Chem.* **1996**, *35*, 1168–1178.

12 A. Bösmann, G. Francio, E. Janssen, M. Solinas, W. Leitner, P. Wasserscheid, *Angew. Chem. Int. Ed. Engl.* **2001**, *40*, 2697–2699.

13 N. Karodia, S. Guise, C. Newlands, J.-A. Andersen, *Chem. Commun.* **1998**, 2341–2342.

14 P. Wasserscheid, R. van Hal, A. Bösmann, *Green Chem.* **2002**, *4*, 400–404.

15 H. Olivier-Bourbigou, L. Magna, *J. Mol. Catal. A: Chemical* **2002**, *2484*, 1–19.

16 R. Sheldon, *Chem. Commun.* **2001**, 2399–2407.

17 C. M. Gordon, *Applied Catalysis A: General* **2001**, *222*, 101–117.

18 P. Wasserscheid, W. Keim, *Angew. Chem., Int. Ed.* **2000**, *39*, 3772–3789.

19 T. Welton, *Chem. Rev.* **1999**, *99*, 2071–2083.

20 J. D. Holbrey, K. R. Seddon, *Clean Products and Processes* **1999**, *1*, 223–226.

21 W. Keim, D. Vogt, H. Waffenschmidt, P. Wasserscheid, *J. Catal.* **1999**, *186*, 481–486.

22 B. Drießen-Hölscher, P. Wasserscheid, W. Keim, *CATTECH*, **1998**, June, 47–52.

23 a) T. Prinz, W. Keim, B. Drießen-Hölscher, *Angew. Chem., Int. Ed. Engl.* **1996**, *35*, 1708–1710; b) C. Dobler, G. Mehltretter, M. Beller, *Angew. Chem., Int. Ed.* **1999**, *38*, 3026–3028.

24 a) R. B. A. Pardy, I. Tkatschenko, J. Chem. Soc., *Chem. Commun.* **1981**, 49–50; b) J. R. Ascenso, M. A. A. F. De, C. T. Carrando, A. R. Dias, P. T. Gomes, M. F. M. Piadade, C. C. Romao, A. Revillon, I. Tkatschenko, *Polyhedron* **1989**, *8*, 2449–2457; c) P. Grenouillet, D. Neibecker, I. Tkatschenko, *J. Organomet. Chem.* **1983**, *243*, 213–22; d) J.-P. Gehrke, R. Taube, E. Balbolov, K. Kurtev; *J. Organomet. Chem.* **1986**, *304*, C4–C6.

25 P. Wasserscheid, C. M. Gordon, C. Hilgers, M. J. Maldoon, I. R. Dunkin, *Chem. Commun.* **2001**, 1186–1187.

26 R. T. Carlin, R. A. Osteryoung, *J. Mol. Catal.* **1990**, *63*, 125–129.

27 Y. Chauvin, S. Einloft, H. Olivier, *Ind. Eng. Chem. Res.* **1995**, *34*, 1149–1155

28 H. Waffenschmidt, P. Wasserscheid, *J. Mol. Catal.*, **2001**, *164*, 61–67.

29 a)A. J. Arduengo, R. L. Harlow, M. Kline, *J. Am. Chem. Soc.* **1991**, *113*, 361–363; b) A. J. Arduengo, H. V. R. Dias, R. L. Harlow, *J. Am. Chem. Soc.* **1992**, *114*, 5530–5534 c) G. T. Cheek,

J. A. Spencer, *9th Int. Symp. on Molten salts*, (C.L. Hussey, D.S. Newman, G. Mamantov, Y. Ito eds.), The Electrochem. Soc., Inc., New York, **1994**, 426–432; d) W. A. Herrmann, M. Elison, J. Fischer, C. Koecher, G. R. J. Artus, *Angew. Chem., Int. Ed. Engl.* **1995**, *34*, 2371–2374; e) D. Bourissou, O. Guerret, F. P. Gabbaï, G. Bertrand, *Chem. Rev.* **2000**, *100*, 39–91.

30 L. Xu, W. Chen, J. Xiao, *Organometallics* **2000**, *19*, 1123–1127.

31 C. J. Mathews, P. J. Smith; T. Welton, A. J. P. White, *Organometallics*, **2001**, *20(18)*, 3848–3850.

32 D. S. McGuinness, K. J. Cavell, B. F. Yates, *Chem. Commun.* **2001**, 355–356.

33 M. Hasan, I. V. Kozhevnikow, M R. H. Siddiqui, C. Fermoni, A. Steiner, N. Winterton, *Inorg. Chem.* **2001**, *40(4)*, 795–800.

34 J. E. L. Dullius, P. A. Z. Suarez, S. Einloft, R. F. de Souza, J. Dupont, J. Fischer, A. D. Cian, *Organometallics* **1998**, *17*, 815–819.

35 J. A. Boon, J. A. Levisky, J. L. Pflug, J. S. Wilkes, *J. Org. Chem.* **1986**, 480–486.

36 a) M. J. Earle, K. R. Seddon, C. J. Adams, G. Roberts, *Chem. Commun.* **1998**, 2097–2098; b) A. Stark, B. L. MacLean, R. D. Singer, *J. Chem. Soc., Dalton. Trans.* **1999**, 63–66.

37 P. N. Davey, C. P. Newman, K. R. Seddon, M. J. Earle, WO 9919288 (to Quest International B.V., Neth.), **1999** [Chem Abstr. **1999**, *130*, 281871].

38 R. J. C. Brown, P. J. Dyson, D. J. Ellis, T. Welton, *Chem. Commun.* **2001**, 1862–1863.

39 C. Hardacre, S. E. J. McMath, J. D. Holbrey, *Chem. Comm.* **2001**, 367–368.

40 P. Wasserscheid, *Ph.D. thesis*, RWTH Aachen, Aachen, **1998**.

41 a) T. B. Scheffler, C. L. Hussey, K. R. Seddon, C. M. Kear, P. D. Armitage, *Inorg. Chem.* **1983**, *22*, 2099–2100; b) T. M. Laher, C. L. Hussey, *Inorg. Chem.* **1983**, *22*, 3247–3251; c) T. B. Scheffler, C. L. Hussey, *Inorg. Chem.* **1984**, *23*, 1926–1932; d) P. B. Hitchcock, T. J. Mohammed, K. R. Seddon, J. A. Zora, C. L. Hussey, E. H. Ward, *Inorg. Chim. Acta* **1986**, *113*, L25–L26.

42 a) D. Appleby, C. L. Hussey, K. R. Seddon, J. E. Turp, *Nature* **1986**, *323*, 614–616; b) A. J. Dent, K. R. Seddon, T. Welton, *J. Chem. Soc. Chem. Commun.* **1990**, 315–316.

43 P. A. Chaloner, M. A. Esteruelas, F. Joó, L. A. Oro, Homogeneous Hydrogenation, Kluwer Academic Publisher, Dordrecht, **1994**.

44 M. Medved, P. Wasserscheid, T. Melin, *Chem.-Ing.-Technik* **2001**, *73*, 715.

45 P. A. Z. Suarez, J. E. L. Dullius, S. Einloft, R. F. de Souza, J. Dupont, Polyhedron **1996**, *15*, 1217–1219.

46 Y. Chauvin, L. Mußmann, H. Olivier, *Angew. Chem.* **1995**, *107*, 2941–2943.

47 P. A. Z. Suarez, J. E. L. Dullius, S. Einloft, R. F. de Souza, J. Dupont, *Inorg. Chim. Acta* **1997**, *255*, 207–209.

48 P. A. Z. Suarez, J. E. L. Dullius, S. Einloft, R. F. de Souza, J. Dupont, *Polyhedron* **1996**, *15*, 1217–1219.

49 P. J. Dyson, D. J. Ellis, D. G. Parker, T. Welton, *Chem. Commun.* **1999**, 25–26.

50 L. A. Müller, J. Dupont, R. F. de Souza, *Macromol. Rapid. Commun.* **1998**, *19*, 409–411.

51 a) B. Drießen-Hölscher, J. Heinen, *J. Organomet. Chem.* **1998**, *570*, 141–146; b) J. Heinen, M. S. Tupayachi, B. Drießen-Hölscher, *Catalysis Today* **1999**, *48*, 273–278.

52 S. Steines, B. Drießen-Hölscher, P. Wasserscheid, *J. Prakt. Chem.* **2000**, *342*, 348–354.

53 J. Dupont, P. A. Z. Suarez, A. P. Umpierre, R. F. De Souza, *J. Braz. Chem. Soc.* **2000**, *11(3)*, 293–297.

54 A. L. Monteiro, F. K. Zinn, R. F. de Souza, J. Dupont, *Tetrahedron Assym.* **1997**, *2*, 177–179.

55 A. Berger, R. F. De Souza, M. R. Delgado, J. Dupont, *Tetrahedron: Asymmetry* **2001**, *12(13)*, 1825–1828.

56 F. Liu, M. B. Abrams, R. T. Baker, W. Tumas, *Chem. Commun.* **2001**, 433–434.

57 R. A. Brown, P. Pollet, E. McKoon, C. A. Eckert, C. L. Liotta, P. G. Jessop, *J. Am Chem. Soc.* **2001**, *123*, 1254–1255.

58 P. J. Dyson, D. J. Ellis, T. Welton, *Can. J. Chem.* **2001**, *79(5/6)*, 705–708.

59 J. Howarth, *Tetrahedron Lett.* **2000**, *41(34)*, 6627–6629.

60 S. V. Ley, C. Ramarao, M. D. Smith, *Chem. Commun.* **2001**, 2278–2279.

61 G. S. Owens, M. M. Abu-Omar, *Chem. Commun.* **2000**, 1165–1166.

62 C. E. Song, E. J. Roh, *Chem. Commun.* **2000**, 837–838.

63 L. Gaillon, F. Bedioui, *Chem. Commun.* **2001**, 1458–1459.

64 R. D. Singer, P. J. Scammells, *Tetrahedron Lett.* **2001**, *42(39)*, 6831–6833.

65 a) E. G. Kuntz, E. Kuntz, DE 2627354 (to Rhone-Poulenc S. A., Fr.), **1976**, [Chem. Abstr. **1977**, *87*, 101944]; b) E. G. Kuntz, CHEMTECH **1987**, *17*, 570–575; c) B. Cornils, W. A. Herrmann, "Aqueous-Phase Organometallic Catalysis", Wiley-VCH, Weinheim, **1998**.

66 P. Wasserscheid, H. Waffenschmidt, *J. Mol. Catal. A: Chem.* **2000**, *164(1–2)*, 61–67.

67 Y. Chauvin, H. Olivier, L. Mußmann, Y. Chauvin, EP 776880 (to Institut Français du Pétrole, Fr.), **1997** [Chem. Abstr. **1997**, *127*, 65507].

68 C. C. Brasse, U. Englert, A. Salzer, H. Waffenschmidt, P. Wasserscheid, *Organometallics* **2000**, *19(19)*, 3818–3823.

69 J. D. Unruh, R. Christenson, *J. Mol. Catal.* **1982**, *14*, 19–34.

70 C. P. Casey, E. L. Paulsen, E. W. Beuttenmueller, B. R. Proft, L. M. Petrovich, B. A. Matter, D. R. Powell, *J. Am. Chem. Soc.* **1997**, *119*, 11817–11825.

71 W. R. Moser, C. J. Papile, D. A. Brannon, R. A. Duwell, *J. Mol. Catal.* **1987**, *41*, 271–292.

72 A. Heßler, O. Stelzer, H. Dibowski, K. Worm, F. P. Schmidtchen, *J. Org. Chem.* **1997**, *62*, 2362–2369.

73 P. Machnitzki, M. Teppner, K. Wenz, O. Stelzer, E. J. Herdtweck, *Organomet. Chem.* **2000**, *602*, 158–169.

74 H. Dibowski, F. P. Schmidtchen, *Angew. Chem., Int. Ed.* **1998**, *37*, 476–478.

75 O. Stelzer, F. P. Schmidtchen, A. Heßler, M. Tepper, H. Dibowski, H. Bahrmann, M. Riedel, DE 19701245 (to Celanese G.m.b.H., Germany), **1998** [Chem. Abstr. **1998**, *129*, 149094].

76 P. Wasserscheid, H. Waffenschmidt, P. Machnitzki, K. Kottsieper, O. Stelzer, *Chem. Commun.* **2001**, 451–452.

77 C. P. Casey, G. T. Whiteker, M. G. Melville, L. M. Petrovich, L. J. A. Gavey, D. R. J. Powell, *J. Am. Chem. Soc.* **1992**, *114*, 5535–5543.

78 M. Kranenburg, Y. E. M. van der Burgt, P. C. J. Kamer, P. W. N. M van Leeuwen, K. Goubitz, J. Fraanje, *Organometallics* **1995**, *14*, 3081–3089.

79 P. W. N. M. van Leeuwen, P. C. J. Kamer, J. N. H. Reek, P. Dierkes, *Chem. Rev.* **2000**, 100, 2741–2770.

80 P. Dierkes, S. Ramdeehul, L. Barloy, A. De Cian, J. Fischer, P. C. J. Kamer, P. W. N. M. van Leeuwen, *Angew. Chem., Int. Ed.* **1998**, *37*, 3116–3118.

81 F. Favre, H. Olivier-Bourbigou, D. Commereuc, L. Saussine, *Chem. Commun.* **2001**, 1360–1361.

82 J. Sirieix, M. Ossberger, B. Betzemeier, P. Knochel, *Synletters* **2000**, 1613–1615.

83 a) K. W. Kottsieper, O. Stelzer, P. Wasserscheid, *J. Mol. Catal. A: Chem.* **2001**, *175(1–2)*, 285–288; b) D. J. Brauer, K. W. Kottsieper, C. Liek, O. Stelzer, H. Waffenschmidt, P. Wasserscheid, *J. Organomet. Chem.* **2001**, *630(2)*, 177–184.

84 M. F. Sellin, P. B. Webb, D. J. Cole-Hamilton, *Chem. Commun.* **2001**, 781–782.

85 D. E. Kaufmann, M. Nouroozian, H. Henze, *Synlett* **1996**, 1091–1092.

86 a) W. A. Herrmann, V. P. W. Böhm, *J. Organomet. Chem.* **1999**, *572*, 141–145; b) V. P. W. Böhm, W. A. Hermann, *Chem. Eur. J.* **2000**, *6*, 1017–1025.

87 S. Bouquillon, B. Ganchegui, B. Estrine, F. Henin, J. Muzart, *J. Organomet. Chem.* **2001**, *634*, 153–156.

88 a) V. Calò, A. Nacci, L. Lopez, A. Napola, *Tetrahedron Lett.* **2001**, *42*, 4701–4703; b) V. Calo, A. Nacci, A. Monopoli, L. Lopez, A. di Cosmo, *Tetrahedron*, **2001**, *57*, 6071–6077.

89 J. Silberg, T. Schareina, R. Kempe, K. Wurst, M. R. Buchmeiser, *J. Organomet. Chem.* **2001**, *622*, 6–18.

90 L. Xu, W. Chen, J. Ross, J. Xiao, *Org. Lett.* **2001**, *3(2)*, 295–297.
91 A. J. Carmichael, M. J. Earle, J. D. Holbrey, P. B. McCormac, K. R. Seddon, *Org. Lett.* **1999**, *1*, 997–1000.
92 R. R. Deshmukh, R. Rajagopal, K. V. Srinivasan, *Chem. Comm.* **2001**, 1544–1545.
93 H. Hagiwara, Y. Shimizu, T. Hoshi, T. Suzuki, M. Ando, K. Ohkubo, C. Yokoyama, *Tetrahedron Lett.* **2001**, *42(26)*, 4349–4351.
94 C. J. Mathews, P. J. Smith, T. Welton, *Chem. Comm.* **2000**, 1249–1250.
95 S. T. Handy, X. Zhang, *Org. Lett.* **2001**, *3(2)*, 233–236.
96 G. Wilke, B. Bogdanovic, P. Hardt, P. Heimbach, W. Keim, M. Kröner, W. Oberkirch, K. Tanaka, E. Steinrücke, D. Walter, H. Zimmermann, *Angew. Chem.* **1966**, *5*, 151–154.
97 a) Y. Chauvin, B. Gilbert, I. Guibard, *J. Chem. Soc. Chem. Commun.* **1990**, 1715–1716; b) Y. Chauvin, S. Einloft, H. Olivier, *Ind. Eng. Chem. Res.* **1995**, *34*, 1149–1155; c) Y. Chauvin, S. Einloft, H. Olivier, FR 93/11,381 (to Institut Français du Pétrole, Fr.), **1996** [Chem. Abstr. **1995**, *123*, 144896c].
98 a) M. Freemantle, *Chem. Eng. News* **1998**, *76(13)*, 32–37; b) E. Burridge, *ECN Chemscope* **1999**, *May*, 27–28; c) H. Olivier, *J. Mol. Catal. A: Chem.* **1999**, *146(1–2)*, 285–289.
99 S. Einloft, F. K. Dietrich, R. F. de Souza, J. Dupont, *Polyhedron* **1996**, *19*, 3257–3259.
100 H. Olivier, P. Laurent-Gérot, *J. Mol. Catal. A: Chem.* **1999**, *148*, 43–48.
101 J. T. Dixon, J. J. C. Grove, A. Ranwell, WO 0138270 (to Sasol Technology (Pty) Ltd, S. Afr.), **2001** [Chem. Abstr. **2001**, *135*, 7150].
102 a) Y. Chauvin, H. Olivier, C. N. Wyrvalski, L. C. Simon, R. F. de Souza, *J. Catal.* **1997**, *165*, 275–278; b) L. C. Simon, J. Dupont, R. F. de Souza, *J. Mol. Catal.* **1998**, *175*, 215–220.
103 W. Keim, B. Hoffmann, R. Lodewick, M. Peukert, G. Schmitt, J. Fleischhauer, U. Meier, *J. Mol. Catal.* **1979**, *6*, 79–97.
104 a) B. Ellis, W. Keim, P. Wasserscheid, *Chem. Commun.* **1999**, 337–338; b) P. Wasserscheid, W. Keim, WO 9847616 (to BP Chemicals), **1997** [Chem. Abstr. **1998**, *129*, 332457].
105 a) M. Eichmann, *dissertation*, RWTH-Aachen, **1999**; b) P. Wasserscheid, M. Eichmann, *Catal. Today* **2001**, *66(2–4)*, 309–316.
106 a) I. Brassat, *Ph.D. thesis*, RWTH Aachen, **1998**; b) I. Brassat, W. Keim, S. Killat, M. Möthrath, P. Mastrorilli, C. Nobile, G. J. Suranna, *Mol. Catal. A: Chem.* **2000**, *43*, 41–58; I. Brassat, U. Englert, W. Keim, D. P. Keitel, S. Killat, G. P. Suranna, R. Wang, *Inorg. Chim. Acta* **1998**, *280*, 150–162.
107 P. Wasserscheid, C. Hilgers, submitted for publication.
108 S. M. Silva, P. A. Z. Suarez, R. F. de Souza, J. Dupont, *Polymer Bull.* **1998**, *41*, 401–405.
109 R. A. Ligabue, R. F. de Souza, J. Dupont, *J. Mol. Catal. A: Chem.* **2001**, *169*, 11–17.
110 a) C. Gurtler, M. Jautelat, EP 1035093 (to Bayer A.G., Germany), **2000** [Chem. Abstr. **2000**, *133*, 237853]; b) R. C. Buijsman, E. van Vuuren, J. G. Sterrenburg, *Org. Lett.* **2001**, *3(23)*, 3785–3787; c) D. Semeril, H. Olivier-Bourbigou, C. Bruneau, P. H. Dixneuf, *Chem. Comm.* **2002**, 146–147.
111 E. Mizushima, T. Hayashi, M. Tanaka, *Green Chem.* **2001**, *2*, 76–79.
112 J. Ross, W. Chen, L. Xu, J. Xiao, *Organometallics* **2001**, *20*, 138–142.
113 S. Toma, B. Gotov, I. Kmentova, E. Solcaniova, *Green Chem.* **2000**, *2*, 149–151.
114 a) W. Chen, L. Xu, C. Chatterton, J. Xiao, *Chem. Commu.*, **1999**, 1247–1248; b) C. de Bellefon, E. Pollet, P. Grenouillet, *J. Mol. Catal.* **1999**, *145*, 121–126.

5.3
Ionic Liquids in Multiphasic Reactions

Hélène Olivier-Bourbigou and Alain Forestière

5.3.1
Multiphasic Reactions: General Features, Scope, and Limitations

While the solubility of organometallic complexes in common organic solvents appears to be an advantage in terms of site availability and tunability, reaction selectivity, and activity, it is a major drawback in terms of catalyst separation and recycling. The quest for new catalyst immobilization or recovery strategies to facilitate reuse is unceasing. Immobilization of the catalyst on a solid support has been widely studied. Except for Ziegler–Natta- and metallocene-type polymerization processes, in which the catalyst is not recycled due to its high activity, this technology has not yet been developed industrially, mainly because of problems of catalyst leaching and deactivation. One successful approach to close the advantage/disadvantage gap between homogeneous and heterogeneous catalysis is multiphasic catalysis [1]. In its simplest version, there are only two liquid phases ("biphasic" catalysis or "two-phase" catalysis). The catalyst is dissolved in one phase (generally a polar phase), while the products and the substrates are found in the other. The catalyst can be separated by decantation and recycled under mild conditions.

It is important to make the distinction between the multiphasic catalysis concept and transfer-assisted organometallic reactions or phase-transfer catalysis (PTC). In this latter approach, a catalytic amount of quaternary ammonium salt $[Q]^+[X]^-$ is present in an aqueous phase. The catalyst's lipophilic cation $[Q]^+$ transports the reactant's anion $[Y]^-$ to the organic phase, as an ion-pair, and the chemical reaction occurs in the organic phase of the two-phase organic/aqueous mixture [2].

The use of multiphasic catalysis has proven its potential in important industrial processes. In 1977, the first large-scale commercial catalytic process to benefit from two-phase liquid/liquid technology was the Shell Higher Olefin Process (SHOP) for oligomerization of ethene into α-olefins, catalyzed by nickel complexes dissolved in diols such as 1,4-butanediol. Subsequently, the advancement in two-phase homogeneous catalysis has been demonstrated by the introduction of biphasic aqueous hydroformylation as an economically competitive large-scale process. The first commercial oxo plant, developed by Ruhrchemie–Rhône–Poulenc for the production of butyraldehyde from propene, came on stream in 1984. This is an example of a gas-liquid-liquid multiphasic system in which the homogeneous rhodium-based catalyst is immobilized in a water phase thanks to its coordination to the hydrophilic trisulfonated triphenylphosphine ligand (TPPTS) [3]. The catalyst separation is more effective and simpler than in classical rhodium processes, but separation of by-products from the catalyst is also an important issue.

Since then, water has emerged as a useful solvent for organometallic catalysis. In addition to the hydroformylation reactions, several other industrial processes employing homogeneous catalysis have been converted to aqueous-phase procedures [4].

5.3.2
Multiphasic Catalysis: Limitations and Challenges

Multiphasic (biphasic) catalysis relies on the transfer of organic substrates into the catalyst phase or on catalysis at the phase boundary. Most organic substrates do not have sufficient solubility in the catalyst phase (particularly in water) to give practical reaction rates in catalytic applications. Therefore, although the use of aqueous-biphasic catalysis has proven its potential in important industrial processes, the current applications of this technique remain limited: firstly to catalysts that are stable in the presence of water, and secondly to substrates that have significant water solubility. Many studies have focused on improving the affinities between the two liquid aqueous/organic phases, either through increasing the lipophilic character of the catalyst phase or even by immobilizing the catalyst on a support. For example, rapid stirring, emulsification, and sonication have all been used to increase the interfacial area. The addition of co-solvents to the aqueous phase has been investigated extensively as a means to improving the solubility of higher olefinic substrates in the catalyst-containing phase. Application of detergents or micellar processes to promote substrate transfer to the interface, or the addition of co-ligands such as PPh_3 – or even of ligands with an amphiphilic character or modified cyclodextrins – also play rate-enhancing roles. The development of supported aqueous-phase catalysis (SAPC), which involves the dissolution of an aqueous-phase complex in a thin layer of water adhering to a silica surface, opens the way to the reactivity of hydrophobic substrates. Although all these techniques can change the solubility of organic substrates in the aqueous phase or favor the concentration of the active center at the interface, they can also cause the leaching of a proportion of the catalyst into the organic phase.

The major advantage of the use of two-phase catalysis is the easy separation of the catalyst and product phases. However, the co-miscibility of the product and catalyst phases can be problematic. An example is given by the biphasic aqueous hydroformylation of ethene to propanal. Firstly, the propanal formed contains water, which has to be removed by distillation, This is difficult, due to formation of azeotropic mixtures. Secondly, a significant proportion of the rhodium catalyst is extracted from the reactor with the products, which prevents its efficient recovery. Nevertheless, the reaction of ethene itself in the water-based Rh-TPPTS system is fast. It is the high solubility of water in the propanal that prevents the application of the aqueous biphasic process [5].

To overcome these limitations, there has been a great deal of investigation of novel methods, one of them focused on the search for alternative solvents [6, 7]. Table 5.3-1 gives different approaches to biphasic catalysis, with some of their respective advantages and limitations.

Although already well known, perfluorinated solvents have only quite recently proved their utility in many organic and catalyzed reactions. The main advantage of

Table 5.3-1: Advantages and limitations of different approaches for multiphasic "homogeneous" catalysis.

Catalyst phase	*Product phase*	*Advantages*	*Limitations*
Water (+co-solvent)	Organic liquid	▪ Easy product separation and catalyst recycling ▪ Lower cost of chemical processes ▪ Lack of toxicity of water	▪ Low reaction rate for poorly water-miscible substrate ▪ Mass transfer limits rate of reaction ▪ Treatment of spent water
Polar solvent	Organic liquid	▪ Solvent effect	▪ Use of volatile organic solvent ▪ Co-miscibility of the two phases
Fluorinated organic solvent	Organic liquid	▪ Temperature dependency of the miscibility of fluorinated phase with organic solvents	▪ Solvent and ligand costs ▪ Product contamination
Water	Supercritical fluids (e.g. CO_2)	▪ Organic co-solvent not needed ▪ High miscibility of CO_2 with gas	▪ Poor solvating ability of supercritical fluids ▪ High investment and operating costs
Ionic liquid	Organic liquid	▪ Tunability of the solubility characteristics of the ionic liquids ▪ Solvent effect	▪ Ionic liquid costs ▪ Disposal of spent ionic liquids
Ionic liquid	Supercritical fluids (e.g. CO_2)	▪ Organic co-solvent not needed ▪ Tunability of the solubility characteristics of the ionic liquids ▪ Presence of CO_2 reduces ionic liquid's viscosity	▪ Ionic liquid costs ▪ High pressure apparatus needed

these solvents is that their miscibility with organic products can be tuned by variation of the temperature. Fluorous-phase catalysis makes possible the association of homogeneous phase catalysis (thus avoiding problems of mass-transfer limitations) and a biphasic separation of the catalyst and reaction mixture [8]. However, these solvents are still relatively expensive and require costly, specially designed ligands to keep the catalyst in the fluorous phase during the separation. In addition, a significant amount of perfluorinated solvent can remain dissolved in the organic phase, and contamination of the products can occur. To date, there are no industrial developments of this technology, due to lack of competitiveness.

Supercritical carbon dioxide ($scCO_2$) has also emerged as a highly promising reaction medium. In combination with homogeneous catalysis, its benefits could be the potential increase of reaction rates (absence of gas–liquid phase boundary, high diffusion rates) and selectivities, and also its lack of toxicity [9]. In combination with water, it has been used in a biphasic system to perform the hydrogenation of cinnamaldehyde. Gas-liquid-liquid mass transfer limitations were ruled out, due to the very high solubility of reactant gas in $scCO_2$ [10]. Although elegant, this approach still appears relatively expensive, especially for the bulk chemical industry. Furthermore, the low solubility of interesting substrates might hamper the commercialization of $scCO_2$ in the fine chemical industry. A very recent and highly interesting development is the combination of an ionic liquid catalyst phase and a product phase containing $scCO_2$. This approach is presented in more detail in Section 5.4.

Further progress in multiphasic catalysis will rely on the development of alternative techniques that allow the reactivity of a broader range of substrates, the efficient separation of the products, and recovery of the catalyst, while remaining economically viable.

5.3.3 Why Ionic Liquids in Multiphasic Catalysis?

Notwithstanding their very low vapor pressure, their good thermal stability (for thermal decomposition temperatures of several ionic liquids, see [11, 12]) and their wide operating range, the key property of ionic liquids is the potential to tune their physical and chemical properties by variation of the nature of the anions and cations. An illustration of their versatility is given by their exceptional solubility characteristics, which make them good candidates for multiphasic reactions (see Section 5.3.4). Their miscibility with water, for example, depends not only on the hydrophobicity of the cation, but also on the nature of the anion and on the temperature.

N,N'-Dialkylimidazolium cations are of particular interest because they generally give low-melting salts, are more thermally stable than their tetraalkylammonium analogues, and have a wide spectrum of physicochemical properties available. For the same $[BMIM]^+$ cation, for example, the $[BF_4]^-$, $[CF_3SO_3]^-$, $[CF_3CO_2]^-$, $[NO_3]^-$, and halide salts all display complete miscibility with water at 25 °C. On cooling the $[BMIM][BF_4]$/water solution to 4 °C, however, a water-rich phase separates. In a similar way, a change of the $[BMIM]^+$ cation for the longer-chain, more hydrophobic $[HMIM]^+$ (1-hexyl-3-methylimidazolium) cation affords a $[BF_4]^-$ salt that shows low co-miscibility with water at room temperature. On the other hand, the $[BMIM][PF_6]$, $[BMIM][SbF_6]$, $[BMIM][NTf_2]$ ($NTf_2 = N(CF_3SO_2)_2$), and $[BMIM][BR_4]$ ionic liquids show very low miscibility with water, but the shorter, symmetrically substituted $[MMIM][PF_6]$ salt becomes water-soluble. One might therefore expect that modification of the alkyl substituents on the imidazolium ring could produce different and very tunable ionic liquid properties.

The influence of the nature of cations and anions on the solubility characteristics of the resulting salts with organic substrates is also discussed in Section 5.3.4. It has

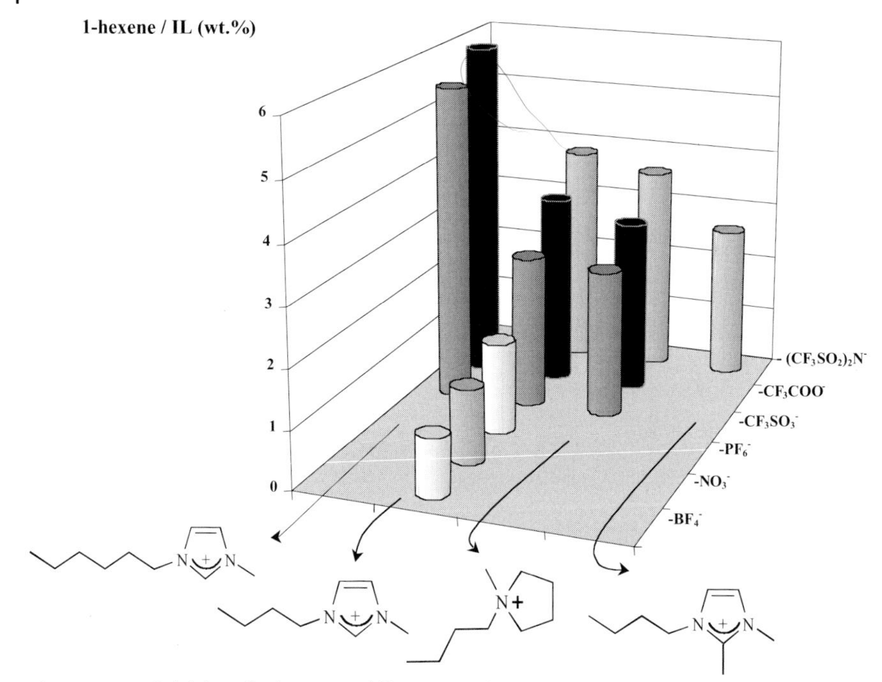

Figure 5.3-1: Solubility of 1-hexene in different ionic liquids as a function of the nature of anions and cations.

been shown (Figure 5.3-1) that increasing the length of the alkyl chain on the imidazolium cation can increase the solubility of 1-hexene, but so can tuning the nature of the anion.

A comparison of the solubility of α-olefins with increasing numbers of carbon atoms in water and in [BMIM][BF_4] (Figure 5.3-2), shows that olefins are at least 100 times more soluble in the ionic liquid than in water.

Addition of co-solvents can also change the co-miscibility characteristics of ionic liquids. As an example, the hydrophobic [BMIM][PF_6] salt can be completely dissolved in an aqueous ethanol mixture containing between 0.5 and 0.9 mole fraction of ethanol, whereas the ionic liquid itself is only partially miscible with pure water or pure ethanol [13]. The mixing of different salts can also result in systems with modified properties (e.g., conductivity, melting point).

One of the key factors controlling the reaction rate in multiphasic processes (for reactions taking place in the bulk catalyst phase) is the reactant solubility in the catalyst phase. Thanks to their tunable solubility characteristics, the use of ionic liquids as catalyst solvents can be a solution to the extension of aqueous two-phase catalysis to organic substrates presenting a lack of solubility in water, and also to moisture-sensitive reactants and catalysts. With the different examples presented below, we show how ionic liquids can have advantageous effects on reaction rate and on the selectivity of homogeneous catalyzed reactions.

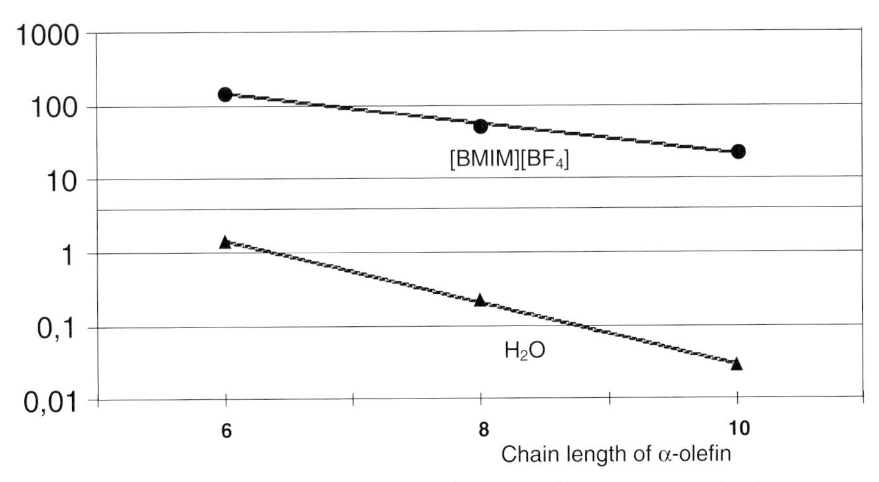

Figure 5.3-2: Comparison of the solubilities of α-olefins with different numbers of carbon atoms in water and in [BMIM][BF$_4$].

5.3.4
Different Technical Solutions to Catalyst Separation through the Use of Ionic Liquids

In general, homogeneous catalysis suffers from complicated and expensive catalyst separation from the products. Homogeneous catalysts are very often unstable at the high temperatures necessary for the distillation of high-boiling products. Multiphasic catalysis makes the separation of products under mild conditions possible. Different technologies to separate the products, and to recycle the catalytic system when using ionic liquids as one of these phases, have been proposed (Table 5.3-2).

The simplest case (Table 5.3-2, case a) is when the ionic liquid is able to dissolve the catalyst, and displays a partial solubility with the substrates and a poor solubility with the reaction products. Under these conditions, the product upper phase, also containing the unconverted reactants, is removed by simple phase decantation, and the ionic liquid containing the catalyst is recycled. This can be illustrated by transition metal-catalyzed olefin transformations into non-polar hydrocarbon products such as olefin oligomerization, hydrogenation, isomerization, metathesis, and acidic olefin alkylation with isobutane. Transition metal catalysts can also be immobilized in ionic liquids with melting points just above room temperature (Table 5.3-2, case b). The reaction occurs in a two-phase liquid–liquid system. By cooling the reaction mixture, the products can be separated by filtration from the “solid” catalyst medium, which can be recycled. The advantages of this technique have been demonstrated for the hydrogenation of 1-hexene catalyzed by ruthenium-phosphine complexes in [BMIM]Cl/ZnCl$_2$ [14] and for the hydroformylation of 1-hexene in the high-melting phosphonium tosylate ionic liquids [15].

Table 5.3-2: Different technologies for multiphasic reactions making use of ionic liquids.

Lower phase (during the reaction)	*Upper phase (during the reaction)*	*Mode of separation catalyst phase/products*	
Ionic liquid + catalyst	**Organic liquid (products + unreacted substrates)**	(a) Decantation (liquid-liquid)	48–50
		(b) Filtration of the ionic liquid on cooling	14, 15
Ionic liquid + catalyst + part of the products	**Organic liquid (part of the products + unreacted substrates)**	(c) Product extraction with an organic co-solvent immiscible with the ionic liquid	18
		(d) Distillation	16
or	**or**		
Ionic liquid + catalyst + products	**No upper phase**	(e) Separation after addition of a co-solvent miscible with the ionic liquid, immiscible with the products	
Ionic liquid + catalyst	**Products + unreacted substrates + CO_2**	(f) Extraction with *sc*CO2	19
Supported ionic liquid + catalyst	**Organic liquid (products) or gas**	(g) Phase separation	21

Thanks to the low vapor pressure of ionic liquids, product distillation without azeotrope formation can reasonably be anticipated if the products are not too high-boiling. One example is the hydroformylation of methyl-3-pentenoate in [BMIM][PF_6] with catalysis by a homogeneous Rh-phosphite system. In the absence of ionic liquid, deactivation of the catalyst is observed. Through the use of the [BMIM][PF_6] salt, the catalyst is stabilized and can be successfully reused after distillation of the products [16]. Nevertheless, this separation technique remains demanding in energy, and the eventual accumulation of high-boiling by-products in the nonvolatile ionic liquid phase can be a problem.

When the products are partially or totally miscible in the ionic phase, separation is much more complicated (Table 5.3-2, cases c–e). One advantageous option can be to perform the reaction in one single phase, thus avoiding diffusional limitation, and to separate the products in a further step by extraction. Such technology has already been demonstrated for aqueous biphasic systems. This is the case for the palladium-catalyzed telomerization of butadiene with water, developed by Kuraray, which uses a sulfolane/water mixture as the solvent [17]. The products are soluble in water, which is also the nucleophile. The high-boiling by-products are extracted with a solvent (such as hexane) that is immiscible in the polar phase. This method

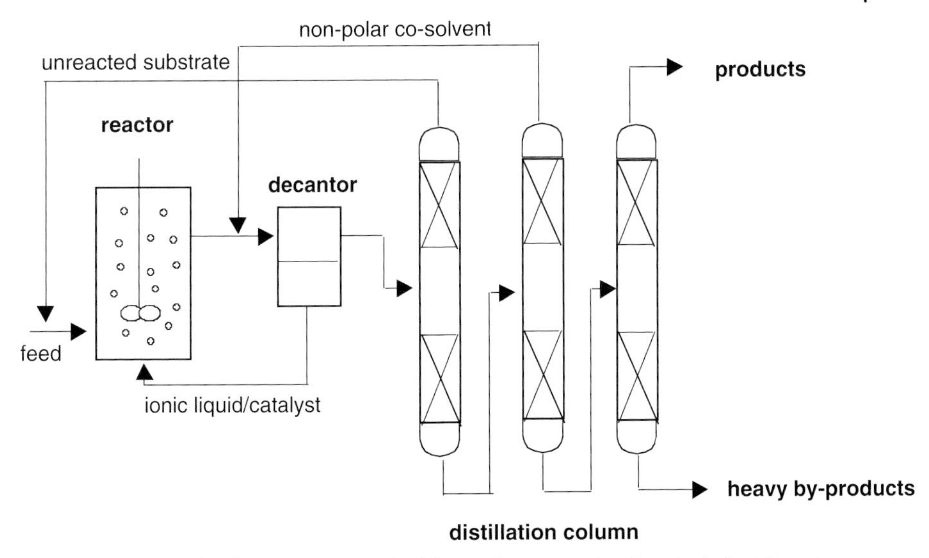

Figure 5.3-3: Example of an extraction method for product separation from ionic liquid/catalyst reaction mixtures.

has the advantage that (i) the catalyst and the products can be separated without heating them, so that thermal deactivation is avoided, and (ii) the extraction is achieved for all the compounds, so that the accumulation of catalyst poisons and high-boiling by-products is minimal. This technology can be applied when ionic liquids are used as the catalyst polar phase (Figure 5.3-3).

A co-solvent that is poorly miscible with ionic liquids but highly miscible with the products can be added in the separation step (after the reaction) to facilitate the product separation. The Pd-mediated Heck coupling of aryl halides or benzoic anhydride with alkenes, for example, can be performed in [BMIM][PF_6], the products being extracted with cyclohexane. In this case, water can also be used as an extraction solvent, to remove the salt by-products formed in the reaction [18]. From a practical point of view, the addition of a co-solvent can result in cross-contamination, and it has to be separated from the products in a supplementary step (distillation). More interestingly, unreacted organic reactants themselves (if they have non-polar character) can be recycled to the separation step and can be used as the extractant co-solvent.

When water-miscible ionic liquids are used as solvents, and when the products are partly or totally soluble in these ionic liquids, the addition of polar solvents, such as water, in a separation step after the reaction can make the ionic liquid more hydrophilic and facilitate the separation of the products from the ionic liquid/water mixture (Table 5.3-2, case e). This concept has been developed by Union Carbide for the hydroformylation of higher alkenes catalyzed by Rh-sulfonated phosphine ligand in the N-methylpyrrolidone (NMP)/water system. Thanks to the presence of NMP, the reaction is performed in one homogeneous phase. After the reaction, water is added in a mixer, followed by efficient phase separation in a settler. One advantage of this process is its flexibility and good performance with respect to the olefin carbon number.

The combination of ionic liquids with supercritical carbon dioxide is an attractive approach, as these solvents present complementary properties (volatility, polarity scale....). Compressed CO_2 dissolves quite well in ionic liquid, but ionic liquids do not dissolve in CO_2. It decreases the viscosity of ionic liquids, thus facilitating mass transfer during catalysis. The separation of the products in solvent-free form can be effective and the CO_2 can be recycled by recompressing it back into the reactor. Continuous flow catalytic systems based on the combination of these two solvents have been reported [19]. This concept is developed in more detail in Section 5.4.

Membrane techniques have already been combined with two-phase liquid catalysis. The main function of this method is to perform fine separation of undesirable constituents from the catalytic system after phase decantation has already performed the coarse separation of the catalyst from the products. This technique can be applied to ionic liquid systems as a promising approach for the selective removal of volatile solutes from ionic liquids [20].

Ionic liquids have already been demonstrated to be effective membrane materials for gas separation when supported within a porous polymer support. However, supported ionic liquid membranes offer another versatile approach by which to perform two-phase catalysis. This technology combines some of the advantages of the ionic liquid as a catalyst solvent with the ruggedness of the ionic liquid-polymer gels. Transition metal complexes based on palladium or rhodium have been incorporated into gas-permeable polymer gels composed of [BMIM][PF_6] and poly(vinylidene fluoride)-hexafluoropropylene copolymer and have been used to investigate the hydrogenation of propene [21].

5.3.5
Immobilization of Catalysts in Ionic Liquids

For the use of ionic liquids in catalytic or organic reactions, two main methodologies have been developed. In the first one, the ionic liquid is both the catalyst and the reaction solvent. This is the case in acid-catalyzed reactions in which Lewis acidic ionic liquids such as acidic chloroaluminates are both active catalytic species and solvents for carbenium ions. In this case, the dissolution of the inorganic Lewis acid (e.g., $AlCl_3$) in the organic phase is not observed. The second approach, discussed in Section 5.2, is when the ionic liquid acts as a "liquid support" for the homogeneous catalyst. In this technology, the catalyst (in general a transition metal complex) is immobilized in the ionic phase and the products form the upper phase, as described in Section 5.3.4. To achieve the development of such an approach in a continuous process, the key point is to immobilize and stabilize the catalyst in the ionic liquid in the presence of an organic second phase with minimum loss of metal. Two approaches have been investigated:

1) the active species is known to be ionic in conventional organic solvents,
2) the active species is characterized as a non-charged complex.

In the first case, one may expect that the catalyst should remain ionic and be retained without modification in the ionic liquid. Different successful examples have been reported, such as olefin and diolefin hydrogenation reactions catalyzed by the cationic $[HRh(PPh_3)_2(diene)][PF_6]$ complexes [22], and aromatic hydrocarbon hydrogenation catalyzed by the $[H_4Ru_4(C_6H_6)_4][BF_4]_2$ cluster [23]. In the presence of hydrogen, this latter complex probably forms the $[H_6Ru_4(C_6H_6)_4][BF_4]_2$ complex, which acts as the effective arene hydrogenation catalyst. Another example is olefin dimerization catalyzed by the cationic [HNi(olefin)][A] (A is a chloroaluminate anion) complexes. These species can be formed by *in situ* alkylation of a nickel(II) salt with an acidic alkylchloroaluminate ionic liquid acting both as the solvent and as the co-catalyst [24]. The cationic $[(methallyl)Ni(Ph_2PCH_2PPh_2(O))][SbF_6]$ complex proved to be stable and active for ethene oligomerization in $[PF_6]^-$-based ionic liquids without the addition of Lewis acid. The high electrophilicity of the Ni center, which is responsible for the activity of the catalyst, is probably not altered by the ionic solvent [25]. In the Suzuki reaction, the active species in $[BMIM][BF_4]$ is believed to be the tricoordinated $[Pd(PPh_3)_2(Ar)][X]$ complex that forms after oxidative addition of the aryl halide to $[Pd(PPh_3)_4]$ [26]. Thanks, therefore, to their low nucleophilicity, ionic liquids do not compete with the unsaturated organic substrate for coordination to the electrophilic active metal center. The different recycling experiments demonstrate the stability of these organometallic complexes in ionic liquids.

Not only cationic, but also anionic, species can be retained without addition of specially designed ligands. The anionic active $[HPt(SnCl_3)_4]^{3-}$ complex has been isolated from the $[NEt_4][SnCl_3]$ solvent after hydrogenation of ethylene [27]. The $PtCl_2$ precursor used in this reaction is stabilized by the ionic salt (liquid at the reaction temperature) since no metal deposition occurs at 160 °C and 100 bar. The catalytic solution can be used repeatedly without apparent loss of catalytic activity.

In the second case, in which the active catalytic species is assumed to be uncharged, leaching of the transition metal in the organic phase can be limited by the use of functionalized ligands. As the triumph of aqueous biphasic catalysis follows the laborious work involved in the development of water-soluble ligands, recent investigations have focused on the synthesis of new ligands with "tailor made" structures for highly active and selective two-phase catalysts and for good solubility in the ionic liquid phase. These ligands are mainly phosphorus ligands with appropriate modifications (Scheme 5.3.1).

Polar groups such as the cationic phenylguanidinium groups **1–3** [28, 29], the imidazolium and pyridinium groups **4** and **5** [30], and the 2-imidazolyl groups **6** and **7** [31] have been reported. A cobaltocinium salt bearing phosphine donors (**8**) has also been described [32]. Phosphites are well known ligands in homogeneous Rh-catalyzed hydroformylation, affording enhanced reaction rates and regioselectivities. Since they are unstable towards hydrolysis, examples of their use in aqueous biphasic catalysis are rare. Ionic liquids offer suitable alternative solvents compatible with phosphites **9** [28].

To date, these functionalized ligands have been investigated on the laboratory scale, in batch operations to immobilize rhodium catalyst in hydroformylation.

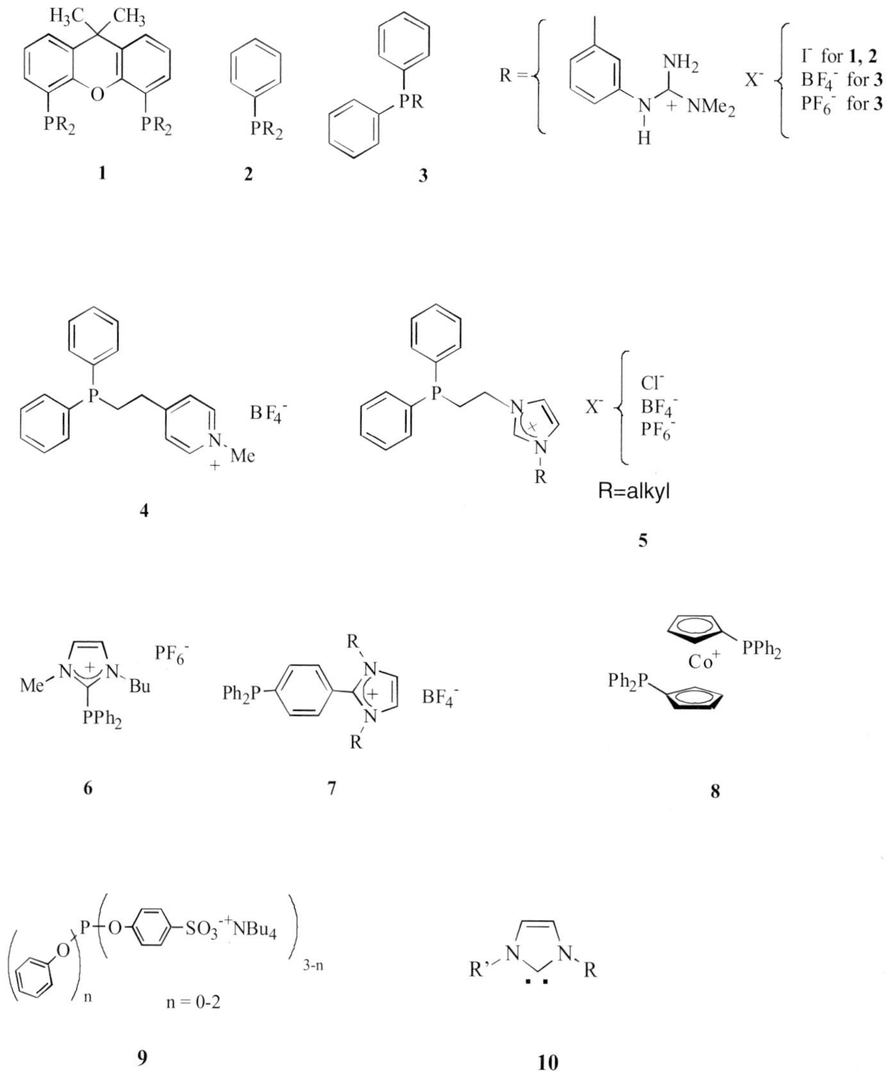

Scheme 5.3-1: Ligands 1-10.

Good rhodium retention results were obtained after several recycles. However, optimized ligand/metal ratios and leaching and decomposition rates, which can result in the formation of inactive catalyst, are not known for these ligands and require testing in continuous mode. As a reference, in the Ruhrchemie–Rhône–Poulenc process, the losses of rhodium are $<10^{-9}$ g Rh per kg *n*-butyraldehyde.

Certain amines, when linked to TPPTS, form ionic solvents liquid at quite low temperatures. Bahrman [33] used these ionic liquids as both ligands and solvents for the Rh catalyst for the hydroformylation of alkenes. In this otherwise interesting

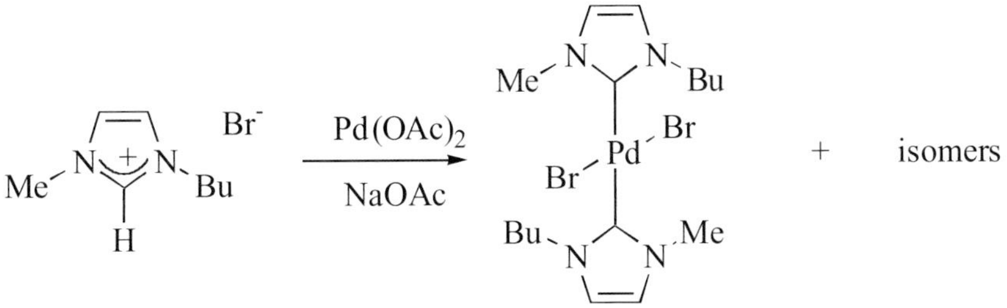

Scheme 5.3-2: Formation of carbene complexes by dialkylimidazolium salt deprotonation.

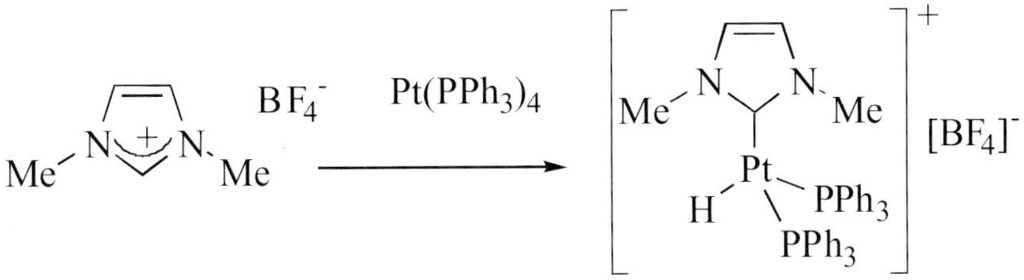

Scheme 5.3-3: Formation of carbene complexes by oxidative addition to Pt(0).

approach, however, the ligand/rhodium ratio, which influences the selectivity of the reaction, is difficult to control.

As well as phosphorus ligands, heterocyclic carbenes ligands **10** have proven to be interesting donor ligands for stabilization of transition metal complexes (especially palladium) in ionic liquids. The imidazolium cation is usually presumed to be a simple inert component of the solvent system. However, the proton on the carbon atom at position 2 in the imidazolium is acidic and this carbon atom can be deprotonated by, for example, basic ligands of the metal complex, to form carbenes (Scheme 5.3-2).

The ease of formation of the carbene depends on the nucleophilicity of the anion associated with the imidazolium. For example, when $Pd(OAc)_2$ is heated in the presence of [BMIM][Br], the formation of a mixture of Pd imidazolylidene complexes occurs. Palladium complexes have been shown to be active and stable catalysts for Heck and other C–C coupling reactions [34]. The highest activity and stability of palladium is observed in the ionic liquid [BMIM][Br]. Carbene complexes can be formed not only by deprotonation of the imidazolium cation but also by direct oxidative addition to metal(0) (Scheme 5.3-3). These heterocyclic carbene ligands can be functionalized with polar groups in order to increase their affinity for ionic liquids. While their donor properties can be compared to those of donor phosphines, they have the advantage over phosphines of being stable toward oxidation.

5.3.6
Scaling up Ionic Liquid Technology from Laboratory to Continuous Pilot Plant Operation

The increasing number of applications that make use of ionic liquids as solvents or catalysts for organic and catalytic reactions emphasizes their key advantages over organic solvents and their complementarity with respect to water or other "green" solvents. For scaling up to large-scale production, however, kinetic models are very often required and have to be developed for an optimum reactor design. In this type of multiphasic (biphasic) catalysis, one important parameter is the location of the reaction: does the reaction take place in the bulk of the liquid, at the interface, or simultaneously at both sites? For a reaction in the bulk of the liquid (e.g., in the ionic liquid), the liquid (and/or gaseous) reactants would first have to dissolve in the catalyst solution phase before the start of the chemical reaction. The reaction rate would therefore be determined by the concentration of the reactants in the catalyst phase. It is important to be able to identify mass transfer limitations that occur when the reaction rate is higher than the mass transfer velocity. In some cases the existence of mass transfer limitations can be used advantageously to control the exothermicity of reactions. For example, a reduction in stirring can be a means to decrease the reaction rate without having to destroy the catalyst. In single-phase homogeneous reactions, catalyst poisons (such as CO or CO_2) are sometimes deliberately injected into the reactor to stop the reaction.

In the aqueous biphasic hydroformylation reaction, the site of the reaction has been much discussed (and contested) and is dependent on reaction conditions (temperature, partial pressure of gas, stirring, use of additives) and reaction partners (type of alkene) [35, 36]. It has been suggested that the positive effects of co-solvents indicate that the bulk of the aqueous liquid phase is the reaction site. By contrast, the addition of surfactants or other surface- or micelle-active compounds accelerates the reaction, which apparently indicates that the reaction occurs at the interfacial layer.

Therefore, important parameters such as phase transfer phenomena (i.e., the solubility of the reactants in the ionic liquid phase), volume ratio of the different phases, and efficiency of mixing so as to provide maximum liquid–liquid interfacial area are key factors in determining and controlling reaction rates and kinetics. Kinetic models have been developed for aqueous biphasic systems and are continuously refined to improve agreements with experimental results. These models might be transferable to biphasic catalysis with ionic liquids, but data concerning the solubilities of liquids (and gases) in these new solvents and the existence of phase equilibria in the presence of organic upper phases have still to be accumulated (see Section 3.4). Very few publications on these topics with respect to ionic liquids are available.

The influence of the concentration of hydrogen in [BMIM][PF_6] and [BMIM][BF_4] on the asymmetric hydrogenation of α-acetamidocinnamic acid catalyzed by rhodium complexes bearing a chiral ligand has been investigated. Hydrogen was found to be four times more soluble in the $[BF_4]^-$-based salt than in the $[PF_6]^-$-based one,

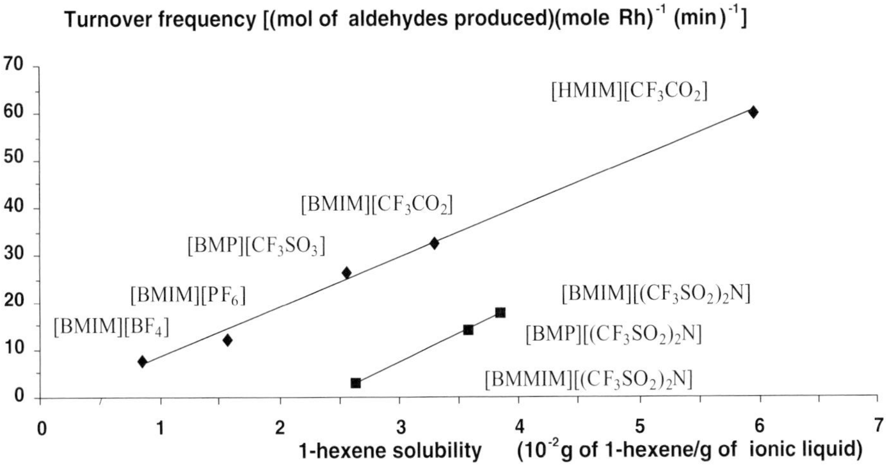

Figure 5.3-4: Turnover frequency of Rh-catalyzed hydroformylation as a function of 1-hexene solubility in ionic liquids. Reactions conditions: $Rh(CO)_2(acac)$ 0.075 mmol, 1-hexene/Rh = 800, TPPTS/Rh = 4, heptane as internal standard, $CO/H_2 = 1$ (molar ratio), P = 2 MPa, T = 80 °C, TOF determined at 25 % conversion of 1-hexene ([BMP]= *N*,*N*-butylmethylpyrolidinium; [BMMIM] = 1-butyl-2,3-dimethylimidazolium)

at the same pressure. This difference in molecular hydrogen concentration in the ionic phase (rather than pressure in the gas phase) has been correlated with the remarkable effect on the conversion and enantioselectivity of the reaction [37].

In the rhodium-catalyzed hydroformylation of 1-hexene, it has been demonstrated that there is a correlation between the solubility of 1-hexene in ionic liquids and reaction rates (Figure 5.3-4) [28].

However, information concerning the characteristics of these systems under the conditions of a continuous process is still very limited. From a practical point of view, the concept of ionic liquid multiphasic catalysis can be applicable only if the resultant catalytic lifetimes and the elution losses of catalytic components into the organic or extractant layer containing products are within commercially acceptable ranges. To illustrate these points, two examples of applications run on continuous pilot operation are described: (i) biphasic dimerization of olefins catalyzed by nickel complexes in chloroaluminates, and (ii) biphasic alkylation of aromatic hydrocarbons with olefins and light olefin alkylation with isobutane, catalyzed by acidic chloroaluminates.

5.3.6.1 Dimerization of alkenes catalyzed by Ni complexes

The Institut Français du Pétrole has developed and commercialized a process, named Dimersol X, based on a homogeneous catalyst, which selectively produces dimers from butenes. The low-branching octenes produced are good starting materials for isononanol production. This process is catalyzed by a system based on a nickel(II) salt, soluble in a paraffinic hydrocarbon, activated with an alkylaluminium chloride derivative directly inside the dimerization reactor. The reaction is sec-

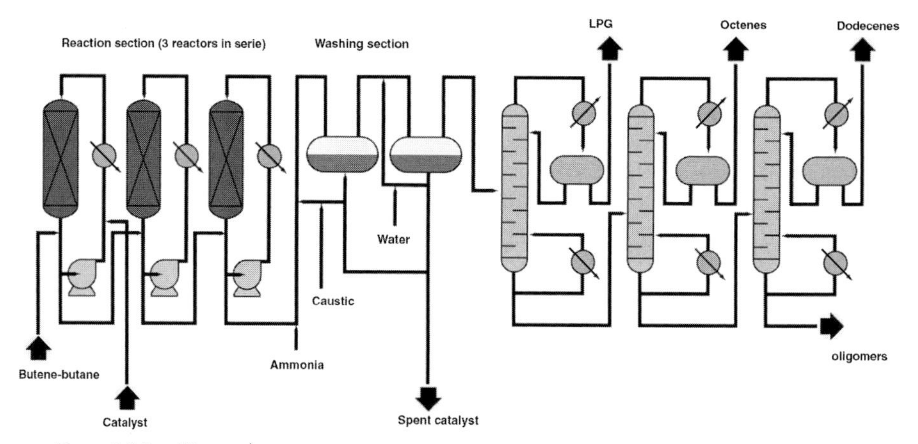

Figure 5.3-5: Dimersol process.

ond order in monomer concentration and first order in catalyst concentration. The butene conversion level is highly dependent on its initial concentration. In its present commercial form, Dimersol X can achieve 80 % conversion of butenes (for 70–75 % butene concentration in the feed) with 85 % octene selectivity. A process flow diagram is shown in Figure 5.3-5.

The reaction takes place at low temperature (40–60 °C), without any solvent, in two (or more, up to four) well-mixed reactors in series. The pressure is sufficient to maintain the reactants in the liquid phase (no gas phase). Mixing and heat removal are ensured by an external circulation loop. The two components of the catalytic system are injected separately into this reaction loop with precise flow control. The residence time could be between 5 and 10 hours. At the output of the reaction section, the effluent containing the catalyst is chemically neutralized and the catalyst residue is separated from the products by aqueous washing. The catalyst components are not recycled. Unconverted olefin and inert hydrocarbons are separated from the octenes by distillation columns. The catalytic system is sensitive to impurities that can coordinate strongly to the nickel metal center or can react with the alkylaluminium derivative (polyunsaturated hydrocarbons and polar compounds such as water).

Despite all the advantages of this process, one main limitation is the continuous catalyst carry-over by the products, with the need to deactivate it and to dispose of wastes. One way to optimize catalyst consumption and waste disposal was to operate the reaction in a biphasic system. The first difficulty was to choose a "good" solvent. *N,N'*-Dialkylimidazolium chloroaluminate ionic liquids proved to be the best candidates. These can easily be prepared on an industrial scale, are liquid at the reaction temperature, and are very poorly miscible with the products. They play the roles both of the catalyst solvent and of the co-catalyst, and their Lewis acidities can be adjusted to obtain the best performances. The solubility of butene in these solvents is high enough to stabilize the active nickel species (Table 5.3-3), the nickel

Table 5.3-3: Solubilities of 1-butene and *n*-butane in the acidic mixture composed of 1-butyl-3-methylimidazolium ([BMIM]) chloride/aluminium chloride/ ethylaluminium dichloride (1:1.22: 0.06 molar ratio) as a function of temperature under atmospheric pressure.

Temperature (°C)	*Solubility of 1-butene (wt %)**	*Solubility of butane (wt %)*
10	4.5	2
20	2	1

*Isomerization of 1-butene into 2-butene is observed

catalyst can be immobilized without the addition of special ligands, and the catalytically active nickel complex is generated directly in the ionic liquid by treatment of a commercialized nickel(II) salt, as used in the Dimersol process, with an alkylaluminium chloride derivative.

The performances of the biphasic system in terms of activity, selectivity, recyclability, and lifetime of the ionic liquid have been evaluated in a continuous flow pilot operation. A representative industrial feed (raffinate II), composed of 70 % butenes (27 % of which is 1-butene) and 1.5 % isobutene (the remaining being *n*-butane and isobutane), enters continuously into a well mixed reactor containing the ionic liquid and the nickel catalyst. Injection of fresh catalyst components can be made to compensate for the detrimental effects of random impurities present in the feed. The reactor is operated full of liquid. The effluent (a mixture of the two liquid phases) leaves the reactor through an overflow and is transferred to a phase separator. The separation of the ionic liquid (density around 1200 g L^{-1}) and the oligomers occurs rapidly and completely (favored by the difference in densities). The ionic liquid and the catalyst are recycled to the reactor. A continuous run has been carried out over a period of 5500 hours. Butene conversion and selectivity were stable, and no addition of fresh ionic liquid was required, demonstrating its stability under the reaction conditions. Relative to the homogeneous Dimersol process, the nickel consumption was decreased by a factor of 10. The octene selectivity was five points higher (90–95 % of the total products). This can be ascribed to the higher solubility of butenes (relative to the octenes) in the ionic liquids, subsequent reactions between octenes and butenes to form trimers thus being disfavored.

Despite the utmost importance of physical limitations such as solubility and mixing efficiency of the two phases, an apparent first-order reaction rate relative to the olefin monomer was determined experimentally. It has also been observed that an increase of the nickel concentration in the ionic phase results in an increase in the olefin conversion.

In the homogeneous Dimersol process, the olefin conversion is highly dependent on the initial concentration of monomers in the feedstock, which limits the applicability of the process. The biphasic system is able to overcome this limitation and promotes the dimerization of feedstock poorly concentrated in olefinic monomer.

The mixing of the two phases proved to be an important parameter for the reaction rate. An increased efficiency of the mixing resulted in an increase in the reaction rate but did not change the dimer selectivity. Elsewhere, batch laboratory experiments showed that no reaction occurred in the organic phase. This could indicate the possibility of the participation of an interfacial reaction.

The ratio of the ionic liquid to the organic phase present in the reactor also plays an important role. A too high level of ionic liquid results in much longer decantation time and causes lower dimer selectivity. To combine efficient decantation and a reasonable size for the settler in the process design, it has been proposed that the separation of the two phases be performed in two distinct settling zones arranged in parallel [38].

A new biphasic process named Difasol has been developed (see Figure 5.3-6). Because of the solubility of the catalyst in the ionic phase and the poor miscibility of the products, the unit is essentially reduced to a continuously stirred tank reactor followed by a phase separator. The heat of the reaction is removed by the circulation of a proportion of the organic phase in a cooler, while the remainder is sent to the washing section. Difasol is ideally suited for use after a first homogeneous dimerization step. This first homogeneous step proved to be the best way to purify the feed of trace impurities. It can be used instead of conventional feedstock pretreatment technologies using adsorbents. Another interesting approach to removal of the impurities from the feed consists of the circulation of the feed to be treated and the ionic liquid already used in the dimerization section as a counter-current [39].

A proposed package consists of a first homogeneous dimerization step, which acts mainly as a pretreatment of the feedstock, and a biphasic component (Figure 5.3-7).

This arrangement ensures more efficient overall catalyst utilization and a significant increase in the yield of octenes. As an example, dimer selectivity in the 90–92 % range with butene conversion in the 80–85 % range can be obtained with a C_4 feed containing 60 % butenes. Thanks to the biphasic technique, the dimerization

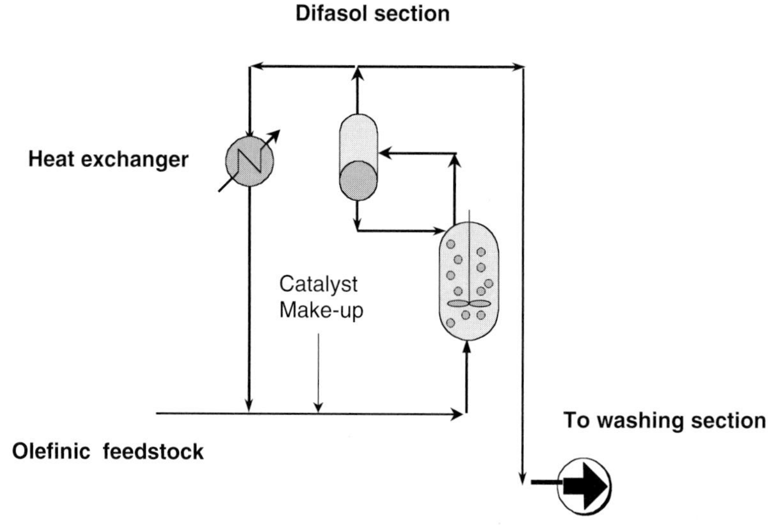

Figure 5.3-6: Difasol reaction section.

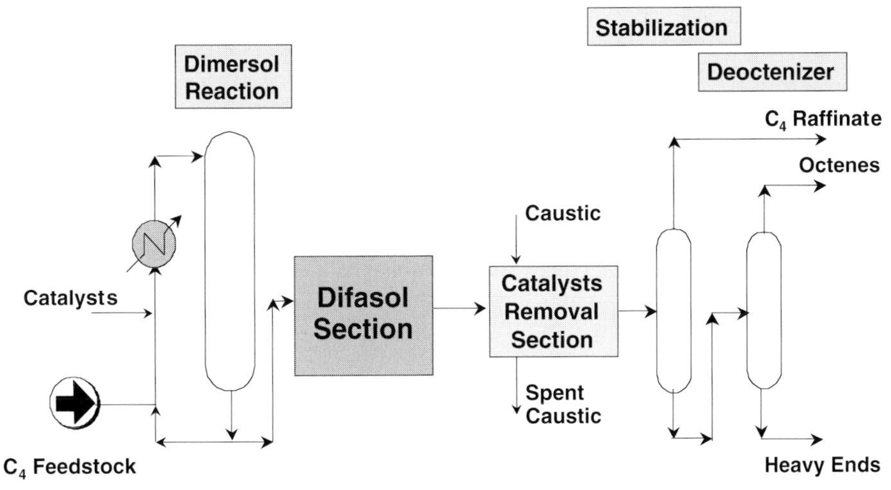

Figure 5.3-7: Process scheme integrating Dimersol and Difasol.

can also be extended to less reactive longer-chain olefins (C_5 feed), allowing the production of decenes and nonenes through co-dimerization of butenes and pentenes.

Since the catalyst is concentrated and operates in the ionic phase, and also probably at the phase boundary, reaction volumes in the biphasic technology are much lower than in the conventional single-phase Dimersol process, in which the catalyst concentration in the reactor is low. As an example, the Difasol reactor volume can be up to 40 times lower than that classically used in the homogeneous process.

A similar catalytic dimerization system has been investigated [40] in a continuous flow loop reactor in order to study the stability of the ionic liquid solution. The catalyst used is the organometallic nickel(II) complex (Hcod)Ni(hfacac) (Hcod = cyclooct-4-ene-1-yl and hfacac = 1,1,1,5,5,5-hexafluoro-2,4-pentanedionato-O,O'), and the ionic liquid is an acidic chloroaluminate based on the acidic mixture of 1-butyl-4-methylpyridinium chloride and aluminium chloride. No alkylaluminium is added, but an organic Lewis base is added to buffer the acidity of the medium. The ionic catalyst solution is introduced into the reactor loop at the beginning of the reaction and the loop is filled with the reactants (total volume 160 mL). The feed enters continuously into the loop and the products are continuously separated in a settler. The overall activity is 18,000 (TON). The selectivity to dimers is in the 98 % range and the selectivity to linear octenes is 52 %.

5.3.6.2 Alkylation reactions

BP Chemicals studied the use of chloroaluminates as acidic catalysts and solvents for aromatic hydrocarbon alkylation [41]. At present, the existing $AlCl_3$ technology (based on "red oil" catalyst) is still used industrially, but continues to suffer from poor catalyst separation and recycling [42]. The aim of the work was to evaluate the $AlCl_3$-based ionic liquids, with the emphasis placed on the development of a clean and recyclable system for the production of ethylbenzene (benzene/ethene alkylation) and synthetic lubricants (alkylation of benzene with 1-decene). The production of linear alkyl benzene (LAB) has also been developed by Akzo [43].

The ethylbenzene experiments were run by BP in a pilot loop reactor similar to that described for the dimerization (Figure 5.3-8).

Ionic liquids operate in true biphasic mode. While the recovery and recyclability of ionic liquid was found to be more efficient than with the conventional $AlCl_3$ catalyst (red oil), the selectivity for the monoalkylated aromatic hydrocarbon was lower. In this gas-liquid-liquid reaction, the solubility of the reactants in the ionic phase (e.g. the benzene/ethene ratio in the ionic phase) and the mixing of the phases were probably critical. This is an example in which the engineering aspects are of the utmost importance.

The use of acidic chloroaluminates as alternative liquid acid catalysts for the alkylation of light olefins with isobutane, for the production of high octane number gasoline blending components, is also a challenge. This reaction has been performed in a continuous flow pilot plant operation at IFP [44] in a reactor vessel similar to that used for dimerization. The feed, a mixture of olefin and isobutane, is pumped continuously into the well stirred reactor containing the ionic liquid catalyst. In the case of ethene, which is less reactive than butene, [pyridinium]Cl/$AlCl_3$ (1:2 molar ratio) ionic liquid proved to be the best candidate (Table 5.3-4).

The reaction can be run at room temperature and provides good quality alkylate (dimethylbutanes are the major products) over a period of three hundred hours.

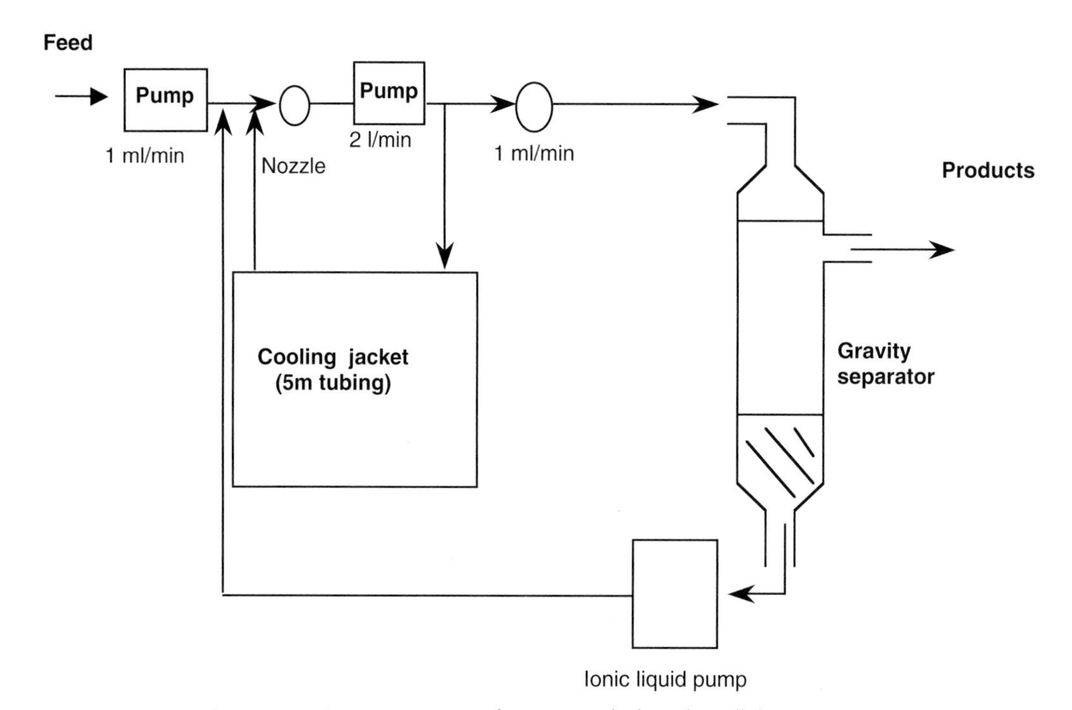

Figure 5.3-8: Loop reactor as used in aromatic hydrocarbon alkylation experiments.

Table 5.3-4: Alkylation of ethylene and 2-butene with isobutane. Semicontinuous pilot-plant results

Operating conditions/ nature of the olefin	*Ethene*		*2-Butene*	
Nature of ionic liquid	[Pyridinium, HCl] /$AlCl_3$ (1:2 molar ratio)		[BMIM][Cl]/$AlCl_3$	
Olefin content in the feed (wt%)	14–20		12–14	
VVH[a] (h^{-1})	0.2		0.35–0.45	
Temperature (°C)	25		5	
Test duration (h)	354		520	
Olefin conversion (wt%)	60–90		>98	
Production (g products/g ionic liquid)	121		172	
Product distribution (wt%)				
	i-C_6[b]	75–90	Light ends[e]:	5–10
	i-C_8[c]	10–17	i-C_8[c]	80–90 (>90 % TMP)
	C_8^+[d]	<5	C_8^+[d]	5–10
MON[f]	90–94		90–95	
RON[f]	98–101		95–98	

[a]Volume of olefin/(volume of ionic liquid.hour). [b]i-C_6 = 2,2- and 2,3-dimethylbutanes, [c]i-C_8 = isooctanes, TMP : trimethylpentanes, [d]C_8^+ = hydrocarbon products with more than eight carbon atoms, [e]Light ends = hydrocarbon products with fewer than eight carbon atoms, [f]RON = research octane number, MON = motor octane number

When butenes are used instead of ethene, lower temperature and fine-tuning of the acidity of the ionic liquid are required to avoid cracking reactions and heavy by-product formation. Continuous butene alkylation has been performed for more than five hundred hours with no loss of activity and stable selectivity. A high level of mixing is essential for a high selectivity and thus for a good quality alkylate. These applications are promising, but efforts are still needed to compete with the existing effective processes based on hydrofluoric and sulfuric acids.

5.3.6.3 Industrial use of ionic liquids

What can drive the switch from existing homogeneous processes to novel ionic liquids technology? One major point is probably a higher cost-effectiveness. This can result from improved reaction rates and selectivity, associated with more efficient catalyst recovery and better environmental compatibility.

The cost of ionic liquids can, of course, be a limiting factor in their development. However, this cost has to be weighed against that of current chemicals or catalysts.

If the ionic liquid can be recycled and if its lifetime is proven to be long enough, then its initial price is probably not the critical point. In Difasol technology, for example, ionic liquid cost, expressed with respect to the octene produced, is lower than that of the catalyst components.

The manufacture of ionic liquids on an industrial scale is also to be considered. Some ionic liquids have already been commercialized for electrochemical devices (such as capacitors) applications [45].

Chloroaluminate laboratory preparations proved to be easily extrapolated to large scale. These chloroaluminate salts are corrosive liquids in the presence of protons. When exposed to moisture, they produce hydrochloric acid, similarly to aluminium chloride. However, this can be avoided by the addition of some proton scavenger such as alkylaluminium derivatives. In Difasol technology, for example, carbon-steel reactors can be used with no corrosion problem.

The purity of ionic liquids is a key parameter, especially when they are used as solvents for transition metal complexes (see Section 5.2). The presence of impurities arising from their mode of preparation can change their physical and chemical properties. Even trace amounts of impurities (e.g., Lewis bases, water, chloride anion) can poison the active catalyst, due to its generally low concentration in the solvent. The control of ionic liquid quality is thus of utmost importance.

As new compounds, very limited research has been done to evaluate the biological effects of ionic liquids. The topical effect of [EMIM]Cl/$AlCl_3$ melts and [EMIM]Cl on the integument of laboratory rat has been investigated. The study reports that [EMIM]Cl is not in itself responsible for tissue damage. However, the chloroaluminate salt can induce tissue irritation, inflammation, and necrosis, due to the presence of aluminium chloride. However, treatments for aluminium chloride and hydrochloric acid are well documented. This study needs to be expanded to the other ionic liquids, and their toxicity need to be investigated [46].

Very few data [47] relating to the disposal of used ionic liquids are available. In Difasol technology, the used ionic liquid is taken out of the production system and the reactor is refilled with fresh catalyst solution.

5.3.7 Concluding Remarks and Outlook

In comparison with classical processes involving thermal separation, biphasic techniques offer simplified process schemes and no thermal stress for the organometallic catalyst. The concept requires that the catalyst and the product phases separate rapidly, to achieve a practical approach to the recovery and recycling of the catalyst. Thanks to their tunable solubility characteristics, ionic liquids have proven to be good candidates for multiphasic techniques. They extend the applications of aqueous biphasic systems to a broader range of organic hydrophobic substrates and water-sensitive catalysts [48–50].

To be applied industrially, performances must be superior to those of existing catalytic systems (activity, regioselectivity, and recyclability). The use of ionic liquid biphasic technology for nickel-catalyzed olefin dimerization proved to be successful, and this system has been developed and is now proposed for commercialization. However, much effort remains if the concept is to be extended to non-chloroaluminate ionic liquids. In particular, the true potential of ionic liquids (and mixtures containing ionic liquids) could be achievable if a substantial body of thermophysical and thermodynamic properties were amassed in order that the best medium for a given reaction could be chosen.

As far as industrial applications are concerned, the easy scale-up of two-phase catalysis can be illustrated by the first oxo aqeous biphasic commercial unit with an initial annual capacity of 100,000 tons extrapolated by a factor of 1:24,000 (batchwise laboratory development → production reactor) after a development period of 2 years [4].

References

1 B. Cornils, W. A. Herrmann in *Applied Homogeneous Catalysis with Organometallic Compounds* (B. Cornils, A. W. Herrmann eds.) Wiley-VCH, Weinheim **2000**, p. 575.

2 V. E. Dehmlow in *Aqueous-Phase Organometallic Catalysis: Concept and Applications* (B. Cornils, A. W. Herrmann eds.), Wiley-VCH, Weinheim **1998**, p. 207.

3 B. Cornils, E. G. Kuntz, in *Aqueous-Phase Organometallic Catalysis: Concept and Applications* (B. Cornils, A. W. Herrmann eds.), Wiley-VCH, Weinheim **1998**, p. 271.

4 B. Cornils, *Org. Process Res. Dev.* **1998**, *2*, 121.

5 J. Herwig, R. Fischer in *Rhodium-catalyzed Hydroformylation in Catalysis by Metal Complexes* (P. W. N. M. van Leewen, C. Claver eds.), Kluwer Academic Publisher, The Netherlands, Vol. 22, **2000**, p. 189.

6 Modern Solvents in Organic Synthesis, in *Topics in current Chemistry*, (P. Knochel Ed), Springer, Berlin, Vol. 206, **1999**.

7 J. N. Reek, P. C. J. Kamer, P. W. N. M. van Leeuwen, *Rhodium-catalyzed Hydroformylation in Catalysis by Metal Complexes* (P. W. N. M. van Leewen, C. Claver eds.), Kluwer Academic Publisher, The Netherlands, Vol. 22, **2000**, p. 253.

8 I. T. Horvath, *Acc. Chem. Res.* **1998**, *31*, 641.

9 P. G. Jessop, T. Ikariya, R. Noyori, *Chem. Rev.* **1999**, *99*, 475.

10 B. M. Bhanage, M. Shirai, M. Arai, Y. Ikushima, *Chem. Commun.* **1999**, 1277.

11 J. G. Huddleston, A. E. Visser, W. M. Reichert, H. D. Willauer, G. A. Broker, R. D. Rogers, *Green Chemistry* **2001**, *3*, 156.

12 H. L. Ngo, K. LeCompte, L. Hargens, A. B. McEwen, *Thermochim. Acta* **2000**, *97*, 357–358.

13 R. P. Sawtloski, A. E. Visser, M. W. Reichert, G. A. Broker, L. M. Farina, J. D. Holbrey, R. D. Rogers, *Chem. Commun.* **2001**, 2070.

14 J. Dupont, P. A. Z. Suarez, A. P. Umpierre, R. F. de Souza, *Catal. Lett.* **2001**, *73*, 211.

15 N. Karodia, S. Guise, G. Newlands, J.-A. Andersen, *Chem. Commun.* **1998**, 2341.

16 W. Keim, D. Vogt, H. Waffenschmidt, P. Wasserscheid, *J. Catal.* **1999**, *186*, 481.

17 N. Yoshimura in *Aqueous-Phase Organometallic Catalysis: Concept and Applications* (B. Cornils, A. W. Herrmann eds.), Wiley-VCH, Weinheim **1998**, p. 408.

18 A. J. Carmichael, M. J. Earle, J. D. Holbrey, P. B. McCormac, K. R. Seddon, *Org. Lett.* **1999**, *1*, 997.

19 M. Freemantle, *Chem. Eng. News* **2001**, 41.

20 T. Schäfer, C. A. Rodrigues, A. M. A. Carlos, J. G. Crespo, *Chem. Commun.* **2001**, 1622.
21 R. T. Carlin, J. Fuller, *Chem Commun.* **1997**, 1345.
22 Y. Chauvin, L. Mussmann, H. Olivier, *Angew. Chem. Int. Ed.* 1995, 34, 2698.
23 P. J. Dyson, D. J. Ellis, D. G. Parker, T. Welton, *Chem. Commun.* **1999**, 25.
24 Y. Chauvin, S. Einloft, H. Olivier, *Ind. Eng. Chem. Res.* **1995**, *34*, 1149.
25 P. Wasserscheid, C. M. Gordon, C. Hilgers, M. J. Muldoon, I. R. Dunkin, *Chem. Commun.* **2001**, 1186.
26 C. J. Mathews, P. J. Smith, T. Welton, *Chem. Commun.* **2000**, 1249.
27 G. W. Parshall, *J. Am. Chem. Soc.* **1972**, *94*, 8716.
28 F. Favre, H. Olivier-Bourbigou, D. Commereuc, L. Saussine, *Chem. Commun.* **2001**, 1360.
29 P. Wasserscheid, H. Waffenschmidt, P. Machnitzki, K. W. Kottsieper, O. Stelzer, *Chem. Commun.* **2001**, 451.
30 D. J. Brauer, K. W. Kottsieper, C. Liek, O. Stelzer, H. Waffenschmidt, P. Wasserscheid, *J. Organomet. Chem.* **2001**, *630*, 177.
31 K. W. Kottsieper, O. Stelzer, P. Wasserscheid, *J. Mol. Catal.* **2001**, *175*, 285.
32 C. C. Brasse, U. Englert, A. Salzer, H. Waffenschmidt, P. Wasserscheid, *Organometallics* **2000**, *19*, 3818.
33 Celanese Chemicals Europe (H. Bahrmann, H. Bohnen) Ger. Offen., DE 19919494 (2000).
34 (a) L. Xu, W. Chen, J. Xiao, *Organometallics* **2000**, *19*, 1123. (b) C. Mathews, P. J. Smith, T. Welton, A. J. P. White, D. J. Williams, *Organometallics,* **2001**, *20*, 3848.
35 O. Wachsen, K. Himmler, B. Cornils, *Catal. Today* **1998**, *42*, 373.
36 Y. Zhang, Z.-S. Mao, J. Chen, *Ind. Eng. Chem. Res.* **2001**, *40*, 4496.
37 A. Berger, R. F. de Souza, M. R. Delgado, J. Dupont, *Tetrahedron: Asymmetry* **2001**, *12*, 1825.
38 Institut Français du Pétrole (C. Bronner, A. Forestière, F. Hugues) US 6 203 712 B1 (**2001**).
39 Institut Français du Pétrole (H. Olivier, D. Commereuc, A. Forestière, F. Hugues) US 6 284 937 (**2001**).
40 P. Wasserscheid, M. Eichmann, Catal. Today **2001**, *66*, 309.
41 BP Chemicals Limited (B. Ellis, F. Hubert, P. Wasserscheid) PCT Int. WO 00/41809 (**2000**).
42 S. E. Knipling, *Petroleum Technology Quarterly* Autumn **2001**, 123.
43 Akzo Nobel (F. G. Sherif, L-J. Shyu, C. Greco, A. G. Talma, C. P. M. Lacroix) WO 98/03454 (**1998**).
44 Y. Chauvin, A. Hirschauer, H. Olivier, *J. Mol. Catal.* **1994**, *92*, 155.
45 K. Xu, M. S. Ding, T. R. Jow, *J. Electrochem. Soc.* **2001**, *148*, A267.
46 W. Mehm, J. B. Nold, C. Randall, B. S. Zernzach, *Aviation, Space, and Environmental Medicine* **1986**, 362.
47 British Nuclear Fuels (A. J. Jeapes, R. C. Thied, K. R. Seddon, W. R. Pitner, D. W. Rooney, J. E. Hatter, T. Welton) PCT Int. WO 01/15175 (**2001**).
48 H. Olivier-Bourbigou, L. Magna, *J. Mol. Catal. A: Chem.* **2002**, *419*, 182–183.
49 P. Wasserscheid, W. Keim, *Angew. Chem. Int. Ed.* **2000**, *39*, 3772.
50 R. Sheldon, *Chem. Commun.* **2001**, 2399.

5.4 Multiphasic Catalysis with Ionic Liquids in Combination with Compressed CO_2

Peter Wasserscheid

5.4.1 Introduction

Ionic liquids are often viewed as promising solvents for "clean processes" and "green chemistry", mainly due to their nonvolatile characters [1, 2]. These two catchphrases encompass current efforts to reduce drastically the amounts of side and coupling products, and also solvent and catalyst consumption in chemical processes. As another "green solvent" concept for chemical reactions, the replacement of volatile organic solvents by supercritical CO_2 (*sc*CO_2) is frequently discussed [3]. *sc*CO_2 combines environmentally benign characteristics (nontoxic, nonflammable) with favorable physicochemical properties for chemical synthesis. Catalyst separation schemes based on the tunable phase behavior of *sc*CO_2 (e.g. CESS process) have been developed [4].

However, ionic liquids and *sc*CO_2 are not competing concepts for the same applications. While ionic liquids can be considered as alternatives for polar organic solvents, the use of *sc*CO_2 can cover those applications in which non-polar solvents are usually used.

With regard to homogeneous transition metal-catalyzed reactions, the two media show complementary strengths and weaknesses. While ionic liquids are known to be excellent solvents for many transition metal catalysts (see Section 5.2), the solubilities of most transition metal complexes in *sc*CO_2 are poor. Usually, special ligand designs (such as phosphine ligands with fluorous "ponytails" [3]) are required to allow sufficient catalyst concentration in the supercritical medium. However, the isolation of the product from the solvent is always very easy in the case of *sc*CO_2, while product isolation from an ionic catalyst solution can become more and more complicated depending on the solubility of the product in the ionic liquid and on the product's boiling point.

In cases in which product solubility in the ionic liquid and the product's boiling point are high, the extraction of the product from the ionic liquid with an additional organic solvent is frequently proposed. This approach often suffers from some catalyst losses (due to some mutual solubility) and causes additional steps in the workup. Moreover, the use of an additional, volatile extraction solvent may nullify the "green solvent" motivation to use ionic liquids as nonvolatile solvents.

Beckman, Brennecke, and their research groups were the first to realize that the combination of *sc*CO_2 and an ionic liquid can offer special advantages. They observed that, although *sc*CO_2 is surprisingly soluble in some ionic liquids, the reverse is not the case, with no detectable ionic liquid solubilization in the CO_2 phase. On the basis of these results they described a method to remove naphthalene quantitatively from the ionic liquid [BMIM][PF_6] by extraction with *sc*CO_2 [5]. Subsequent work by Brennecke's team has applied the same procedure to the extraction of a large variety of different solutes from ionic liquids, without observation of any ionic liquid contamination in the isolated substances [6].

Research efforts aiming to quantify the solubility of CO_2 in ionic liquids revealed a significant influence of the ionic liquid's water content on the CO_2 solubility. While water-saturated [BMIM][PF_6] (up to 2.3 wt% water) has a CO_2 solubility of only 0.13 mol fraction, 0.54 mol fraction CO_2 dissolves in dry [BMIM][PF_6] (about 0.15 wt% water) at 57 bar and 40 °C [7]. Kazarian et al. used ATR-IR to determine the solubility of CO_2 in [BMIM][PF_6] and [BMIM] [BF_4]. They reported a solubility of 0.6 mol fraction CO_2 in [BMIM][PF_6] at 68 bar and 40 °C [8].

5.4.2
Catalytic Reaction with Subsequent Product Extraction

The first application involving a catalytic reaction in an ionic liquid and a subsequent extraction step with *sc*CO_2 was reported by Jessop et al. in 2001 [9]. These authors described two different asymmetric hydrogenation reactions using [Ru(OAc)$_2$(tolBINAP)] as catalyst dissolved in the ionic liquid [BMIM][PF_6]. In the asymmetric hydrogenation of tiglic acid (Scheme 5.4-1), the reaction was carried out in a [BMIM][PF_6]/water biphasic mixture with excellent yield and selectivity. When the reaction was complete, the product was isolated by *sc*CO_2 extraction without contamination either by catalyst or by ionic liquid.

In a similar manner, the asymmetric hydrogenation of isobutylatropic acid to afford the anti-inflammatory drug ibuprofen has been carried out (Scheme 5.4-2). Here, the reaction was carried out in a [BMIM][PF_6]/MeOH mixture, again followed by product extraction with *sc*CO_2 (see Section 5.2.4.1 for more details on these hydrogenation reactions).

5.4.3
Catalytic Reaction with Simultaneous Product Extraction

More recently, Baker, Tumas, and co-workers published catalytic hydrogenation reactions in a biphasic reaction mixture consisting of the ionic liquid [BMIM][PF_6] and *sc*CO_2 [10]. In the hydrogenation of 1-decene with Wilkinson's catalyst [$RhCl(PPh_3)_3$] at 50 °C and 48 bar H_2 (total pressure 207 bar), conversion of 98 %

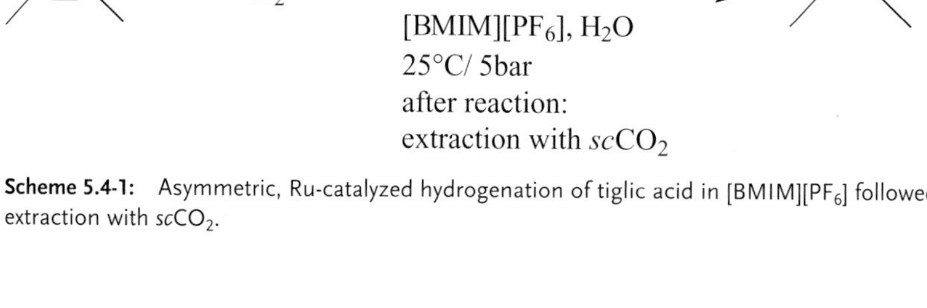

CO_2H + H_2 → CO_2H (*)

Ru(O_2CMe)$_2$(tolBINAP)
[BMIM][PF_6], H_2O
25°C/ 5bar
after reaction:
extraction with *sc*CO_2

Scheme 5.4-1: Asymmetric, Ru-catalyzed hydrogenation of tiglic acid in [BMIM][PF_6] followed by extraction with *sc*CO_2.

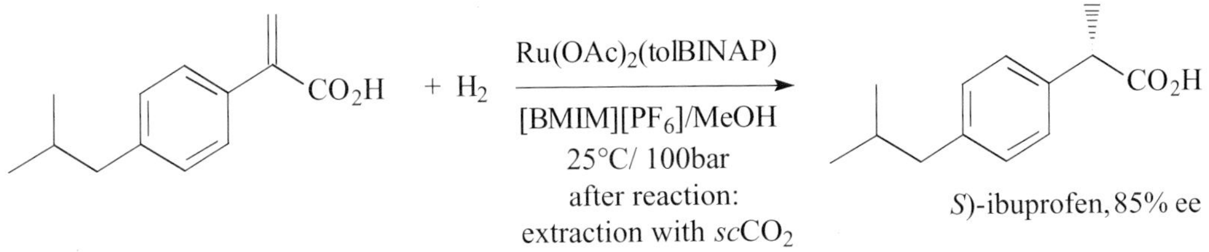

Ru(OAc)$_2$(tolBINAP)
[BMIM][PF$_6$]/MeOH
25°C/ 100bar
after reaction:
extraction with $scCO_2$

S)-ibuprofen, 85% ee

Scheme 5.4-2: Synthesis of ibuprofen by asymmetric, Ru-catalyzed hydrogenation in [BMIM][PF$_6$] with product isolation by subsequent extraction with $scCO_2$.

after 1 h was reported, corresponding to a turnover frequency (TOF) of 410 h^{-1}. Under identical conditions, the hydrogenation of cyclohexene proceeded with 82 % conversion after 2 h (TOF = 220 h^{-1}). The isolated ionic catalyst solution could be recycled in consecutive batches up to four times. The fact that a biphasic hydrogenation of 1-decene can be successfully achieved is not, however, any special benefit of the unconventional [BMIM][PF$_6$]/$scCO_2$ biphasic system. In fact, no reactivity advantage with the use of $scCO_2$ in place of a more common alkane solvent for such a biphasic system can be concluded from the reported results.

5.4.4 Catalytic Conversion of CO_2 in an Ionic Liquid/$scCO_2$ Biphasic Mixture

In the same paper [10], the authors described the [RuCl$_2$(dppe)]-catalyzed (dppe = Ph_2P-$(CH_2)_2$-PPh_2) hydrogenation of CO_2 in the presence of dialkylamines to obtain *N,N*-dialkylformamides. The reaction of di-*n*-propylamine in the [BMIM][PF$_6$]/$scCO_2$ system resulted in complete amine conversion to provide the desired *N,N*-di-*n*-propylformamide with high selectivity. This compound showed very high solubility in the ionic liquid phase, and complete product isolation by extraction with $scCO_2$ proved to be difficult. However, product extraction with $scCO_2$ became possible once the ionic catalyst solution had become completely saturated with the product.

5.4.5 Continuous Reactions in an Ionic Liquid/Compressed CO_2 System

Cole-Hamilton and co-workers demonstrated the first flow apparatus for a continuous catalytic reaction using the biphasic system [BMIM][PF$_6$]/$scCO_2$ [11]. They investigated the continuous Rh-catalyzed hydroformylation of 1-octene over periods of up to 33 h using the ionic phosphine ligand [PMIM]$_2$[PhP($C_6H_4SO_3$)$_2$] ([PMIM] = 1-methyl-3-propylimidazolium). No catalyst decomposition was observed during the period of the reaction, and Rh leaching into the $scCO_2$/product stream was less than 1 ppm. The selectivity for the linear hydroformylation product was found to be stable over the reaction time (*n*/*iso* = 3.1).

During the continuous reaction, alkene, CO, H_2, and CO_2 were separately fed into the reactor containing the ionic liquid catalyst solution. The products and unconverted feedstock were removed from the ionic liquid still dissolved in *sc*CO_2. After decompression the liquid product was collected and analyzed. A schematic view of the apparatus used by Cole-Hamilton et al. is given in Figure 5.4-1.

Obviously, the motivation to perform this hydroformylation reaction in a continuous flow reactor arose from some problems during the catalyst recycling when the same reaction was first carried out in repetitive batch mode. In the latter case, Cole-Hamilton et al. observed a continuous drop of the product's *n/iso* ratio from 3.7 to 2.5 over the first nine runs. Moreover, the isomerization activity of the system increased during the batch-wise recycling experiments, and Rh leaching became significant after the ninth run. The authors concluded from ^{31}P NMR investigations that ligand oxidation due to contamination of the systems with air (during the opening of the reactor for recycling) had resulted in the formation of $[RhH(CO)_4]$ as the active catalytic species. This compound is known to show more isomerization activity and a lower *n/iso* ratio than the phosphine-modified catalyst system. Moreover, $[RhH(CO)_4]$ is also known to display some solubility in *sc*CO_2, which explains the observed leaching of rhodium into the organic layer.

All the problems associated with the batch-wise catalyst recycling could be convincingly overcome by application of the continuous operation mode described above. The authors concluded that continuous flow *sc*CO_2/ionic liquid biphasic systems provided a method for continuous flow homogeneous catalysis with integrated separation of the products from the catalyst and from the reaction solvent. Most interestingly, this unusual continuous biphasic reaction mode enabled the quantitative separation of relatively high boiling products from the ionic catalyst solution under mild temperature conditions and without use of an additional organic extraction solvent.

Slightly later, and independently of Cole-Hamilton's pioneering work, the author's group demonstrated in collaboration with Leitner et al. that the combination of a suitable ionic liquid with compressed CO_2 can offer much more potential for homogeneous transition metal catalysis than only being a new procedure for easy product isolation and catalyst recycling. In the Ni-catalyzed hydrovinylation of

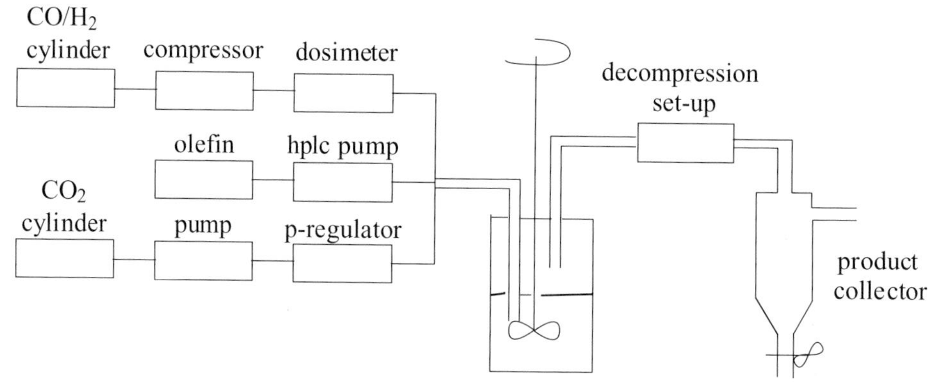

Figure 5.4-1: Continuous flow apparatus as used for the hydroformylation of 1-octene in the biphasic system $[BMIM][PF_6]$/*sc*CO_2

styrene it was possible to activate, tune, and immobilize the well-known Wilke complex by use of this unusual biphasic system (Scheme 5.4-3). Obviously, this reaction benefits from this special solvent combination in a new and highly promising manner.

Hydrovinylation is the transition metal-catalyzed co-dimerization of alkenes with ethene yielding 3-substituted 1-butenes [12]. This powerful carbon–carbon bond-forming reaction can be achieved with high enantioselectivity by the use of Wilke's complex as a catalyst precursor [13]. In conventional solvents, this pre-catalyst needs to be activated with a chloride abstracting agent, such as $Et_3Al_2Cl_3$. Leitner et al. reported the use of Wilke's complex in compressed CO_2 (under liquid and under supercritical conditions) after activation with alkali salts of weakly coordinating anions such as Na[BARF] ($[BARF]^- = [(3,5\text{-}(CF_3)_2C_6H_3)_4B]^-$) [14].

At first, the reaction was investigated in batch mode, by use of different ionic liquids with weakly coordinating anions as the catalyst medium and compressed CO_2 as simultaneous extraction solvent. These experiments revealed that the activation of Wilke's catalyst by the ionic liquid medium was clearly highly dependent on the nature of the ionic liquid's anion. Comparison of the results in different ionic liquids with $[EMIM]^+$ as the common cation showed that the catalyst's activity drops in the order $[BARF]^- > [Al\{OC(CF_3)_2Ph\}_4]^- > [(CF_3SO_2)_2N]^- > [BF_4]^-$. This trend is consistent with the estimated nucleophilicity/coordination strength of the anions.

Interestingly, the specific environment of the ionic solvent system appears to activate the chiral Ni-catalyst beyond a simple anion-exchange reaction. This becomes obvious from the fact that even the addition of a 100-fold excess of $Li[(CF_3SO_2)_2N]$ or $Na[BF_4]$ in pure, compressed CO_2 produced an at best moderate activation of Wilke's complex in comparison to the reaction in ionic liquids with the corresponding counter-ion (e.g., 24.4 % styrene conversion with 100-fold excess of $Li[(CF_3SO_2)_2N]$, in comparison to 69.9 % conversion in $[EMIM][(CF_3SO_2)_2N]$ under otherwise identical conditions).

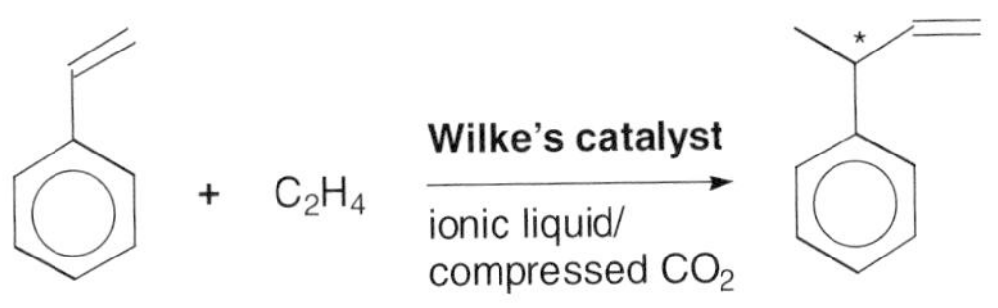

+ C_2H_4 —Wilke's catalyst / ionic liquid/ compressed CO_2→

Wilke's catalyst:

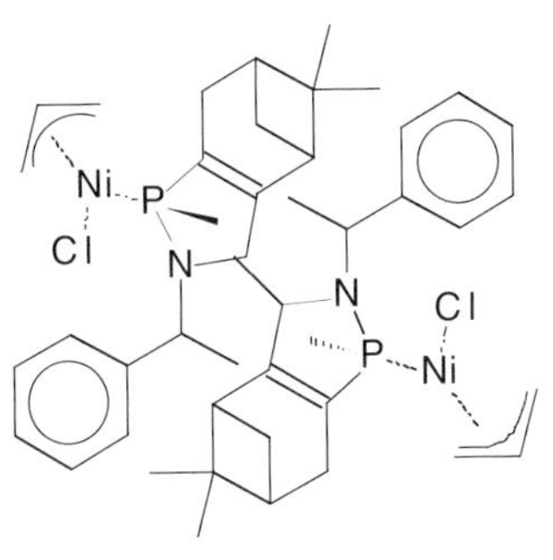

Scheme 5.4-3: The enantioselective hydrovinylation of styrene with Wilke's catalyst.

In the biphasic batch reaction the best reaction conditions were found for the system [EMIM][$(CF_3SO_2)_2N$]/compressed CO_2. It was found that increasing the partial pressure of ethylene and decreasing the temperature helped to suppress the concurrent side reactions (isomerization and oligomerization), 58 % conversion of styrene (styrene/Ni = 1000/1) being achieved after 1 h under 40 bar of ethylene at 0 °C with 3-phenyl-1-butene being detected as the only product and with a 71 % *ee* of the *R* isomer.

However, attempts to reuse the ionic catalyst solution in consecutive batches failed. While the products could readily be isolated after the reaction by extraction with *sc*CO_2, the active nickel species deactivated rapidly within three to four batchwise cycles. The fact that no such deactivation was observed in later experiments with the continuous flow apparatus described below (see Figure 5.4-2) clearly indicate the deactivation of the chiral Ni-catalyst being mainly related to the instability of the active species in the absence of substrate.

In the continuous hydrovinylation experiments, the ionic catalyst solution was placed in the reactor R, where it was in intimate contact with the continuous reaction phase entering from the bottom (no stirring was used in these experiments). The reaction phase was made up in the mixer from a pulsed flow of ethylene and a continuous flow of styrene and compressed CO_2.

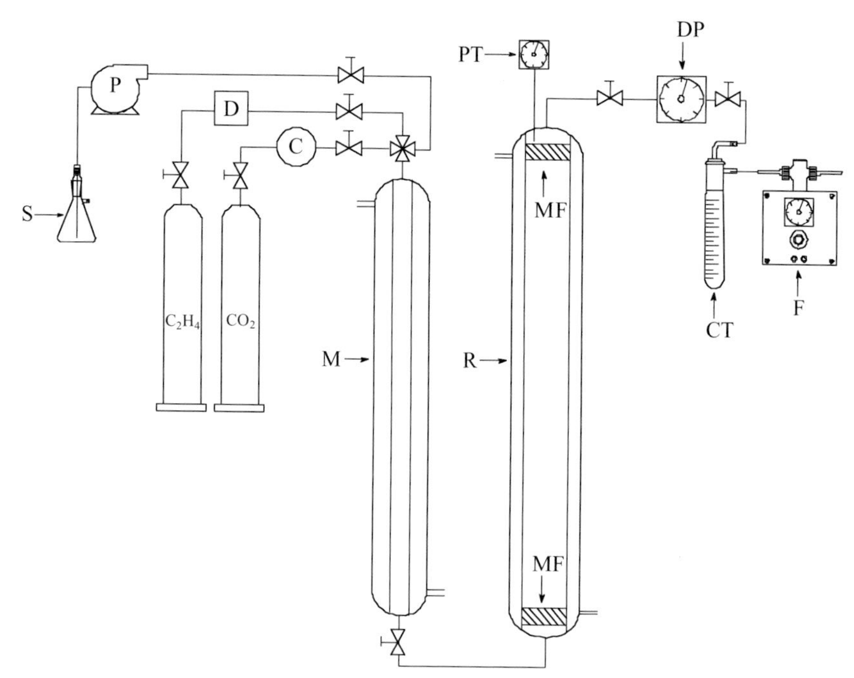

Figure 5.4-2: Schematic view of the continuous flow apparatus used for the enantioselective hydrovinylation of styrene in the biphasic [EMIM][$(CF_3SO_2)_2N$] system. The components are labeled (alphabetically) as follows: *C*: compressor, *CT*: cold trap, *D*: dosimeter, *DP*: depresurizer, *F*: flow-meter, *M*: mixer, *MF*: metal filter, *P*: HPLC pump, *PT*: pressure transducer and thermocouple, *R*: reactor, *S*: styrene.

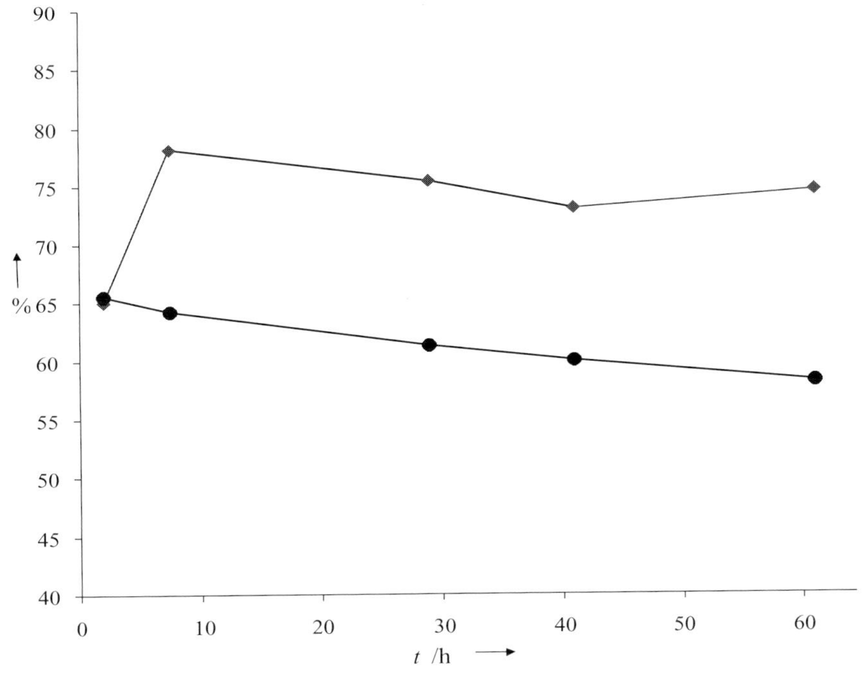

Figure 5.4-3: Lifetime study of Wilke's catalyst in the hydrovinylation of styrene, activated and immobilized in the $[EMIM][(CF_3SO_2)_2N]$/compressed CO_2 system (● *ee*; ♦ conversion).

Figure 5.4-3 shows the results of a lifetime study for Wilke's catalyst dissolved, activated, and immobilized in the $[EMIM][(CF_3SO_2)_2N]$/compressed CO_2 system. Over a period of more than 61 h, the active catalyst showed remarkably stable activity while the enantioselectivity dropped only slightly. These results clearly indicate – at least for the hydrovinylation of styrene with Wilke's catalyst – that an ionic liquid catalyst solution can show excellent catalytic performance in continuous product extraction with compressed CO_2.

5.4.6 Concluding Remarks and Outlook

The combination of ionic liquids and compressed CO_2 – at opposite extremes of the volatility and polarity scales – offers a new and intriguing immobilization technique for homogeneous catalysis.

In comparison with catalytic reactions in compressed CO_2 alone, many transition metal complexes are much more soluble in ionic liquids without the need for special ligands. Moreover, the ionic liquid catalyst phase provides the potential to activate and tune the organometallic catalyst. Furthermore, product separation from the catalyst is now possible without exposure of the catalyst to changes of temperature, pressure, or substrate concentration.

In contrast to the use of pure ionic liquid, the presence of compressed CO_2 greatly decreases the viscosity of the ionic catalyst solution, thus facilitating mass transfer during the catalytic reaction. Moreover, high-boiling products with some solubility in the ionic liquid phase can now be removed without use of an additional organic solvent. Finally, the use of compressed CO_2 as the mobile phase allows a reactor design very similar to a classical fixed bed reactor [15]. Thus, the combination of ionic liquids and compressed CO_2 provides a new and highly attractive approach, benefiting from the advantages of both homogeneous and heterogeneous catalysis. Moreover, this approach promises to overcome some of the well known limitations of conventional biphasic catalysis (catalyst immobilization, feedstock solubility in the catalytic phase, solvent cross-contamination, mass transfer limitation). In particular, the combination of nonvolatile ionic liquids with non-hazardous CO_2 offers fascinating new possibilities for the design of environmentally benign processes.

References

1 M. Freemantle, *Chem. Eng. News* **1998**, *76(13)*, 32.

2 a) M. Freemantle, *Chem. Eng. News* **1999**, *77(1)*, 23; b) D. Bradley, *Chem. Ind.* **1999**, 86; c) M. Freemantle, *Chem. Eng. News* **2000**, *78(20)*, 37.

3 a) P. G. Jessop, W. Leitner (eds.) *"Chemical Synthesis Using Supercritical Fluids"*, Wiley-VCH, Weinheim, **1999**; b) For recent reviews, see a) P. G. Jessop, T. Ikariya, R. Noyori, *Science*, **1995**, *269*, 1065–1069; c) P. G. Jessop, T. Ikariya, R. Noyori, *Chem. Rev.* **1999**, 99, 475–493; d) M. Poliakoff, S. M. Howdle, S. G. Kazarian, *Angew. Chem. Int. Ed. Engl.* **1995**, *34*, 1275–1295.

4 a) G. Franciò, K. Wittmann, W. Leitner, *J. Organomet. Chem.* **2001**, *621*, 130–142; b) S. Kainz, A. Brinkmann, W. Leitner, A. Pfaltz, *J. Am. Chem. Soc.* **1999**, *121*, 6421–6429; c) D. Koch, W. Leitner, *J. Am. Chem. Soc.* **1998**, *120*, 13,398–13,404.

5 L. A. Blanchard, D. Hancu, E. J. Beckman, J. F. Brennecke, *Nature* **1999**, *299*, 28–29.

6 L. A. Blanchard, J. F. Brennecke, *Ind. Eng. Chem. Res.* **2001**, *40*, 287–292.

7 L. A. Blanchard, Z. Gu, J. F. Brennecke, *J. Phys. Chem. B* **2001**, *105*, 2437.

8 S. G. Kazarian, B. J. Biscoe, T. Welton, *Chem. Commun.* **2000**, 2047–2048.

9 R. A. Brown, P. Pollett, E. McKoon, C. A. Eckert, C. L. Liotta, P. G. Jessop, *J. Am. Chem. Soc.* **2001**, *123*, 1254–1255.

10 F. Liu, M. B. Abrams, R. T. Baker, W. Tumas, *Chem. Commun.* **2001**, 433–434.

11 a) M. F. Sellin, P. B. Webb, D. J. Cole-Hamilton, *Chem. Commun.* **2001**, 781–782; b) D. J. Cole-Hamilton, M. F. Sellin, P. B. Webb WO 0202218 (to the University of St. Andrews) 2002 [Chem. Abstr. **2002**, *136*, 104215].

12 For reviews, see a) P. W. Jolly, G. Wilke, *Applied Homogenous Catalysis with Organic Compounds 2*, (B. Cornils, W. A. Herrman eds.), Wiley-VCH, **1996**, 1024–1048; b) T. V. RajanBabu, N. Nomura, J. Jin, B. Radetich, H. Park, M. Nandi, *Chem. Eur. J.* **1999**, *5*, 1963–1968.

13 G. Wilke, J. Monkiewicz, H. Kuhn, DE 3618169 (to Studiengesellschaft Kohle m.b.H., Germany), **1987** [Chem. Abstr. **1988**, *109*, P6735].

14 A. Wegner, W. Leitner, *Chem. Commun.* **1999**, 1583–1584.

15 For the use of CO_2 as a mobile phase in classical heterogeneous catalysis, see: W. K. Gray, F. R. Smail, M. G. Hitzler, S. K. Ross, M. Poliakoff, *J. Am. Chem. Soc.* **1999**, *121*, 10,711–10,718.

6
Inorganic Synthesis

Frank Endres and Tom Welton

6.1
Directed Inorganic and Organometallic Synthesis

Tom Welton

Although a great deal of excitement has surrounded the use of ionic liquids as solvents for organic synthesis, the rational synthesis of inorganic and organometallic compounds in ionic liquids has remained largely unexplored.

6.1.1
Coordination Compounds

Some halogenometalate species have been observed to have formed spontaneously during spectroelectrochemical studies in ionic liquids. For example, $[MoCl_6]^{2-}$ (which is hydrolyzed in water, is coordinated by solvent in polar solvents, and has salts that are insoluble in non-polar solvents) can only be observed in basic $\{X(AlCl_3) < 0.5\}$ chloroaluminate ionic liquids [1]. However, this work has been directed at the measurement of electrochemical data, rather than exploitation of the ionic liquids as solvents for synthesis [2]. It has been shown that the tetrachloroaluminate ion will act as a bidentate ligand in acidic $\{X(AlCl_3) > 0.5\}$ chloroaluminate ionic liquids, forming $[M(AlCl_4)_3]^-$ ions [3]. This was also the result of the spontaneous formation of the complexes, rather than a deliberate attempt to synthesize them.

The only reports of directed synthesis of coordination complexes in ionic liquids are from oxo-exchange chemistry. Exposure of chloroaluminate ionic liquids to water results in the formation of a variety of aluminium oxo- and hydroxo-containing species [4]. Dissolution of metals more oxophilic than aluminium will generate metal oxohalide species. Hussey et al. have used phosgene ($COCl_2$) to deoxochlorinate $[NbOCl5]^{2-}$ (Scheme 6.1-1) [5].

$$[NbOCl_5]^{2-} \underset{O^{2-}}{\overset{COCl_2}{\rightleftharpoons}} [NbCl_6]^-$$

Scheme 6.1-1: Nb(V) oxo-exchange chemistry in a basic [EMIM]Cl/$AlCl_3$ ionic liquid.

$$[VO_2Cl_2]^- \underset{PhIO}{\overset{triphosgene}{\rightleftharpoons}} [VOCl_4]^{2-} \underset{O_2}{\overset{triphosgene}{\rightleftharpoons}} [VCl_6]^{3-}$$

Scheme 6.1-2: Vanadium oxo-exchange chemistry in a basic [EMIM]Cl/$AlCl_3$ ionic liquid

Triphosgene (bis(trichloromethyl)carbonate) has been used to deoxochlorinate $[VOCl_4]^{2-}$ to $[VCl_6]^{3-}$ and $[VO_2Cl_2]^-$ to $[VOCl_4]^{2-}$ [6]. In both these cases the deoxochlorination was accompanied by spontaneous reduction of the initial products (Scheme 6.1-2).

6.1.2
Organometallic Compounds

With the enthusiasm currently being generated by the (so-called) stable carbenes (imidazolylidenes) [7], it is surprising that there are few reports of imidazolium-based ionic liquids being used to prepare metal imidazolylidene complexes. Xiao et al. have prepared bis(imidazolylidene)palladium(II) dibromide in [BMIM]Br [8]. All four possible conformers are formed, as shown in Scheme 6.1-3.

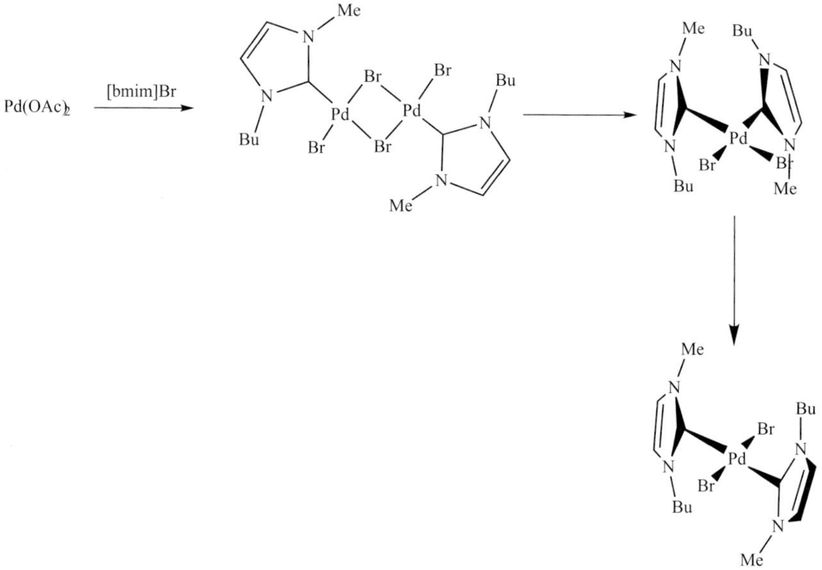

Scheme 6.1-3: The formation of bis(1-butyl-3-methylimidazolylidene)palladium(II) dibromide in [BMIM]Br

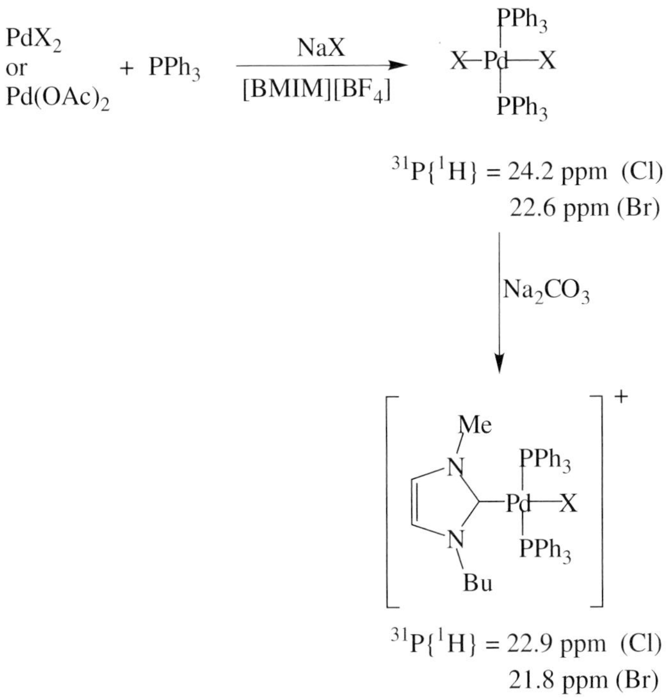

$$PdX_2 \text{ or } Pd(OAc)_2 + PPh_3 \xrightarrow[\text{[BMIM][BF}_4\text{]}]{NaX} trans\text{-}[PdX_2(PPh_3)_2]$$

$^{31}P\{^1H\}$ = 24.2 ppm (Cl)
22.6 ppm (Br)

Na_2CO_3

$[(\text{1-Bu-3-Me-imidazolylidene})Pd(PPh_3)_2X]^+$

$^{31}P\{^1H\}$ = 22.9 ppm (Cl)
21.8 ppm (Br)

Scheme 6.1-4: The formation of (1-butyl-3-methylimidazolylidene)bis(triphenylphosphine)palladium(II) chloride in [BMIM][BF_4].

In the presence of triphenylphosphine and four equivalents of chloride, (1-butyl-3-methylimidazolylidene)bis(triphenylphosphine)palladium(II) chloride is formed (Scheme 6.1-4).

Singer and co-workers have investigated the acylation reactions of ferrocene in ionic liquids made from mixtures of [EMIM]I and aluminium(III) chloride (Scheme 6.1-5) [9, 10]. The ionic liquid acts both as solvent and as source of the Friedel–Crafts catalyst. In mildly acidic $\{X(AlCl_3) > 0.5\}$ [EMIM]I/$AlCl_3$, the monoacetylated ferrocene was obtained as the major product. In strongly acidic [EMIM]I/$AlCl_3$ $\{X(AlCl_3) = 0.67\}$ the diacylated ferrocene was the major product. Also, when R = alkyl, the diacetylated product was usually the major product, but for R = Ph, the monoacetylated product was favored.

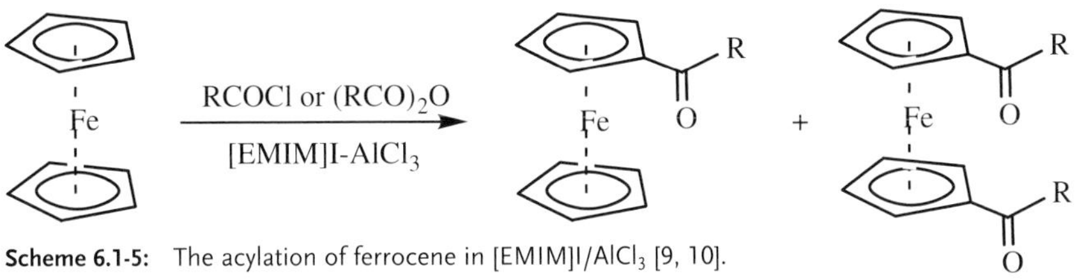

$$Fe(C_5H_5)_2 \xrightarrow[\text{[EMIM]I-AlCl}_3]{\text{RCOCl or (RCO)}_2\text{O}} (C_5H_5)Fe(C_5H_4COR) + Fe(C_5H_4COR)_2$$

Scheme 6.1-5: The acylation of ferrocene in [EMIM]I/$AlCl_3$ [9, 10].

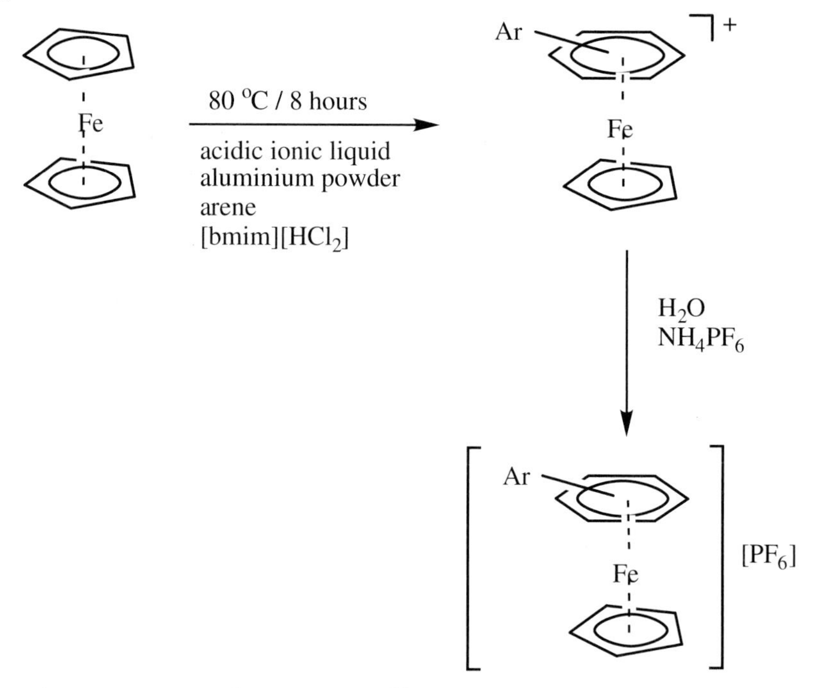

Scheme 6.1-6: Arene exchange reactions of ferrocene in $[BMIM]Cl/AlCl_3$.

In another study relying on chloroaluminate chemistry, the Fisher–Hafner-type ligand-exchange reactions of ferrocene were investigated (Scheme 6.1-6) [11]. Again, the acidic ionic liquids acted as combinations of solvent and catalyst. In these reactions it was necessary to add $[BMIM][HCl_2]$ as a proton source, to generate the cyclopentadiene leaving group.

The strong halide-abstracting properties of acidic $\{X(AlCl_3) = 0.67\}$ $[BMIM]Cl/AlCl_3$ have been used for the synthesis of the "piano stool" complexes $[Mn(CO)_3(\eta^6\text{-arene})]^+$ (Scheme 6.1-7) [12].

In all of the above cases the products were isolated by the destruction of the chloroaluminate ionic liquids by addition to water.

6.1.3
Other Reactions

The only other report of the use of an ionic liquid to prepare an inorganic material is that of the formation of a silica aerogel in $[EMIM][(CF_3SO_2)_2N]$ [13]. Formic acid was added to tetramethylorthosilicate in the ionic liquid, yielding a gel that cured over a period of three weeks (Scheme 6.1-8). Here, it was the nonvolatile nature of the ionic liquid, preventing the loss of solvent during the curing process, that was exploited. The ionic liquid was retrieved from the aerogel by extraction with acetonitrile.

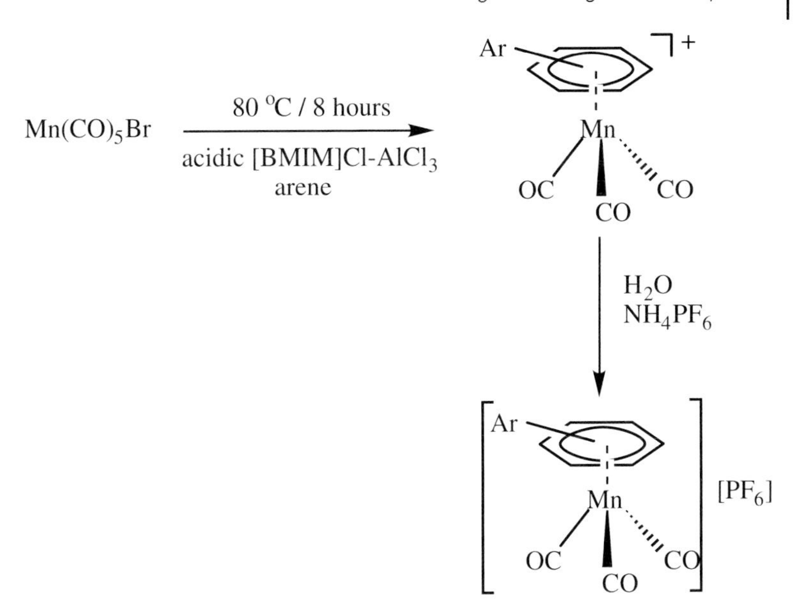

Scheme 6.1-7: The synthesis of $[Mn(CO)_3(\eta^6\text{-arene})]^+$ "piano stool" complexes in $[BMIM]Cl/AlCl_3$.

$$2\,HC(O)OH + (CH_3O)_4Si \xrightarrow[\text{[EMIM][(CF}_3\text{SO}_2)_2\text{N]}]{} SiO_2 + 2\,CH_3OH + 2\,HC(O)OCH_3$$

Scheme 6.1-8: The formation of a SiO_2 aerogel in $[EMIM][(CF_3SO_2)_2N]$.

6.1.4
Outlook

There is no doubt that inorganic and organometallic synthesis in ionic liquids is lagging behind organic synthesis. This is not due to any lack of importance. If, for instance, ionic liquids are to find use in biphasic catalysis, a point will arrive at which the ionic liquid layer can no longer be recycled. It will only be through an understanding of the chemistry of the dissolved catalysts, deliberately prepared to be difficult to remove, that they will be transformable into materials capable of being extracted from the ionic liquids.

Ionic liquids hold as much promise for inorganic and organometallic synthesis as they do for organic synthesis. Their lack of vapor pressure has already been exploited [13], as have their interesting solubility properties. The field can only be expected to accelerate from its slow beginnings.

References

1 T. B. Scheffler, C. L. Hussey, K. R. Seddon, C. M. Kear and P. D. Armitage, *Inorg. Chem.* **1984**, *23*, 1926.
2 C. L. Hussey, *Pure Appl. Chem.* **1988**, *60*, 1763.
3 A. J. Dent, K. R. Seddon and T. Welton, *J. Chem. Soc., Chem. Commun.* **1990**, 315.
4 T. Welton, *Chem. Rev.* **1999**, *99*, 2071.
5 I. W. Sun, E. H. Ward and C. L. Hussey, *Inorg. Chem.* **1987**, *26*, 4309.
6 A. J. Dent, A. Lees, R. J. Lewis, and T. Welton, *J. Chem. Soc., Dalton Trans.* **1996**, 2787.
7 (a) W. A. Herrmann and C. Kocher, *Angew. Chem. Int. Ed.* **1997**, *36*, 2163; (b) D. Bourissou, O Guerret, F. P. Gabbai and G. Bertrand, *Chem. Rev.* **2000**, *100*, 39.
8 L. Xu, W. Chen, and J. Xiao, *Organometallics* **2000**, *19*, 1123.
9 J. K. D. Surette, L. Green, and R. D. Singer, *Chem. Commun.* **1996**, 2753.
10 A. Stark, B. L. MacLean, and R. D. Singer, *J. Chem. Soc., Dalton Trans.* **1999**, 63.
11 Paul J. Dyson, Martin C. Grossel, N. Srinivasan, T. Vine, T. Welton, D. J. Williams, A. J. P. White, and T. Zigras, *J. Chem. Soc., Dalton Trans.* **1997**, 3465.
12 D. Crofts, P. J. Dyson, K. M. Sanderson, N. Srinivasan, and T. Welton, *J. Organomet. Chem.* **1999**, *573*, 292
13 S. Dai, Y. H. Ju, H. J. Gao, J. S. Lin, S. J. Pennycook, and C. E. Barnes, *Chem. Commun.* **2000**, 243.

6.2 Making of Inorganic Materials by Electrochemical Methods

Frank Endres

6.2.1 Electrodeposition of Metals and Semiconductors

6.2.1.1 General considerations

Electrodeposition is one of the main fields in electrochemistry, both in industrial processes and in basic research. In principal, all metals and semiconductors can be obtained by electrolysis of the respective salts in aqueous or organic solutions and molten salts, respectively. As well as electrowinning of the elements, electrocoating of materials for corrosion protection is an important field in industry and in basic research. With the help of the Scanning Tunneling Microscope, a great deal of work on the nanometer scale has been done over the past 15 years. Insight into how the initial stages of metal deposition influence the bulk growth has been obtained. Furthermore, the role of brighteners, added to solutions to make shining deposits, can now be understood; they seem to adsorb at growing clusters and force the metal to grow layer-by-layer instead of in the form of clusters [1]. Aqueous solutions, however, are unsuitable for the electrodeposition of less noble elements because of their limited electrochemical windows. For light, refractory, and rare earth metals, water fails as a solvent because hydrogen evolves long before deposition of the metal. Ionic liquids are ideal solvents for such purposes though, because they have – depending on their compositions – wide electrochemical windows combined with good solubilities of most metal salts and semiconductor compounds [2]. Many technical processes, such as the electrowinning of the rare earth and refractory metals Mg, Al, and several others, are performed in high-temperature molten salts. These systems are highly corrosive and sometimes make it difficult to find materials that will withstand chemical attack by the melts. The design of electrochemical cells for low-melting ionic liquids, in contrast, is much easier. These combine, more or less, the advantages of classical molten salts and those of aqueous media. Thanks to their wide electrochemical windows, several metals and alloys conventionally accessible from high-temperature molten salts – such as Al and its alloys, La, etc. – can also be deposited at room temperature. Furthermore, metals obtainable from aqueous media can in most cases also be deposited from ionic liquids, often with superior quality since hydrogen evolution does not occur. Pd is a good example, since deposits from aqueous solutions can contain varying amounts of hydrogen, which can make the deposits rather brittle. However, shining, even, nanosized Pd deposits can easily be obtained from ionic liquids. These features and their good ionic conductivities of between 10^{-3} and 10^{-2} $(\Omega cm)^{-1}$ make the ionic liquids interesting solvents for electrodeposition.

Section 6.2.1 offers literature data on the electrodeposition of metals and semiconductors from ionic liquids and briefly introduces basic considerations for electrochemical experiments. Section 6.2.2 describes new results from investigations of process at the electrode/ionic liquids interface. This part includes a short introduction to *in situ* Scanning Tunneling Microscopy.

6.2.1.2 **Electrochemical equipment**

Any redox couple has a defined electrode potential on the electrochemical potential scale. In aqueous solutions, for example, a silver wire immersed into a solution containing Ag^{+} ions with the activity 1 has a value of +799 mV vs. the normal hydrogen electrode. At more positive potentials a Ag electrode will dissolve, at more negative values Ag will deposit from the ions. If one applies a certain voltage between two electrodes, an electrochemical reaction may occur, depending on the applied voltage. If one wants to know the processes involved, for example during electrodeposition, it is necessary to know the electrode potential relative to a reference electrode exactly. This can be measured by means of a third electrode immersed in the solution. This is the three-electrode setup, with the working electrode of interest (WE), the reference electrode (RE), and the counter-electrode (CE). However, any current I(EC) that flows through the cell will influence the electrode potentials, so a stable value would be hard to obtain. A potentiostat allows precise control of the potential of the working electrode with respect to a reference electrode. It always applies the desired value U(setpoint) to the working electrode, usually responding to changes within microseconds. A simplified setup based on an operational amplifier (OPA), where the working electrode is connected to ground, is presented in Figure 6.2-1.

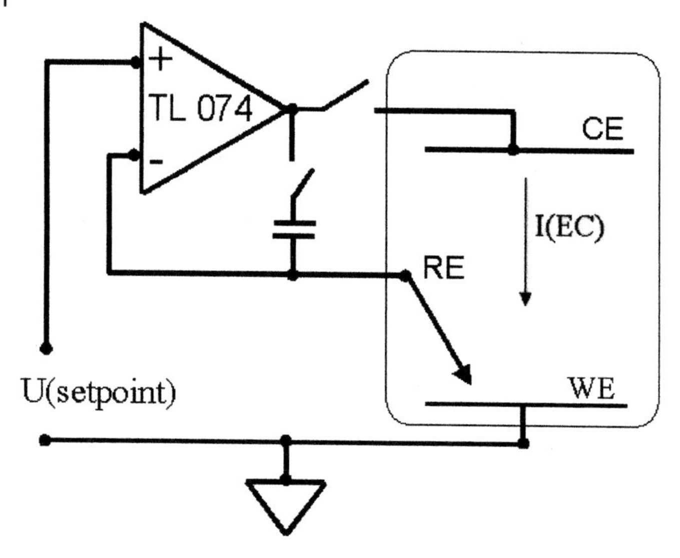

Figure 6.2-1: Simplified circuit of a potentiostat with working electrode (WE) on ground. Reference electrode (RE) and potentiostatic setpoint are fed to the inverting and noninverting input of an operational amplifier. The counter-electrode (CE) is connected to the output of the operational amplifier. I(EC): electrochemical current.

The reference electrode (RE) is connected to the inverting input of an operational amplifier (for example: Texas Instruments TL 074), and the setpoint is applied between ground and the noninverting input of the operational amplifier. For electronic reasons Equation 6.2-1 applies.

$$U(CE) = F * [\, U(RE) - U(setpoint) \,] \qquad (6.2\text{-}1)$$

F: amplification factor

As F has typical values from ca. 10^6–10^7, it follows U(RE) = U(setpoint).

Good electrode contacts are required because of the high amplification factor, as any fluctuation would result in strong oscillations at the output of the OPA. To prevent such problems, a capacitor of roughly 1 μF can be inserted between the reference electrode and the counter-electrode, thus damping such oscillations. U(setpoint) can be a constant voltage or any externally generated signal. In cyclic voltammetry, for example, a linearly varying potential is applied between an upper and a lower limit. If electrode reactions occur in the applied potential range, a current flows and can be plotted against the electrode potential, thus giving Cyclic Voltammograms (CVs). The currents are limited by kinetics or transport limitation, depending on the system, so that peak currents are observed, and these can be evaluated to provide insight into the electrochemical processes [3].

6.2.1.3 Electrodeposition of less noble elements

Aluminium electrodeposition Al electrodeposition from chloroaluminate ionic liquids has been investigated by several authors on different substrates by classical electrochemical methods, such as cyclic voltammetry, potential step experiments, and *ex situ* techniques [4–7]. In all cases, Al deposition was only observed in the acidic regime and the quality of the deposits was reported to be superior to those obtained from organic solutions. Deposition on substrates such as glassy carbon, tungsten, and platinum is preceded by a nucleation step and is electrochemically quasireversible (i.e., it is not solely diffusion-controlled; the charge-transfer reaction also plays an important role). On Pt, however, there are some hints from the published electrochemical data for underpotential phenomena. The bulk deposits of Al are rather granular and the current density has an influence on the size of the clusters, with a tendency to smaller crystals with higher current densities. If dry toluene [4] or benzene [8] are added to the liquid, mirror-bright deposits have been reported. It is likely that the organic molecules play the role of brighteners. Such effects have been known for a long time, and organic molecules such as crystal violet are widely used in aqueous electroplating processes to deposit shining layers of Cu, Ag, etc. [9]. The miscibility of the chloroaluminates with toluene, xylene, and other organic solvents has the further advantage that the liquid can easily be washed away from the samples after the electrodeposition had been performed, so that clean substrates can be prepared as easily as from aqueous solutions.

Although these chloroaluminate-based liquids will most probably not replace high-temperature molten salts for Al electrowinning purposes, they could become important in electroplating of Al and several Al alloys. We recently succeeded, with the aid of special electrochemical techniques and special bath compositions, in preparing high quality deposits of nanocrystalline metals such as Al with grain sizes down to only several nanometers [10]. Such nanocrystalline deposits are interesting as coatings for corrosion protection, for example.

Electrodeposition of less noble elements and aluminium alloys In technical processes, elements such as the alkali, the alkaline earth, the refractory, or the rare earth metals are obtained by high-temperature molten salt electrolysis [11, 12]. Eutectics of alkali halides are used as solvents in many cases, while in some cases – La and Ce, for example – the metal halides can be electrolyzed directly. The temperatures vary from about 450 °C to more than 1000 °C. On the one hand, these are pretty difficult experimental conditions: on the other hand, a high electronic conductivity is observed in many cases as soon as metal is deposited. Na dissolves easily in liquid NaCl, for example, and a nonmetal/metal transition is observed with rising Na content [13]. As a consequence, the current efficiency during electrolysis can reduce enormously due to a partial electronic short-circuit. Consequently, it would be interesting to apply low-melting ionic liquids for electrowinning or electroplating of these elements. To date, only a few examples have been deposited in elemental form from low-melting ionic liquids. In most cases the chloroaluminate systems were employed, and some aluminium alloys with interesting properties have been reported in the literature.

Sodium and Lithium

Both sodium [14] and lithium [15] electrodeposition have been achieved in neutral chloroaluminate ionic liquids containing protons. These elements are interesting for Na- or Li-based secondary batteries in which the metals would serve directly as the anode material. The electrodeposition is not possible in basic or acidic chloroaluminate ionic liquids; only proton-rich NaCl- or LiCl-buffered neutral chloroaluminate liquids are feasible for the electrodeposition. The protons enlarge the electrochemical window towards the cathodic regime so that the alkali metal electrodeposition becomes possible. For Na, the proton source was dissolved HCl, introduced either in the gas phase or as [EMIM][HCl_2]. For Li electrodeposition, triethanolamine hydrogen dichloride was also employed as proton source. Reversible deposition and stripping was reported for both alkali metals, on tungsten and stainless steel substrates, respectively.

Gallium

Elemental gallium can be electrodeposited both from chloroaluminate [16] and from chlorogallate [17] ionic liquids. In the latter case, [EMIM]Cl was mixed with $GaCl_3$, thus giving an ionic liquid that was studied for GaAs thin film electrodeposition. In the chloroaluminates, Ga can be deposited from Lewis acidic systems. It was found that the electroreduction from Ga(III) first gives rise to Ga(I), and the elemental Ga then forms from Ga(I) upon further reduction. On glassy carbon the electrodeposition involves instantaneous three-dimensional nucleation with diffusion-controlled growth of the nuclei. No alloying with Al was reported if deposition of Ga was performed in the Ga(I) diffusion regime. Reproducible electrodeposition of Ga is a promising route for production of binary and ternary compound semiconductors. A controlled electrodeposition of GaX quantum dots (X = P, As, Sb) would be very attractive for nanotechnology.

Iron

The electrodeposition of iron was investigated in neutral and acidic chloroaluminates [18, 19]. Although iron can be deposited from certain aqueous solutions, ionic liquids offer the advantage of depositing it in high quality in elemental form without side reactions such as hydrogen evolution or oxidation by water if the potential control is switched off. This is an interesting feature, especially for nanotechnology, as iron is a magnetic material. It has been reported that elemental Fe can be reversibly deposited on several substrates, including tungsten or glassy carbon in acidic chloroaluminate liquids, although the electrode potential for its electrodeposition is very close to the Al deposition potential. The reduction can be performed, prior to the deposition of the element, either from $FeCl_3$ that is reduced to $FeCl_2$, or directly from $FeCl_2$. The fact that the electrode potentials for Al and Fe deposition are close together makes it possible to deposit Fe-Al alloys with interesting properties.

Aluminium alloys with iron, cobalt, nickel, copper, and silver

The bulk deposition of alloys from Al with Fe, Co, Ni, Cu, and Ag was recently investigated with electrochemical and *ex situ* analytical techniques [20]. The alloys were prepared under near steady-state, diffusion-controlled conditions. For $CoAl_x$, $FeAl_x$, and $CuAl_x$, compositions with $x \approx 1$ were obtained, while kinetic phenomena complicated a reliable analysis for $NiAl_x$. In the case of $AgAl_x$, the authors reported that analysis was precluded by a dendritic growth of the deposits. All of the alloy systems displayed complex electrodissolution, and the nature of the oxidation process was different for the alloys produced in specific potential regimes. However, one has to keep in mind that classical electrochemistry and *ex situ* analysis give mainly integral information on the deposits. Nanometer resolution in *ex situ* methods is not yet a straightforward procedure. Nevertheless, although the alloy deposition is obviously complicated, the results are quite interesting for the electrodeposition of thin alloy films, as alloys of Al with Fe, Ni, or Co could perhaps give magnetic nanostructures more stable than the respective elements.

Aluminium alloys with niobium and tantalum

Nb and Ta can be obtained in elemental form from high-temperature molten salts. Nb and Ta are widely used as coatings for corrosion protection, since they – like Al – form thin oxide layers that protect the underlying material from being attacked. In technical processes, several high-temperature molten salts are employed for electrocoating, and the morphology of the deposit is strongly influenced by the composition of the baths. Some attempts have been made to deposit Nb and Ta from ionic liquids [21, 22]. In [21] the authors focused on the electrodeposition of $AlNb_x$ alloys from room-temperature ionic liquids containing both $AlCl_3$ and chlorides of Nb. The authors reported that they obtained Nb contents of up to 29 wt-% in the deposits, at temperatures between 90 and 140 °C. In [22], chloroaluminate liquids were employed at room temperature and $AlNb_x$ films could only be obtained if $NbCl_5$ was prereduced in a chemical reaction. The authors reported that Nb powder is the most effective reducing agent for this purpose. Similar preliminary results have been obtained for Ta electrodeposition. Although it seems to be difficult to deposit pure Nb and Ta in low-melting ionic liquids, the alloys with Al could have quite interesting properties.

Aluminium alloys with titanium

Titanium is an interesting material for corrosion protection and lightweight construction. Hitherto it could only be deposited in high quality from high-temperature molten salts, although attempts have been made to deposit it from organic solutions and even aqueous media. In general, bulk electrodeposition of Ti is complicated because traces of water immediately form passivating oxide layers on Ti. Deposition of the element has also not yet been successful in ionic liquids. In a recent article [23], however, it was reported that $AlTi_x$ alloys can be obtained from chloroaluminates. The corrosion resistance of the layers is reported to be superior to that of Al itself and seems to become even better with increasing Ti content. However, Ti was not deposited in elemental form without codeposition of Al. In chloroaluminates,

Al is more noble than Ti, and so at room temperature only codeposits and alloys can be obtained. Furthermore, kinetic factors also play a role in the electrodeposition of the element.

Aluminium alloys with chromium

The electrodeposition of Cr in acidic chloroaluminates was investigated in [24]. The authors report that the Cr content in the $AlCr_x$ deposit can vary from 0 to 94 mol %, depending on the deposition parameters. The deposit consists both of Cr-rich and Al-rich solid solutions as well as intermetallic compounds. An interesting feature of these deposits is their high-temperature oxidation resistance, the layers seeming to withstand temperatures of up to 800 °C, so coatings with such an alloy could have interesting applications.

Lanthanum and aluminium-lanthanum alloys

It was quite recently reported that La can be electrodeposited from chloroaluminate ionic liquids [25]. Whereas only $AlLa_x$ alloys can be obtained from the pure liquid, the addition of excess LiCl and small quantities of thionyl chloride ($SOCl_2$) to a $LaCl_3$-saturated melt allows the deposition of elemental La, but the electrodissolution seems to be somewhat kinetically hindered. This result could perhaps be interesting for coating purposes, as elemental La can normally only be deposited in high-temperature molten salts, which require much more difficult experimental or technical conditions. Furthermore, La and Ce electrodeposition would be important, as their oxides have interesting catalytic activity as, for instance, oxidation catalysts. A controlled deposition of thin metal layers followed by selective oxidation could perhaps produce catalytically active thin layers interesting for fuel cells or waste gas treatment.

6.2.1.4 Electrodeposition of metals that can also be obtained from water

As already mentioned above, most of the metals that can be deposited from aqueous solutions can also be obtained from ionic liquids. One could reasonably raise the question of whether this makes sense, as aqueous solutions are much easier to handle. However, there are two properties of the ionic liquids that are superior to those of aqueous solutions. Firstly, their electrochemical windows are much wider, so that side reactions during electrodeposition can easily be prevented. Whereas palladium, for example, can give brittle deposits in aqueous media due to hydrogen evolution and dissolution of hydrogen in the metal, shining, nanosized deposits can be obtained in ionic liquids [10]. Secondly, the temperature can be varied over a wide range, in some cases more than 400 °C. In general, variation of the temperature has a strong effect on the kinetics of the deposition and on the surface, as well as on the interface mobility of the deposits. Although there are no systematic studies on temperature variation upon electrodeposition in ionic liquids, this is an attractive research field, as there would be a certain link to the classical high-temperature molten salts.

Indium and antimony The electrodeposition of In on glassy carbon, tungsten, and nickel has been reported [26]. In basic chloroaluminates, elemental indium is formed in one three-electron reduction step from the $[InCl_5]^{2-}$ complex, but In(I) species have also been reported [27]. The overpotential deposition involves progressive three-dimensional nucleation on a finite number of active sites on carbon and tungsten, while on nickel progressive three-dimensional nucleation is observed. Electrodeposition of In from acidic melts is reported not to occur, but liquids based on $InCl_3$ and organic salts were successfully used to deposit InSb [27, 28]. Sb electrodeposition on tungsten, platinum, and glassy carbon has been reported by Osteryoung et al. [29, 30]. The metal can be deposited from acidic melts, but partly irreversible behavior is observed. Unpublished data by C. Hussey [31] confirm this interesting behavior. Both passivation phenomena and reactions of the deposits with the metal substrates could play a role, and so *in situ* STM studies would be of great interest in order to elucidate the processes at the electrode surface. If the electrode processes were known in detail, definite InSb layers or nanosized InSb quantum dots could perhaps be made by simultaneous electrodeposition (see below). InSb is a direct semiconductor, and quantum dots of InSb, made under ultra-high vacuum conditions, have already been successfully studied for laser applications [32]. Quantum dots are widely under investigation nowadays and this is a rapidly growing research field. Definite electrodeposition from ionic liquids would be an important contribution.

Tellurium and cadmium Electrodeposition of Te has been reported [33]: in basic chloroaluminates the element is formed from the $[TeCl_6]^{2-}$ complex in one four-electron reduction step. Furthermore, metallic Te can be reduced to Te^{2-} species. Electrodeposition of the element on glassy carbon involves three-dimensional nucleation. A systematic study of the electrodeposition in different ionic liquids would be of interest because – as with InSb – a defined codeposition with cadmium could produce the direct semiconductor CdTe. Although this semiconductor can be deposited from aqueous solutions in a layer-by-layer process [34], variation of the temperature over a wide range would be interesting since the grain sizes and the kinetics of the reaction would be influenced.

Electrodeposition of Cd has also been reported [35, 36]. In [35], $CdCl_2$ was used to buffer neutral chloroaluminate liquids from which the element could be deposited. In an interesting recent work [36], a $[EMIM][BF_4]$ ionic liquid with added [EMIM]Cl was successfully used to deposit Cd. It is formed on platinum, tungsten, and glassy carbon from $CdCl_4{}^{2-}$ in a quasireversible two-electron reduction process. This result is promising, as Te might perhaps also be deposited from such an ionic liquid, thus possibly giving a system for direct CdTe electrodeposition.

Copper and silver Electrodeposition of Cu in the chloroaluminate liquids has been widely investigated. It has been reported to be deposited from acidic liquids only, and it also shows some interesting deviations from its behavior in aqueous solutions. If $CuCl_2$ is added to an acidic liquid, Cu(II) undergoes two one-electron reduction steps on glassy carbon and on tungsten [37, 38]; in the first step Cu(I) is formed, in the second step the metal deposits. At high overvoltages for the deposition, an alloying with Al begins [39]. The electrodeposition of Cu from a $[BF_4]^-$ liquid has been investigated [40] on polycrystalline tungsten, on platinum, and on glassy carbon. On Pt, UPD phenomena were reported, whereas on tungsten and glassy carbon only OPD was apparent. *Ex situ* analysis proved that the deposit was composed solely of copper.

The electrodeposition of Ag has also been intensively investigated [41–43]. In the chloroaluminates – as in the case of Cu – it is only deposited from acidic solutions. The deposition occurs in one step from Ag(I). On glassy carbon and tungsten, three-dimensional nucleation was reported [41]. Quite recently it was reported that Ag can also be deposited in a one-electron step from tetrafluoroborate ionic liquids [43]. However, the charge-transfer reaction seems to play an important role in this medium and the deposition is not as reversible as in the chloroaluminate systems.

Nickel and cobalt Nickel and cobalt have been intensively investigated in aqueous solutions. Both of these metals are interesting for nanotechnology, as magnetic nanostructures can be formed in aqueous solutions [44]. However, their bulk electrodeposition is accompanied by hydrogen evolution. Both elements can also be deposited from acidic chloroaluminate liquids [45, 46]. The main literature interest is devoted to alloys with aluminium, as such deposits also show magnetic behavior. Recent in situ STM studies have shown that on Au(111) in the underpotential regime, one Ni monolayer exhibiting an 8×8 Moiré superstructure is formed. Furthermore, island growth along the steps starts in the UPD regime [47]. In situ scanning tunneling spectroscopy has shown that the tunneling barrier is significantly reduced on going from Ni to Ni Al_x clusters [48].

Palladium and gold Palladium electrodeposition is of special interest for catalysis and for nanotechnology. It has been reported [49] that it can be deposited from basic chloroaluminate liquids, while in the acidic regime the low solubility of $PdCl_2$ and passivation phenomena complicate the deposition. In our experience, however, thick Pd layers are difficult to obtain from basic chloroaluminates. With different melt compositions and special electrochemical techniques at temperatures up to 100 °C we succeeded in depositing mirror-bright and thick nanocrystalline palladium coatings [10].

Gold electrodeposition has been reported [50, 51] from chloroaluminate-based liquids and from a liquid made of an organic salt and $AuCl_3$. Although high quality gold can be electrodeposited from aqueous solutions, the latter result is especially interesting with respect to the deposition of unusual alloys between gold and less noble elements.

Zinc and tin The electrodeposition of Zn [52] has been investigated in acidic chloroaluminate liquids on gold, platinum, tungsten, and glassy carbon. On glassy carbon only three-dimensional bulk deposition was observed, due to the metal's underpotential deposition behavior. At higher overvoltages, codeposition with Al has been reported. As Zn is widely used in the automobile industry for corrosion protection, a codeposition with Al could also be interesting for selected applications. Tin has been electrodeposited from basic and acidic chloroaluminate liquids on platinum, gold, and glassy carbon [53]. On Au the deposition starts in the UPD regime and, from the electrochemical data, one monolayer was reported. Furthermore there seems to be some evidence for alloying between Sn and Au. On glassy carbon three-dimensional growth of Sn occurs.

6.2.1.5 Electrodeposition of semiconductors

Many studies on semiconductor electrodeposition have in the past been performed in different solutions, such as aqueous media, organic solutions, molten salts, and also a few ionic liquids. A good overview on the topic in general is presented in ref. [54]. To date, however, industrial procedures have not yet been established. In addition to bulk deposits of semiconductors for photovoltaic applications, thin layers or quantum dots would be of great interest both in basic research and in nanotechnology. Hitherto, most basic studies on semiconductor formation and characterization have been performed under UHV conditions. Molecular Beam Epitaxy is a widely used method for such purposes. In technical processes, Chemical or Physical Vapor Deposition are still the methods of the greatest importance. Although high quality deposits can be obtained, such processes are cost-intensive and the layers are consequently expensive. A simple and cheaper electrodeposition would surely be of commercial interest. Work by Stickney [34] has shown that Electrochemical Atomic Layer Epitaxy (ECALE) in aqueous media is a suitable deposition method for compound semiconductors with qualities comparable to those made by vacuum techniques. In special electrochemical polarization routines the elements of a compound semiconductor are successively deposited one onto the other, layer-by-layer. Unfortunately, direct electrodeposition of CdTe, CdSe, and others is difficult for kinetic reasons, and in many cases the elements are codeposited together with the desired semiconductor in varying amounts at room temperature. Variation of the temperature can strongly affect the quality of the electrodeposits [55]. In general, direct deposition of a compound semiconductor would be interesting as it would be less time-consuming than the elegant ECALE process. Although there are only a few articles on semiconductor electrodeposition from ionic liquids, these media are interesting for such studies for several reasons: the acidity can be varied over wide ranges, they have low vapor pressures, and as a consequence – depending on the system – the temperature can be varied over several hundred degrees, so that kinetic barriers in compound formation can be overcome. Furthermore, because of the wide electrochemical windows, it is possible to obtain compounds that are inaccessible from aqueous solutions, one example being GaAs. For GaSb, InSb, InP, and ternary compound semiconductors, electrodeposition from ionic liquids could be interesting, especially if higher temperatures were applied. As well as the compound semiconductors, elemental semiconductors can also be obtained from ionic liquids. Si and Ge are widely used as wafer material for different electronic applications, and junctions of n- and p-doped Si are still interesting for photovoltaic applications. Controlled electrodeposition of both elements and their mixtures would also surely be attractive for nanotechnology, as Ge quantum dots made under UHV conditions show interesting photoluminescence.

GaAs The direct electrodeposition of GaAs from ionic liquids has been studied mainly by two groups. Wicelinski et al. [56] used an acidic chloroaluminate liquid at 35–40 °C to codeposit Ga and As. However, it was reported that Al underpotential deposition on Ga occurs. Verbrugge and Carpenter employed an ionic liquid based on $GaCl_3$ to which $AsCl_3$ had been added [17, 57]. Unfortunately, the quality of the deposits in these studies was not convincing, and both pure arsenic and gallium could be found in the deposits. Nevertheless, this route is promising for the electrodeposition of Ga-based semiconductors, as thermal annealing could improve the quality of the deposits.

InSb The principal of InSb electrodeposition is the same as for GaAs. An ionic liquid based on $InCl_3$ is formed, to which $SbCl_3$ is added [27, 28]. At 45 °C, InSb can be directly electrodeposited, but elemental In and Sb are also reported to co-deposit. The In/Sb ratio depends strongly on the deposition potential. Despite some problems, the authors of these studies are optimistic that ionic liquids based on $GaCl_3$ and $InCl_3$ may also be useful for depositing ternary compound semiconductors such as AlGaAs and InGaSb.

ZnTe The electrodeposition of ZnTe was published quite recently [58]. The authors prepared a liquid that contained $ZnCl_2$ and [EMIM]Cl in a molar ratio of 40:60. Propylene carbonate was used as a co-solvent, to provide melting points near room temperature, and 8-quinolinol was added to shift the reduction potential for Te to more negative values. Under certain potentiostatic conditions, stoichiometric deposition could be obtained. After thermal annealing, the band gap was determined by absorption spectroscopy to be 2.3 eV, in excellent agreement with ZnTe made by other methods. This study convincingly demonstrated that wide band gap semiconductors can be made from ionic liquids.

Germanium In situ STM studies on Ge electrodeposition on gold from an ionic liquid have quite recently been started at our institute [59, 60]. In these studies we used dry [BMIM][PF_6] as a solvent and dissolved GeI_4 at estimated concentrations of 0.1–1 mmol l^{-1}, the substrate being Au(111). This ionic liquid has, in its dry state, an electrochemical window of a little more than 4 V on gold, and the bulk deposition of Ge started several hundreds of mV positive from the solvent decomposition. Furthermore, distinct underpotential phenomena were observed. Some insight into the nanoscale processes at the electrode surface is given in Section 6.2.2.3.

6.2.2 Nanoscale Processes at the Electrode/Ionic Liquid Interface

6.2.2.1 General considerations

In situ STM studies on electrochemical phase formation from ionic liquids were started in the author's group five years ago. On the one hand there was no knowledge of the local processes of phase formation in ionic liquids and molten salts. On the other hand – thanks to their wide electrochemical windows – these systems offered access to elements that cannot be obtained from aqueous solutions. In the rapidly growing field of nanotechnology, in which semiconductor nanostructures will play an important role, we see great opportunity for electrodeposition of nanostructures from ionic liquids. It is known that germanium quantum dots on silicon made by Molecular Beam Epitaxy under UHV conditions display an interesting photoluminescence around 1 eV [61]. Furthermore, lasers based on compound semiconductor quantum dots such as InSb have been discussed in the literature [32]. Although UHV conditions are straightforward in basic research, a possible nanotechnological process would be relatively complicated and presumably expensive. An electrochemical routine would therefore be preferred if comparable results could be obtained. For this purpose, the electrochemical processes and other factors influencing the deposition and the stability of the structures have to be understood on the nanometer scale.

6.2.2.2 The scanning tunneling microscope

The main technique employed for *in situ* electrochemical studies on the nanometer scale is the Scanning Tunneling Microscope (STM), invented in 1982 by Binnig and Rohrer [62] and combined a little later with a potentiostat to allow electrochemical experiments [63]. The principle of its operation is remarkably simple, a typical simplified circuit being shown in Figure 6.2-2.

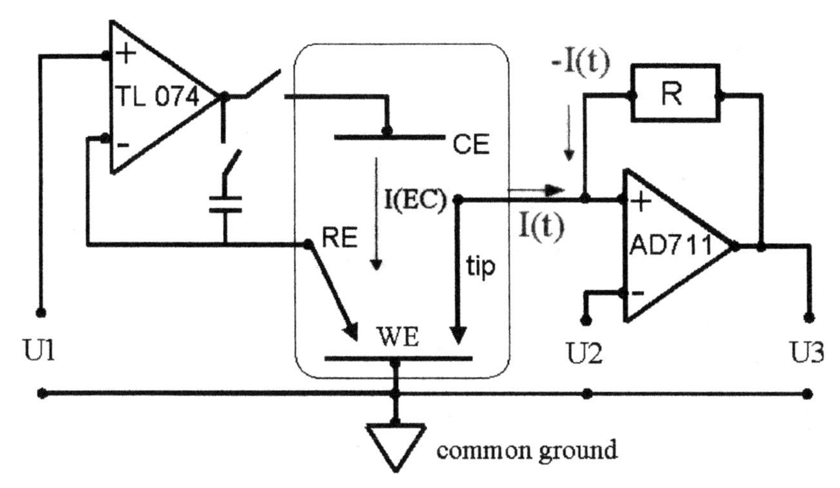

Figure 6.2-2: Simplified circuit of an electrochemical STM setup. In addition to the potentiostat (see Figure 6.2.1), an STM preamplifier is added, to which the tip is connected. U1: potentiostatic setpoint, U2: tunneling voltage, I(t): tunneling current, U3 = –R I(t).

The left side is essentially identical to the potentiostat circuit presented in Section 6.2.1.2.. The right side is the preamplifier of the STM. The atomically sharp metal tip is located roughly 1 nm over an electronically conductive substrate, here the working electrode (WE) of interest. This is done by computer control, with the help of step motors and micrometer screws as well as piezoelectric elements, to which the tip is connected. If a potential U2 is applied between tip and sample (typically 5–500 mV) a tunneling current I(t) flows, with typical values between 0.1 and 10 nA, depending on the distance. This current is transformed into a voltage U3 (= –R*I(t)) that can be further processed. The tunneling current is strongly dependent on the distance (d) and is a function of the electronic density of states of tip D (tip) and sample D (sample). For a strongly simplified case one obtains to a first approximation Equation 6.2-2 [64]:

$$I(\text{tunnel}) = f(U\ \text{bias}) * D(\text{tip}) * D(\text{sample}) * \exp[-\ \text{const.} * d] \qquad (6.2\text{-}2)$$

Because of this strong distance dependence, local height changes can in principle be detected in the picometer range. There are two modes of operation. In the "Constant Height Mode" the tip is scanned over the surface at a constant height and the local changes of the tunneling current are acquired. In the "Constant Current Mode" the distance between the tip and the sample is kept constant by feedback electronics. It works such that U3 is amplified and finally fed back through an adding amplifier to the piezo control. It is also clear that the STM tip acts as an electrode in the electrochemical cell. As soon as a voltage is applied to the tip, a current can flow. Such Faradaic currents (the deposition of metal, or hydrogen or oxygen evolution, for example) can easily reach some hundreds of nanoamperes. Macroscopically this is negligible but, as the tunneling currents are only some nanoamperes, the tip has to be insulated – with the exception of its very end – by a paint or by glass. Hence, the Faradaic currents can be reduced down to the picoampere range, making stable tunneling conditions under electrochemical conditions possible.

6.2.2.3 **Results**

Aluminium electrodeposition on Au(111) The processes during electrodeposition of aluminium have been investigated on the nanometer scale [65]. As already pointed out in preceding sections, Al is an important metal for various applications in technical processes. In order to obtain insight into the growth of the metal and to understand the initial stages of the phase formation, *in situ* STM experiments were performed under electrochemical conditions during electroreduction of $AlCl_3$ in an acidic [BMIM]Cl/$AlCl_3$ ionic liquid. This liquid is extremely corrosive and we had to build our own STM heads to allow measurements under inert gas conditions. The cyclic voltammogram on Au(111) shows several UPD processes and one OPD process, as can be seen in Figure 6.2-3.

Gold oxidation starts at electrode potentials > +1200 mV vs. Al/$AlCl_3$, first at the steps between different terraces. At higher potentials pits are formed, rapidly resulting in complete disintegration of the substrate.

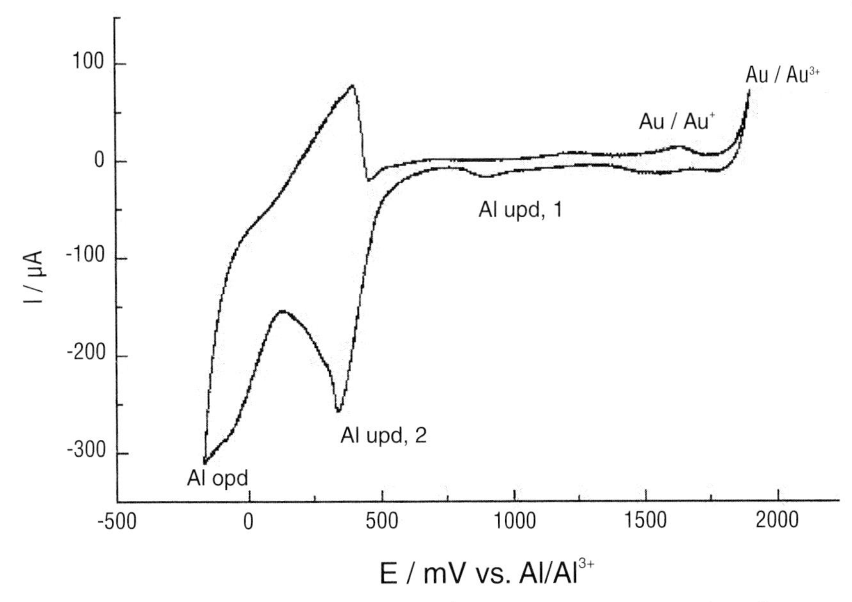

Figure 6.2-3: Cyclic voltammogram of acid $[BMIM]^+Cl^-/AlCl_3$ on Au(111): Au oxidation begins at electrode potentials > +1.2 V vs. $Al/AlCl_3$. UPD processes are observed at about +900 and +400 mV before bulk deposition of Al starts (see also [65]).

The following electrode processes at the surfaces have been identified during electrodeposition. Upon a potential step from E > +1000 mV to E = +950 mV vs. $Al/AlCl_3$, two-dimensional islands formed irreversibly on the surface. Their height was 250 ± 20 pm, indicative of gold islands. We attributed this observation to the formation of Au–Al compounds followed by expulsion of surplus Au atoms to the surface. At an electrode potential of +400 mV, small Al islands with an averaged height of 230 ± 20 pm started growing (Figure 6.2-4).

At +100 mV vs. $Al/AlCl_3$, clusters of up to 1 nm in height formed. When a potential step to +1100 mV vs. $Al/AlCl_3$ was performed, the clusters dissolved immediately, but both holes and gold islands of up to two monolayers in height remained on the surface. It is likely that strong alloying between Au and Al took place both in the surface and in the deposited clusters (Figure 6.2-5).

In the overpotential deposition regime we observed that nanosized Al was deposited in the initial stages. Furthermore, a transfer of Al from the scanning tip to the Al covered substrate was observed. We accidentally succeeded in an indirect tip-induced nanostructuring of Al on growing Al (Figure 6.2-6).

These results are quite interesting. The initial stages of Al deposition result in nanosized deposits. Indeed, from the STM studies we recently succeeded in making bulk deposits of nanosized Al with special bath compositions and special electrochemical techniques [10]. Moreover, the preliminary results on tip-induced nanostructuring show that nanosized modifications of electrodes by less noble elements are possible in ionic liquids, thus opening access to new structures that cannot be made in aqueous media.

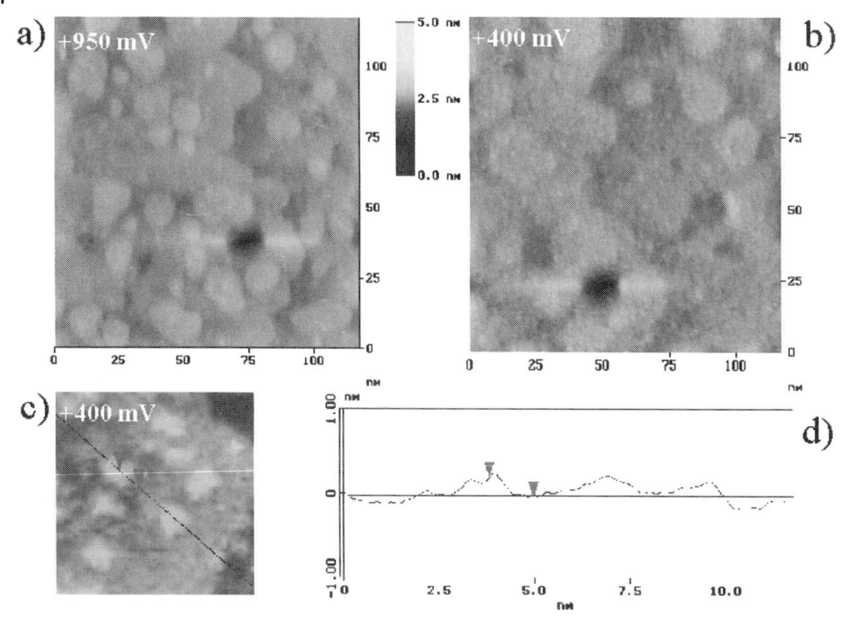

Figure 6.2-4

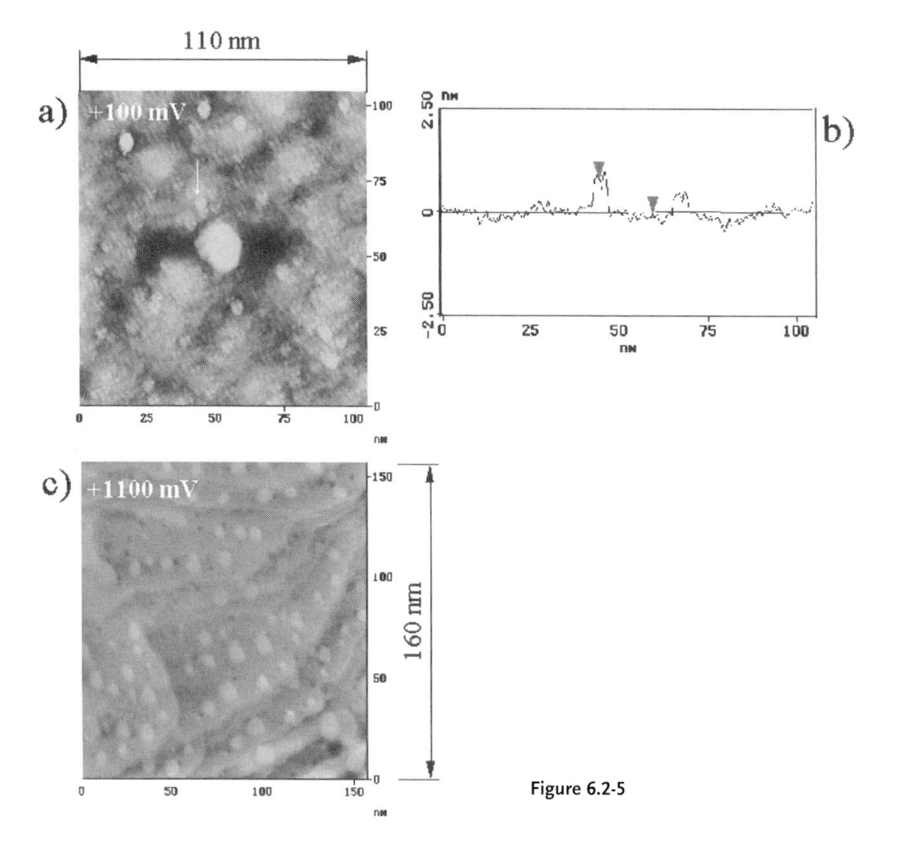

Figure 6.2-5

Figure 6.2-4: Underpotential phenomena during Al reduction in acidic $[BMIM]^+Cl^-/AlCl_3$ on Au(111). At +950 mV vs. $Al/AlCl_3$, two-dimensional islands with a height of 250 ± 20 pm form (a). At +400 mV (b), two-dimensional Al islands with an averaged height of 230 ± 20 pm are reversibly deposited. In (c) the islands are shown with a higher resolution, while (d) shows a typical height profile.

Figure 6.2-5: Underpotential phenomena during Al reduction in acidic $[BMIM]^+Cl^-/AlCl_3$ on Au(111): nanoclusters with heights of up to 1 nm form at +100 mV vs. $Al/AlCl_3$ (a); a typical height profile is shown in (b). Upon a potential step to +1100 mV vs. $Al/AlCl_3$ the clusters dissolve immediately and leave holes in the surfaces as well as small Au islands (c): alloying between Al and Au is very likely.

Copper electrodeposition on Au(111) Copper is an interesting metal and has been widely investigated in electrodeposition studies from aqueous solutions. There are numerous publications in the literature on this topic. Furthermore, technical processes to produce Cu interconnects on microchips have been established in aqueous solutions. In general, the quality of the deposits is strongly influenced by the bath composition. On the nanometer scale, one finds different superstructures in the underpotential deposition regime if different counter-ions are used in the solutions. A co-adsorption between the metal atoms and the anions has been reported. In the underpotential regime, before the bulk deposition begins, one Cu monolayer forms on Au(111) [66].

In situ STM experiments in an acidic $[BMIM]Cl/AlCl_3$ ionic liquid with either CuCl or $CuCl_2$ as copper sources have recently been carried out in the author's group [67]. The motivation was based mainly on two facts: copper in chloroaluminates is deposited from Cu(I), and Cu^+ can furthermore be regarded as a naked cation because there is no distinct solvation shell, unlike in aqueous media. As a consequence, distinct deviations from the behavior in aqueous solutions were expected. The cyclic voltammogram on Au(111) displayed three UPD processes, followed by three-dimensional Cu growth in the OPD regime. At potentials > +1000 mV vs. Cu/Cu^+, gold oxidation starts first at the steps, at higher electrode potentials bulk oxidation of gold begins (Figure 6.2-7).

The following two pictures (Figure 6.2-8a and b) were acquired at +500 mV and at +450 mV vs. Cu/Cu^+ and show that at +450 mV vs. Cu/Cu^+ monolayer high Cu clusters nucleate at the steps between different Au terraces. Thus, the pair of shoulders in the cyclic voltammogram is correlated with this surface process.

If the electrode potential is further reduced to +350 mV, a hexagonal superstructure with a periodicity of 2.4 ± 0.2 nm is observed. With respect to the interatomic distances in the Au(111) structure at the surface, this corresponds – within the error limits – to an 8 × 8 superstructure (Figure 6.2-9).

The integrated charge would correspond to 0.7 ± 0.1 Cu monolayers. Thus, either a less closely packed Cu layer or an anion co-adsorption that can both lead to a Moiré superstructure are probed; in the solution investigated $[Al_2Cl_7]^-$ is the predominant anion. At +200 mV vs. Cu/Cu^+ the superstructure disappears and a completely closed Cu monolayer is observed, with a charge corresponding to 1.0 ± 0.1 Cu monolayers.

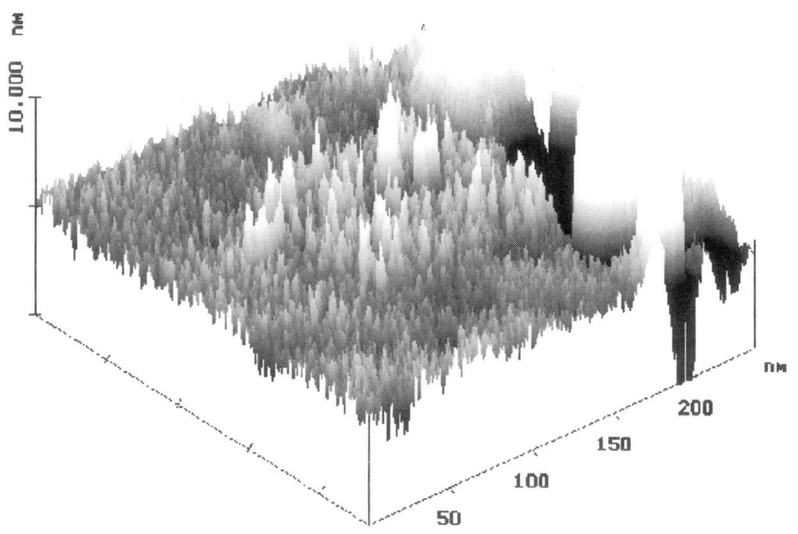

Figure 6.2-6: The initial stages of Al overpotential deposition result in nanosized deposits. A jump to contact transfer of Al from the scanning tip to the growing Al was observed (picture from [65] – with permission of the Pccp owner societes).

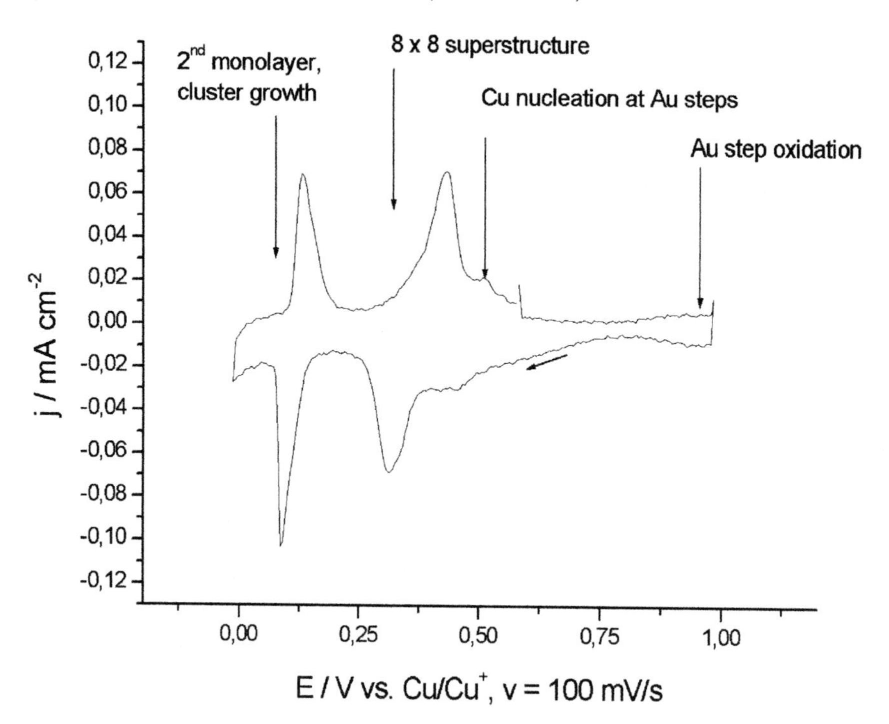

Figure 6.2-7: Cyclic voltammogram of CuCl in acidic $[BMIM]^{+}Cl^{-}/AlCl_3$ on Au(111): three UPD processes are observed, correlated with decoration of Au steps by copper, formation of an 8×8 superstructure followed by a Cu monolayer. Before the bulk deposition a second monolayer grows together with clusters

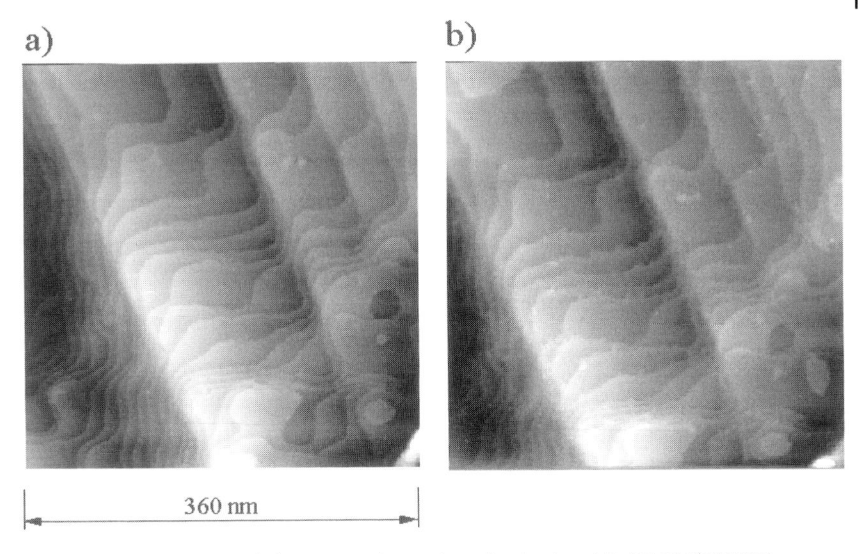

Figure 6.2-8: Underpotential phenomena during Cu reduction in acidic $[BMIM]^+Cl^-/AlCl_3$ on Au(111): a potential step from +500 mV vs. Cu/Cu^+ (a) to +450 mV results in the growth of small Cu islands at the steps of the gold terraces (b) (picture from [66] – with permission of the Pccp owner societes).

Unlike the case in aqueous solutions, the growth of a second 200 ± 20 pm high monolayer at +50 mV was observed, together with clusters of heights up to 1 nm (Figure 6.2-10).

This result is quite surprising, as no second Cu monolayer has yet been reported in aqueous solutions, nor have clusters up to 1 nm in height in the UPD regime. It

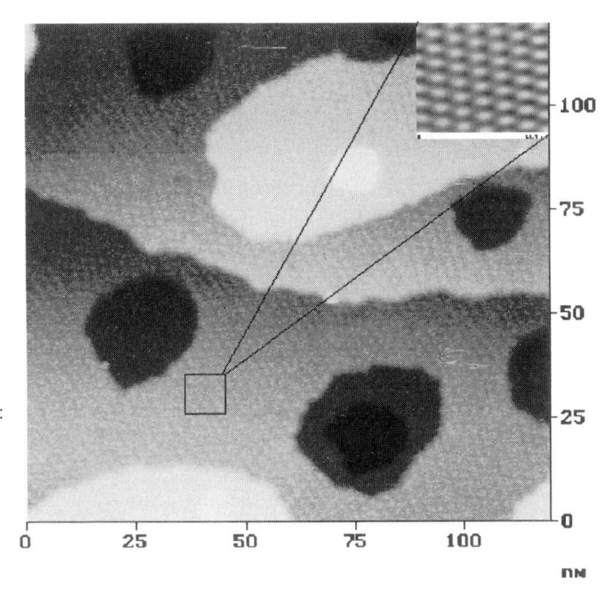

Figure 6.2-9: Underpotential phenomena during Cu reduction in acidic $[BMIM]^+Cl^-/AlCl_3$ on Au(111): at +350 mV an 8 × 8 superstructure is observed; the integrated charge would correspond to 0.7 ± 0.1 Cu monolayers (picture from [66] – with permission of the Pccp owner societes).

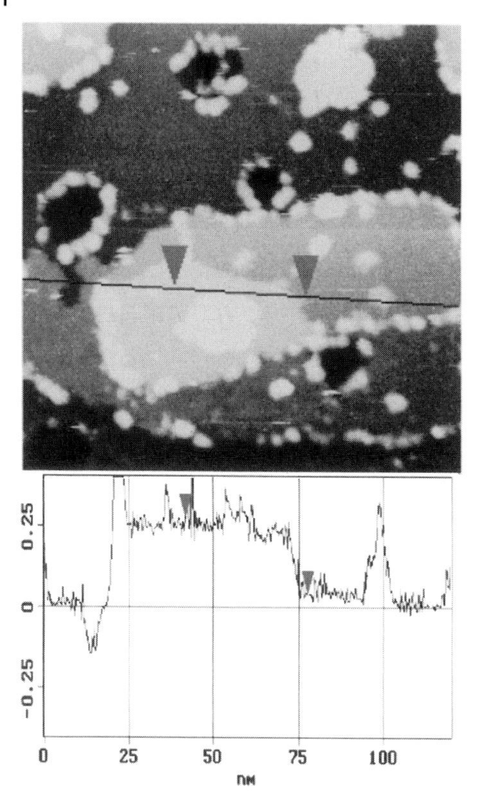

Figure 6.2-10: Underpotential phenomena during Cu reduction in acidic $[BMIM]^{+}Cl^{-}/AlCl_3$ on Au(111): at +50 mV a second monolayer with a height of 200 ± 20 pm grows, together with a pronounced deposition of clusters containing Cu and perhaps also a small amount of Al (picture from [66] – with permission of the Pccp owner societes).

cannot be excluded completely here that the clusters as well as the second monolayer could contain a small amount of Al. Such an underpotential alloying could stabilize the clusters enormously, especially as several stoichiometric Al-Cu compounds are known.

In the OPD regime, finally, the Cu bulk phase starts growing.

Germanium electrodeposition on Au(111) As a third example of *in situ* STM results, the electrodeposition of germanium should be mentioned here [59,60]. Germanium is an elemental semiconductor with a bandgap of 0.67 eV. In contrast to those of metals, furthermore, its crystal structure is determined by the tetrahedral symmetry of the Ge atoms, so that the diamond structure is thermodynamically the most stable. As the chemistry of Si and Ge are quite similar, such experiments could also give some insight into deposition process of the less noble Si. Germanium is hard to obtain in aqueous solutions, as its deposition potential is very close to that of hydrogen evolution. However, the ionic liquid $[BMIM][PF_6]$ (and others) can easily be prepared with water levels below 20 ppm and is therefore ideally suited for such electrodeposition studies. The pure liquid shows only capacitive behavior on Au(111), as can be seen in the cyclic voltammogram (Figure 6.2-11), acquired with a scan rate of 1 mV/s under inert gas conditions.

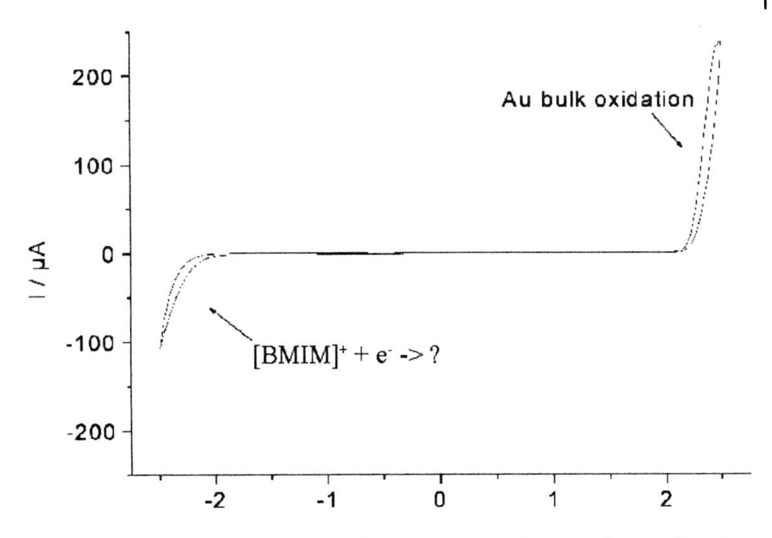

Figure 6.2-11: Cyclic voltammogram of dry $[BMIM]^+ PF_6^-$ on Au(111): between the anodic and the cathodic limits only capacitive currents flow: an electrochemical window of a little more than 4 V is obtained (picture from [59] – with permission of the Pccp owner societes).

If GeI_4 is added in an estimated concentration between 0.1 and 1 mmol dm^{-3}, several processes are observed in the cyclic voltammogram (Figure 6.2-12).

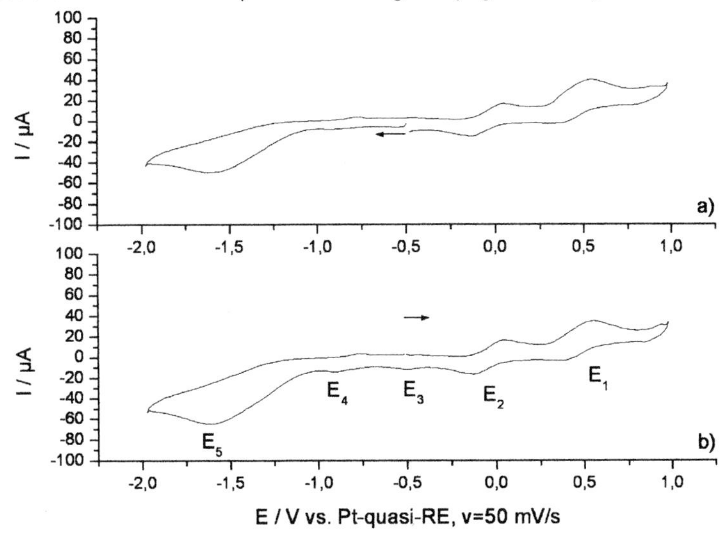

Figure 6.2-12: Cyclic voltammogram of 0.1 – 1 mmol dm^{-3} GeI_4 on gold in dry $[BMIM]^+PF_6^-$, starting at –500 mV towards cathodic (a) and anodic (b) regime. Two quasireversible (E_1 and E_2) and two apparently irreversible (E_4 and E_5) diffusion-controlled processes are observed. E_3 is correlated with the growth of two-dimensional islands on the surface, E_4 and E_5 with the electrodeposition of germanium, E_2 with gold step oxidation, and E_1 probably with the iodine/iodide couple. Surface area: 0.5 cm^2 (picture from [59] – with permission of the Pccp owner societes).

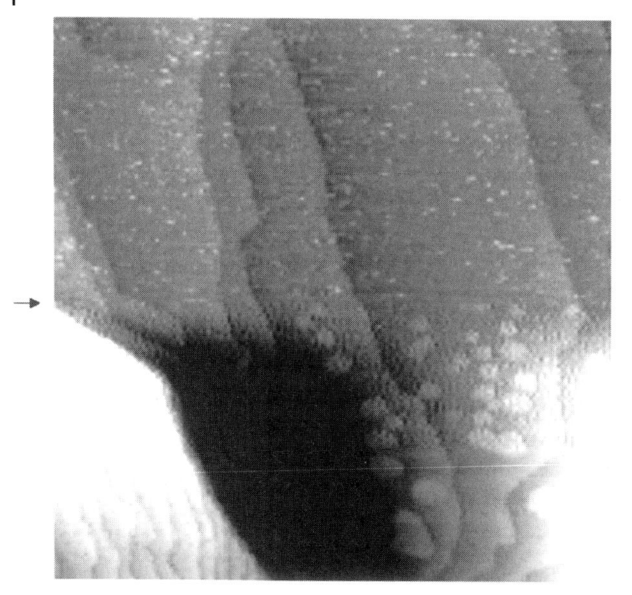

200 x 200 nm²

Figure 6.2-13: UPD phenomena of Ge on Au(111) in dry $[BMIM]^+$ PF_6^-: two-dimensional islands with an average height of 250 ± 20 pm start growing upon a potential step from the open circuit potential to –500 mV vs. the Pt quasi reference electrode (picture from [59] – with permission of the Pccp owner societes).

The electrode process at –500 mV on this potential scale is correlated to the growth of 250 ± 20 pm high islands. They grow immediately upon a potential step from the open circuit potential to –500 mV (arrow in Figure 6.2-13).

They form a monolayer that is rich in defects, but no second monolayer is observed. The interpretation of these results is not straightforward; from a chemical point of view both the electrodeposition of low-valent Ge_xI_y species and the formation of Au-Ge or even $Au_xGe_yI_z$ compounds are possible. A similar result is obtained if the electrodeposition is performed from $GeCl_4$. There, 250 ± 20 pm high islands are also observed on the electrode surface. They can be oxidized reversibly and disappear completely from the surface. With GeI_4 the oxidation is more complicated, because the electrode potential for the gold step oxidation is too close to that of the island electrodissolution, so that the two processes can hardly be distinguished. The gold step oxidation already occurs at +10 mV vs. the former open circuit potential, at +485 mV the oxidation of iodide to iodine starts.

In the reductive regime, a strong, apparently irreversible, reduction peak is observed, located at –1510 mV vs. the quasi reference electrode used in this system. With *in situ* STM, a certain influence of the tip on the electrodeposition process was observed. The tip was therefore retracted, the electrode potential was set to –2000 mV, and after two hours the tip was reapproached. The surface topography that we obtained is presented in Figure 6.2-14.

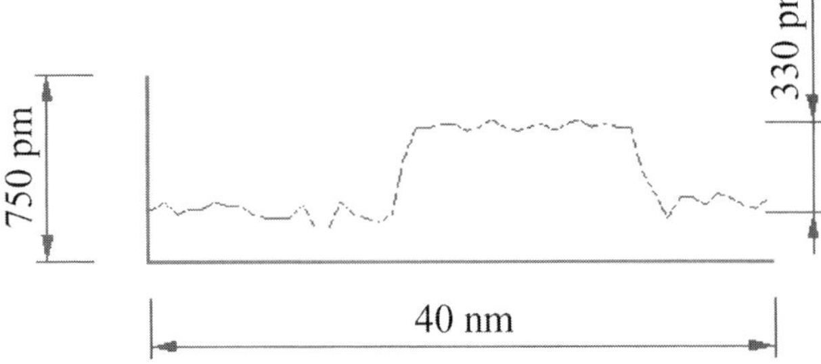

Figure 6.2-14: Bulk deposition of Ge on Au(111) in dry $[BMIM]^+$ PF_6^-: at –2000 mV terraces with an average height of 330 ± 30 pm, indicative of Ge(111) bilayers, are obtained.

The surface consists of terraces of a height of 330 ± 30 pm. Within error limits, this is the value that would be expected for Ge(111) bilayers. Furthermore, we were able to observe that the electrodeposition gave rise to a less ordered surface structure with nanoclusters, transforming over a timescale of about 1 hour into a layered structure. With $GeBr_4$ a transformation of clusters into such a layered surface was only partly seen; with $GeCl_4$ this transformation could not be observed.

The oxidation of the deposited germanium is also a complicated process; we found that mainly chemical oxidation by GeI_4 takes place, together with some electrooxidation. It is likely that kinetic factors play a dominant role.

If the germanium layers are partly oxidized by a short potential step to –1500 mV, random worm-like nanostructures form, healing in a complex process if the electrode potential is set back to more negative values (Figure 6.2-15).

As well as electrodissolution and electrodeposition, periphery and surface diffusion play important roles.

Unfortunately, only thin films of about 20 nanometers in thickness could be obtained with GeI_4. An *ex situ* analysis was difficult, because of experimental limitations, but XPS clearly showed that elemental Ge was also obtained, besides

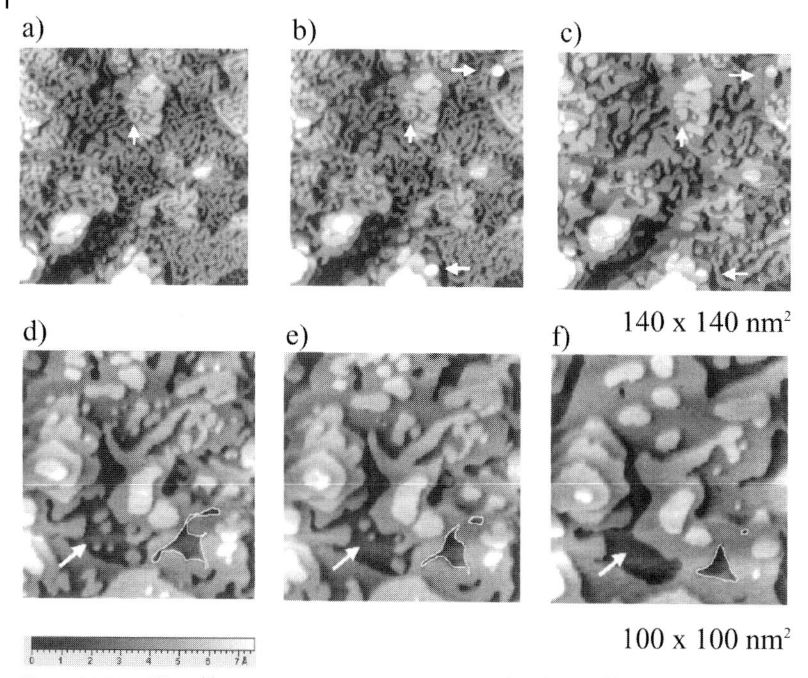

Figure 6.2-15: Wormlike nanostructures in Ge(111) can be obtained by partial oxidation; they heal in a complex process comprising electrodeposition/electrodissolution and periphery diffusion. Ring-like defects transform to points as predicted by Grayson's theorem (vertical arrow, a: 0 min, b: 8 min, c: 20 min – time with reference to a), electrodeposition of clusters occurs (horizontal arrows). Furthermore, the clusters can also dissolve (arrows in d–f) and pinch-off phenomena are observed (manually surrounded structures, d: 0 min, e: 12 min, f: 52 min – time with reference to d) (picture from [60] – with permission of the Pccp owner societes.

Ge(IV). The observation of Ge(IV) in the XPS analysis is most probably the result of chemical attack by ambient oxygen. Such an attack has also been reported for Ge on Au made by Physical Vapor Deposition [68].

6.2.3 Summary

This section presents an insight into the electrodeposition of metals, alloys, and semiconductors from ionic liquids. Besides environmental considerations, these media have the great advantage that they give access by electrodeposition to elements that cannot be obtained from aqueous solutions. Not only could technical procedures and devices profit, but interesting insights into the nanoscale processes during electrodeposition of elements such as germanium, silicon, etc. are also possible, especially for semiconductor nanostructures that will be important in nanotechnology. The ionic liquids give access to a great variety of elements and compounds. Therefore, electrodeposition in ionic liquids is an important contribution to nanotechnology. Perhaps it will be possible in future to establish nanoelectrochemical processes, for instance to make nanochips by this methodology. In any case, many more studies will be necessary if the deposition processes is to be understood on the nanometer scale.

References

1 Nanoscale probes of the solid/liquid interface, NATO ASI series E 288 (A.A. Gewirth and H. Siegenthaler eds.), Kluwer Academic Publishers, Dordrecht **1995**.

2 Chemistry of nonaqueous solutions: current progress (G. Mamantov and A.I. Popov eds.), VCH, Weinheim, **1994**.

3 A.J. Bard and L.R. Faulkner, "Electrochemical Methods: Fundamentals and Applications", 2nd edition, John Wiley & Sons, Inc., **2001**, ISBN: 0-471-04372-9.

4 S. Takahashi, K. Akimoto, I. Saeki, *Hyomen Gijutsu* **1989**, *40*, 134.

5 P.K. Lai, M. Skyllas-Kazacos, *J. Electroan. Chem.* **1988**, *248*, 431.

6 Y. Zhao, T.J. VanderNoot, *Electrochim. Acta* **1997**, *42*, 1639.

7 M.R. Ali, A. Nishikata, T. Tsuru, *Indian J. Chem. Technol.* **1999**, *6*, 317.

8 Q. Liao, W.R. Pitner, G. Stewart, C.L. Hussey, *J. Electrochem. Soc.* **1997**, *144*, 936

9 J. Fischer, US patent 2828252 (**1958**).

10 M. Bukowski, F. Endres, H. Natter, and R. Hempelmann, patent pending DE-101 08 893.0-24 (08.29.**2002**)

11 D. Wei, M. Okido, *Curr. Top. Electrochem.* **1997**,*5*, 21.

12 I. Galasiu, R. Galasiu, J. Thonstad, *Nonaqueous Electrochem.* **1999**, 461.

13 W. Freyland, *Zeitschrift für Physikalische Chemie* **1994**, *184*, 139.

14 G.E. Gray, P.A. Kohl, J. Winnick, *J. Electrochem. Soc.* **1995**, *142*, 3636.

15 B.J. Piersma, *Proc. Electrochem. Soc.*, **1994**, *94*, 415.

16 P.-Y. Chen, Y.-F. Lin, I.-W. Sun, *J. Electrochem. Soc.* **1999**, *146*, 3290.

17 M.W. Verbrugge, M.K. Carpenter, *AIChE J.* **1990**, *36*, 1097.

18 C. Nanjundiah, K. Shimizu, R.A. Osteryoung, *J. Electrochem. Soc.* **1982**, *11*, 2474.

19 M. Lipsztajn, R.A. Osteryoung, *Inorg. Chem.* **1985**, *24* , 716.

20 R.T. Carlin, H.C. De Long, J. Fuller, P.C. Trulove, *J. Electrochem. Soc.* **1998**, *145*, 1598.

21 N. Koura, T. Kato, E. Yumoto, *Hyomen Gijutsu*, **1994**, *45*, 805.

22 G.T. Cheek, H.C. De Long, P.C. Trulove, *Proc. Electrochem. Soc.* **2000**, *99-41*, 527.

23 N. Guo, J. Guo, S. Xiong, *Fushi Kexue Yu Fanghu Jishu* **1998**, *10*, 290.

24 M.R. Ali, A. Nishikata, T. Tsuru, *Electrochim. Acta* **1997**, *42*, 2347.

25 T. Tsuda, Y. Ito, *Proc. Electrochem. Soc.* **2000**, *99-41*, 100.

26 J.S-Y. Liu, I.-W. Sun, *J. Electrochem. Soc.* **1997**, *144*, 140.

27 M.K. Carpenter, M.W. Verbrugge, *J. Mater. Res.* **1994**, *9*, 2584.

28 M.K. Carpenter, M.W. Verbrugge, US patent 92-926103 (**1993**).

29 M. Lipsztjan, R.A. Osteryoung, *Inorg. Chem.* **1985**, *24*, 3492.

30 D.A. Habboush, R.A. Osteryoung, *Inorg. Chem.* **1984**, *23*, 1726.

31 C.L. Hussey, personal communication.

32 A.F. Tsatsulnikov, S.V. Ivanov, P.S. Kopev, I.L. Krestnikov, A.K. Kryganovskii, N.N. Ledentsov, M.V. Maximov, B.Y. Meltser, P.V. Nekludov, A.A. Suvorova, A.N. Titkov, B.V. Volovik, M. Grundmann, D. Bimberg, Z.I. Alferov, A. F. Ioffe, *Microelectron. Eng.* **1998**, *43/44*, 85.

33 E. G.-S. Jeng, I.-W. Sun, *J. Electrochem. Soc.* **1997**, *144*, 2369.

34 B. Gregory, J.L. Stickney *J. Electroan. Chem.* **1994** *365*, 87.

35 M.A.M Noel, R.A. Osteryoung, *J. Electroan. Chem.* **1996**, *293*, 139.

36 P.-Y. Chen, I.-W. Sun, *Electrochim. Acta* **2000**, *45*, 3163.

37 C.L. Hussey, L.A. King, R.A. Carpio, *J. Electrochem. Soc.* **1979**, *126*, 1029.
38 C. Nanjundiah, R.A. Osteryoung, *J. Electrochem. Soc.* **1983**, *130*, 1312.
39 B.J. Tierney, W.R. Pitner, J.A. Mitchell, C.L. Hussey, *J. Electrochem. Soc.* **1998**, *145*, 3110.
40 P.-Y. Chen, I.-W. Sun, *Proc. Electrochem. Soc.* **1998**, *98-11*, 55.
41 X.-H. Xu, C.L. Hussey, *J. Electrochem. Soc.* **1992**, *139*, 1295.
42 F. Endres and W. Freyland, *J. Phys. Chem. B* **1998**, *102*, 10,229.
43 Y. Katayama, S. Dan, T. Miura, T. Kishi, *J. Electrochem. Soc.* **2001**, *148*, C102.
44 W. Schindler, D. Hofmann, J. Kirschner, *J. Electrochem. Soc.* **2001**, *148*, C124–C130.
45 J.A. Mitchell, W.R. Pitner, C.L. Hussey, G.R. Stafford, *J. Electrochem. Soc.* **1996**, *143*, 3448.
46 W.R. Pitner, C.L. Hussey, G.R. Stafford, *J. Electrochem. Soc.* **1996**, *143*, 130.
47 C.A. Zell, W. Freyland, F. Endres, *Progress in Molten Salt Chemistry* **2000**, *1*, 597, Elsevier, ISBN: 2-84299-249-0.
48 C.A. Zell, W. Freyland, *Chem. Phys. Lett.* **2001**, *337*, 293.
49 H.C. De Long, J.S. Wilkes, R.T. Carlin, *J. Electrochem. Soc.* **1994**, *141*, 1000.
50 X.-H. Xu, C.L. Hussey, *Proc. Electrochem. Soc.* **1992**, *16*, 445.
51 E.R. Schreiter, J.E. Stevens, M.F. Ortwerth, R.G. Freeman, *Inorg. Chem.* **1999**, *38*, 3935.
52 W.R. Pitner, C.L. Hussey, *J. Electrochem. Soc.* **1997**, *144*, 3095.
53 X.-H. Xu, C.L. Hussey, *J. Electrochem. Soc.* **1993**, *140*, 618.
54 R.K. Pandey, S.N. Sahu, S. Chandra, "*Handbook of Semiconductor Electrodeposition*", Marcel Dekker, Inc., **1996**, ISBN 0-8247-9701-9.
55 A. Raza, R. Engelken, B. Kemp, A. Siddiqui, O. Mustafa, *Proc. Arkansas Acad. Sci.* **1995**, *49*, 143
56 S.P. Wicelinski, R.J. Gale, *Proc. Electrochem. Soc.* **1987**, *134*, 262.
57 M.K. Carpenter, M.W. Verbrugge, *J. Electrochem. Soc.* **1987**, *87-7*, 591.
58 M.-C. Lin, P.-Y. Chen, I.-W. Sun, *J. Electrochem. Soc.* **2001**, *148(10)*, C653.
59 F. Endres, C. Schrodt, *Phys. Chem. Chem. Phys.* **2000**, *24*, 5517.
60 F. Endres, *Phys. Chem. Chem. Phys.* **2001**, *3*, 3165.
61 O. Leifeld, A. Beyer, E. Müller, D. Grützmacher, K. Kern, *Thin Solid Films* **2000**, *380*, 176.
62 G. Binnig, H. Rohrer, *Helv. Phys. Acta* **1982**, *55*, 726.
63 R. Sonnenfeld, P.K. Hansma, *Science* **1986**, *232*, 211.
64 "*Scanning Tunneling Microscopy and Spectroscopy: Theory, Techniques and Applications*" (D.A. Bonnell ed.), John Wiley & Sons, **2000**, ISBN: 0-471-24824-X.
65 C.A. Zell, F. Endres, W. Freyland, *Phys. Chem. Chem. Phys.* **1999**, *1*, 697.
66 T. Will, M. Dietterle, D.M. Kolb in ref. 1, p. 137–162.
67 F. Endres, A. Schweizer, *Phys. Chem. Chem. Phys.* **2000**, *2*, 5455.
68 S. Ingrey, B. MacLaurin, *J. Vac. Sci. Technol., A* **1984**, *2*, 358.

7
Polymer Synthesis in Ionic Liquids

Adrian J. Carmichael and David M. Haddleton

7.1
Introduction

Ambient-temperature ionic liquids have received much attention in both academia and industry, due to their potential as replacements for volatile organic compounds (VOCs) [1–3]. These studies have utilized the ionic liquids as direct replacements for conventional solvents and as a method to immobilize transition metal catalysts in biphasic processes.

Many organic chemical transformations have been carried out in ionic liquids: hydrogenation [4, 5], oxidation [6], epoxidation [7], and hydroformylation [8] reactions, for example. In addition to these processes, numerous synthetic routes involve a carbon-carbon (C–C) bond-forming step. As a result, many C–C bond-forming procedures have been studied in ambient-temperature ionic liquids. Among those reported are the Friedel–Crafts acylation [9] and alkylation [10] reactions, allylation reactions [11, 12], the Diels–Alder reaction [13], the Heck reaction [14], and the Suzuki [15] and Trost–Tsuji coupling [16] reactions.

The C–C bond-forming reaction that has received most attention in ionic liquids is the dimerization of simple olefins (such as ethene, propene, and butene) [17–20]. An existing commercial procedure, the Dimersol process [21], is widely used for the dimerization of simple olefins, producing approximately 3×10^6 tonnes per annum. The technology in this process has been adapted to function under biphasic conditions by use of the ternary ionic liquid system [BMIM][Cl-$AlCl_3$-$EtAlCl_2$] (where BMIM is 1-butyl-3-methylimidazolium) as a solvent for the nickel catalysts [17, 18, 22], Improvements in catalyst activity, better selectivity, and simple removal of pure products allowing easy recycling of the ionic liquid and catalyst are the benefits offered over the conventional process.

In these reactions the system is tuned, for example by adjustment of the reaction temperature and time and modification of the catalyst structure to maximize the quantity of the desired dimers produced, and to minimize the production of higher molecular weight oligomers and polymers. In other reactions it is the opposite that is true: higher weight products are desired. The most industrially useful C–C bond-forming reaction is addition polymerization. This is used to obtain polymers that are used for a multitude of applications: in, for example, coatings, detergents, adhesives, plastics, etc. The use of ambient-temperature ionic liquids as solvents for the preparation of polymers has received little attention, especially in comparison with studies concerning their use in other synthetic areas. This chapter surveys the polymerization reactions carried out in ionic liquids thus far.

7.2 Acid-catalyzed Cationic Polymerization and Oligomerization

Strong Brønsted acids that have non-nucleophilic anions (such as $HClO_4$ and CF_3CO_2H) are capable of initiating cationic polymerization with vinyl monomers that contain an electron-donating group adjacent to a carbon-carbon double bond, (such as vinyl ethers, isobutylene, styrene, and dienes). Lewis acids are also used as initiators in cationic polymerization with the formation of high molecular weight polymers. These Lewis acids include metal halides (such as $AlCl_3$, BF_3, and $SbCl_5$), organometallic species (such as $EtAlCl_2$), and oxyhalides (such as $POCl_3$). Lewis acids are often used in the presence of a proton source (such as H_2O, HCl, or MeOH) or a carbocation source (such as *t*BuCl), which produces an acceleration in the rate of polymerization [23].

The chloroaluminate(III) ionic liquids – [EMIM][Cl-$AlCl_3$], for example (where EMIM is 1-ethyl-3-methylimidazolium) – are liquid over a wide range of $AlCl_3$ concentrations [24]. The quantity of $AlCl_3$ present in the ionic liquid determines the physical and chemical properties of the liquid. When the mole fraction, $X(AlCl_3)$, is below 0.5, the liquids are referred to as basic. When $X(AlCl_3)$ is above 0.5, the liquids are referred to as acidic, and at an $X(AlCl_3)$ of exactly 0.5 they are referred to as neutral.

Studies have shown that when protons (from HCl as the source) are dissolved at ordinary temperatures and pressures in the acidic ionic liquid [EMIM][Cl-$AlCl_3$] ($X(AlCl_3) = 0.55$) they are superacidic, with a strength similar to that of a liquid HF/Lewis acid mixture [25]. The precise Brønsted acidity observed depends on proton concentration and on ionic liquid composition. The ambient-temperature chloroaluminate(III) ionic liquids are extremely sensitive to moisture, reacting exothermically to give chlorooxoaluminate(III) species and generating HCl. Since moisture is ever present, even in the most carefully managed systems, chloroaluminate(III) ionic liquids generally possess superacidic protons. In addition, acidic chloroaluminate(III) ionic liquids contain Lewis acid species [26] ($[Al_2Cl_7]^-$, for example), so it is unsurprising that, with the combination of these factors, acidic chloroaluminate(III) ionic liquids catalyze the cationic oligomerization and polymerization of olefins.

Studies on the dimerization and hydrogenation of olefins with transition metal catalysts in acidic chloroaluminate(III) ionic liquids report the formation of higher molecular weight fractions consistent with cationic initiation [17, 20, 27, 28]. These studies ascribe the occurrence of the undesired side reaction to both the Lewis acid- and the proton-catalyzed routes. Attempts to avoid these side reactions resulted in the preparation of alkylchloroaluminate(III) ionic liquids and buffered chloroaluminate(III) ionic liquids [17, 20, 28].

Attempts to bring the benefits of ionic liquid technology, drawing on this inherent ability of the chloroaluminate(III) ionic liquids, to catalysis of cationic polymerization reactions, as opposed to their minimization, were patented by Ambler et al. of BP Chemicals Ltd. in 1993 [29]. They used acidic [EMIM][Cl-$AlCl_3$] ($X(AlCl_3) =$ 0.67) for the polymerization of butene to give products that have found application as lubricants. The polymerization could be carried out by bubbling butene through the ionic liquid, the product forming a separate layer that floated upon the ionic liquid and was isolated by a simple process. Alternatively, the polymerization could be carried out by injecting the ionic liquid into a vessel charged with butene. After a suitable settling period, the poly(butene) was isolated in a similar fashion. The products from these reactions are best described as oligomers as opposed to polymers, as the product is still in the liquid form. Chain transfer to impurities, ionic liquid, monomer, and polymer will terminate the propagation reaction, resulting in the low-mass products.

Synthesis of higher molecular weight polymers by cationic polymerization requires the formation of charged centers that live for long enough to propagate without chain transfer or termination. For this to occur, stabilization of the propagating species by solvation is generally required. In addition, low temperatures are usually employed, in an attempt to reduce side reactions that destroy the propagating centers. Use of a pure isobutene feedstock gives poly(isobutene) with properties that depend upon the reaction temperature. As the temperature is reduced, the molecular weight of the product is reported to increase dramatically, which is a result of the rates of the side reactions and the rate of polymerization being reduced (Table 7.2-1) [29].

Ionic liquid-catalyzed polymerization of butene is not limited to the use of pure alkene feedstocks, which can be relatively expensive. More usefully, the technology can be applied to mixtures of butenes, such as the low-value hydrocarbon feedstocks raffinate I and raffinate II. The raffinate feedstocks are principally C4 hydrocarbon mixtures rich in butenes. When these feedstocks are polymerized in the presence of acidic chloroaluminate(III) ionic liquids, polymeric/oligomeric products with

Table 7.2-1: Polymerization of isobutene in the acidic ionic liquid [EMIM]Cl/$AlCl_3$ ($X(AlCl_3) =$ 0.67) [29].

Reaction temperature (°C)	*Yield (% w/w)*	*Molecular weight of product (g mol^{-1})*
–23	26	100,000[a]
0	75	3000 and 400[b]

[a] Polystyrene equivalents. [b] Bimodal.

molecular weights higher than those obtained by conventional processes are produced, even though higher reaction temperatures are used. With the ionic liquid-catalyzed process, although isobutene conversion is much higher than *n*-butene conversion, the produced polymers have a much higher incorporation of *n*-butenes than would be possible from conventional cationic polymerization processes (Table 7.2-2) [29].

The ionic liquid process has a number of advantages over traditional cationic polymerization processes such as the Cosden process, which employs a liquid-phase aluminium(III) chloride catalyst to polymerize butene feedstocks [30]. The separation and removal of the product from the ionic liquid phase as the reaction proceeds allows the polymer to be obtained simply and in a highly pure state. Indeed, the polymer contains so little of the ionic liquid that an aqueous wash step can be dispensed with. This separation also means that further reaction (e.g., isomerization) of the polymer's unsaturated ω-terminus is minimized. In addition to the ease of isolation of the desired product, the ionic liquid is not destroyed by any aqueous washing procedure and so can be reused in subsequent polymerization reactions, resulting in a reduction of operating costs. The ionic liquid technology does not require massive capital investment and is reported to be easily retrofitted to existing Cosden process plants.

Further development of the original work in which [EMIM][Cl-$AlCl_3$] ($X(AlCl_3)$ = 0.67) was used as the ionic liquid has found that replacement of the ethyl group attached to the imidazolium ring with alkyl groups of increasing length (e.g., octyl, dodecyl, and octadecyl) produces increased catalytic activity towards the oligomerization of the olefins in the ionic liquid. Thus, the longer the alkyl chain, the greater the degree of polymerization achieved [31]. This provides an additional method for altering the product distribution. Increased polymer yield with the raffinate I feedstock was achieved by the use of an [EMIM][Cl-$AlCl_3$] ionic liquid containing a small proportion of the quaternary ammonium salt [NEt_4]Cl. The ternary ionic liquid [NEt_4]Cl/[EMIM][Cl-$AlCl_3$] of mole ratio 0.08:0.25:0.67, used under the same reaction conditions as the binary ionic liquid [EMIM][Cl-$AlCl_3$] ($X(AlCl_3)$ = 0.67), produced ~70 % of a polymer/oligomer mixture, as opposed to ~40 % polymer/oligomer produced with the original binary system. Both systems produced oligomers with M_n = 1,000 g mol^{-1} [32]. These examples demonstrate the capability to tune the ionic liquids' properties by changing the ancillary substituents. This allows the solvent to be adapted to the needs of the reaction, as opposed to altering the reaction to the needs of the solvent.

Raffinate I feedstock Olefin fraction	*Concentration (% w/w)*	*Reacted (% w/w)*
Isobutene	46	91
1-Butene	25	47
trans-2-Butene	8	34
cis-2-Butene	3	37

Table 7.2-2: Polymerization of Raffinate I in the acidic ionic liquid [EMIM]Cl/$AlCl_3$ ($X(AlCl_3)$ = 0.67): conversion of the individual components [29].

This technology has been utilized by BP Chemicals for the production of lubricating oils with well defined characteristics (for example, pour point and viscosity index). It is used in conjunction with a mixture of olefins (i.e., different isomers and different chain length olefins) to produce lubricating oils of higher viscosity than obtainable by conventional catalysis [33]. Unichema Chemie BV have applied these principals to more complex monomers, using them with unsaturated fatty acids to create a mixture of products [34].

Apart from one mention in the original patent of the synthesis of a high molecular weight poly(isobutene) (see Table 7.2-1) [29], the remaining work has until recently been concerned with the preparation of lower weight oligomers. In 2000, Symyx Technologies Inc. protected a method for the production of high molecular weight poly(isoolefin)s without the use of very low temperatures [35]. Symyx used the [EMIM][Cl-$AlCl_3$] ionic liquid to produce poly(isobutene)s with weight average molecular weights (M_ws) in excess of 100,000 g mol^{-1} which are of use in the automotive industry due to their low oxygen permeability and mechanical resilience (Table 7.2-3). The table shows that polymers with molecular weights higher than half a million are obtained at temperatures as high as –40 °C. As would be expected, when the temperature is increased the molecular weight decreases. In all cases the yield is less than 50 %. If the reaction is performed under biphasic conditions, reducing the concentration of isobutene and adding ethylaluminium(III) dichloride, however, the reaction yield becomes quantitative, (Table 7.2-4). This shows that, in addition to the use of temperature to control the molecular weight of the product, control can also be achieved through the quantity of ethylaluminium(III) dichloride added to the reaction: the more alkylaluminum(III) that is added, the lower the molecular weight of the product. It might be expected that the ethylaluminium(III) dichloride would act as a proton scavenger, which should stop the polymerization, thus it seems it acts either/both as a strong Lewis acid or/and as an alkylating agent promoting polymerization.

For the results reported in both Table 7.2-3 and Table 7.2-4, the only reported detail concerning the ionic liquid was that it was [EMIM][Cl-$AlCl_3$]. No details of the aluminium(III) chloride content were forthcoming. As with most of the work presented in this chapter, data are taken from the patent literature and not from peer reviewed journals, and so many experimental details are not available. This lack of clear reporting complicates issues for the synthetic polymer chemist. Simpler and cheaper chloroaluminate(III) ionic liquids prepared by using cations derived from the reaction between a simple amine and hydrochloric acid (e.g., $Me_3N \cdot HCl$ and

Table 7.2-3: Polymerization of isobutene to high molecular weight poly(isobutene)s in the ionic liquid [EMIM]Cl/$AlCl_3$ [35].

Quantity of ionic liquid (μl)	*Quantity of isobutene (μl)*	*Temperature (°C)*	*Yield (%)*	M_w *(g mol^{-1})*
10	483	–40	38	526,000
10	483	–30	33	302,000
10	483	–20	45	128,000

Table 7.2-4: Polymerization of isobutene to high molecular weight poly(isobutene)s in the ionic liquid [EMIM]Cl/$AlCl_3$ under biphasic conditions [35].

Quantity of ionic liquid (µl)	*Quantity of hexane (µl)*	*Quantity of isobutene (µl)*	*Quantity of $EtAlCl_2$ (µl)*	*Yield (%)*	M_w *g mol^{-1})*
50	321	25	11	100	276,000
50	310	25	23	100	235,000
50	298	25	34	100	186,000

Conditions: temperature = –30 °C; [$EtAlCl_2$] = 1 M solution in hexane.

$Bu_2NH \cdot HCl$) have successfully been used in the polymerization of isobutene and styrene [36]. Although these ionic liquids have much higher melting points than their imidazolium analogues, they are liquid at temperatures suitable for their use in the preparation of low molecular weight oligomers (i.e., 1000 to 4000 g mol^{-1}). This reduces one of the barriers to exploitation of the technology, the relatively high expense of the imidazolium halide salts.

7.3 Free Radical Polymerization

Free radical polymerization is a key method used by the polymer industry to produce a wide range of polymers [37]. It is used for the addition polymerization of vinyl monomers including styrene, vinyl acetate, tetrafluoroethylene, methacrylates, acrylates, (meth)acrylonitrile, (meth)acrylamides, etc. in bulk, solution, and aqueous processes. The chemistry is easy to exploit and is tolerant to many functional groups and impurities.

The first use of ionic liquids in free radical addition polymerization was as an extension to the doping of polymers with simple electrolytes for the preparation of ion-conducting polymers. Several groups have prepared polymers suitable for doping with ambient-temperature ionic liquids, with the aim of producing polymer electrolytes of high ionic conductance. Many of the prepared polymers are related to the ionic liquids employed: for example, poly(1-butyl-4-vinylpyridinium bromide) and poly(1-ethyl-3-vinylimidazolium bis(trifluoromethanesulfonyl)imide [38–41].

Noda and Watanabe [42] reported a simple synthetic procedure for the free radical polymerization of vinyl monomers to give conducting polymer electrolyte films. Direct polymerization in the ionic liquid gives transparent, mechanically strong and highly conductive polymer electrolyte films. This was the first time that ambient-temperature ionic liquids had been used as a medium for free radical polymerization of vinyl monomers. The ionic liquids [EMIM][BF_4] and [BP][BF_4] (BP is *N*-butylpyridinium) were used with equimolar amounts of suitable monomers, and polymerization was initiated by prolonged heating (12 hours at 80 °C) with benzoyl peroxide. Suitable monomers for this purpose were those that dissolved in the ionic liquid solvent to give transparent, homogeneous solutions (Table 7.3-1), with unsuitable monomers phase-separating and therefore not being subjected to polymerization. Of all the monomers found to give transparent homogeneous solutions, only vinyl acetate failed to undergo polymerization. In all other polymerizations, with the exception of that of 2-hydroxyethyl methacrylate (HEMA), the polymer was insoluble in the ionic liquid and phase-separated. The compatibility of HEMA with the ionic liquids resulted in its use for the preparation of polymer electrolyte films, which were found to be highly conductive. For film formation, the reaction mixtures were simply spread between glass plates and heated; no degassing procedures were carried out. Analysis of the films found that the amount of unreacted monomer was negligible, indicating fast polymerization. No characterization of the polymers, or indeed analysis of the polymerization reactions, was reported in any of the reactions described by Noda and Watanabe [42].

More recent studies by May and by ourselves have looked into the kinetics and the types of polymers formed by the free radical polymerization reactions of vinyl monomers with ambient-temperature ionic liquids as solvents [43, 44]. The free radical polymerization of methyl methacrylate (MMA) in [BMIM][PF_6], initiated by 2,2'-azobisisobutyronitrile (AIBN) at 60 °C, proceeds rapidly, causing a large increase in viscosity that hampers efficient stirring of the reaction mixture. The polymerization reactions produce poly(methyl methacrylate) (PMMA) with very high molecular weights (see Table 7.3-2) [44]. In comparison with a free radical polymerization in a conventional organic solvent, toluene in this case, both conversion and M_n are increased by approximately one order of magnitude. This could be due to one of two reasons. Firstly, the rate of bimolecular termination either by disproportionation or by combination could be suppressed, which might be due in part to the large increase in viscosity of the reaction medium. In this case termination would be dominated by chain transfer either to solvent or to monomer. The molecular weights observed are consistent with this explanation. Alternatively, the rate constant of propagation, k_p, may be increased in the ionic liquid due to local environment effects.

Table 7.3-1: Compatibility of the ionic liquids [EMIM][BF_4] and [BP][BF_4] with monomers and their polymers [42].

	[EMIM][BF_4]		*[NBPY][BF_4]*	
	Monomer	*Polymer*	*Monomer*	*Polymer*
Methyl methacrylate	X	–	O	X
Acrylonitrile	O	X	O	X
Vinyl acetate	O	no reaction	O	no reaction
Styrene	X	–	X	–
2-Hydroxyethyl methacrylate	O	ρ	O	ρ

Legend: O, transparent homogenous solution; X, phase-separated; ρ, translucent gel.

Table 7.3-2: Free radical polymerization of MMA in the ionic liquid [BMIM][PF_6] [44].

Reaction media	*[AIBN] (w/v %)*	*Conversion (%)*	*M_n (g mol^{-1})*	*PDi*
[BMIM][PF_6]	1	25	669,000	1.75
[BMIM][PF_6]	2	27	600,000	1.88
[BMIM][PF_6]	4	36	416,000	2.22
[BMIM][PF_6]	8	56	240,000	2.59
Toluene	1	3	58,300	1.98

Conditions: temperature = 60 °C; time = 20 min; 20 % v/v monomer in ionic liquid.

The effects of increasing the concentration of initiator (i.e., increased conversion, decreased M_n, and broader PDi) and of reducing the reaction temperature (i.e., decreased conversion, increased M_n, and narrower PDi) for the polymerizations in ambient-temperature ionic liquids are the same as observed in conventional solvents. May et al. have reported similar results and in addition used ^{13}C NMR to investigate the stereochemistry of the PMMA produced in [BMIM][PF_6]. They found that the stereochemistry was almost identical to that for PMMA produced by free radical polymerization in conventional solvents [43]. The homopolymerization and copolymerization of several other monomers were also reported. Similarly to the findings of Noda and Watanabe, the polymer was in many cases not soluble in the ionic liquid and thus phase-separated [43, 44].

7.4 Transition Metal-catalyzed Polymerization

The previous sections show that certain ionic liquids, namely the chloroaluminate(III) ionic liquids, are capable of acting both as catalyst and as solvent for the polymerization of certain olefins, although in a somewhat uncontrolled manner, and that other ionic liquids, namely the non-chloroaluminate(III) ionic liquids, are capable of acting as solvents for free radical polymerization processes. In attempts to carry out polymerization reactions in a more controlled manner, several studies have used dissolved transition metal catalysts in ambient-temperature ionic liquids and have investigated the compatibility of the catalyst towards a range of polymerization systems.

7.4.1 Ziegler–Natta Polymerization of Ethylene

Ziegler–Natta polymerization is used extensively for the polymerization of simple olefins (such as ethylene, propene, and 1-butene) and is the focus of much academic attention, as even small improvements to a commercial process operated on this scale can be important. Ziegler–Natta catalyst systems, which in general are early transition metal compounds used in conjunction with alkylaluminium compounds, lend themselves to study in the chloroaluminate(III) ionic liquids, especially those of acidic composition.

During studies into the behavior of titanium(IV) chloride in chloroaluminate(III) ionic liquids, Carlin et al. carried out a brief study to investigate whether Ziegler–Natta polymerization was possible in an ionic liquid [45]. They dissolved $TiCl_4$ and $EtAlCl_2$ in [EMIM][Cl-$AlCl_3$] ($X(AlCl_3)$ = 0.52) and bubbled ethylene through for several minutes. After quenching, poly(ethylene) with a melting point of 120–130 °C was isolated in very low yield, thus demonstrating that Ziegler–Natta polymerization works in these liquids, albeit not very well.

The same ionic liquid was employed, giving higher yields of poly(ethylene), with bis(η-cyclopentadienyl)titanium(IV) dichloride in conjunction with $Me_3Al_2Cl_3$ as catalyst [46]. However, the catalytic activities are still low when compared to other homogeneous systems, which may be attributed to, among other things, low solubility of ethylene in the ionic liquids or the presence of alkylimidazole impurities that coordinate and block the active titanium sites. In chloroaluminate(III) ionic liquids of basic composition, no catalysis was observed. This was ascribed to the formation of the inactive $[Ti(\eta\text{-}C_5H_5)_2Cl_3]^-$ species. In comparison, the zirconium and hafnium analogues $[Zr(\eta\text{-}C_5H_5)_2Cl_2]$ and $[Hf(\eta\text{-}C_5H_5)_2Cl_2]$ showed no catalytic activity towards the polymerization of ethylene either in acidic or in basic ionic liquids. This is presumably due to the presence of stronger M-Cl bonds that preclude the formation of a catalytically active species.

7.4.2 Late Transition Metal-catalyzed Polymerization of Ethylene

The surge in development of late transition metal polymerization catalysts has been due, in part, to the need for systems that can copolymerize ethylene, and related monomers, with polar co-monomers under mild conditions. Late transition metals have lower oxophilicity than early transition metals, and therefore a higher tolerance for a wider ranger of functional groups (e.g., -COOR and -COOH groups) [47]. A recent study reports the use of the nickel complex **1** (Figure 7.4-1) for the homopolymerization of ethylene in an ambient-temperature ionic liquid [48]. Compound **1** was used under mild biphasic conditions with the ternary ionic liquid [BMIM][Cl-$AlCl_3$-$EtAlCl_2$] (1.0:1.0:0.32, $X(Al)$ = 0.57) and toluene, producing poly(ethylene) which was easily isolated from the reaction mixture by decanting the upper toluene layer. This allowed the ionic liquid and **1** to be recycled for use in further polymerizations. Before reuse, however, trimethylaluminium(III) was added to overcome the loss of free alkylaluminium species into the separated organic phase [48]. The characteristics of the isolated poly(ethylene) depend upon several reaction conditions. On increasing the reaction temperature from –10 to +10 °C, the melting point decreases from 123 to 85 °C due to a greater amount of chain branching, and a decrease in the M_w from 388,000 to 280,000 g mol^{-1} also results. Reuse of the catalyst/ionic liquid solution also has an effect, with subsequent reactions giv-

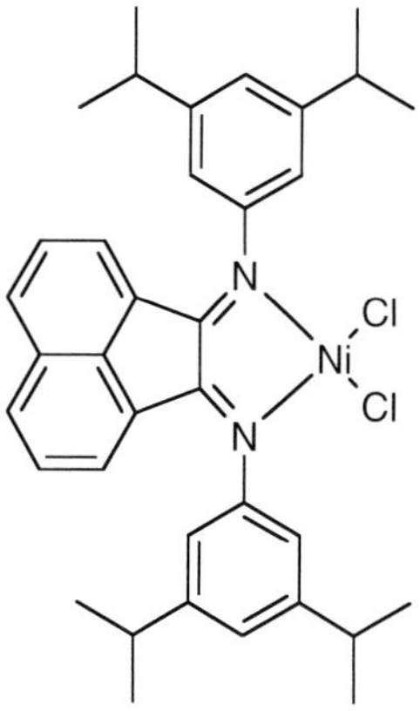

Figure 7.4-1: Nickel catalysts used for the polymerization and oligomerization of ethylene in ambient-temperature ionic liquids [48, 49].

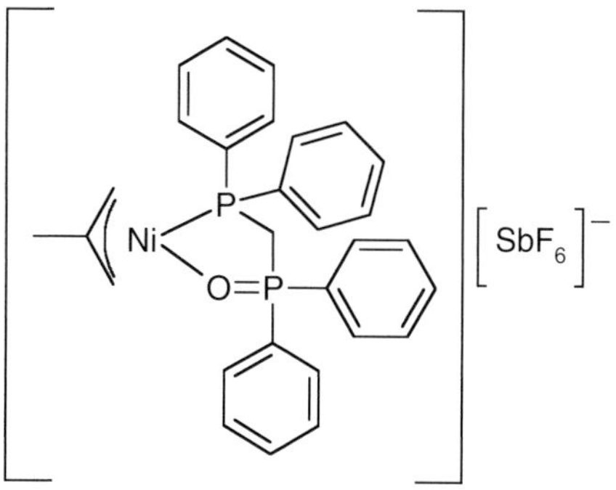

ing a progressive shift from crystalline to amorphous polymer, with a period that gives rise to bimodal product distributions. This change is due to the changing composition of the ionic liquid, as fresh co-catalyst is added after each polymerization run, giving rise to the formation of different active species.

A related study used the air- and moisture-stable ionic liquids [RMIM][PF_6] (R = butyl-decyl) as solvents for the oligomerization of ethylene to higher α-olefins [49]. The reaction used the cationic nickel complex **2** (Figure 7.4-1) under biphasic conditions to give oligomers of up to nine repeat units, with better selectivity and reactivity than obtained in conventional solvents. Recycling of the catalyst/ionic liquid solution was possible with little change in selectivity, and only a small drop in activity was observed.

7.4.3
Metathesis Polymerization

Acyclic diene molecules are capable of undergoing intramolecular and intermolecular reactions in the presence of certain transition metal catalysts: molybdenum alkylidene and ruthenium carbene complexes, for example [50, 51]. The intramolecular reaction, called ring-closing olefin metathesis (RCM), affords cyclic compounds, while the intermolecular reaction, called acyclic diene metathesis (ADMET) polymerization, provides oligomers and polymers. Alteration of the dilution of the reaction mixture can to some extent control the intrinsic competition between RCM and ADMET.

Gürtler and Jautelat of Bayer AG have protected methods that use chloroaluminate(III) ionic liquids as solvents for both cyclization and polymerization reactions of acyclic dienes [52]. They employed the neutral ionic liquid [EMIM][Cl-$AlCl_3$]

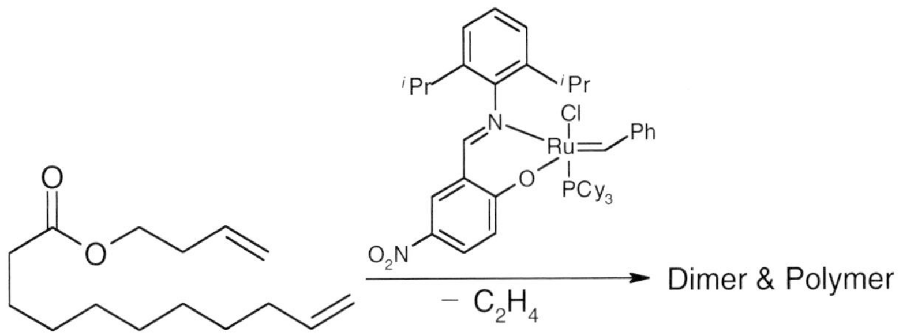

Figure 7.4-2: Acyclic diene metathesis polymerization (ADMET) reaction carried out in the neutral ionic liquid [EMIM]Cl/$AlCl_3$ ($X(AlCl_3)$ = 0.5) [52].

($X(AlCl_3)$ = 0.5) to immobilize a ruthenium carbene complex for biphasic ADMET polymerization of an acyclic diene ester (Figure 7.4-2). The reaction is an equilibrium processes, and so removal of ethylene drives the equilibrium towards the products. The reaction proceeds readily at ambient temperatures, producing mostly polymeric materials but also ~10 % dimeric material.

7.4.4 Living Radical Polymerization

As discussed in Section 7.3, conventional free radical polymerization is a widely used technique that is relatively easy to employ. However, it does have its limitations. It is often difficult to obtain predetermined polymer architectures with precise and narrow molecular weight distributions. Transition metal-mediated living radical polymerization is a recently developed method that has been developed to overcome these limitations [53, 54]. It permits the synthesis of polymers with varied architectures (for example, blocks, stars, and combs) and with predetermined end groups (e.g., rotaxanes, biomolecules, and dyes).

A potential limitation to commercialization of this technology is that relatively high levels of catalyst are often required. Indeed, it is common that one mole equivalent is required for each growing polymer chain to achieve acceptable rates of polymerization, making catalyst removal and reuse problematic. In order to overcome this problem, a range of approaches have been reported, including supported catalysts [55], fluorous biphase reactions [56], and more recently the use of ionic liquids [57, 44]. It was found that copper(I) bromide in conjunction with *N*-propyl-2-pyridylmethanimine as ligand catalyzes the living radical polymerization of MMA in the neutral ionic liquid [BMIM][PF_6]. The reaction progressed in a manner consistent with a living polymerization: that is, good first-order kinetic behavior and evolution of number-average molecular weight (M_n) with time were observed, and a final product with low M_n and PDi values was obtained [57]. Polymerization in the ionic liquid proceeded much more rapidly than that in conventional organic solvents; indeed, polymerization occurred at 30 °C in [BMIM][PF_6] at a rate comparable to that found in toluene at 90 °C.

The cationic nature of the copper(I) catalyst means that it is immobilized in the ionic liquid. This permits the PMMA product to be obtained, with negligible copper contamination, by a simple extraction procedure with toluene (in which the ionic liquid is not miscible) as the solvent. The ionic liquid/catalyst solution was subsequently reused.

The technique of copper(I) bromide-mediated living radical polymerization is compatible with other ambient-temperature ionic liquids. It proceeds smoothly in hexyl- and octyl-3-methylimidazolium hexafluorophosphate and tetrafluoroborate ionic liquids. However, use of [BMIM][BF_4] for the polymerization of MMA generates a product with a bimodal product distribution. Figure 7.4-3 shows the trace, together with a trace from a similar reaction carried out in [BMIM][PF_6] [44]. The mass distribution for [BMIM][PF_6] shows a single, narrow, low molecular weight peak consistent with living radical polymerization, whereas the mass distribution for [BMIM][BF_4] shows a similar peak, but also an additional peak that is broad and at high molecular weight. This high molecular weight peak is consistent with the results observed for conventional free radical polymerization in ionic liquids, as discussed in Section 7.3. This anomalous result can be explained in terms of the synthetic method used to prepare the ionic liquids. Of all the ionic liquids used, [BMIM][BF_4] was the only one in the study that was miscible with water. It was therefore the only one not subjected to an aqueous workup, and so was contaminated with halide salts [58]. The halide salts might poison the catalyst, with subsequent polymerization proceeding by two different mechanisms. Alternatively, it might be that, under living polymerization conditions, the terminal halide atom on the propagating polymer chain does not fully separate from the polymer during propagation, creating a "caged-radical" that undergoes propagation. Under appropriate conditions, separation occurs, resulting in irreversible homolytic fission and the production of free radicals. Conventional free radical polymerization ensues in competition with the atom-transfer mechanism, giving high conversion and high-mass polymer alongside the low-mass polymer from the living mechanism. This

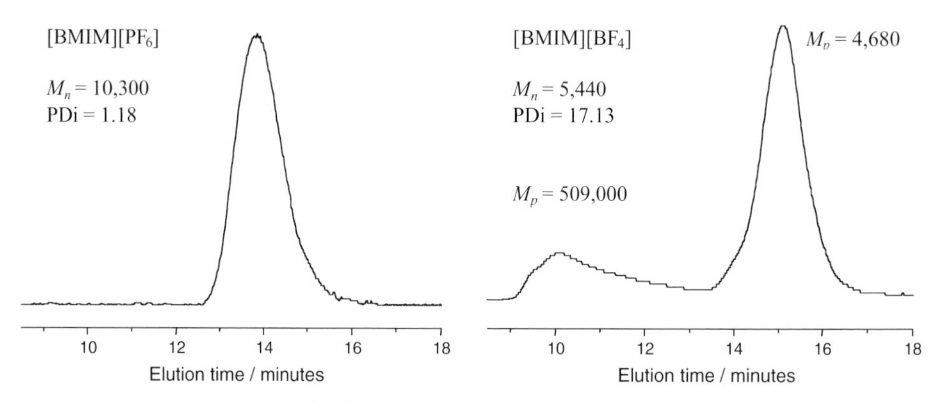

Figure 7.4-3: SEC traces for the Cu(I)Br-mediated living radical polymerization of MMA in the ionic liquids [BMIM][X] (X = [PF_6] or [BF_4]) [44].

implies that the rate of termination in conventional radical propagation is drastically reduced, maybe by coordination with the cation or anion from the ionic liquid, which also prevents recombination with the halide atom.

In a related study, Kubisa has investigated the Atom-transfer Radical Polymerization (ATRP) of acrylates in [BMIM][PF_6] [59]. The solubility of the monomer in the ionic liquid chosen depends very much upon the substituent on the monomer. Homogeneous polymerization of methyl acrylate gave living polymerization, with narrow polydispersity polymers and good molecular weight control. Higher order acrylates gave heterogeneous reactions, with the catalyst remaining in the ionic liquid phase. Although deviations from living polymerization behavior were observed, butyl acrylate showed controlled polymerization. The same group is currently extending this work and also looking at cationic vinyl polymerization and various ring-opening polymerization reactions.

7.5 Preparation of Conductive Polymers

Electronically conducting polymers have a number of potential applications, including as coatings for semiconductors [60], in electrocatalysis [61], and as charge-storage materials [62]. Of these, poly(*para*-phenylene) (PPP), the simplest of the poly(arene) classes, possesses properties that include excellent thermal stability, high coke number, and good optical and electrophysical characteristics [63]. For PPP to be utilizable in devices and advanced materials, it should have a high relative molecular mass (M_r), a homogenous structure, and good submolecular packing.

Poly(*para*-phenylene) can be prepared by a variety of chemical routes, but the polymers obtained are generally of low quality due to low masses and the occurrence of polymerization through 1,2-linkages, resulting in a disruption of molecular packing. They are obtained as powders and are often contaminated with oxygen and chlorine products and catalyst residues. The chemical synthesis of PPP can be carried out in ambient-temperature ionic liquids. The oxidative dehydropolycondensation of benzene was carried out in the acidic ionic liquid [BP]Cl/$AlCl_3$ ($X(AlCl_3) = 0.67$) with $CuCl_2$ as the catalyst [64, 65]. This gave PPP with relative molecular masses considerably higher than those obtained in conventional solvents, and M_r could be tuned by varying the benzene concentration. The high M_r values observed were attributed to greater solubility of PPP in the ionic liquid, permitting a greater degree of polymerization before phase separation occurred. The electrochemical synthesis of PPP reduces many of the disadvantages of the chemical route. The same group carried out the electrochemical polymerization of benzene in the same ionic liquid, preparing PPP as conductive films that were flexible and transparent. The films were prepared with very high relative molecular masses, with degrees of polymerization of up to 200 being observed [65, 66]. The electrochemical polymerization of benzene to PPP has not been carried out exclusively in [BP][Cl-$AlCl_3$] ionic liquids. Other reports use [BP]Cl/$AlCl_2$(OEt),

[CTP]Cl/$AlCl_3$ (CTP is *N*-cetylpyridinium), and [EMIM][Cl-$AlCl_3$], with the best results having been observed in the traditional aluminium(III) chloride ionic liquids [67–69].

The electrochemical oxidation of fluorene in [EMIM][Cl-$AlCl_3$] ionic liquids of acidic or neutral compositions gives poly(fluorene) films that are more stable and have less complicated electrochemical behavior than those prepared in acetonitrile, the usual solvent. Basic ionic liquids cannot be used, as chloride ions are more easily oxidized than fluorene [70]. A number of aromatic compounds containing heteroatoms, such as pyrrole, aniline, and thiophene, can also be oxidized electrochemically in chloroaluminate(III) ionic liquids to give polymer films [71–74]. In ionic liquids of acidic composition, electrochemical polymerization of the nitrogen- and sulfur-containing compounds is either more difficult or not possible at all, due to the formation of adducts with $AlCl_3$ [75]. Any interactions between benzene and $AlCl_3$ are not significant enough to influence its polymerization to PPP [69].

7.6 Conclusions

It is readily apparent that the volume of research concerning polymerization of any type in ionic liquids is sparse. It is not immediately clear why this is the case, and the field has not really started as yet. Ionic liquid technology has brought a number of benefits to polymer synthesis. For example, the application of chloroaluminate(III) ionic liquids as both solvent and catalyst for the cationic polymerization of olefins has provided a system that not only produces cleaner polymers than traditional processes but permits the recovery and reuse of the ionic liquid solvent/catalyst. Ionic liquids have allowed the preparation of high molecular weight conducting polymers such as poly(*para*-phenylene), and have been useful for the immobilization of transition metal polymerization catalysts, thus offering a potential solution to a problem that prevents the commercialization of transition metal-mediated living radical polymerization.

The use of neutral ionic liquids for free radical polymerization highlights one of their problems: their relatively high viscosity. The viscosity of the reaction mixture has a significant effect on the outcome of polymerization reactions, and these liquids can have viscosities much higher than those of conventional organic solvents. The free radical polymerization of MMA in [BMIM][PF_6] generates polymers with high molecular weights, which, when combined with the reduced fluidity of the ionic liquid, causes the reaction mixture to set after a very short time. This problem can be avoided in polymerization reactions if phase-separation of the product occurs, as with, for example, the free radical polymerization of MMA in [BP][BF_4], or, if the reaction is operated under biphasic conditions, for example, the reported ADMET polymerization of an acyclic diene ester.

As well as viscosity, other factors to be aware of include the purity of the ionic liquids. The presence of residual halide ions in neutral ionic liquids can poison transition metal catalysts, while different levels of proton impurities in chloroaluminate(III) ionic liquids can alter the product distribution of the reaction. The reduced temperatures required for many polymerization reactions in ionic liquids, together with the reduced solubility of oxygen in ionic liquids compared to that in conventional solvents, means that two of the most common quenching methods are reduced in effectiveness. If detailed studies are being carried out, in particular kinetic studies, it is necessary to stop further reaction completely so that accurate data may be obtained.

The controlled synthesis of polymers, as opposed to their undesired formation, is an area that has not received much academic interest. Most interest to date has been commercial, and focused on a narrow area: the use of chloroaluminate(III) ionic liquids for cationic polymerization reactions. The lack of publications in the area, together with the lack of detailed and useful synthetic information in the patent literature, places hurdles in front of those with limited knowledge of ionic liquid technology who wish to employ it for polymerization studies. The expanding interest in ionic liquids as solvents for synthesis, most notably for the synthesis of discrete organic molecules, should stimulate interest in their use for polymer science.

Even within the small numbers of studies conducted to date, we are already seeing potentially dramatic effects. Free radical polymerization proceeds at a much faster rate and there is already evidence that both the rate of propagation and the rate of termination are effected. Whole polymerization types – such as ring-opening polymerization to esters and amides, and condensation polymerization of any type (polyamides, polyesters, for example) – have yet to be attempted in ionic liquids. This field is in its infancy and we look forward to the coming years with great anticipation.

References

1 J. D. Holbrey, K. R. Seddon, *Clean Products and Processes* **1999**, *1*, 223–236.
2 T. Welton, *Chem. Rev.* **1999**, *99*, 2071–2083.
3 P. Wasserscheid, W. Keim, *Angew. Chem., Int. Ed. Engl.* **2000**, *39*, 3772–3789.
4 C. J. Adams, M. J. Earle, K. R. Seddon, *J. Chem. Soc., Chem. Commun.* **1999**, 1043–1044.
5 Y. Chauvin, L. Mussmann, H. Olivier, *Angew. Chem., Int. Ed. Engl.* **1995**, *34*, 2698–2700.
6 J. Howarth, *Tetrahedron Lett.* **2000**, *41*, 6627–6629.
7 G. S. Owens, M. M. Abu-Omar, *J. Chem. Soc., Chem. Commun.* **2000**, 1165–1166.
8 C. C. Brasse, U. Englert, A. Salzer, H. Waffenschmidt, P. Wasserscheid, *Organometallics* **2000**, *19*, 3818–3823.
9 C. J. Adams, M. J. Earle, G. Roberts, K. R. Seddon, *J. Chem. Soc., Chem. Commun.* **1998**, 2097–2098.
10 C. E. Song, W. H. Shim, E. J. Roh, J. H. Choi, *J. Chem. Soc., Chem. Commun.* **2000**, 1695–1696.
11 W. Chen, L. Xu, C. Chatterton, J. Xiao, *J. Chem. Soc., Chem. Commun.* **1999**, 1247–1248.
12 C. M. Gordon, A. McCluskey, *J. Chem. Soc., Chem. Commun.* **1999**, 1431–1432.
13 M. J. Earle, P. B. McCormac, K. R. Seddon, *Green Chemistry* **1999**, *1*, 23–25.

14 A. J. Carmichael, M. J. Earle, J. D. Holbrey, P. B. McCormac, K. R. Seddon, *Org. Lett.* **1999**, *1*, 997–1000.
15 C. J. Mathews, P. J. Smith, T. Welton, *J. Chem. Soc., Chem. Commun.* **2000**, 1249–1250.
16 C. de Bellefon, E. Pollet, P. Grenouillet, *J. Mol. Catal. A-Chemical* **1999**, *145*, 121–126.
17 Y. Chauvin, B. Gilbert, I. Guibard, *J. Chem. Soc., Chem. Commun.* **1990**, 1715–1716.
18 Y. Chauvin, S. Einloft, H. Oliver, *Ind. Eng. Chem. Res.* **1995**, *34*, 1149–1155.
19 S. Einloft, F. K. Dietrich, R. F. de Souza, J. Dupont, *Polyhedron* **1996**, *15*, 3257–3259.
20 B. Ellis, W. Keim, P. Wasserscheid, *J. Chem. Soc., Chem. Commun.* **1999**, 337–338.
21 D. Commereuc, Y. Chauvin, G. Leger, J. Gaillard, *Revue de L'Institut Français du Pétrole* **1982**, *37*, 639–649.
22 D. Commereuc, Y. Chauvin, F. Hugues, L. Saussine, A. Hirschauer, *French Patent* **1988**, FR 2,611,700.
23 G. Odian, *Principles of Polymerization*, Wiley, New York, **1991**.
24 A. A. Fannin, D. A. Floreani, L. A. King, J. S. Landers, B. J. Piersma, D. J. Stech, R. L. Vaughn, J. S. Wilkes, J. L. Williams, *J. Phys. Chem.* **1984**, *88*, 2614–2621.
25 G. P. Smith, A. S. Dworkin, R. M. Pagni, S. P. Zingg, *J. Am. Chem. Soc.* **1989**, *111*, 5075–5077.
26 J. A. Boon, J. A. Levisky, J. L. Pflug, J. S. Wilkes, *J. Org. Chem* **1986**, *51*, 480–483.
27 P. A. Suarez, J. E. L. Dullius, S. Einloft, R. F. de Souza, J. Dupont, *Polyhedron* **1996**, *15*, 1217–1219.
28 P. Wasserscheid, W. Keim, *World Patent* **1998**, WO 98/47616.
29 P. W. Ambler, P. K. G. Hodgson, N. J. Stewart, *European Patent* **1993**, EP 0558187.
30 K. Weissermel, H. -J. Arpe, *Industrial Organic Chemistry*, VCH, Weinheim, **1997**.
31 A. A. K. Abdul-Sada, P. W. Ambler, P. K. G. Hodgson, K. R. Seddon, N. J. Stewart, *World Patent* **1995**, WO 95/21871.
32 A. A. K. Abdul-Sada, K. R. Seddon, N. J. Stewart, *World Patent* **1995**, WO 95/21872.
33 P. M. Atkins, M. R. Smith, B. Ellis, *European Patent* **1997**, EP 0791643.
34 G. Roberts, C. M. Lok, C. J. Adams, K. R. Seddon, M. J. Earle, J. Hamill, *World Patent* **1998**, WO 98/07679.
35 V. Murphy, *World Patent* **2000**, WO 00/32658.
36 F. G. Sherif, L. J. Shyu, C. P. M. Lacroix, A. G. Talma, *US Patent* **1998**, US 5731101.
37 G. Moad, D. H. Solomon, *The Chemistry of Free Radical Polymerization*, Pergamon, Oxford, **1995**.
38 M. Watanabe, S. Yamada, N. Ogata, *Electrochimica Acta* **1995**, *40*, 2285–2288.
39 J. Fuller, A. C. Breda, R. T. Carlin, *J. Electrochem. Soc* **1997**, *144*, L67–L70.
40 M. Hirao, K. Ito-Akita, H. Ohno, *Polym. Adv. Technol.* **2000**, *11*, 534–538.
41 H. Ohno, *Electrochimica Acta* **2001**, *46*, 1407–1411.
42 A. Noda, M. Watanabe, *Electrochimica Acta* **2000**, *45*, 1265–1270.
43 H. Zhang, L. Bu, M. Li, K. Hong, J. W. Mays, R. D. Rogers, *ACS Symposium series chapter, in press* **2001**.
44 A. J. Carmichael, D. A. Leigh, D. M. Haddleton, *ACS Symposium series chapter, in press* **2002**.
45 R. T. Carlin, R. A. Osteryoung, J. S. Wilkes, J. Rovang, *Inorg. Chem* **1990**, *29*, 3003–3009.
46 R. T. Carlin, J. S. Wilkes, *J. Mol. Cat.* **1990**, *63*, 125–129.
47 S. D. Ittel, L. K. Johnson, M. Brookhart, *Chem. Rev.* **2000**, *100*, 1169–1203.
48 M. F. Pinheiro, R. S. Mauler, R. F. de Souza, *Macromol. Rapid Commun.* **2001**, *22*, 425–428.
49 P. Wasserscheid, C. M. Gordon, C. Hilgers, M. J. Muldoon, I. R. Dunkin, *J. Chem. Soc., Chem. Commun.* **2001**, 1186–1187.
50 A. Fürstner, *Angew. Chem., Int. Ed. Engl.* **2000**, *39*, 3012–3043.
51 M. R. Buchmeiser, *Chem. Rev.* **2000**, *100*, 1565–1604.

52 C. Gürtler, M. Jautelat, *European Patent* **2000**, EP 1035093.

53 K. Matyjaszewski, *J. Macromol. Sci., Pure Appl. Chem.* **1997**, *10*, 1785–1801.

54 K. Matyjaszewski, J. Xia, *Chem. Rev.* **2001**, *101*, 2921–2990.

55 D. M. Haddleton, D. Kukulj, A. P. Radigue, *J. Chem. Soc., Chem. Commun.* **1999**, 99–100.

56 D. M. Haddleton, S. G. Jackson, S. A. F. Bon, *J. Am. Chem. Soc.* **2000**, *122*, 1542–1543.

57 A. J. Carmichael, D. M. Haddleton, S. A. F. Bon, K. R. Seddon, *J. Chem. Soc., Chem. Commun.* **2000**, 1237–1238.

58 K. R. Seddon, A. Stark, M. J. Torres, *Pure Appl. Chem.* **2000**, *12*, 2275–2287.

59 T Biedron, P Kubisa, *Macromol. Rapid Commun.* **2001**, *22*, 1237–1242.

60 A. J. Frank, K. Honda, *J. Phys. Chem.* **1982**, *86*, 1933–1935.

61 R. A. Bull, F. R. Fran, A. J. Bard, *J. Electrochem. Soc.* **1983**, *130*, 1636–1638.

62 B. J. Feldman, P. Burgmayer, R. W. Murray, *J. Am. Chem. Soc.* **1985**, *107*, 872–878.

63 P. Kovacic, M. B. Jones, *Chem. Rev.* **1987**, *87*, 357–379.

64 V. M. Kobryanskii, S. A. Arnautov, *J. Chem. Soc., Chem. Commun.* **1992**, 727–728.

65 V. M. Kobryanskii, S. A. Arnautov, *Synth. Met.* **1993**, *55*, 1371–1376.

66 S. A. Arnautov, V. M. Kobryanskii, *Macromol. Chem. Phys.* **2000**, *201*, 809–814.

67 S. A. Arnautov, *Synth. Met.* **1997**, *84*, 295–296.

68 D. C. Trivedi, *J. Chem. Soc., Chem. Commun.* **1989**, 544–545.

69 L. M. Goldenberg, R. A. Osteryoung, *Synth. Met.* **1994**, *64*, 63–68.

70 L.. Janiszewska, R. A. Osteryoung, *J. Electrochem. Soc.* **1988**, *135*, 116–122.

71 P. G. Pickup, R. A. Osteryoung, *J. Am. Chem. Soc.* **1984**, *106*, 2294–2299.

72 J. Tang, R. A. Osteryoung, *Synth. Met.* **1991**, *45*, 1–13.

73 L.. Janiszewska, R. A. Osteryoung, *J. Electrochem. Soc.* **1987**, *134*, 2787–2794.

74 R. T. Carlin, R. A. Osteryoung, *J. Electrochem. Soc.* **1994**, *141*, 1709–1713.

75 T. A. Zawodzinski, L. Janiszewska, R. A. Osteryoung, *J. Electroanal. Chem.* **1988**, *255*, 111–117.

8
Biocatalytic Reactions in Ionic Liquids

Udo Kragl, Marrit Eckstein, and Nicole Kaftzik

8.1
Introduction

Biocatalytic reactions and production processes have been established as useful tools for several decades. The Reichstein process for the oxidation of D-sorbitol to L-sorbose by the use of whole microorganisms, which is still in use, was introduced as early as 1934 [1]. Several years ago, BASF introduced a lipase-catalyzed process for the kinetic resolution of chiral amines [2]. During the history of biocatalysis, alternative reaction conditions have been investigated with the goals of overcoming such problems as substrate solubility, selectivity, yield, or catalyst stability. Some progress has been made through the use of organic solvents [3, 4], the addition of high salt concentrations [5], and the use of microemulsions [6] or supercritical fluids [7]. Recently the methods of gene technology – site-directed mutagenesis and directed evolution – have added new and powerful tools for the development of better biocatalysts [8, 9]. It was thus unsurprising that researchers in the field of biocatalysis have begun to focus on ionic liquids as novel solvents in order to find new solutions to known problems.

In this chapter, we try to summarize the work so far reported in this field. We first give a short introduction into the different forms of biocatalytic reactions, highlighting some special properties of biocatalysts.

8.2
Biocatalytic Reactions and their Special Needs

Biotechnological processes may be divided into fermentation processes and biotransformations. In a fermentation process, products are formed from components in the fermentation broth, as primary or secondary metabolites, by microorganisms or higher cells. Product examples are amino acids, vitamins, or antibiotics such as penicillin or cephalosporin. In these cases, co-solvents are sometimes used for *in situ* product extraction.

The term "biotransformation" or "biocatalysis" is used for processes in which a starting material (precursor) is converted into the desired product in just one step. This can be done by use either of whole cells or of (partially) purified enzymes. Product examples range from bulk chemicals (such as acrylamide) to fine chemicals and chiral synthons (chiral amines or alcohols, for example). There are several books and reviews dealing with the use of biotransformations either at laboratory or at industrial scales [1, 10–13].

Biocatalysts in nature tend to be optimized to perform best in aqueous environments, at neutral pH, temperatures below 40 °C, and at low osmotic pressure. These conditions are sometimes in conflict with the need of the chemist or process engineer to optimize a reaction with respect to space-time yield or high product concentration in order to facilitate downstream processing. Furthermore, enzymes and whole cells are often inhibited by products or substrates. This might be overcome by the use of continuously operated stirred tank reactors, fed-batch reactors, or reactors with *in situ* product removal [14, 15]. The addition of organic solvents to increase the solubility of substrates and/or products is a common practice [16].

Generally, there are three ways to use organic solvents or ionic liquids in a biocatalytic process:

1. as a pure solvent,
2. as a co-solvent in aqueous systems, or
3. in a biphasic system.

When either the organic solvent or the ionic liquid is used as pure solvent, proper control over the water content, or rather the water activity, is of crucial importance, as a minimum amount is necessary to maintain the enzyme's activity. For ionic liquids, a reaction can be operated at constant water activity by use of the same methods as established for organic solvents [17]. [BMIM][PF_6] or [BMIM][$(CF_3SO_2)_2N$], for example, may be used as pure solvents and in biphasic systems. Water-miscible ionic liquids, such as [BMIM][BF_4] or [MMIM][$MeSO_4$], can be used in the second case.

It should be noted that, despite the success of the application of conventional organic solvents, there is no general rule as to which solvent is "enzyme friendly". To a certain extent, the log P concept, based on the distribution coefficient between water and octanol, can be used as guideline [18]. In general, solvents with a log P value greater than 3, such as xylene (3.1) or hexane (3.9), are less deactivating than those with a low log P value, such as ethanol (–0.24). Certainly, the hydrophilicity of the co-solvent is important, as it allows interaction with and breaking of hydrogen bonds that stabilize the tertiary structure of the protein. Such interactions are very likely to occur with ionic liquids as well. Surprisingly, enzymes and even whole cells are active in various ionic liquids, as shown in Section 8.3. So far, ionic liquids have not been treated according to the log P concept. However, the polarities of ionic liquids have been investigated by different groups [19–22]. The polarities of different ionic liquids such as [BMIM][PF_6] or [EMIM][$(CF_3SO_2)_2N$] are similar to those of polar solvents such as ethanol or *N*-methylformamide. On Reichardt's normalized polarity scale, ranging from 0 for tetramethylsilane to 1 for water, ionic liquids have polarities around 0.6. Toluene (0.1) and MTBE (0.35) are less polar [22, 21] (for more details on the polarity of ionic liquids see Section 3.5). Both of these solvents are commonly used as water-immiscible solvents in enzyme catalysis. When used with whole cells, organic solvents often damage the cell membrane. So far, only little is known about the toxicity of ionic liquids, although LD_{50} values of 1400 mg kg^{-1} in female Wistar rats have recently been reported for 3-hexoyloxymethyl-1-methylimidazolium tetrafluoroborate [23]. From this the authors concluded that tetrafluoroborates could be used safely.

When starting our first experiments with available ionic liquids, in screening programs to identify suitable systems, we encountered several difficulties such as pH shifts or precipitation. More generally, the following aspects should be taken into account when ionic liquids are used with biocatalysts:

- In some cases, impurities in the ionic liquids resulted in dramatic pH shifts, causing enzyme inactivation. This could sometimes be overcome simply by titration or higher buffer concentrations. In other cases, purification of the ionic liquid or an improved synthesis might be necessary.
- Enzymatic reactions are often performed in aqueous buffer solution; addition of increasing amounts of ionic liquids sometimes caused precipitates of unknown composition.
- To maintain enzymatic activity a minimal amount of water has to be present, best described by the water activity. However, water present in the reaction system may cause hydrolysis of some ionic liquids.
- Some enzymes require metal ions – such as cobalt, manganese or zinc – for their activity; if these are removed by the ionic liquid by complexation, enzyme inactivation may occur.
- Impurities or the ions of the liquid themselves may act as reversible or irreversible enzyme inhibitors.
- For kinetic investigations and for activity measurements, either photometric assays or – because of the higher complexity of the reactants converted by biocatalysts – HPLC methods can often be used. Here the ionic liquid itself or impurities may interfere with the analytical method.
- Unlike in the case of conventional organic solvents, most research groups prepare the ionic liquids themselves. This may be the reason why different results are sometimes obtained with the same ionic liquids. Park and Kazlauskas performed a washing procedure with aqueous sodium carbonate and found improved reaction rates, but this might also be related to a more precisely defined water content/water activity in the reaction system [22].

As with organic solvents, proteins are not soluble in most of the ionic liquids when they are used as pure solvent. As a result, the enzyme is either applied in immobilized form, coupled to a support, or as a suspension in its native form. For production processes, the majority of enzymes are used as immobilized catalysts in order to facilitate handling and to improve their operational stability [24–26]. As support, either inorganic materials such as porous glass or different organic polymers are used [27]. These heterogeneous catalyst particles are subject to internal and external mass transport limitations, which are strongly influenced by the viscosity of the reaction medium. For [BMIM][$(CF_3SO_2)_2N$], a dynamic viscosity of 52 mPa s at 20 °C has been reported [19]. For comparison, MTBE has a viscosity of only 0.34 mPa s. The viscosity can be reduced to a large extent by increasing the temperature or by addition of small amounts of an organic solvent [28] (for more information on viscosity of ionic liquids, see Section 3.2). This important aspect of the use of ionic liquids in biocatalysis warrants further study.

8.3 Examples of Biocatalytic Reactions in Ionic Liquids

Thanks to their special properties and potential advantages, ionic liquids may be interesting solvents for biocatalytic reactions to solve some of the problems discussed above. After initial trials more than 15 years ago, in which ethylammonium nitrate was used in salt/water mixtures [29], results from the use of ionic liquids as pure solvent, as co-solvent, or for biphasic systems have recently been reported. The reaction systems are summarized in Tables 8.3-1 and 8.3-2, below. Table 8.3-1 compiles all biocatalytic systems except lipases, which are shown separately in 8.3-2. Some of the entries are discussed in more detail below.

8.3.1 Whole-cell Systems and Enzymes other than Lipases in Ionic Liquids

In 1984, Magnuson et al. (Entry 1) investigated the influence of ethylammonium/water mixtures on enzyme activity and stability [29]. At low [H_3NEt][NO_3] concentrations, an increased activity of alkaline phosphatase was found. The same ionic liquid was used by Flowers and co-workers, who found improved protein refolding after denaturation (Entry 2) [30].

So far only two groups have reported details of the use of ionic liquids with whole-cell systems (Entries 3 and 4) [31, 32]. In both cases, [BMIM][PF_6] was used in a two-phase system as substrate reservoir and/or for *in situ* removal of the product formed, thereby increasing the catalyst productivity. Scheme 8.3-1 shows the reduction of ketones with bakers' yeast in the [BMIM][PF_6]/water system.

The recovery of *n*-butanol from a fermentation broth in a similar way has been investigated by *in situ* extraction with [BMIM][PF_6] (Entry 5) [33].

In the first publication describing the preparative use of an enzymatic reaction in ionic liquids, Erbeldinger et al. reported the use of the protease thermolysin for the synthesis of the dipeptide Z-aspartame (Entry 6) [34]. The reaction rates were comparable to those found in conventional organic solvents such as ethyl acetate. Additionally, the enzyme stability was increased in the ionic liquid. The ionic liquid was recycled several times after the removal of non-converted substrates by extraction with water and product precipitation. Recycling of the enzyme has not been reported. It should be noted, however, that according to the log P concept described in the previous section, ethyl acetate – with a value of 0.68 – may interfere with the pro-

Table 8.3-1: Whole-cell systems and enzymes other than lipases in ionic liquids.

Entry	Biocatalyst	Ionic liquid	Reaction system	Ref.
1	Alkaline phosphatase *E. coli*	$[H_3NEt][NO_3]$	Enzyme activity and stability assayed by hydrolysis of *p*-nitrophenol phosphate	29
2	Hen egg white lysozyme	$[H_3NEt_3][NO_3]$	Protein renaturation	30
3	Whole cells of *Rhodococcus* R312	$[BMIM][PF_6]$/buffer (two-phase) $[BMIM][PF_6]$	Biotransformation of 1,3-dicyanobenzene; extraction of erythromycin	 31
4	Whole cells of bakers' yeast	$[BMIM][PF_6]$/buffer (two-phase)	Reduction of ketones	32
5	Whole cells of yeast	$[BMIM][PF_6]$/buffer (two-phase)	Recovery of *n*-butanol from fermentation broth	33
6	Thermolysin	$[BMIM][PF_6]$	Synthesis of *Z*-aspartame	34
7	α-Chymotrypsin	$[OMIM][PF_6]$ $[BMIM][PF_6]$	Transesterification of *N*-acetyl-L-phenylalanine ethyl ester with 1-propanol	35
8	α-Chymotrypsin	$[EMIM][BF_4]$ $[EMIM][(CF_3SO_2)_2N]$ $[BMIM][BF_4]$ $[BMIM][PF_6]$ $[MTOA] [(CF_3SO_2)_2N]$	Transesterification of *N*-acetyl-L-tyrosine ethyl ester with 1-propanol	36
9	β-Galactosidase subtilisin	$[BMIM][BF_4]$/buffer (one-phase)	Hydrolytic activity	37
10	*β-Galactosidase Bacillus circulans*	$[MMIM][MeSO_4]$/buffer (one-phase)	Synthesis of *N*-acetyl-lactosamine	38
11	*Peptide amidase* *β-Galactosidase Bacillus circulans*	$[BMIM][MeSO_4]$ $[MMIM][MeSO_4]$	Amidation of H-Ala-Phe-OH; Synthesis of lactose by reverse hydrolysis	39
12	Formate dehydrogenase	$[MMIM][MeSO_4]$ $[4-MBP][BF_4]$	Regeneration of NADH	41

Scheme 8.3-1

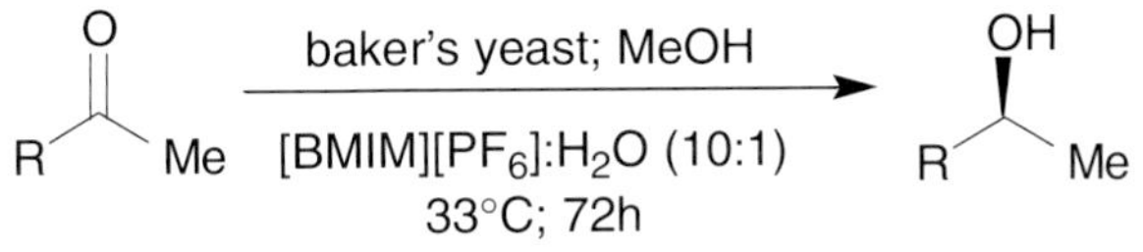

R–C(=O)–Me → (baker's yeast; MeOH / [BMIM][PF_6]:H_2O (10:1) / 33°C; 72h) → R–CH(OH)–Me

R = -C_4H_9 (yield 22%, ee_S 95%)
R = -CH_2-COOEt (yield 75%, ee_S 84%)

[32]

tein in an undesired way. The commercial production process for aspartame uses the soluble enzyme in an aqueous system [1].

The protease α-chymotrypsin has been used for transesterification reactions by two groups (Entries 7 and 8) [35, 36]. N-Acetyl-l-phenylalanine ethyl ester and N-acetyl-l-tyrosine ethyl ester were transformed into the corresponding propyl esters (Scheme 8.3-2).

Laszlo and Compton used [OMIM][PF_6] and [BMIM][PF_6] and compared the results with those obtained with other organic solvents such as acetonitrile or hexane (Entry 7) [35]. They also investigated the influence of the water content on enzyme activity, as well as on the ratio of transesterification and hydrolysis. They found that, as with polar organic solvents, a certain amount of water was necessary to maintain enzymatic activity. For both ionic liquids and organic solvents, the rates were of the same order of magnitude. No data concerning the recycling of the enzyme or its stability were given.

Iborra and co-workers (Entry 8) examined the transesterification of *N*-acetyl-L-tyrosine ethyl ester in different ionic liquids and compared their stabilizing effect relative to that found with 1-propanol as solvent [36]. Despite the fact that the enzyme activity in the ionic liquids tested reached only 10 to 50 % of the value in 1-propanol, the increased stability resulted in higher final product concentrations. Fixed water contents were used in both studies.

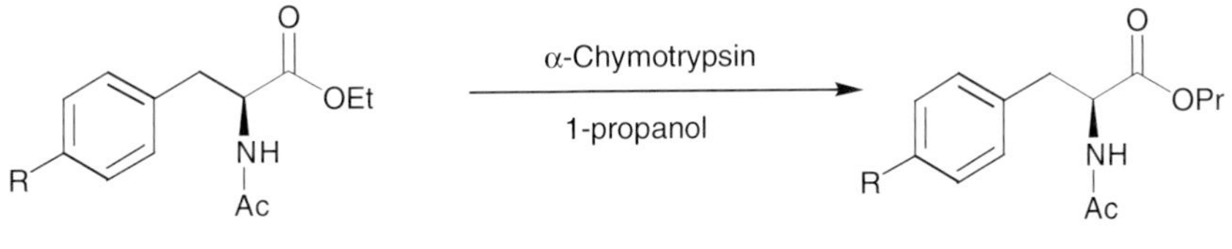

(R-C₆H₄)CH₂–CH(NHAc)–C(=O)OEt → (α-Chymotrypsin / 1-propanol) → (R-C₆H₄)CH₂–CH(NHAc)–C(=O)OPr

[35]: ionic liquid (up to 1.0% v/v H_2O); 40°C
R= H
N-acetyl-L-phenylalanine ethyl ester

[36]: ionic liquid (2% v/v H_2O); 50°C
R= OH
N-acetyl-L-tyrosine ethyl ester

Scheme 8.3-2

Husum et al. found that the hydrolytic activities of β-galactosidase from *E. coli* and the protease subtilisin in a 50 % aqueous solution of the water-miscible ionic liquid [BMIM][BF_4] were comparable to those in 50 % aqueous solutions of ethanol or acetonitrile (Entry 9) [37].

We have studied transfer galactosylation with β-galactosidase from *Bacillus circulans* for the synthesis of *N*-acetyl-lactosamine, starting from lactose and *N*-acetyl-glucosamine (Entry 10) [38]. When the reaction is performed in an aqueous system, the problem of this approach is the secondary hydrolysis of the product by the same enzyme. As a consequence, yields are less than 30 %, and it is important to separate enzyme and product when the maximum yield is obtained. Through the addition of 25 % v/v of [MMIM][$MeSO_4$] as a water-miscible co-solvent, the secondary hydrolysis of the product formed is effectively suppressed, resulting in a doubling of the yield to almost 60 %! Kinetic studies demonstrated that the enzyme activity was not influenced by the presence of the ionic liquid. The enzyme is stable under the conditions employed, allowing its repeated use after filtration with a commercially available ultrafiltration membrane. Further studies to explain the observed effect through analysis of the water activity are underway.

Entries 7, 8, and 10 describe so-called kinetically controlled syntheses starting from activated substrates such as ethyl esters or lactose. In two reaction systems it was possible to demonstrate that ionic liquids can also be useful in a thermodynamically controlled synthesis starting with the single components (Entry 11) [39]. In both cases, as with the results presented in entry 6, the ionic liquids were used with addition of less than 1 % water, necessary to maintain the enzyme activity. The yields observed were similar or better than those obtained with conventional organic solvents.

In order to broaden the field of biocatalysis in ionic liquids, other enzyme classes have also been screened. Of special interest are oxidoreductases for the enantioselective reduction of prochiral ketones [40]. Formate dehydrogenase from *Candida boidinii* was found to be stable and active in mixtures of [MMIM][$MeSO_4$] with buffer (Entry 12) [41]. So far, however, we have not been able to find an alcohol dehydrogenase that is active in the presence of ionic liquids in order to make use of another advantage of ionic liquids: that they increase the solubility of hydrophobic compounds in aqueous systems. On addition of 40 % v/v of [MMIM][$MeSO_4$] to water, for example, the solubility of acetophenone is increased from 20 mmol L^{-1} to 200 mmol L^{-1}.

8.3.2
Lipases in Ionic Liquids

The majority of enzymes reported so far to be active in ionic liquids belong to the class of lipases, the "work horses" of biocatalysis [11]. Designed in nature to work at aqueous/organic interfaces for the cleavage of fats and oils, making the cleavage products accessible as nutrients, lipases in general tolerate and are active in pure organic solvents. This concept has been pioneered by Klibanov and co-workers [16, 42].

Table 8.3-2: Lipases in ionic liquids.

Entry	*Biocatalyst*	*Ionic liquid*	*Reaction system*	*Ref.*
13	*Lipase* *Candida antarctica*	[BMIM][PF_6] [BMIM][BF_4]	Alcoholysis, amminolysis, perhydrolysis	43
14	Screening of eight lipases and two esterases	10 different ionic liquids	Kinetic resolution of (*R,S*)-1-phenylethanol	44
15	Lipases *Candida antarctica* *Pseudomonas cepacia*	[EMIM][PF_6] [BMIM][PF_6]	Kinetic resolution of sec. alcohols	45
16	*Lipase* *Candida antarctica*	[EMIM][BF_4] [BMIM][$(CF_3SO_2)_2N$] [BMIM][PF_6] [EMIM][$(CF_3SO_2)_2N$]	Synthesis of butyl butyrate by transesterification	46
17	Lipases *Candida antarctica* *Pseudomonas cepacia* *Candida rugosa* porcine liver	[BMIM][PF_6] [BMIM][CF_3SO_3] [BMIM][BF_4] [BMIM][$(CF_3SO_2)_2N$] [BMIM][SbF_6]	Kinetic resolution of allylic alcohols	47
18	Lipases *Pseudomonas cepacia* *Candida antarctica*	Several ionic liquids; washing with aqueous sodium carbonate	Kinetic resolution of (*R,S*)-1-phenylethanol; acylation of β-glucose	22
19	Three lipases	[BMIM][PF_6] [BMIM][BF_4]	Synthesis of simple esters	37
20	Lipase *Pseudomonas sp.*	[BMIM][$(CF_3SO_2)_2N$]	Kinetic resolution of (*R,S*)-1-phenylethanol; influence of water activity and temperature	38 48

The report from Sheldon and co-workers was the second publication demonstrating the potential use of enzymes in ionic liquids and the first one for lipases (Entry 13) [43]. They compared the reactivity of *Candida antarctica* lipase in ionic liquids such as [BMIM][PF_6] and [BMIM][BF_4] with that in conventional organic solvents. In all cases the reaction rates were similar for all of the reactions investigated: alcoholysis, ammoniolysis, and perhydrolysis.

Lipases and esterases are often used for kinetic resolution of racemates, variously by hydrolysis, esterification, or transesterification of suitable precursors. Scheme 8.3-3 illustrates the principal for the resolution of a secondary alcohol by esterification with vinyl acetate.

The kinetic resolution of 1-phenylethanol was investigated in our group for a set of eight different lipases and two esterases in ten ionic liquids with MTBE as reference (Entry 14) [44]. Vinyl acetate was used for the transesterification. No activity was observed for the esterases, but for the lipases from *Pseudomonas sp.* and *Alcaligenes sp.*, an improved enantioselectivity was observed in [BMIM] [$(CF_3SO_2)_2N$] as solvent, in comparison to MTBE. The best results were obtained for *Candida antarctica* lipase B in [BMIM][CF_3SO_3], [BMIM][$(CF_3SO_2)_2N$], and [OMIM][PF_6]. Almost no activity was observed in [BMIM][BF_4] and [BMIM][PF_6], contrary to findings of other groups. This might be due to the quality of the ionic liquids we were using at that time. Other groups investigating the same system observed good activities in these ionic liquids (Entries 15–18) [22, 45–47], Park and Kazlauskas even demonstrating the influence of additional washing steps on the enzyme activity [22]. All groups reported excellent enantioselectivities. In addition to our own work, several groups reported the repeated use of the lipase after the workup procedure. In all cases the remaining substrates and formed products were extracted by use either of ether or of hexane. As a consequence of the use of these conditions, there was a slight reduction in enzyme activity after each cycle.

One particular feature of ionic liquids lies in their solvation properties, not only for hydrophobic compounds but also for hydrophilic compounds such as carbohydrates. Park and Kazlauskas reported the regioselective acylation of glucose in 99 % yield and with 93 % selectivity in [MOEMIM][BF_4] (MOE = $CH_3OCH_2CH_2$), values much higher than those obtained in the organic solvents commonly used for this purpose (Entry 18) [22] (Scheme 8.3-4).

Further studies of *Pseudomonas sp.* lipase revealed a strong influence of the water content of the reaction medium (Entry 20) [48]. To be able to compare the enzyme activity and selectivity as a function of the water present in solvents of different polarities, it is necessary to use the water activity (a_w) in these solvents. We used the

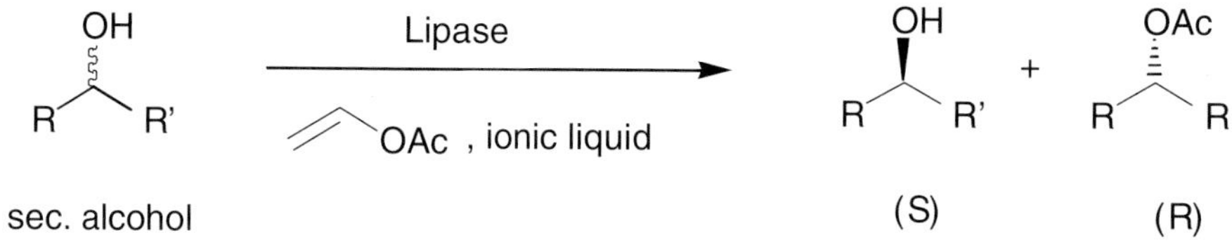

OH R R' sec. alcohol — Lipase, OAc , ionic liquid → OH R R' (S) + OAc R R' (R)

Scheme 8.3-3

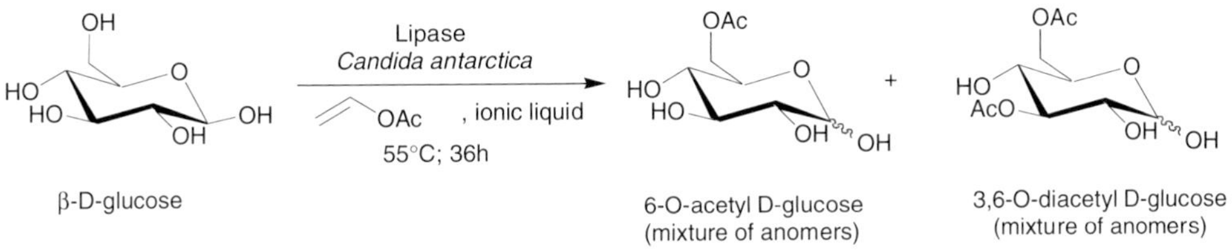

Lipase
Candida antarctica
OAc, ionic liquid
55°C; 36h

β-D-glucose → 6-O-acetyl D-glucose (mixture of anomers) + 3,6-O-diacetyl D-glucose (mixture of anomers)

example: 1-Methoxyethyl-3-methylimidazolium ([MOEMIM]) [BF_4] dissolves ~5mg/ml glucose at 55°C
yield 99%; selectivity: 93% 6-O-acetyl D-glucose **[22]**

Scheme 8.3-4

method of water activity equilibration over saturated salt solutions [49] and were able to demonstrate that, in contrast to MTBE, which is commonly used for this type of reaction, the enantioselectivity of the lipase was less influenced either by the water content or by the temperature when the reaction was performed in [BMIM][$(CF_3SO_2)_2N$].

8.4
Conclusions and Outlook

The results reported so far clearly demonstrate the potential of ionic liquids as solvents for biotransformations. The possible variations for tailor-made solvents may have an impact similar to that of the pioneering work of Klibanov in the use of enzymes in pure organic solvents [42]. Further studies are necessary to identify the reasons for the effects observed, such as better stability, selectivity, or suppression of side reactions. Because of their ionic nature, ionic liquids might interact with charged groups in the enzyme, either in the active site or at its periphery, causing changes in the enzyme's structure. To use ionic liquids in biocatalytic reactions in some cases requires special properties or purities, in order – for example – to avoid changes in the pH of the reaction medium.

When ionic liquids are used as replacements for organic solvents in processes with nonvolatile products, downstream processing may become complicated. This may apply to many biotransformations in which the better selectivity of the biocatalyst is used to transform more complex molecules. In such cases, product isolation can be achieved by, for example, extraction with supercritical CO_2 [50]. Recently, membrane processes such as pervaporation and nanofiltration have been used. The use of pervaporation for less volatile compounds such as phenylethanol has been reported by Crespo and co-workers [51]. We have developed a separation process based on nanofiltration [52, 53] which is especially well suited for isolation of nonvolatile compounds such as carbohydrates or charged compounds. It may also be used for easy recovery and/or purification of ionic liquids.

There is still a long way to go before ionic liquids can become commonly used in biocatalysis. This will require:

- demonstration of stability and recyclability over prolonged periods of times under the reaction conditions applied,
- investigation of mass transport limitations for biocatalysts immobilized on heterogeneous supports, and
- the development of suitable methods for product isolation if they are of limited or no volatility.

References

1 A. Liese, K. Seelbach, C. Wandrey, *Industrial Biotransformations*, Wiley-VCH, Weinheim, **2000**.
2 F. Balkenhol, K. Ditrich, B. Hauer, W. Lander, *J. Prakt. Chem.* **1987**, *339*, 381.
3 G. Carrea, S. Riva, *Angew. Chem. Int. Ed.* **2000**, *39*, 2226.
4 J. M. S. Cabral, M. R. Aires-Barros, H. Pinheiro, D. M. F: Prazeres, *J. Biotechnol.* **1997**, *59*, 133.
5 A. M. Blinkorsky, Y. L. Khmelnitzky, J. S. Dordick, *J. Am. Chem. Soc.* **1999**, *116*, 2697.
6 B. Orlich, R. Schomäcker, *Biotechnol. Bioeng.* **1999**, *65*, 357.
7 T. Hartmann, E. Schwabe, T. Scheper, "*Enzyme catalysis in supercritical fluids*" in R. Patel, *Stereoselective Biocatalysis*, Marcel Dekker, **2000**, 799.
8 U. T. Bornscheuer, M. Pohl, *Curr. Opin. Chem. Biol.* **2001**, *5*, 137.
9 F. Arnold, *Nature* **2001**, *409*, 253.
10 K. Faber, *Biotransformations in Organic Chemistry*, Springer, Berlin, **2000**.
11 U. T. Bornscheuer, R. J. Kazlauskas, *Hydrolases in Organic Synthesis*, Wiley-VCH, Weinheim, **1999**.
12 R. Patel, *Stereoselective Biocatalysis*, Marcel Dekker, **2000**.
13 A. Schmid, J. S. Dordick, B. Hauer, A. Kiener, M. Wubbolts and B. Witholt, *Nature* **2001**, *409*, 258.
14 J. E. Bailey and D. F. Ollis, *Biochemical Engineering Fundamentals*, McGraw–Hill, New York, **1986**.
15 U. Kragl, and A. Liese, in *Encyclopedia of Bioprocess Technology*, edited by M. C. Flickinger and S. W. Drew, Wiley, New York, 454, **2000**.
16 A. Klibanov, *Nature* **2001**, *409*, 241.
17 R. H. Valivety, P. J. Halling, *Biochim. Biophys. Acta* **1992**, *1118*, 218.
18 C. Laane, S. Boeren, R. Hilhorst and C. Veeger, in *Biocatalysis in organic media* (C. Laane, J. Tramper, M. D. Lilly eds.), Elsevier, Amsterdam, **1987**.
19 P. Bonhote, A. P. Dias, K. Papageorgiou, M. Grätzel, *Inorg. Chem.* **1996**, *35*, 1168.
20 A. J. Carmichael, K. R. Seddon, *J. Phys. Org. Chem.* **2000**, *13*, 591.
21 S. Aki, J. Brenneke, A. Samanta, *Chem. Comm.* **2001**, 413.
22 S. Park, R. J. Kazlauskas, *J. Org. Chem.* **2001**, *66*, 8395.
23 J. Pernak, A. Czepukowicz, R. Pozniak, *Ind. Eng. Chem. Res.* **2001**, *40*, 2379.
24 U. Kragl, L. Greiner and C. Wandrey, in *Encyclopedia of Bioprocess Technology* (M. C. Flickinger, S. W. Drew eds.), Wiley, New York, 1064, **2000**.
25 W. Tischer, V. Kasche, *Trends Biotechnol.* **1999**, 17, 326.
26 E. Katchalski-Katzir, D. M. Kraemer, *J. Mol. Cat. B* **2000**, *10*, 157.
27 W. Keim and B. Drießen-Hölscher in *Handbook of Heterogeneous Catalysis* (G. Ertl, H. Knözinger, J. Weitkamp eds.), Wiley-VCH, Weinheim, 231, **1997**.
28 P. Wasserscheid, W. Keim, *Angew. Chem. Int. Ed.* **2000**, *112*, 3926.
29 D. K. Magnuson, J. W. Bodley, D. F. Evans, *J. Solution Chem.* **1984**, *13*, 583.
30 C. A. Summers, R. A. Flowers, *Protein Science* **2000**, *9*, 2001.
31 S. G. Cull, J. D. Holbrey, V. Vargas-Mora, K. R. Seddon, G. J. Lye, *Biotechnol. Bioeng.* **2000**, *69*, 227.
32 J. Howarth, P. James, J. Dai, *Tetrahedron Lett.* **2001**, *42*, 7517.
33 A. G. Fadeev, M. M. Meagher, *Chem. Comm.* **2001**, 295.

34 M. Erbeldinger, A. J. Mesiano, A. J. Russel, *Biotechnol. Prog.* **2000**, *16*, 1129.

35 J. A. Laszlo, D. L. Compton, *Biotechnol. Bioeng.* **2001**, *75*, 181.

36 P. Lozano, T. de Diego, J.-P. Guegan, M. Vaultier, J. L. Iborra, *Biotechnol. Bioeng.* **2001**, *75*, 563.

37 T. L. Husum, C. T. Jorgensen, M. W. Morten, O. Kirk, *Biocatal. Biotransform.* **2001**, *19*, 331.

38 U. Kragl, N. Kaftzik, S. H. Schöfer, M. Eckstein, P. Wasserscheid, C. Hilgers, *CHIMICA OGGI/Chemistry Today* **2001**, *7/8*, 22–24.

39 N. Kaftzik, S. Neumann, M.-R. Kula, U. Kragl, submitted.

40 M. R. Kula, U. Kragl, "*Dehydrogenases in the synthesis of chiral compounds*" in R. Patel, *Stereoselective Biocatalysis*, Marcel Dekker, **2000**, 839.

41 N. Kaftzik, *unpublished results.*

42 A. M. Klibanov, *CHEMTECH* **1986**, *16*, 354.

43 R. Madeira Lau, F. van Rantwijk, K. R. Seddon, R. A. Sheldon, *Org. Lett.* **2000**, *2*, 4189.

44 S. Schöfer, N. Kaftzik, P. Wasserscheid, U. Kragl, *Chem. Comm.* **2001**, 425.

45 K.-W. Kim, B. Song, M.-Y. Choi, M.-J. Kim, *Org. Lett.* **2001**, *3*, 1507.

46 P. Lozano, T. de Diego, D. Carrie, M. Vaultier, J. L. Iborra, *Biotechnol. Lett.* **2001**, *23*, 1529.

47 T. Itoh, E. Akasaki, K. Kudo, S. Shirakami, *Chem. Lett.* **2001**, 262.

48 M. Eckstein, P. Wasserscheid, U. Kragl, *Biotechnol. Lett.* **2002**, *24*, 763.

49 H. L Goderis, G. Ampe, M. P. Feyten, B. L. Fouwe, W. M. Guffens, S. M van Cauwenberg, P. P. Tobback, *Biotechnol. Bioeng.* **1987**, *30*, 258.

50 L. A. Blanchard, J. F. Brennecke, *Ind. Eng. Chem. Res.* **2001**, *40*, 287.

51 T. Schäfer, C. M. Rodrigues, C. A. M. Afonso, J. G. Crespo, *Chem. Comm.* **2001**, 1622.

52 G. Dudziak, S. Fey, L. Hasbach, U. Kragl, *J. Carbohydr. Chem.* **1999**, *18*, 41.

53 J. Kröckel, unpublished results.

9
Outlook

Peter Wasserscheid and Tom Welton

It has been our intention in the eight preceding chapters to provide the essential information for a deep understanding of the nature of ionic liquids as well as a comprehensive review of all different synthetic applications that have so far benefited from ionic liquid technology. For some areas the use of ionic liquids seems to be still in its infancy and – despite some promising results – absolute proof of superiority over existing technology is still lacking. In other areas, however, substantial advantages in the replacement of common catalysts or solvents with ionic liquids have already been demonstrated. We now wish to look to the future. However, it is difficult, and probably foolish, to try to predict what will be discovered in the next few years. In fact, the most exciting part of any new science is its ability to cause surprises. So we have taken the approach of trying to answer the questions that we are most commonly asked when telling people about ionic liquids for the first time.

What is going to be the first area of broad, commercial ionic liquid application? This is probably the question most frequently asked of everybody who is active in developing ionic liquid methodology. The answer is not easy to give. Some petrochemical processes are ready to be licensed or are in pilot plant development (as described in Section 5.2), but there is still some time needed to bring these applications on stream and to claim a broad replacement of existing technologies by ionic liquids in this area. For some non-synthetic applications, in contrast, the lead time from the first experiments to full technical realization is much shorter.

For example, Novasina S.A. (www.novasina.com), a Swiss company specializing in the manufacture of devices to measure humidity in air, has developed a new sensor based on the non-synthetic application of an ionic liquid. The new concept makes simple use of the close correlation between the water uptake of an ionic liquid and its conductivity increase. In comparison with existing sensors based on polymer membranes, the new type of ionic liquid sensor shows significantly faster response times (up to a factor of 2.5) and less sensitivity to cross contamination (with alcohols, for example). Each sensor device contains about 50 μl of ionic liquid, and the new sensor system became available as a commercial product in 2002. Figure 9-1 shows a picture of the sensor device containing the ionic liquid, and Figure 9-2 displays the whole humidity analyzer as commercialized by Novasina S.A..

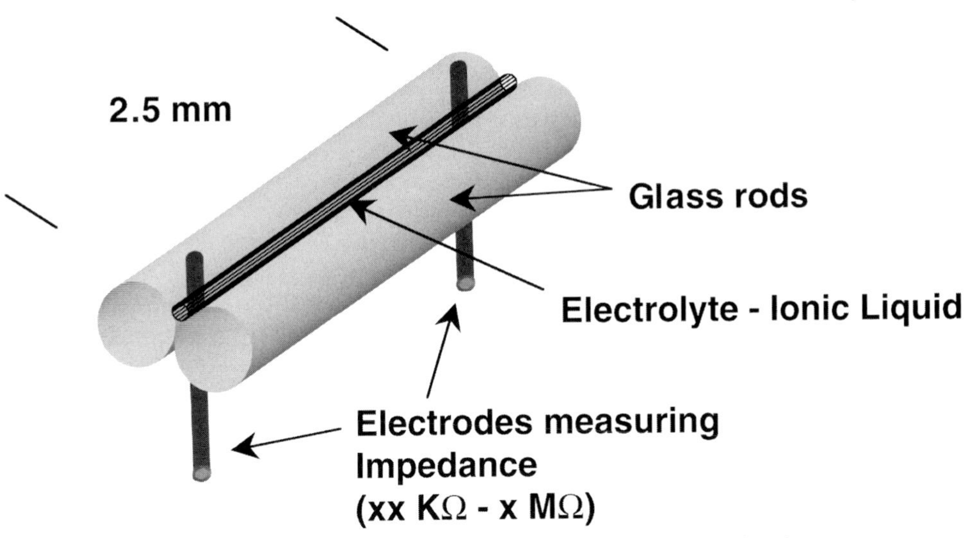

Figure 9-1: Sensor device for measurement of relative humidity, containing an ionic liquid as electrolyte (with permission of Novasina S.A.)

This is only one of some very promising potential non-synthetic applications of ionic liquids that have emerged recently. Many others – some more, some less fully documented in patent or scientific literature – have been published. Table 9-1 gives a few examples, showing that most of the non-synthetic applications of ionic liquids can be grouped into three areas. Electrochemical applications benefit from the wide electrochemical window of ionic liquids and/or from the distinct variation of con-

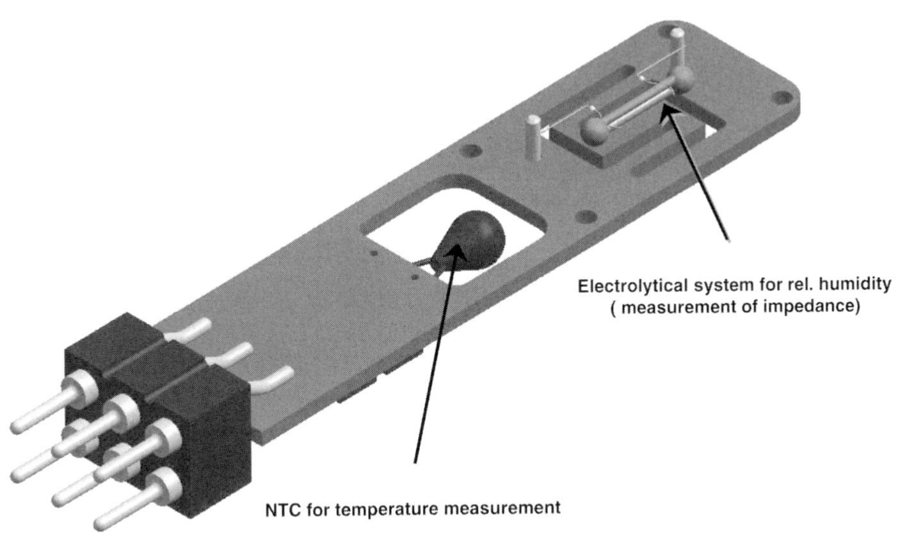

Figure 9-2: High-precision CC-1 measurement cell for measurement of relative humidity and temperature including an ionic liquid as "sensing" electrolyte, as commercialized by Novasina S.A. (with permission of Novasina S.A.).

ductivity if substances are dissolved in them. Analytical applications often profit from the special solubility properties of ionic liquids. Applications in which ionic liquids are used as novel "engineering fluids" are based on their solubility properties, their thermal properties, their mechanical properties, or the special mixture of all these that is provided by some ionic liquids. All applications displayed in Table 9-1 make use – to a greater or lesser extent – of the nonvolatile character of the ionic liquids.

Actually, it is quite likely that the first area of broader technical ionic liquid use will indeed be a non-synthetic application. Why? Certainly not because non-synthetic applications have shown more potential, more performance, or more possibilities, but because many of these are relatively simple, with clearly defined technical targets. The improvement over existing technology is often based on just one or a very few specific properties of the ionic liquid material, whereas for most synthetic appli-

Table 9-1: Non-synthetic application of ionic liquids – selected examples and references.

Application	*Research group*	*Reference*
Electrochemical applications		
Ionic liquids as active component in sensors	Dai et al.	1
Ionic liquids as electrolyte in batteries	Caja et al.	2
Electrodeposition of metals from ionic liquids	Endres	3
Analytic applications		
Ionic liquid as matrix for mass spectroscopy	Armstrong et al.	4
Ionic liquid as stationary phase for gas chromatography	Armstrong et al.	5, 6
"Engineering liquids"		
Ionic liquid as heat carrier and thermofluid	Wilkes et al.	7
Ionic liquid as lubricant	Liu et al.	8
Ionic liquid as antistatic	Pernak et al.	9
Ionic liquids as liquid crystals	Seddon, Holbrey, Gordon et al.	10, 11
Ionic liquids as solvents for extraction	*Ionic liquid/aqueous*	
	Rogers et al.	12, 13, 14, 15
	Dietz et al.	16
	Dai et al.	17
	Ionic liquid/hydrocarbon	
	Jess et al.	18
Ionic liquid as active layer in supported liquid phase membranes for gas separation	Melin, Wasserscheid, et al.	19, 20

cations a complex mixture of physicochemical properties in dynamic mixtures has to be considered. So the question of why non-synthetic applications of ionic liquids today look so promising with regard to their technical development can be answered in that these are just quicker and easier to develop, since they do not require the same degree of knowledge about the complex nature of the ionic liquid material.

At this stage of development, knowledge of ionic liquid properties is patchy, to say the least. For some applications only limited, very specific information is needed to allow the translation of a research project into technical reality (mostly non-synthetic applications). For others (mostly synthetic applications), a lot more detailed information, skills, and data are required to make the technology feasible. This process takes time, even though the ever growing ionic liquid community has already added a lot of information to the ionic liquid "toolbox".

Several of the examples in Table 9-1 are looking quite promising for technical realization on a short to medium timescale. Other ideas are still in their infancy, and there is still a lot of potential for the development of other new non-synthetic applications of ionic liquids in the years to come.

How does one identify a promising non-synthetic application for ionic liquid technology? We basically expect that, in all non-synthetic, high value-adding applications, in which the application of an ionic liquid achieves some unique and superior performance of a technical device, ionic liquid technology may have a very good chance of quick and successful introduction.

In this book we have decided to concentrate on purely synthetic applications of ionic liquids, just to keep the amount of material to a manageable level. However, we think that synthetic and non-synthetic applications (and the people doing research in these areas) should not be treated separately for a number of reasons. Each area can profit from developments made in the other field, especially concerning the availability of physicochemical data and practical experience of development of technical processes using ionic liquids. In fact, in all production-scale chemical reactions some typically non-synthetic aspects (such as the heat capacity of the ionic liquid or product extraction from the ionic catalyst layer) have to be considered anyway. The most important reason for close collaboration by synthetic and non-synthetic scientists in the field of ionic liquid research is, however, the fact that in both areas an increase in the understanding of the ionic liquid material is the key factor for successful future development.

Why is lack of understanding still the major limitation for the development of ionic liquid methodology? After having read the preceding eight chapters you will probably agree that ionic liquids are complex liquid materials. The detailed study of ionic liquids is still in its infancy, and there has simply been insufficient time to accumulate large amounts of good quality data on a wide range of liquids. Also, the fact that we are beginning to understand more about the basic nature of these materials in their pure form still does not answer the question of what happens to substances dissolved in the ionic liquid. This, though, is what all chemical reactions in ionic liquids are about. To give a very simplistic idea of this important point we can consider that a pure ionic liquid may be regarded – more or less – as big packages of cations and anions (see self-diffusion measurements and electrical conductivity measurements in Chapters 3 and 4). However, a very dilute solution of an ionic liquid in a molecular solvent (or substrate or product) will probably be much more like ion-pairs dissolved in the solvent. Is this still an ionic liquid then? Probably not. Can it still have some of the typical ionic liquid features (such as activation, solvation of ionic species, etc.)? Maybe. This leads to questions such as: what is the critical concentration of the ionic liquid in such a solution (e.g., with the substrate/product during a chemical reaction) for the system to display ionic liquid-like behavior? Or how do the physicochemical properties of the pure ionic liquid change in the reaction mixture when reactants are dissolved in the medium?

A few examples from the literature should illustrate this aspect further. Seddon et al., for example, have described the great influence of relatively small amounts of impurities on the physicochemical properties of ionic liquids [21]. Chauvin found that traces of Cl^- ion impurities prevented rhodium-catalyzed hydrogenation of olefins [22], whereas Welton found that the same impurities were needed in order to allow the Suzuki reaction to proceed [23]. Song et al. reported significant activation of an Mn(salen) complex in a solution consisting of 20 volume% of ionic liquid in CH_2Cl_2 versus pure CH_2Cl_2 [24]. Wasserscheid et al. found that the strength of diastereomeric interactions between a chiral ionic liquid and a chiral substrate was strongly dependent on the concentration of the substrate in the ionic liquid [25].

Of course, these concentration effects will be highly dependent on the nature of the substrate dissolved in the ionic liquid, as well as on the nature of the ionic liquid's cation and anion. Given the enormous opportunity to vary these last two, it becomes clear that a detailed understanding of the role of the ionic liquid in reaction mixtures is far from complete. Clearly, this limited understanding is currently restricting our opportunities to benefit from the full potential of an ionic liquid solvent in a given synthetic application.

One frequently discussed idea by which to overcome the lack of available data and understanding on a short time range is to pick one, "universal" ionic liquid and to study this one in very great detail, instead of developing many new systems (combined with an obvious lack in detailed information about these).

Is there a "universal" ionic liquid at the present state of development? The answer is clearly no. Many of the ionic liquids commonly in use have very different physical and chemical properties (see Chapter 3) and it is absolutely impossible that one type of ionic liquid could be used for all synthetic applications described in Chapters 5–8. In view of the different possible roles of the ionic liquid in a given synthetic application (e.g., as catalyst, co-catalyst, or innocent solvent) this point is quite obvious. However, some properties, such as nonvolatility, are universal for all ionic liquids. So the answer becomes, if the property that you want is common to all ionic liquids, then any one will do. If not, you will require the ionic liquid that meets your needs.

Nevertheless, a certain process of focussing can be expected in the future. The authors expect that this process will give rise to two different groups of ionic liquids that will be routinely used throughout academia and industry.

The first group is expected to fall under the definition of "bulk ionic liquids". This means a class of ionic liquids that is produced, used, and somehow consumed in larger quantities (>100 liter ionic liquid consumption per application unit per year). Applications for these ionic liquids are expected to be as solvents for organic reactions, homogeneous catalysis and biocatalysis, and other synthetic applications with some ionic liquid consumption: heat carriers, lubricants, additives, new surfactants, new phase-transfer catalysts, extraction solvents, solvents for extractive distillation, antistatics, etc. These "bulk ionic liquids" would be relatively cheap (around €30 per liter), halogen-free (e.g., for easy disposal of spent ionic liquid) and toxicologically well characterized (a preliminary study about the acute toxicity of a non-chloroaluminate ionic liquid has recently been published [9]) . We expect that, of all ionic liquids meeting these requirements, only a very limited number of candidates will be selected for the described applications. However, these candidates will become well characterized and – because of their larger production quantities – readily available.

On the other hand, we also anticipate a wider range of highly specialized ionic liquids that will be produced and consumed in smaller quantities (<100 liter ionic liquid consumption per application unit per year). Fields of applications for these highly specialized ionic liquids are expected to be as special solvents for organic synthesis, homogeneous catalysis, biocatalysis and all other synthetic applications with very low ionic liquid consumption (due, for example, to very efficient multiphasic operation), catalytically active ionic liquids with low catalyst consumption, analytic devices (stationary or mobile phases for chromatography, matrixes for MS, etc.), sensors, batteries, electrochemical baths for electrodeposition, etc. This group will contain all sophisticated and relatively expensive ionic liquids, such as task-specific ionic liquids, chiral ionic liquids, expensive fluorine-containing anions, etc. Here we expect that the ionic liquid will be designed and optimized for the best performance in each specific high-value-adding application. Consequently, only scientists' imaginations will limit the number of ionic liquids used in this group.

Which type of reaction should be studied in an ionic liquid? This is another frequently asked question, which is of course closely related to the question of which ionic liquid to use. As mentioned before, not all chemistry will make sense in all types of ionic liquid.

We are far here from aiming to advise anybody about future research projects. The only message that we would like to communicate is that a chemical reaction is not necessarily surprising or important because it somehow works as well in an ionic liquid. One should look for those applications in which the specific properties of the ionic liquids may allow one to achieve something special that has not been possible in traditional solvents. If the reaction can be performed better (whatever *you* may mean by that) in another solvent, then use that solvent. In order to be able to make that judgement, it is imperative that we all include comparisons with molecular solvents in our studies, and not only those that we know are bad, but those that are the best alternatives.

What reaction can be carried out in an ionic liquid that is not possible in organic solvents or water? Many convincing examples have been described in Chapters 5–8.

These should not be repeated here. To identify new examples, the easiest way is probably to start from a detailed understanding of the special properties of the ionic liquid material and to identify promising research fields from this point. Two successful examples from the past should illustrate this approach in more detail.

The fact that ionic liquids with weakly coordinating anions can combine, in a unique manner, relatively high polarity with low nucleophilicity allows biphasic catalysis with highly electrophilic, cationic Ni-complexes to be carried out for the first time [26].

The wide electrochemical windows of ionic liquids, in combination with their ability to serve as solvents for transition metal catalysts, opens up new possibilities for a combination of electrochemistry and transition metal catalysis. A very exciting first example has recently been published by Bedioui et al. [27].

There is still a lot of potential for new and somehow unique synthetic chemistry in ionic liquids, but understanding is crucial to develop the right ideas. We are still at the very beginning. A lot of exciting chemistry is still to be done in ionic liquids!

References

1 S. Dai, C. Liang, *Abstracts of Papers, 222nd ACS National Meeting*, Chicago, IL, United States, August 26–30 **2001**, ANYL.

2 J. Caya, T. D. J. Dunstan, D. M. Ryan, V. Katovic, *Proc. Electrochem Soc.* **2000**, *99-41* (Molten Salts XII), 150.

3 F. Endres, *Phys. Chem. Chem. Phys.* **2001**, *3*, 3165–3174.

4 D. W. Armstrong, L.-K. Zhang, L. He, M. L. Gross, *Anal. Chem.* **2001**, *73(15)*, 3679–3686.

5 D. W. Armstrong, L. He, Y.-S. Liu, *Anal. Chem.* **1999**, *71(17)*, 3873–3876.

6 A. Berthod, L. He, D. W. Armstrong, Chromatographia **2001**, *53(1/2)*, 63–68.

7 M. L. Mutch, J. S. Wilkes, *Proc. Electrochem. Soc.* **1998**, *98-11* (Molten Salt XI), 254.

8 C. Ye, W. Liu, Y. Chen, L. Yu, *Chem. Commun.* **2001**, 2244–2245.

9 J. Pernak, A. Czepukowicz, R. Pozniak, *Ind. Eng. Chem. Res.* **2001**, *40*, 2379–2383.

10 J. D. Holbrey, K. R. Seddon, *J. Chem. Soc., Dalton Trans.* **1999**, *13*, 2133–2140.

11 C. M. Gordon, J. D. Holbrey, A. R. Kennedy, K. R. Seddon, *J. Mater. Chem.* **1998**, *8(12)*, 2627–2736.

12 J. G. Huddleston, R. D. Rogers, *Chem. Commun.* **1998**, 1765–1766.

13 A. E. Visser, R. P. Swatloski, S. T. Griffin, D. H. Hartman, R. D. Rogers, *Sep. Technol.* **2001**, *36*, 785–804.

14 A. E. Visser, R. P. Swatloski, W. M. Reichert, J. H. Davis, Jr., R. D. Rogers, R. Mayton, S. Sheff, A. Wierzbicki, *Chem. Commun.* **2001**, 135–136.

15 A. E. Visser, R. P. Swatloski, M. W. Reichert, S. T. Griffin, R. D. Rogers, *Ind. & Eng. Res.* **2000**, *39(10)*, 3596–3604.

16 M. L. Dietz, J. A. Dzielawa, *Chem. Commun.* **2001**, 2124–2125.

17 S. Dai, Y. H. Ju, C. E. Barnes, *J. Chem. Soc., Dalton Trans.* **1999**, *8*, 1201–1202.

18 A. Bösmann, L. Datsevitch, A. Jess, A. Lauter, C. Schmitz, P. Wasserscheid, *Chem. Commun.* **2001**, 2494–2495.

19 M. Medved, P. Wasserscheid, T. Melin, *Proceedings of the 8th Aachener Membran Kolloquium*, **2001**, II-123.

20 M. Medved, P. Wasserscheid, T. Melin, *Chem.-Ing.-Techn.* **2001**, *73*, 715.

21 K. R. Seddon, A. Stark, M.-J. Torres, *Pure Appl. Chem.* **2000**, *72*, 2275–2287.

22 Y. Chauvin, L. Mussmann, H. Olivier, *Angew. Chem., Int. Ed. Engl.* **1995**, *34*, 2698–2700.

23 C. Mathews, P. J. Smith, T. Welton, *Chem. Commun.* **2000**, 1249–1250.

24 E. C. Song, E. J. Roh, *Chem. Commun.* **2000**, 837–838.

25 P. Wasserscheid, A. Bösmann, C. Bolm, *Chem. Commun.* **2002**, 200–201.

26 P. Wasserscheid, C. M. Gordon, C. Hilgers, M. J. Muldoon, I. R. Dunkin, *Chem. Commun.* **2001**, 1186–1187.

27 L. Gaillon, F. Bedioui, *Chem. Commun.* **2001**, 1458–1459.

Index

c

d

u

v

w

x

y

z

Larsen, Christian L
Growth and government in Sacramento [by] Christian L. Larsen [and others] Indiana University Press [1966]
238 p. maps. (Metropolitan action studies, no. 4)

1. Sacramento metropolitan area--Pol. & govt. 2. Metropolitan government--U.S. --Case studies.

METROPOLITAN ACTION STUDIES NO. 4

Growth and Government in Sacramento

THE AUTHORS

James R. Bell, Professor of Government and Coordinator of Public Administration Curricula, Sacramento State College

Leonard D. Cain, jr., Professor and Head, Department of Sociology, Sacramento State College

Lyman A. Glenny, Executive Director, Board of Higher Education, State of Illinois

William H. Hickman, Professor and Head, Department of Economics, Sacramento State College

Irl A. Irwin, Associate Professor of Psychology, Sacramento State College

Christian L. Larsen, Professor and Head, Department of Government and Police Science, Sacramento State College

METROPOLITAN ACTION STUDIES NO. 4

Growth and Government in Sacramento

CHRISTIAN L. LARSEN

JAMES R. BELL

LEONARD D. CAIN, JR.

LYMAN A. GLENNY

WILLIAM H. HICKMAN

IRL A. IRWIN

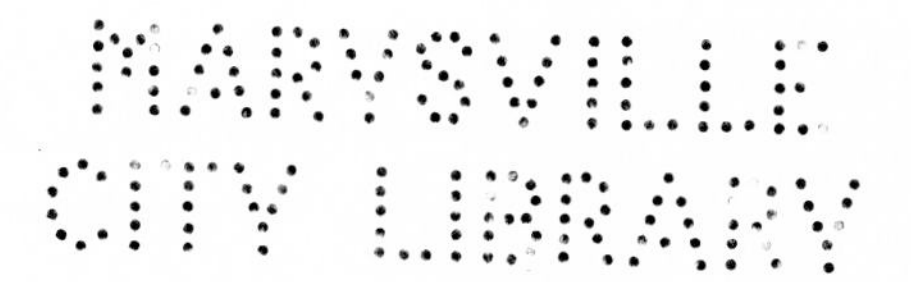

BLOOMINGTON • INDIANA UNIVERSITY PRESS • LONDON

Manufactured in the United States of America
Library of Congress catalog card number: 65-19708

Contents

Tables, Maps

TABLES

MAPS

Foreword

The most difficult and complex problems of domestic public policy of our time and nation are associated with the changing characteristics of the urban community. Congestion, deterioration, and delinquency in the older cities; ugliness and sprawl in the suburbs; the tensions accompanying social and economic segregation as the exploding populations group and regroup themselves into communities which only inadequately provide for common purposes and the public welfare—these are pervasive and endemic in current American society.

The institutions and arrangements for confronting the basic public policy questions lag behind the needs of changing social patterns. There is often no appropriate forum, no viable governmental entity, through which community decisions can be considered and reached. Nearly all American cities, faced with institutional inadequacies, have struggled with the need to adjust their govermental units and agencies to cope with the problems of urban services and urban direction.

This is the fourth in a group of case studies of the efforts of particular American cities to adapt and adjust their governmental institutions to the new metropolitan problems. *Decisions in Syracuse* (1961), *The Miami Metropolitan Experiment* (1963), and *The Milwaukee Metropolitan Study Commission* (1965), all published by the Indiana University Press, have preceded it. The four cities described in these case studies differ significantly, yet there is much in common in their problems and in their reactions to the problems. Sacramento, capital city of the state which is now the most populous in the union, is frequently overshadowed in national image by its two giant metropolitan neighbors, Los Angeles and San Francisco, but it is itself clearly representative of the new and rapidly growing cities of the West. Much newer and more dynamic in growth than Syracuse or Milwaukee, it is not yet as large as Miami. Its approaches to the governmental problems of rapid urban expansion are not the same as those in the other three cities, but its use of citizen study committees and of outside consultants, of petitions and referenda, have obvious counterparts in the other cities.

These case studies were supported by a series of grants from the Ford Foundation; a grant was also made to Indiana University to enable it to attempt some coordination of the individual studies. The investigations were conducted, in each instance, by a group of social scientists at a university or college in the metropolitan community involved. The authors of this study were members of the faculty at Sacramento State College.

Through this and the other volumes of the series, we hope that a better understanding of the processes of governmental adaptation to social change, and of community decision-making in general, will be promoted.

Indiana University YORK WILLBERN

Preface

In 1959 the Sacramento State College Foundation received a grant from the Ford Foundation to study the governmental reorganization attempts of the Sacramento metropolitan area. The study was proposed and executed by three political scientists, one economist, one sociologist, and one psychologist—all from the Sacramento State College faculty. Professor Larsen, a political scientist, was designated by the group to direct the study.

The grant for this study was one of several made for the study of particular communities and their efforts toward solving specific metropolitan area problems. The studies are intended to serve both comparative and historical purposes. The specific purpose of the grant was to "capture and record" experiences of metropolitan areas. The study was also to be interdisciplinary and to be completed within a two-year period.

The goals and research methods of the Sacramento study were, of course, affected by the terms and purposes of the grant.[1] To have examined in detail all aspects of the governmental units, the personal attributes of public officeholders, the total pattern of interest groupings, and all other possible influences on the reorganization movements not only would have been beyond the resources of time and money available to the research committee, but could well have buried the significance of the primary purpose of "capturing and recording." The committee had to decide which areas to study in order to "capture" the salient influences in the reorganization attempts. Three different classifications of groups could be identified in these attempts: the special interest groups such as realtors, merchants, suburban newspapers, school officials, and local government officials and employees; the civic organizations interested in good government such as the League of Women Voters, neighborhood improvement clubs, and various study committees; and the people active in the movements themselves. In the final analysis, we abandoned depth studies of groups and sample polls of residents in favor of presenting the third group—the people who were active in the reorganization movements—in sharp focus. The group also decided that, in order to present a meaningful picture of attempted governmental changes, the description of events ought to be augmented by a study of the leadership power structure in the area. The following statement served as a guide throughout the study:

> To capture and record the history of efforts since the mid-1940's to secure modification of the governmental forms serving the Sacramento metropolitan area. To attempt to identify, describe, and analyze the forces, opinions, attitudes, obstacles, and associated factors at work in the community relative to the changes which have occurred and the changes which have been proposed as to local governmental arrangements. To describe and evaluate the effectiveness of community-wide leadership and organization as they apply to the metropolitan area.

The major source of information for these reorganization movements was extensive personal interviews with a total of

96 residents of the Sacramento area. These persons were selected because of their activity in one or more specific movements for change or because they were generally acknowledged to be influential in the metropolitan area as a whole. The authors also relied upon newspaper accounts and official documents of groups involved in the reorganization attempts (see Appendix).

Some economic, political, and social background is included when necessary to clarify an action or attitude, but the main emphasis is on description and analysis of the action groups connected with reorganization efforts. Interest groups and "good government" organizations were not studied directly, but were evaluated by the activists in the movements, in interviews and questionnaires. Persons who had not participated in reorganization attempts but who held power and influence in the community were also studied. The reorganization efforts are presented in chronological order and in some sense re-enact the give-and-take between the participants in the movements. The interpretations and evaluations of the authors are, for the most part, postponed until the concluding chapters.

The study committee has not recommended an appropriate governmental structure for the metropolitan area; this was not the purpose of the grant. Observations and suggestions from persons who were interviewed are included, but no single comprehensive solution is suggested and no solution mentioned is endorsed by the study group.

The authors are most grateful to the following organizations and individuals: the Ford Foundation for the generous grant which made this study possible; all the interviewees, who so graciously gave us time from busy schedules and whose contributions to the study cannot be overrated; Mrs. Doris Hollister, Dr. James Cowan, Mr. Walter Isenberg, and Dr. Hubert J. McCormick, for allowing the use of the official minutes and the newspaper histories of the organizations with which they were associated; graduate assistants Thomas Cuny, John Lopes, and Robert Curry, who did an excellent job of locating and organizing materials; Michael Cullivan and John Berg for excellent services in preparing the maps; Miss Nancy Ryan and Miss Nina McCoy for providing capable typing and stenographic assistance, often under the most trying conditions; and Dr. Ernest G. Miller, whose editorial services gave added emphasis to the central theme of this study.

CHRONOLOGICAL OUTLINE OF GOVERNMENTAL REORGANIZATION EVENTS

September-October, 1955—Publication of 19 articles in *Sacramento Bee* on "Sacramento's Community Crisis."

November, 1955—Sacramento Metropolitan Area Advisory Committee (SMAAC) authorized.

February 1, 1956—First meeting of SMAAC.

May 9, 1956—Public Administration Service employed by SMAAC.

September, 1956—Publication of the report of the Urban Government Committee of the Greater North Area Chamber of Commerce entitled *Problems of Local Government in the Greater North Area of Sacramento County*.

November 13, 1956—PAS issued the first preliminary report.

February 19, 1957—PAS issued the second preliminary report.

May 22, 1957—PAS final report released.

November 12, 1957—SMAAC issued its final report relative to the PAS report.

December 18, 1957—Decision made to terminate SMAAC and to create a new committee—the Metropolitan Government Committee (MGC).

January 13, 1958—Final meeting of SMAAC.

February 19, 1958—First meeting of MGC.

June 17, 1958—Proposal to incorporate a big new city made public.

July 3, 1958—Sacramento City Council approval of the proposal to annex the Hagginwood-Del Paso Heights area requested.

November 18, 1958—Effort to incorporate a big new city abandoned.

December 18, 1958—Sacramento City Council approval of the proposal to annex Arden-Arcade requested.

January 20, 1959—Hagginwood-Del Paso Heights annexation proposal defeated by voters.

March 5, 1959—MGC adopted the proposal of mass annexation plus city-county consolidation of some services.

June, 1959—MGC issued its final report entitled *Government Reorganization for Metropolitan Sacramento.*

September 29, 1959—Proposed Arden-Arcade annexation defeated by voters.

June 2, 1964—Voters of North Sacramento approve merger with Sacramento.

METROPOLITAN ACTION STUDIES NO. 4

Growth and Government in Sacramento

1

Metropolitan Sacramento

> Sacramento, not only the Valley's chief city, but the even busier capital of a rich, progressive . . . state, is in the midst of the most phenomenal expansion of all. It is in the midst of it, but curiously apart.[1]

Sacramento, the "Camellia Capital of the World," spreads eastward from the confluence of the Sacramento and American rivers and dominates the northern part of the 450-mile long great Central Valley that lies between the Sierra Nevada to the east and the Coast Range to the west. Sacramento County, which by U.S. Bureau of the Census definition was also the Sacramento metropolitan area when this study was undertaken, has grown to a population of over one-half million people. Between 1950 and 1960 the population increased by more than 225,000. The communities surrounding the capital city have grown faster than the city of Sacramento has. Sacramento has been cited as "a classic example of a small American city that is subject to rather than master of today's powerful new social, economic, and political pressures. . . ."[2] Although the outlying communities have strong ties to the central city, Sacramento, they continue to be governed as small, separate units.

Unless otherwise indicated, in this study Sacramento County is considered as being the Sacramento metropolitan area, since the U.S. Bureau of the Census used this definition of a metropolitan area at the time this study was begun. If any part of a county is considered part of a metropolitan area, then the entire county is so considered. It should be noted, however, that certain areas of neighboring counties are a part of the Sacramento community in a social and an economic sense whereas the southern part of Sacramento County really is not. Certainly, the portion of Yolo County lying just across the Sacramento River to the west of Sacramento city is a part of the Sacramento community. Urban developments along the Roseville freeway have brought the city of Roseville (Placer County) and its surrounding area into the Sacramento metropolitan area in the same sense. This hardly can be said of Isleton, a small incorporated municipality many miles to the south and west of Sacramento city but contained within Sacramento County. An awareness of this difference between the county boundaries and a realistic definition of the metropolitan community was indicated by the Public Administration Service in its survey, *The Government of Metropolitan Sacramento,* published in 1957. However, the Public Administration Service decided not to cross county boundary lines in its study, on the ground that California constitutional complications and political considerations made any approach to metropolitan problems other than one based on the county virtually impossible to put into effect.

During the late 1950's several attempts were made to reorganize the governmental structure of the metropolitan area.

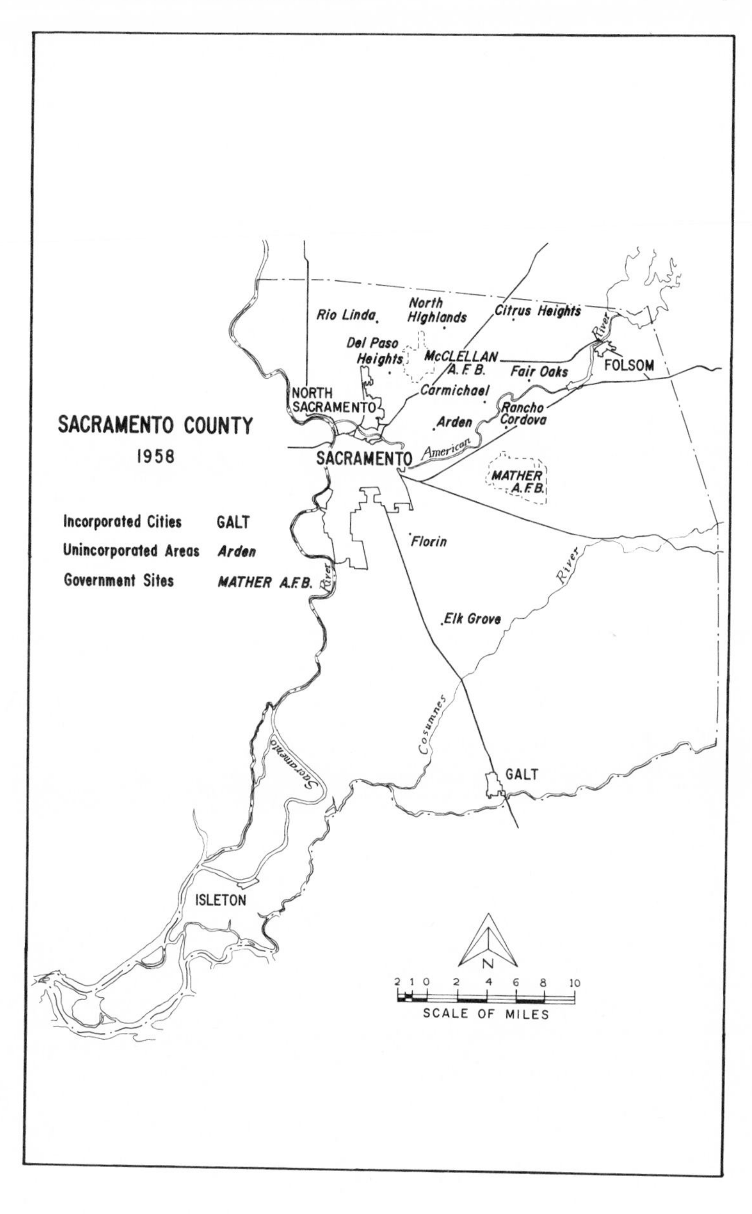
SACRAMENTO COUNTY
1958
Incorporated Cities GALT
Unincorporated Areas Arden
Government Sites MATHER A.F.B.
Rio Linda
North Highlands
Citrus Heights
Del Paso Heights
McCLELLAN A.F.B.
Fair Oaks
FOLSOM
NORTH SACRAMENTO
Carmichael
Rancho Cordova
Arden
SACRAMENTO
American
River
MATHER A.F.B.
Florin
River
Elk Grove
Cosumnes
Sacramento
River
GALT
ISLETON
N
2 1 0 2 4 6 8 10
SCALE OF MILES

Two groups were established to study the problem of governing in the entire area. Two large areas outside the city of Sacramento considered and voted on the question of annexation to the city of Sacramento. One area attempted to incorporate itself as one large city, separate from the city of Sacramento. Because of their related, often overlapping nature, these reorganizational attempts may be viewed as parts of a single, wide-ranging response to the problem of governance in the Sacramento metropolitan area. This study is essentially a record of these movements and an analysis of the citizens' response to them.

Since none of these attempts to consolidate governments in the Sacramento area was accepted by the citizenry at the time (the Hagginwood-Del Paso Heights area annexation has been achieved in more recent years), the conclusions about governmental reorganization so far indicate concrete achievements to have been quite limited. An important problem dealt with in this study, however, is why these attempts at change failed. Answers to this question cannot be found without an understanding of the nature of the communities involved and a recognition of the leading individuals and organizations in the area. Through study of the records of the groups involved in the reorganization efforts, through newspaper accounts, and through extensive interviews with community leaders and leaders in the movements, we attempt to paint a broad picture of the arena in which decisions about governance in the Sacramento metropolitan area have been made.

THE SETTING

Sacramento, the capital of California, is the chief city in a large and rich agricultural area located at the confluence of two of the largest rivers in northern California. Its steady and rather spectacular growth can be attributed largely to the burgeoning state government and to the expansion of federal field offices. The city has experienced a stable government devoted to city interests and committed, at least on the record, to maintaining a good level of municipal services while holding down property taxes. The city has been accused of being smug, parochial, and conservative, and this attitude may be traced in part to the rural background of the central city. This insular point of view is countered by the incipient urbanism resulting primarily from many annexations, which have substantially increased Sacramento's population and have extended its geographical area. In spite of the fact that the population outside the city is growing faster, Sacramento remains the largest distinct governmental, business, and residential entity in the metropolitan area.

The postwar growth of the metropolitan area has increased the size and strengthened the separate identity of several unincorporated business and residential clusters that have grown up in Sacramento County. In a sense they have grown separately, and in some respects, competitively. Their facilities for retail trade and services appeal to the suburbanites. Since the suburban trend after World War II, many new neighborhood centers have emerged in the county, which have been groping for some sense of their identity and seeking an influential role commensurate with their size and economic importance. Each of these subcommunities, with its loosely defined and overlapping geographical boundaries, claims a degree of loyalty or attachment from its residents. For instance, the northeast area includes Arden, Arcade, Country Club Centre, Town and Country, North Highlands, Citrus Heights, Orangevale, and others. These groupings may be further subdivided into loyalty associations of improvement clubs and park or other special districts. Several of these larger neighborhood clusters exceed in population any of the cities in the county except Sacramento.

While the people in these areas consider themselves an important part of Sacramento County, the interview results in this study show that they do not associate their area with any of the cities or share a general sense of community in any tangible, meaningful form. One of their characteristics is rivalry, among themselves and with the city of Sacramento. These localized loyalties are not sufficiently strong to promote incorporation as separate cities but, at the same time, attachments have been strong enough to prevent annexation or incorporation into larger units.

THE CITY OF SACRAMENTO

The city of Sacramento, the largest and first of the five (now four since the merger of North Sacramento with Sacramento) municipalities to be incorporated (August 1, 1849), was granted a charter by the state legislature on February 27, 1850. Its present charter dates back to June 30, 1921. The original boundaries of the city, which encompassed 4.5 square miles, remained unchanged for 65 years. With the annexation of the Oak Park area southeast of Sacramento in 1911, the city limits were extended by 9.4 square miles. No other territory was annexed until 1946, but from that year to 1955, 27 annexations, totaling approximately 9.5 square miles, took place.

Possibly because of the work of the Sacramento Metropolitan Area Advisory Committee and the Public Administration Service, the nationally known consulting firm employed by the committee, no annexation occurred during 1956 and 1957. Several annexations (and one merger) have taken place since then, adding about 69.29 square miles to the city. As of June, 1965, its total area was approximately 92.75 square miles.

Sacramento's annexations, for the most part, have extended

the city's boundaries eastward to the American River and southward for several miles. The annexation, as uninhabited territory, of the new site for the California State Fair in 1959 represented the first acquisition of territory across the American River. This extended the area of Sacramento city to that side of the American River where most of the people of Sacramento County's unincorporated area live. In early 1961 Sacramento's first inhabited area across the river, Northgate, was annexed, and the additional annexation in May, 1961, of approximately 8.5 square miles of contiguous uninhabited territory opened the gate for further expansion into the North Area.

OTHER INCORPORATED AREAS

The city of Sacramento was the county's only incorporated municipality until 1923, when Isleton, a small city in the delta area of the southwestern portion of the county, was incorporated. The census reports for Isleton show a population decline from 1,837 in 1940 to 1,039 in 1960. The municipality has had three small annexations and has a total area of between one-fourth and one-third of a square mile.

North Sacramento was incorporated in 1924. Annexations to the city of Sacramento had resulted in North Sacramento being completely surrounded by Sacramento by the time merger of the two municipalities was approved by North Sacramento voters in June of 1964. North Sacramento got off to a shaky start, when a petition to disincorporate was circulated during the first year of its existence. Its first annexation, in 1939, plus others in the post World War II period, enlarged North Sacramento's area from .75 to 6.58 square miles by the time of the merger. Its 1960 population of 12,922 represented an increase of about 100 per cent over the 1950 figure.

In April, 1946, Folsom became Sacramento County's fourth incorporated municipality. It is situated on the banks of the

American River, approximately 25 miles to the east and north of the city of Sacramento. Until 1959, when it annexed territory across the American River and up to the Placer County line, Folsom lay entirely on the south side of the river. Folsom has actively annexed in recent years, and by January, 1965, it had an area of approximately 7.5 square miles. Its 1960 population was 3,925, compared with 1,690 in 1950.

In August, 1946, Galt, which is located about 30 miles southeast of the city of Sacramento, became the fifth incorporated municipality in Sacramento County. A few small annexations to Galt have been made, and its total area is approximately 1.35 square miles. Its population has risen somewhat—from 1,333 in 1950 to 1,868 in 1960.

The incorporated cities of Isleton, Folsom, and Galt have their own civic identity and pride, and outwardly, at least, they have few ties with Sacramento or with each other. They are self-reliant, and jealous of their municipal status and prerogatives.

SACRAMENTO COUNTY AND URBANIZED UNINCORPORATED AREAS

Sacramento was far from being a metropolitan area in 1850, when the county was created.[3] The entire population of the county numbered only 9,087. Prior to World War II, Sacramento County was agricultural and rural, typical of counties throughout the country. Mushrooming suburban tracts that have grown up since that time have encircled the incorporated communities and now occupy much of the northern half of the county. The population increased by over 225,000 between 1950 and 1960, and the growth of Sacramento County since the middle 1940's had equaled its growth for all the preceding years (see Table 1). In the 1960 census it ranked eighth among the California counties.[4]

The county has become increasingly urban in function and,

since the mid-1950's, urban in attitude. The area of the county is approximately 983 square miles, and only about 102 of these lay within municipal boundaries as of June, 1965. Although the 1950 census reported more people living within municipalities than outside, by 1960 the proportions had changed—211,421 of Sacramento County's residents lived in the incorporated municipalities, and 291,357 lived outside. Most of the latter reside in the northern one-third of Sacramento County, bordering upon the city of Sacramento.

A new county manager and a "reformed" board of supervisors have provided the sprawling suburban area with some municipal-type services—generally through a hodgepodge of overlapping special districts. While each district meets a need of some segment of the population, in the aggregate these districts are unplanned and disorderly. According to the Public Administration Service, there were 157 special districts (exclusive of school districts), created under 27 separate laws, as of February, 1957. The few special districts that have been abolished or consolidated since 1957 have been offset by the creation of new districts. Only Los Angeles County exceeds Sacramento in number of special districts.[5]

Compared with many metropolitan areas, Sacramento's governmental pattern might not seem overwhelmingly complicated, nor so difficult to streamline. There are only four incorporated municipalities in the county, containing less than one-ninth of the total county area. The great increase in population has taken place primarily during the past 20 years; the total population of the unincorporated areas has surpassed the total population of the incorporated areas only in the past ten years. The people outside the cities have had little time to develop interests and loyalties that might conflict with efforts at governmental consolidation.

Consolidation or unification has been difficult nevertheless.

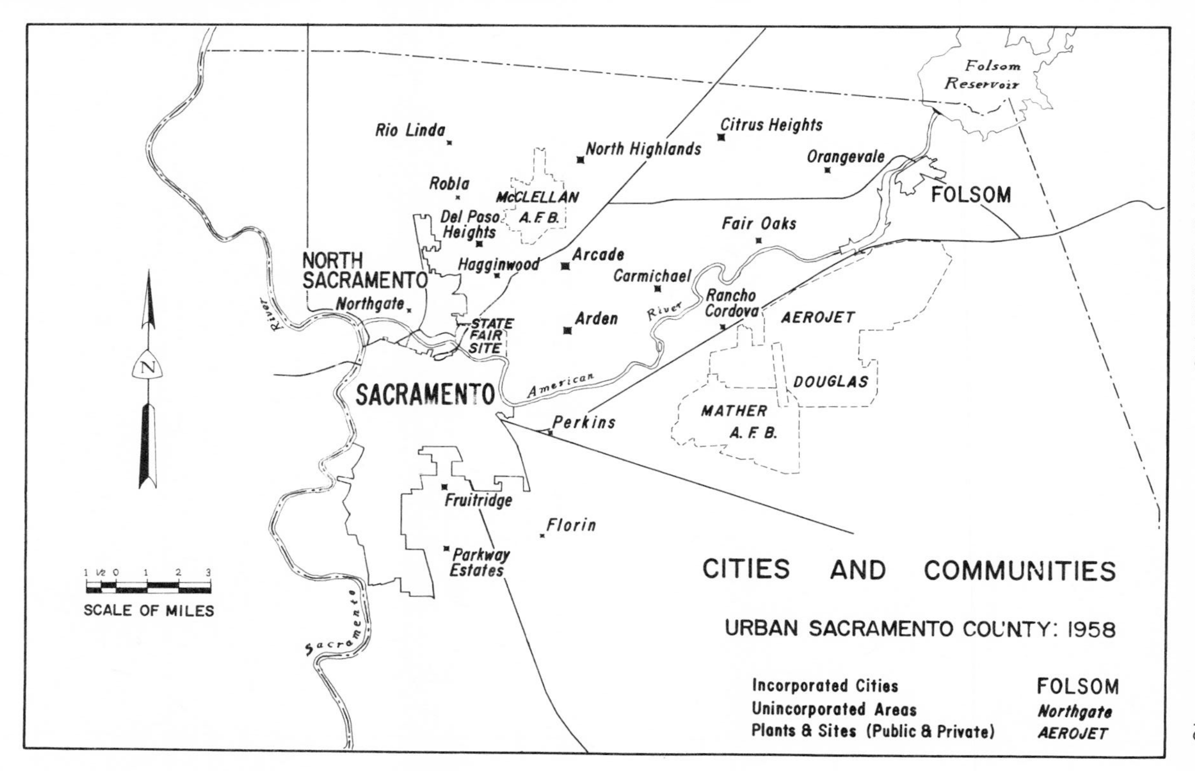

Folsom Reservoir
Rio Linda
North Highlands
Citrus Heights
Orangevale
FOLSOM
Robla
McCLELLAN A.F.B.
Del Paso Heights
Fair Oaks
Arcade
NORTH SACRAMENTO
Hagginwood
Carmichael
Northgate
River
Rancho Cordova
AEROJET
Arden
River
STATE FAIR SITE
DOUGLAS
American
SACRAMENTO
MATHER A.F.B.
Perkins
Fruitridge
Florin
Parkway Estates
1 1/2 0 1 2 3
SCALE OF MILES
Sacramento
CITIES AND COMMUNITIES
URBAN SACRAMENTO COUNTY: 1958
Incorporated Cities FOLSOM
Unincorporated Areas Northgate
Plants & Sites (Public & Private) AEROJET

Many distinct but unincorporated communities, once agricultural, have developed, originally around the crossroads store, the rural church, the country school, and the fire station. During the many years that these communities were small in area and population, a strong feeling of separate identity grew up. As they became a part of the urbanized area of metropolitan Sacramento some residents retained this attachment. Among these, Carmichael, Fair Oaks, Orangevale, Citrus Heights, and Florin are all now part of the urban complex extending north and east from the city of Sacramento.

Also, ordinarily urban expansion would take in first those areas adjacent to the central city, since municipal-type services could be extended to such new developments without great difficulty. But the areas adjacent to Sacramento had problems of drainage, seepage, and river overflow. Real estate developers found it more expedient to begin their subdivisions a few miles beyond the boundaries of the city of Sacramento. In addition, the first residential subdivisions were built close to the major suburban employers such as McClellan Air Force Base, Mather Air Force Base, Aerojet-General Corporation, and Douglas Aircraft. The territory between these outlying residential developments and the municipal boundaries of Sacramento gradually became urbanized, but in the meantime a number of the outlying communities such as North Highlands, Rancho Cordova, Parkway Estates, Arden, and Arcade developed a sense of separate identity.

As population increased, the unincorporated areas created special districts to provide municipal-type services such as police protection, water, and sewage. In California the special district may perform only those functions set forth in the state enabling law. The city of Sacramento, however, under its home rule charter, has broader authority, covering "all municipal and police powers necessary to the complete and efficient administration of the municipal government, although such powers may not be herein expressly enumerated. . . ."[6]

These special districts have a greater claim on the citizen than does the county. Even though the county may provide some municipal services directly, and even though respect for county government has risen considerably in the past few years, the county government is not likely ever to command sufficient loyalty to create a "community." The citizen looks not to the county but to the special district that he joins with his immediate friends and neighbors for essential governmental services.

TABLE 1

Population of City of Sacramento and Sacramento County, 1850-1960

	County	*Per cent of increase*	*City*	*Per cent of increase*
1850	9,087		6,830	
1860	24,142	165.68	13,785	102.13
1870	26,830	11.13	16,283	18.12
1880	34,390	28.18	21,420	31.55
1890	40,339	17.30	26,386	23.18
1900	45,915	13.82	29,282	10.98
1910	67,806	47.80	44,696	45.81
1920	91,029	34.25	65,908	47.46
1930	141,999	55.99	93,750	42.24
1940	170,333	19.95	105,958	13.02
1950	277,140	62.70	137,572	29.83
1960	502,778	81.42	191,667	39.32

Source: United States Bureau of the Census.

OCCUPATIONS AND INCOME LEVELS

Sacramento's unique economic characteristics are important aspects of the metropolitan setting in the late 1950's. In California as a whole, manufacturing and trade combined account for about 45 per cent of total employment, with manufacturing slightly in the lead. Manufacturing employment is increasing most rapidly. Far behind come finance and government employment.[7]

In the Sacramento area, however, government employment and government-contract employment predominate. Together they account for over 40 per cent of total employment, with approximately 20,000 employees each in federal, state, and local government, and government-contract employment. Employment in wholesale and retail trade trails far behind in second place. Employment in service occupations follows and then manufacturing. The only other occupations providing sizable volumes of employment are transportation, communications, and utilities and contract construction. Much of the Federal government employment and the employment in manufacturing (Aerojet-General and Douglas Aircraft, for example) is related closely to military preparedness programs.[8] The income per household in Sacramento County, $7,408, ranks well above the California average of $6,950. In 1959 per capita income for California was $2,250 while in Sacramento County the figure was $2,319. The city of Sacramento has a higher income level—$2,646 per capita and $8,132 per household.[9]

Sacramento, then, is a "government city" with a relatively high income level. The economy of the city, and of the entire metropolitan area is dominated by governmental activity, either directly or indirectly.

PRESSURES FOR REORGANIZATION AND REFORM

THE *Sacramento Bee*

In September and October, 1955, the *Sacramento Bee,* the major metropolitan daily, carried a series of 19 lengthy articles on "Sacramento's Community Crisis." Some of the titles included: "Growth Makes Orphans of Cityless Fringes Bordering Sacramento," "Rising County Taxes, Assessments Plague Fringe

Area Dwellers," "Suburban Residents Find Little Independence in District Government," "Taxpayers Can Benefit from Consolidation of Duplicating Functions," "Merger of City, County as Fringe Problem Solution Merits Study," "Mass Annexation to Sacramento Is Proposed as Fringe Solution," "Citizens Commission Is First Step Toward Fringe, City Solution."

The series painted an over-all picture of an "urbanopolis, a lush jungle of subdivisions, shopping centers, commercial strips, septic tanks, sewer lines. . . ." The plight of the fringe-area residents, faced with heavier county taxation and increased costs for their municipal-type services, was described and the city's more favorable financial position was stressed.

One article debunked the idea that residents of the unincorporated areas are more independent and have more "grass roots" government than their neighbors in the city, noting the unrepresentative character of district government. Various schemes for providing better governmental services to the suburbs—sale of services by the city, incorporation, urban county services, functional consolidation, city-county merger, etc.—were analyzed. Although the *Bee* favored mass annexation of suburban areas to the city of Sacramento, the concluding article called for the creation of a citizens' committee to study the problem and recommend a proper solution.

The articles, together with a few *Sacramento Bee* editorials on the subject, were combined into a free booklet entitled *Sacramento: A Crisis of Growth.*

URBAN GOVERNMENT COMMITTEE OF THE GREATER NORTH AREA CHAMBER OF COMMERCE

Even before 1960, when the Greater North Area Chamber of Commerce became a part of the Sacramento City-County Chamber of Commerce, the North Area organization claimed to serve all of Sacramento County north of the American River.

James R. Cowan, who later played a prominent role both in the Sacramento Metropolitan Area Advisory Committee and in the Metropolitan Government Committee surveys, was president in 1955-1956 when the North Area chamber planned a program which included a study of the adequacy of local government in the North Area. The chamber created the Urban Government Committee; Mr. Cowan appointed the membership of the committee and assumed the chairmanship.

About two weeks after the *Bee* had published the final article in its series, Cowan proposed that certain kinds of questions and problems be studied:

> Shall the metropolitan North Area agree to systematic annexation by the city of Sacramento?
>
> Shall the unincorporated communities be encouraged to incorporate?
>
> Shall a "big city" be incorporated north of the American River? What are all the facts and aspects of incorporation as a solution?
>
> Shall a combined city-county government be considered?
>
> How should the city of North Sacramento grow?
>
> Is functional consolidation an answer?
>
> How can equitable representation be provided to all segments of the area under various proposals for governmental change?
>
> How can recommendations be best utilized for study by other community and civic groups?
>
> What are the legal problems and steps toward an orderly procedure for annexation and/or incorporation?[10]

While the Urban Government Committee did not find answers for many of these questions, it did issue a report, approximately ten months later (September, 1956), suggesting governmental arrangements for metropolitan Sacramento.[11] Detailed recommendations were avoided because of the work of the Sacramento Metropolitan Area Advisory Committee, the purpose of which was to make specific recommendations. Four interim conclusions were presented by the Urban Government Committee: a streamlined governmental unit, probably on a county-wide basis, should be supported by the North Area as a solution to the general problem; in any revised system of local government, representation should be on a community or area basis; flexible tax rates should be used in order to recognize the varying types and demands for municipal services in different areas; and until SMAAC completed its study, consolidation of districts with overlapping functions in the North Area and the county as a whole should continue.

SACRAMENTO AREA PLANNING ASSOCIATION

The Sacramento Area Planning Association was organized on October 19, 1953, apparently through the initiative of the League of Women Voters. The association disbanded in late 1959. The major objective of the Sacramento Area Planning Association was to promote planning in the Sacramento metropolitan area—recreation facilities, community beautification, traffic flow and parking, land use, and more efficient governmental units. The group was concerned about the political pressure and favoritism evident in zoning decisions and in the issuance of building permits; it called for a county master plan, reorganization of the tax structure, and more effective and influential planning commissions in the area.

The association sponsored discussions in 1956 of population growth and governmental problems in metropolitan Sacramento. A few of its members, some of whom were also members of the North Area's Urban Government Committee, advocated city-county merger.

SACRAMENTO COUNTY GRAND JURY

Considerable credit probably belongs to the grand jury for initiating major efforts at reorganizing local government in metropolitan Sacramento. The grand jury had been reporting upon unsatisfactory conditions over a period of several years—for example, in the 1953 report, the home for the aged, the Franklin road camp, and the county jail were criticized. The 1954 grand jury report dealt with the gross inadequacies in the county welfare department and the county civil service commission and personnel office. This report revealed some underlying reasons for the failure of the county welfare department to fulfill its responsibilities:

> Our examination has presented a picture of a county board of supervisors which has failed to recognize that Sacramento County has grown enormously within the past few years, and to take steps to meet the welfare problems that have increased with this growth. The Sacramento County Board of Supervisors, although responsible to the people for the conduct of the welfare department, have taken little active interest in and have been given little knowledge of the manner in which public assistance is administered in this county. . . .
>
> The information obtained from these two investigations just completed indicates to what extent our Board of Supervisors and several of our officials have failed to meet the responsibilities of their office. They have not recognized that Sacramento City and County have grown over the past several years into a large metropolitan area, or they have ignored this fact. . . .
>
> It is regrettable to find conditions existing in our county government such as enumerated in the two reports. It is especially regrettable that the taxpayer had to be put to the expense of two investigations to get facts the Supervisors should have known. . . .

The *Sacramento Bee*, in late February, 1955, called the grand jury report a devastating indictment of the administration of Sacramento County government and stated that a completely new concept of county government was needed. The year 1955 could be "a year of defeat or of unparalleled opportunity for the future of the 369,000 persons in the city and county of Sacramento." Sacramento County, it stated, had one of the finest county charters and yet was one of the worst run counties in the state.

It is not possible to determine the precise degree to which these several pressures for reorganization and reform caused subsequent attempts at governmental change in the Sacramento metropolitan area. But altogether the *Bee* articles, the activities of the Urban Government Committee, the urgings of the Sacramento Area Planning Association, and the reports of the grand jury constituted a persuasive call to action. Before the year 1955 had come to a close, the decision to establish the Sacramento Metropolitan Area Advisory Committee had been made. In January, 1956, the committee began the task of studying the problems of governmental organization in the metropolis. Two years later SMAAC expired, the recommendations in its final report unfulfilled. The Metropolitan Government Committee was established to carry forward the study of metropolitan Sacramento's governmental problems. During the year and a half in which MGC worked, however, the metropolis experienced a movement to create a big new city northeast of the city of Sacramento, a movement to annex to the city of Sacramento a sizable area to the north (Hagginwood-Del Paso Heights), and still another movement to annex a major area to the northeast (Arden-Arcade) to Sacramento. In essence, the SMAAC and MGC studies were phases of one continuous effort, while the incorporation and two annexation movements were separate attempts at governmental change. But in a real sense each of these movements affected, and was affected by, one or more of the others. In the end they were related actions and reactions to the problem of government in metropolitan Sacramento.

2

Citizens' Committees For Metropolitan Study

SACRAMENTO METROPOLITAN AREA ADVISORY COMMITTEE

At the October, 1955, organizational meeting of the Urban Government Committee of the Greater North Area Chamber of Commerce the difficulty of separating the North Area for study was discussed. A truly metropolitan study was needed. On October 26, 1955, the *Sacramento Bee* suggested that the mayor and city manager of Sacramento, the Sacramento county executive, and the chairman of the county board of supervisors constitute themselves a committee to study fringe-area problems. Approximately one week later the city council and the county board of supervisors agreed to establish a thirteen-member committee to study the problems of fringe areas and special districts in the county. Shortly, however, this plan was modified to provide for a larger committee (Sacramento Metropolitan Area Advisory Committee) to study the problems of the entire metropolitan area. Nine members each were to be named by the city of Sacramento and by Sacramento County, and three members were to be named by North Sacramento.

The three participating governmental jurisdictions adopted a common resolution setting forth functional and organizational guidelines for SMAAC: to "study and make recommendations to interested public agencies regarding the broad problems of metropolitan area organization and growth, with attention to incorporation, the formation and/or consolidation of special districts, annexation or any other possible solution to the problems of governmental organization. The Committee shall act in an advisory capacity to the city councils and to the county board of supervisors." The resolution authorized the expenditure of funds that might be appropriated and the employment of consultants, and provided that certain local public officials serve as "ex-officio consultant members."[1]

By mid-January, 1956, all three jurisdictions had appointed their members. Selected from a list of 41 nominees given by councilmen and by civic, professional, and business organizations to represent the city of Sacramento were:

> A field representative for the State Board of Equalization, and president of the county employees union (H. E. Johnson)
> A member of the executive board of the Sacramento Women's Council and former chairman of the Sacramento City Planning Commission (Mrs. Arnold Waybur)
> A physician and board member of the County Medical Society (Orland Wiseman)
> An attorney and member of the Sacramento City Employees Retirement Board (Philip C. Wilkins)
> An attorney and member of the Sacramento City Transit Authority (Alvin Landis)
> An attorney who had just been defeated in the election for city council (Kneeland H. Lobner)

A partner in a large local real estate firm and past president of Sacramento Junior Chamber of Commerce (Thomas W. Yeates)
An administrative official of Sacramento State College (H. J. McCormick)
The owner of a large general construction company (Milton J. Heller)

Sacramento County representatives, chosen from the unincorporated area of the county, included:

A superintendent of a large elementary school district (James R. Cowan)
A business manager of a construction and general laborers' union (Percy F. Ball)
A president of a large general trucking concern (A. F. Dredge, Jr.)
A real estate subdivider and former president of the Sacramento Real Estate Board (Roland Federspiel)
An attorney (Edward McDonell)
A partner in a real estate firm (Fred W. Speich)
An attorney (John J. Wells)
An investment broker and resident partner of a brokerage firm (Edwin Witter)
Manager of the local Campbell Soup Company plant (V. A. Glidden)

There is no specific information on how the county representatives were selected. Except for Glidden, who moved out of the area in 1957 and was replaced by Elmer E. Nelson, of Aerojet-General Corporation, the city and county members remained constant throughout the life of SMAAC.

Only one of North Sacramento's original representatives on SMAAC served to the end, and the manner of their selection is not known. Signing the final SMAAC report for North Sacramento were:

A proprietor of a men's clothing store (Sammy Powell)
An owner of a motor parts company (J. H. Lanphier, Jr.)

The president of a large general construction company (Frank Erickson)

How did SMAAC get started? Who initiated action? Why did they think something should be done? Eight individuals involved in SMAAC were interviewed at length and their responses to such questions varied in interpretation but indicated general agreement that the *Sacramento Bee* was a prime mover in bringing about action. They said, in effect, that the *Bee* articles, in pinpointing the problems of the metropolitan area, caused the council and board of supervisors to take action. Comments ranged from "appointment of SMAAC was a way to stop other action which they [the boards] could not control" to the observation that the *Bee* articles forced the governing boards to give up their reluctance to act. In fact, one of the *Bee* articles in the "Crisis of Growth" series recommended the establishment of a citizens' study committee on metropolitan problems. A few individuals gave credit to the Sacramento Area Planning Association and the Greater North Area Chamber of Commerce Urban Government Committee for sparking initial interest from which the *Bee* took its cue.

In the mid-1950's, the *Bee* articles had presented the only comprehensive identification of governmental and economic problems of the area. They alone made specific recommendations for the appointment of a study committee and clearly assigned responsibility for action to the Sacramento city council and the Sacramento county board of supervisors. There is no evidence that any action would have been taken then or in the foreseeable future without this impetus.

While the Sacramento city council received nominations from a variety of sources for its appointees to SMAAC, the interviewees did not know what procedure had been followed in North Sacramento and Sacramento County. The members evidently were not selected to reflect a particular point of view, either that of the selecting body or of the public. None of those interviewed admitted knowing of any persons or groups who "unofficially" advised the appointing bodies. A few interviewees felt, however, that SMAAC membership did not constitute a cross section of community leaders, particularly of community leaders from the city of Sacramento.

Those interviewed reported that there was little community-wide interest shown in the activities of SMAAC. Several organizations that had had a long-time interest in metropolitan affairs followed SMAAC's work. Both the League of Women Voters and the Sacramento Area Planning Association were represented at most SMAAC meetings. The Sacramento City-County Chamber of Commerce, the Greater North Area Chamber of Commerce, and the Junior League were said to have shown interest. The interviewees knew of no overt opposition to SMAAC's efforts until the final report was issued.

INITIAL ORGANIZATION AND OPERATION OF SMAAC

The first meeting of SMAAC was held on February 1, 1956. James R. Cowan was elected temporary chairman, and Thomas W. Yeates was elected temporary secretary. Both were later named to the permanent posts, along with a vice-chairman, Fred W. Speich. A steering committee, consisting of the officers plus four others appointed by the chairman, was authorized to propose organizational and procedural changes as the work of SMAAC progressed.[2]

The interviewees without exception emphasized the "hands off" policy taken by the three governmental bodies with regard to SMAAC's choice of officers, its procedures, and its final decisions. They also stressed that there was a minimum of internal politics among the SMAAC membership. Since the meetings were public and the press was always present, most SMAAC news was straight reporting. The chairman sent out only a small number of press releases. The chairman was, in effect, the official spokesman for the group and he made most of the public speeches regarding SMAAC and its work.

Although the resolution creating SMAAC authorized the employment of consultants, no funds were provided. The major financial assistance ultimately given by the sponsoring governments, $67,000, was used for the work of an outside consultant.

CONTRACT WITH PUBLIC ADMINISTRATION SERVICE

During its early orientation meetings SMAAC heard local officials speak on the problems and services of their jurisdictions. The central question in these meetings, as reported by several interviewees, was whether SMAAC should be an active study group and gather and weigh evidence, debate alternatives, and make findings and recommendations, or whether it should delegate the study, analysis, and recommendations to a consulting staff and act only as judges of whether these recommendations ought to be accepted, rejected, or modified. The most serious difference of opinion within SMAAC arose over this question.

In an orientation meeting a speaker from the University of California (Berkeley) recommended that SMAAC employ a consulting firm and then decide as to the general orientation and objectives of the consultant's work. SMAAC should keep in close contact with the survey group, review its work, make all decisions resulting from the survey, and work for the adoption of whatever recommendations might result.[3]

In early April a subcommittee appointed for the purpose presented the names of five firms or individuals who might be considered for employment as consultants. Only Public Administration Service and Harold Wise and Associates accepted the invitation to bid on the survey. These firms made their presentations to SMAAC on May 9, and the Public Administration Service was chosen for the job by a secret ballot vote of 13 to 3.

On May 15, 1956, the steering committee and PAS worked out a $67,000 proposal for the study, $55,000 of the amount being allocated for salaries. The PAS agreed to study existing governments in metropolitan Sacramento; to analyze existing governmental services, costs, and modes of financing; and to make recommendations for improvement. It would issue a final report within about ten months after the beginning of the survey. The contract, between PAS and the three sponsoring governments, was soon approved. The city of Sacramento and the county each agreed to pay 49 per cent of the total, and North Sacramento was to pay the remaining 2 per cent.

SMAAC's relationship to the work of the PAS and to the metropolitan government problem in general was not clear in the contract, and, in fact, was never worked out satisfactorily. SMAAC itself did comparatively little by way of actual study. About three months after the survey started, the chairman of SMAAC complained about poor attendance at meetings. Members did not want to come to meetings just to approve minutes of the previous meeting. "We've got to have some meat."[4] Even when PAS was about to issue its final report, SMAAC members disagreed over the role of the committee in handling the report.

SMAAC AT WORK

SMAAC met regularly, usually once a month, while the PAS was making its survey. In addition to the steering committee, which was active in guiding the group throughout, SMAAC created an aims and objectives committee and a public relations committee. PAS worked apart from SMAAC in developing the survey report although the PAS group reported from time to time on what other metropolitan areas had done or were doing to solve problems similar to those of Sacramento. There is nothing in SMAAC minutes or in the newspapers to indicate that PAS discussed or gave a preview of its findings and recommendations to the committee.

SMAAC deliberated over whether all changes in governmental structure should be discouraged until the PAS report had been completed. SMAAC went on record at its April meeting as opposing the incorporation of any new municipalities or the formation of any new community service districts until SMAAC's work had been completed. It opposed any annexations "to existing cities of established communities," but not necessarily all annexations. When the problem arose again in later meetings, some members did not oppose "normal growth patterns" but did want to "prevent the formation of islands of development that will make any final solution to our metropolitan problem impossible." In November, 1956, SMAAC continued to oppose new incorporations, but decided not to take a stand on other specific governmental changes because of "the present danger of misinterpretation of any stand the Committee might take."

SMAAC's public relations committee suggested the development of a Metropolitan Information Panel which would mobilize civic organizations in support of the final recommendations. But the Metropolitan Information Panel never became an effective organization, mainly because it was undertaken too late. As a result, no organized support existed after PAS had completed its survey and SMAAC had been disbanded.

Poor attendance by a few members caused SMAAC some concern. The North Sacramento representatives were particularly lax and were sent written reminders of their responsibility. SMAAC apparently did not resort to the extreme measures set out in its rules—that any member who failed to attend three consecutive meetings without cause was automatically dropped from the membership.

PAS FINAL REPORT

Two preliminary reports were issued by the PAS before its final report was made. The first, released November 13, 1956, gave details of the fact-finding project, which was almost complete. It dealt with the number of governmental units in Sacramento County and gave an analysis of the special districts in the county.[5] In the second report, released on February 19, 1957, the costs for governmental services, based on total local taxes and service charges, were compared for a standard house in the city of Sacramento and for one situated in each of six communities outside the city. The taxes and service costs were greater for all of the six communities. A second part of the report discussed the advantages and disadvantages of various possible governmental arrangements for metropolitan areas in general, but there was no statement as to what the PAS would recommend for metropolitan Sacramento.[6]

On May 22, 1957, PAS presented its final report to SMAAC and the public. During preceding months there had been much debate over how and to whom the report should be made. The uncertain relationship between SMAAC and PAS was clearly reflected in this disagreement over procedure. Some of the committee members felt that PAS should keep them informed of the progress and direction of its work so that the final recommendations of SMAAC would be coordinated with those of the consultant they had employed. Others wanted PAS to be free of any influence, even that of SMAAC. The latter group prevailed and the committee agreed that, since under contract PAS was solely responsible for its findings, SMAAC should not know details of the survey until the final report was made public.

This decision was made on February 19, 1957, but the manner of presentation was debated by SMAAC members, and also by newspaper editors, right up to the eve of the presentation of the report. At the March 19 meeting, SMAAC adopted a motion that "there shall be no formal presentation by PAS until the report has been fully and freely discussed by the committee, and that PAS encumber the funds necessary for consultation after discussion by the committee." In spite of this, the PAS report was presented formally without any prior study by SMAAC. The administrative assistant employed locally by PAS kept the office open for several months thereafter, but no further official consultation by PAS with SMAAC or local government officials took place.

As a concession to SMAAC, PAS agreed that city and county officials should not be invited to the luncheon at which the PAS report was presented. A number of SMAAC members decided that the presence of city and county officials would "clothe the report with an aura of finality" and leave SMAAC no choice but to accept the report. These public officials were, of course, the official parties to the contract with PAS and had appropriated the money to pay for the report.

In broad outline, PAS proposed a merger of the five incorporated municipalities and the county into a single unit—the city-county of Sacramento—with the boundaries of the existing county. This "metro" was to be divided into five boroughs: the city of Sacramento would compose two boroughs; the unincorporated area to the north and east, two more; and the remainder of the county, one. A metropolitan council composed of eleven members serving four-year, staggered terms would be chosen as follows: one councilman elected in each of the boroughs and six others elected at large. A mayor would be elected for a two-year term by the councilmen from among their group. A metropolitan manager with complete responsibility for the administrative side of the government would be selected by the council. The manager's powers would be comparable to those of the city manager of Sacramento. Each borough would have an advisory council, which in time might assume some local responsibilities. Areas not receiving municipal-type services would be designated as rural service areas and would have lower tax rates than the rest of the metro.

The PAS consultants apparently felt confident that this major governmental change could be accomplished under existing constitutional provisions.[7] They relied upon a provision in the California constitution that permitted a city of at least 50,000 population to withdraw from the county, taking additional territory with it if desired, and set itself up as a new city-county. The report proposed that the city of Sacramento withdraw from Sacramento County and take the remainder of the county with it. But PAS expressed some uncertainty over the legal basis for this drastic change: "Added guidance on constitutionality and on the proper legal procedures for consolidation should be secured by the Sacramento Metropolitan Area Advisory Committee from the sources already mentioned [Sacramento City Attorney, Sacramento County Counsel, State Legislative Counsel] and from the Attorney General. All possible legal doubts should be quickly resolved." Nevertheless, the legal discussion on consolidation concluded on this optimistic note: "If the people of Sacramento want metropolitan government it is not likely that the state legislature, special interest groups, local officials, or any combination of factors will prevent their realization of this goal."

Most local government officials, when asked by newsmen for their reactions to the PAS plan, avoided taking a stand. They believed in principle in the merits of consolidation, but they could not pass judgment on the specific recommendations until they had read the PAS report carefully. The *Sacramento Union*—an independent morning daily in the Sacramento metropolitan area having a total circulation of about one-third that of the *Bee*—reported the reaction of a councilman of North Sacramento. He was quoted as saying, "I have always been against any consolidation, as I don't believe we would benefit in any way beyond what we have now." The mayor pro tempore of Isleton, quoted in the *Sacramento Bee,* said, "I think it's all right for the metropolitan area up there, but I don't go for it for Isleton, 40 miles removed from the scene of activities. We certainly can't be said to benefit this far down here. From what I've seen of it, I don't like it." These were the only direct statements positively rejecting the PAS proposal at this stage.

Various groups immediately undertook to study the PAS report. The Urban Government Committee of the Greater North Area Chamber of Commerce pointed out that the PAS recommendations were similar to the previous suggestions of their group. The Arden-Arcade District Council of Improvement Clubs and the Carmichael Chamber of Commerce also appointed committees to study the report.

SMAAC AND THE PAS REPORT

While SMAAC decided not to study the PAS report before its release to the general public, it did not intend to "rubber stamp" the recommendations. Those members of SMAAC who wanted to study the report did convince the group that, after release of the PAS report, a few days should be allowed for individual study by SMAAC members. A series of meetings for more detailed study would follow.

Before considering what occurred in those meetings, it may be well to review the elapsed time since the decision to create SMAAC in late 1955. SMAAC held its first meeting on February 1, 1956. By mid-May, 1956, arrangements had been completed for employing the Public Administration Service, which made a report one year later, in May, 1957.

At the presentation luncheon the chairman reminded SMAAC members that the report was a research project first of all: "The committee will go into every single phase of it before deciding its recommendations." On May 27, 1957, SMAAC began a series of meetings on the PAS report. One member proposed that the group begin by voting general approval of the PAS plan, subject to later modifications of specific details, but a majority insisted that this would be premature. The group then turned to the issue that proved to be a major stumbling block for the PAS plan—the legal question of how the plan could be put into effect. The procedure apparently favored by PAS and theoretically possible under existing constitutional provisions would fail upon an adverse vote in any one of the incorporated municipalities. It was soon clear that there would be an adverse vote in one or more of them. Consequently, the recommended procedure was never seriously considered. The state legislature could be asked to enact legislation to implement the authorization for city-county consolidation already contained in the constitution. However, because of constitutional restrictions upon special legislation, such a statute would have to be applicable to all communities in the state, and would undoubtedly be too controversial to win approval in both houses.[8] A legislative subcommittee was appointed to study the legal aspects.

As the weeks went by, SMAAC members became convinced that only an amendment to the state constitution could insure the adoption of the PAS plan.[9] The PAS report had mentioned a constitutional amendment as one method of securing legal authority for the metropolitan plan, but since the plan could be put into effect without change in either state law or state constitution, the PAS people hoped that "these longer procedures can be avoided. . . ." No serious effort to secure an amendment was ever made in the legislature.

During SMAAC's deliberations doubts arose over the legality of PAS's plan for urban and rural tax differential zones that would permit different tax rates depending on the services received. In June, 1957, when SMAAC asked for an opinion regarding the constitutionality of such zones, the county counsel declared they were unconstitutional. The city attorney gave the same opinion and California's state attorney general concurred in a ruling on October 1, 1957.[10] SMAAC ultimately decided to ask for a constitutional amendment authorizing such special tax zones.

In addition to the legal obstacles, the group discussed whether it should approve the plan in principle or wrestle with specific details. One faction argued that SMAAC should use the report as a basis for preparing a detailed plan of government for metropolitan Sacramento. Other members felt that a charter commission ought to be allowed to work out the details. The issue was never really resolved.

SMAAC's own report, 12 printed pages approved on September 11, 1957, presented a brief history of its organization, followed by a summary of the PAS recommendations. SMAAC expressed agreement in principle with nearly all of the recommendations, but raised objection to having the membership of the first metro council chosen from the present members of the legislative bodies of the city of Sacramento, Sacramento County, and North Sacramento. In spite of this and other specific differences, the SMAAC report stated that matters of detail should be left to a charter commission.

To implement the PAS plan, SMAAC recommended a state constitutional amendment that applied only to Sacramento County. Under this amendment city-county merger could be adopted by a county-wide majority vote. It also should authorize varying tax rates based upon differences in services. In the meantime, an unofficial charter commission, appointed by the county and the five incorporated municipalities, should undertake at once to draft a proposed charter for the new metropolitan government which would be ready for consideration by the voters as soon as the constitutional amendment had been adopted. The amendment would give a permanent legal basis to the unofficial charter commission. SMAAC would extend beyond January 1, 1958, so that it could assist with the drafting of the constitutional amendment.[11]

On November 12, 1957 (about two years after it had been created), SMAAC presented this final report to the officials of the city of Sacramento, Sacramento County, and North Sacramento. Representatives from Folsom, Isleton, and Galt were also invited. One hundred fifty interested citizens attended the meeting. According to the minutes, the audience was equally divided between persons from within and from outside Sacramento city.

The minutes of this meeting are brief. Presentation talks were made by several SMAAC members, and there were questions from the officials and the audience concerning water, schools, and taxes. The chairman of the county board of supervisors, the mayor of Sacramento, and the mayor of North Sacramento, in commending SMAAC on its work, promised study and some sort of action on the part of the governments they represented. They suggested a joint meeting of the three governments about December 1 to consider the SMAAC final report further and to decide whether the life of SMAAC should be extended.

POPULAR REACTION TO THE PAS PLAN

Few positive rejections of the PAS plan were expressed immediately after its release; neither had there been strongly expressed approval. During the six months that SMAAC was studying the plan, the newspapers of the Sacramento area were strongly supporting it. The *Bee* evidently preferred mass annexation plus functional consolidation, but gave the PAS plan much publicity and insisted editorially that the people should have the opportunity of voting upon it. The *Union,* which had not given as much attention to metropolitan problems and proposals for change as the *Bee,* supported the PAS plan in editorials. The community newspapers and the shoppers' news publications of the general Sacramento area were enthusiastic about the plan.

Several organizations in the county also favored the plan. The Urban Government Committee of the Greater North Area Chamber of Commerce, which had studied the PAS plan during this period, presented a resolution to SMAAC endorsing the plan, subject to certain modifications. The Urban Government Committee continued throughout the remainder of its own existence to voice support for the PAS plan. Support also was given by the Carmichael Chamber of Commerce, which helped to sponsor essay contests on the PAS proposals in the area's schools. The Sacramento Area Council of Chambers of Commerce requested that the board of supervisors approve the PAS plan. According to a survey made by the *Sacramento Union,* only the Isleton Chamber of Commerce among the chambers in the area had taken a stand in opposition to the PAS plan.[12] The League of Women Voters in the area voted approval of the plan.

There was no evidence, however, of support from the public in general. The average citizen did not seem to feel that there were serious problems to be considered or that any action or change was needed. Very little support for the PAS plan came from public officials. Most county and municipal officials spoke only in general terms and avoided committing themselves, although a few went on record in opposition to the plan. Only one of Sacramento County's three representatives in the state legislature was willing to introduce the proposed amendment in the legislature. He would do this only if SMAAC, the county board of supervisors, and the city councils of all five municipalities made such a request.[13] At the public meeting in which SMAAC's final report was presented, one of the Sacramento County legislators was heard to say to a representative from one of the smaller cities: "Don't you worry. We'll kill this —— ——— thing!"[14]

In November, about one month before SMAAC officially ceased to exist, the Interim Committee on Municipal and County Government of the state assembly held a one-day hearing on the SMAAC-PAS proposals to decide what state legislation should be proposed. No action was taken, and the view of the assembly committee was perhaps best expressed in this statement made by one of its members, a representative from the Sacramento area:

> . . . I believe that the members of the Committee [SMAAC] who were here today have benefited a great deal by the questions you have asked, and I think that it makes them realize all the more the inadequacy of the study that they have come up with at the present time. In other words, it is quite obvious that considerable more study must be given to the problem, and the details that can't be assumed to be filled in later on. . . . The details have got to be presented in the plan at the time the plan is submitted to this committee, or submitted to either the City Council of Sacramento, North Sacramento, or the County Board of Supervisors.[15]

SMAAC IS DISSOLVED

Within two or three days after SMAAC's final report, the *Sacramento Bee* polled the members of the county board of supervisors, the Sacramento city council, and the North Sacramento city council, asking whether they favored extending the life of SMAAC. A majority of the supervisors and Sacramento city councilmen indicated that they did. North Sacramento councilmen would not comment. The North Sacramento city council, on November 18, voted unanimously to cut its ties with SMAAC as of January 1. The city council had no "functional right" to approve a plan under which persons not residing in the city could vote the city out of existence. Further, the council had received no direction from the voters of North Sacramento.

The chairman of SMAAC at first stated that North Sacramento's action came as no surprise. North Sacramento sentiment had consistently been against any plan of governmental reorganization that would modify its status as an independent city. He pointed to the poor attendance record of North Sacramento's representatives to SMAAC. The next day, however, the chairman announced that negotiations were under way to compromise the differences between North Sacramento and SMAAC. The proposed constitutional amendment would contain a provision that the governmental merger would apply to Sacramento and to North Sacramento only after "a majority of the electorate in each city voted approval."[16]

In the meantime, Sacramento County and the city of Sacramento continued to weigh the question of SMAAC's future. On November 25 the county board of supervisors voted that all of the requests for continuation of SMAAC be "taken under advisement." One supervisor stated that a new committee might be created in place of SMAAC to draft the necessary laws and constitutional amendments. On November 26, representatives of SMAAC met with the Sacramento city council to discuss the future of SMAAC. SMAAC's chairman made a case for SMAAC's continuation and suggested that another advisory group be appointed to make "a two or three year study of charter provisions for the new government." The city council decided that it should have another meeting with SMAAC and also a joint meeting with the county board of supervisors before arriving at a decision.

Some SMAAC members disagreed with the chairman and felt that SMAAC had outlived its usefulness and should be disbanded. It would be futile, they said, to expect "the political boys in office," who traditionally oppose governmental change, to favor SMAAC's continued existence.[17]

When the Sacramento city council met with SMAAC representatives on December 9, the council postponed decision. The county board of supervisors and representatives of North Sacramento, Folsom, Galt, and Isleton were invited to the next council meeting, on December 18, to decide the issue. The fate of SMAAC seemed to have been settled already. SMAAC representatives spoke in terms of a smaller committee. Sacramento's mayor and its city manager talked more and more about a new committee. The manager wanted to be sure that any new committee would not keep the sponsors in the dark until the issuance of a final report.

At the December 18 meeting, continuation of SMAAC received slight consideration, even though the majority of county supervisors and the Sacramento city councilmen had recently gone on record in favor of the continuation of SMAAC. A new committee composed of 15 members representing Sacramento County, the city of Sacramento, North Sacramento, Folsom, Galt, and Isleton was to be formed. The Sacramento county board of supervisors, Sacramento city council, and four Isleton city councilmen present agreed that the new committee should study only the merger plan and "problems related to it." Representatives of Folsom, Galt, and North Sacramento said they would take the matter before their full councils before announcing a stand. One newspaper reporter interpreted the Isleton vote as "the first crack in the wall the smaller cities were setting up against any change of the county's government structure."[18]

When the Sacramento Metropolitan Area Advisory Committee went out of existence, no action had been taken on its recommendations by the sponsoring governmental bodies. In the months intervening between SMAAC's final report and the committee's demise, it became clear that neither the city of Sacramento nor Sacramento County was going to press for further consideration of the SMAAC-PAS plan. They apparently felt that there was little community support for a consolidated government and that the legal and procedural obstacles to forming such a government were indeed imposing. City-county merger meant the death of the larger cities and seemed to have no advantages for the smaller, more rural communities of Isleton and Galt. The governing bodies probably would have preferred to let SMAAC's report lie without action of any kind. At the same time, the local newspapers were constantly reminding them that they had spent $67,000 of tax money for the report. It could hardly be ignored or swept under the civic rug. Their solution to this problem was typically American—form a new committee.

FORMATION OF THE METROPOLITAN GOVERNMENT COMMITTEE

On December 27, 1957, the Sacramento city council approved a resolution creating the Metropolitan Government Committee (MGC). Sacramento County was to be represented on MGC by the county executive plus four members appointed by the county board of supervisors and living in the unincorporated areas. The city manager plus four city residents appointed by the city council would represent the city. North Sacramento's city manager plus one person appointed by its city council were made members, and the city councils of Folsom, Galt, and Isleton were to appoint one member each. The original resolution provided that four of Sacramento County's members and four members from the city of Sacramento must be former members of SMAAC. The city council made this permissive and set a termination date for MGC of June 30, 1959.[19]

As finally adopted by the county board of supervisors and the cities of North Sacramento, Isleton, Galt, and Folsom, the resolution stated that the creation of MGC was not to be construed as an endorsement of SMAAC's recommendations, and that membership was to be made up of six persons each from Sacramento County and the city of Sacramento, two each from Folsom, Isleton, and Galt, and three from North Sacramento. The duties and responsibilities of MGC were stated as follows:

> It shall study the problems of government in the Sacramento Metropolitan Area and possible solutions to those problems, including the recommendations made by the Public Administration Service and the Sacramento Metropolitan Area Advisory Committee.
>
> It shall keep the respective City Councils and the Board of Supervisors regularly informed through quarterly reports concerning the progress of its studies, and shall take such other appropriate means to keep the Councils and the Board of Supervisors informed as it may deem necessary and proper.
>
> It shall act only in an advisory capacity to the respective City Councils and the Board of Supervisors.
>
> It shall not officially represent the City Councils and the Board of Supervisors unless specifically authorized.

"Metropolitan Area" was defined as including all of the county of Sacramento. Any legislation, or constitutional amendment, that MGC felt was necessary should be prepared and submitted to the governing bodies of areas affected for their consideration. Unless specifically authorized by all of the governing bodies in the county, MGC could not recommend such legislation or constitutional amendment to other legislative bodies or to other organizations.

The cost of MGC was to be shared equally by Sacramento County and the city of Sacramento. The four smaller municipalities were to pay nothing. All fiscal matters were to be handled by the appropriate officers of the Sacramento County government. MGC's operation was under the strict control of the city of Sacramento and the county. All of the sponsoring governments could strongly influence its conclusions.

Four of Sacramento County's representatives on MGC had been members of SMAAC: James R. Cowan, Elmer E. Nelson, Fred W. Speich, and Thomas W. Yeates. The two other members named by the county were a rancher and former state assemblyman who was also a member of the county board of education, Dwight Stephenson, and the county executive, M. D. Tarshes. The Sacramento city council, on January 30, 1958, approved the amended resolution and named its six members. Four of these had been members of SMAAC: Milton J. Heller, Alvin Landis, H. J. McCormick, and Philip C. Wilkins. Also named were an assistant general manager of the Sacramento Municipal Utility District, Carl L. Richey, and the city manager, Bartley W. Cavanaugh. North Sacramento named an investment firm executive, R. O. Mapes; an architectural draftsman and member of the city planning commission, Malcolm O. Mau; and the city manager, Homer H. Jack. Folsom selected Marjorie Handy, a member of its city council, and the city attorney, Louis A. Boli. A retired canning firm employee, Emil Evers, and a member of the city council, Emil Schock, represented Galt. Isleton named the city superintendent of schools, Alwyn Amerman, and an oil company representative, Glenn L. Maxey. Nineteen of the 21 persons originally appointed to MGC signed the final report. Elmer E.

Nelson of Sacramento County, who resigned before work of MGC was completed, was succeeded by Percy F. Ball, a former member of SMAAC who remained to sign the final MGC report. Marjorie Handy from Folsom attended some of the early meetings, but did not participate in MGC activities thereafter or sign the final report.

Cowan, SMAAC's chairman, expressed reservations about what he viewed as the overrepresentation of the small cities. Although these cities had a total population of only 16,000, nine out of the 21 members came from them.

> This will definitely weaken the committee. It's not that the small cities shouldn't have the right to speak their piece, but I don't believe the bare fact that a community is incorporated should give it that much larger voice. . . . How about Arcade with its 45,000 people, or Carmichael with 26,000 or North Highlands with more than 20,000? This is a basic injustice to the proposal.[20]

At the final meeting of SMAAC, less than a week later, the SMAAC chairman seemed to have reconciled himself to the membership arrangement. He said: "In thinking about this, it seems to me that the small cities have been a big hurdle for our merger plan and perhaps it will be easier to adjust their problems into the plan if they have this large number of representatives."[21]

Thirteen members of MGC were interviewed. Most stated categorically that MGC was the idea of either the Sacramento city council or the Sacramento County board of supervisors or both. Most also believed that a general dissatisfaction with SMAAC's report had spurred these official bodies to act. SMAAC had been "99 per cent theory—not practicable." "Insufficient attention had been given to alternatives to city-county consolidation, and a further review was needed." "Consolidation was impossible without a constitutional amendment and the smaller cities were opposed to it." One interviewee thought that the county board had strongly supported the creation of MGC because, although it favored consolidation, the supervisors wanted the county to play a more important role than that outlined in the SMAAC-PAS plan. The city and the county of Sacramento were still under pressure from the newspapers to justify the $67,000 spent for the SMAAC report. Something had to be done. One person, thoroughly disillusioned, said, "The last step in killing off SMAAC and the easiest was—study it to death."

There was no evidence from the interviewees that community opinion was widely solicited in selecting the MGC members. The selection of some former SMAAC members was to be expected. Participation by city and county officials in all of MGC's activities should avoid recurrence of the lack of liaison between the committee and the sponsoring bodies, as had been the case with SMAAC, and so the appointment of the two city managers, a city attorney, and the county executive was not surprising. Views expressed at various stages of the interviews emphasized that representatives of the small municipalities participated in MGC simply to insure that their autonomy and power of self-determination would in no way be infringed. The fact that most of these people did not play an active role indicates that they were satisfied that the interests of their cities were not in danger or that they did not believe the problems being discussed were their concern.

The respondents noted a general lack of interest in MGC's work among organizations and interest groups. The League of Women Voters, the Sacramento Area Planning Association, the chambers of commerce, the Junior League, and a few other groups sent representatives to the meetings from time to time, but no direct support was given to MGC as such. Most of these organizations were committed to follow the study because of their previous interest and work toward some kind of area-wide planning. As for opposition, two respondents stated that the Urban Government Committee of the Greater North Area Chamber of Commerce was basically opposed to MGC's work and conclusions. The Urban Government Committee had gone on record as favoring SMAAC's plan for city-county consolidation and was suspicious of, if not opposed to, MGC's work, which might not approve or effectively implement the merger. Two respondents cited the opposition of local, nonmunicipal fire departments to MGC's desire to push for mass annexation and consolidation of functions. Such actions would have brought an end to many of these small departments. Generally, MGC attracted lukewarm support, at best, from outside organizations, and some potentially strong opposition from others.

EARLY WORK OF MGC

The first meeting of MGC was held February 19, 1958. After election of temporary officers—all of whom were formerly members of SMAAC and were later named to the permanent posts—and the decision to meet semimonthly, the meeting was given over to general discussion. The county executive expressed the hope that members of MGC would not approach their assignment with fixed opinions about the SMAAC proposals or about any other proposals for governmental reorganization. One member suggested that MGC should get experts and also the public into its discussions.[22]

As with SMAAC, organizational and procedural decisions of the committee were arrived at by democratic processes in public meetings. The interviewees stressed that there was a minimum of internal politics among the members. The sponsoring governmental bodies maintained the same "hands off" policy during MGC's life, but the membership on MGC of the appointed executives of the county and the two largest cities created a significant difference between MGC and SMAAC. While the influence of these executives is impossible to assess with precision, there is little doubt that the effect of the presence of the Sacramento city manager and the county executive was important. The press attended all meetings of MGC and its activities were fully reported. The chairman gave occasional press releases, and in this sense he was MGC's official spokesman.

Sacramento County paid all of MGC's expenses and billed the city of Sacramento for its proportionate share. The total cost of MGC was approximately $20,000. None of those interviewed felt that MGC's work was inadequately financed or was not given proper support by the city and the county of Sacramento.

A study program of MGC was discussed at the second meeting. Assessment and collection of taxes, inequality of tax rates, duplication of services by special districts, and overlapping boundaries of special districts were considered, but the group did not agree on a program of study. The members decided to invite the Sacramento county planner, the Sacramento city planner, the chairman or some other member of the planning commissions of Sacramento County and the municipalities of Sacramento, North Sacramento, Folsom, and Galt, and the mayor of Isleton to their next meeting to discuss present and future plans for the total area.[23]

The minutes for the next five meetings show that MGC paralleled SMAAC in its early meetings. The group listened to representatives of various governmental agencies explain the nature and operations of their organizations. The meetings were poorly attended and, on one occasion, the chairman was instructed to write a letter to each member stressing the importance of attendance and promptness. After having heard from all of the representatives, MGC decided to hire a full-time research secretary. A subcommittee was appointed to determine the qualifications for the job and to screen the applicants.

INCORPORATION PROPOSAL INFLUENCES MGC

On the evening of June 17, 1958, a proposal to incorporate a large new city northeast of Sacramento, which is the subject of the next chapter, was made public. The ineffectiveness of MGC so far was certainly one of a number of reasons for this proposal. The MGC chairman announced a few days thereafter that he would recommend that MGC give immediate attention to a study of the advantages of separate incorporation of the northeast area versus the advantages of annexation to Sacramento. This recommendation was adopted on June 25. According to newspaper accounts, the chairman emphasized that MGC was not abandoning plans to study city-county merger and other reorganization possibilities, but that problems growing out of the incorporation movement now under way must be studied first. The committee planned to devote six months to this study. The Sacramento city manager promised to provide the needed cost figures.

In mid-August a research director was appointed. (He was forced to resign within a few months because of illness.) MGC was divided into four subcommittees, of which the county executive and the two city managers were ex officio members, to study incorporation versus annexation. The subcommittees held meetings throughout the summer. Their studies were based primarily on a comparison of costs of services under existing arrangements, under separate incorporation, and under annexation to the city of Sacramento.

In September the chairman repeated MGC's ultimate obligation to analyze and evaluate the SMAAC-PAS recommendations, even though the incorporation movement compelled the committee to give full attention to that for the time being. He stated that if MGC decided to support city-county merger it would be obligated to draft a constitutional amendment and submit it to the respective governing bodies for consideration. When it was brought out in subsequent discussion that about 90 per cent of the county's population was living in about 25 per cent of the county's area, someone suggested that perhaps MGC should concern itself only with this 25 per cent instead of the entire county. In the light of MGC's final report, this suggestion was prophetic.

The report on incorporation versus annexation was submitted to MGC at its January, 1959, meeting, although the new city incorporation effort had been abandoned in November. The report concluded that separate incorporation, with a level of services equal to that of the city of Sacramento, would require a municipal property tax rate of $2.26 in the new municipality. In the case of mass annexation an over-all municipal property tax rate of only $1.73 would be needed to achieve a uniform level of services. Present city taxpayers would pay about two-thirds of the cost of raising the level of services in the northeast area. Over a period of years, however, the taxpayers of the central city would benefit from increased land valuation in the northeast area.[24] The report was given to a subcommittee for further study, corrections, and revisions. In a meeting on February 4, 1959, this subcommittee decided to recommend mass annexation of all of the area expected to be urban in character by 1980. County and city planners would draw the exact boundaries, but not, as suggested by some, the boundaries for "stage annexations" also. Folsom and North Sacramento could join if they wished. The subcommittee also agreed to recommend functional consolidation of city and county services. The schools were left to the county committee on school organization. In early March the subcommittee made the additional recommendation that the Arden-Arcade annexation movement—begun late in 1958 and considered in a later chapter—be endorsed as an important step toward implementation of mass annexation.[25]

The subcommittee report was accepted in essence by MGC. The *Sacramento Bee*, in an editorial entitled "Big Annexation Proposal Best Meets the Realities" (March 6, 1959), argued that city-county merger was not feasible at present and the MGC plan offered those people now served by many special districts the opportunity to change to efficient, integrated government. The MGC plan does not kill merger, but is a logical step toward it. On April 1, the MGC chairman urged the Sacramento city council to get together with the county to study what governmental functions could logically be consolidated. The county board of supervisors had already instructed the county executive to undertake this determination on behalf of the county. Although one member of the county board of supervisors asked the board in April to go on record in favor of the MGC plan, the matter was not brought to a vote.

FATE OF MGC RECOMMENDATIONS

The final report of MGC, a document of only 31 pages, presents four alternative solutions to metropolitan area problems of government—separate incorporation, federation, city-county separation, and city-county consolidation. All are found unsatisfactory or unfeasible, although the report states that city-county consolidation may "ultimately be the best solution." The committee recommended action on its subcommittee's plan for mass annexation of the area expected to be urban in character by 1980 and for approval of the pending Arden-Arcade annexation. MGC also recommended concurrent studies by the city and county governing bodies leading to consolidation of such city-county functions as planning, tax assessment, recreation, sewers, road construction, storm drainage, and water. After enlargement of the city through mass annexation, a charter commission, representing all areas, should study the adequacy of the city charter under the new conditions.[26]

The *Sacramento Bee* called for action. After three years of study and at a cost of more than $70,000, a fair and feasible plan for the city and county had been developed. The pending Arden-Arcade annexation was a part of this plan. The city and county must show good faith by getting to work on functional consolidation. Two months later the *Bee* took the Sacramento city council to task for dragging its feet on this matter. A resolution to study the feasibility of functional consolidation was, upon introduction, simply referred to the city attorney for study. The *Bee* insisted that the council get at the business of functional consolidation in order to counter the "land grab" charges voiced by opponents of the Arden-Arcade annexation. The voters in Arden-Arcade needed to be assured that their annexation would be part of an over-all plan for the area. The election on the issue was only five weeks away.[27]

The Urban Government Committee still urged city-county merger, but the North Area chamber overrode its committee's negative recommendation and did finally endorse the Arden-Arcade annexation as a part of MGC's plan for functional consolidation. The Sacramento City-County Chamber of Commerce, although it had not been active in governmental reorganization efforts, endorsed the Arden-Arcade annexation early. The two chambers agreed to work to achieve annexation to the city of Sacramento of the remaining unincorporated areas of metropolitan Sacramento if the annexation succeeded. If the annexation was defeated, they would wait the 12-month period required by law and then cooperate to bring about mass annexation.

The overwhelming defeat of the Arden-Arcade proposal was

a serious blow to the entire MGC plan and to these incipient efforts by organizations to act upon governmental problems. The SMAAC and MGC plans have been revived periodically but never acted upon. On January 2, 1960, the *Sacramento Bee* printed an editorial stating that MGC recommendations cannot implement themselves. Since representatives of the city of Sacramento and of Sacramento County had signed the MGC report, a commission representative of both the city and the county should be set up to put the plan into action. The planning directors, the engineers, and the chief executives of the two governments were proposed as members of such a commission. Again in January, 1964, the executive officer of the newly created Sacramento County Local Agency Formation Commission indicated that the commission would probably give serious attention to SMAAC-MGC plans.[28]

SMAAC AND MGC EVALUATED BY SELECTED MEMBERS

Why have the SMAAC-PAS plan and the MGC recommendations lain dormant? The 16 individuals close to the movement (five of whom served on both committees) who were interviewed almost unanimously cited lack of community support as a major reason. Several felt that more support from the city and county, from service clubs, chambers of commerce, neighborhood clubs, home builders and real estate groups, and from central political organizations all would have been helpful. A few felt that the lack of coordination among SMAAC, PAS, and the sponsoring bodies had discouraged support for the plan. Beyond this general recognition of lack of community-wide interest, the interviewees did not seem to be able to judge how much and from what source support for their work came. The *Bee*, they all agreed, was in favor of the SMAAC study, and most felt that the paper supported the work of MGC also. The editorial policy of the paper had always viewed mass annexation as the best solution to governmental problems.

Besides the *Bee*, the individuals in SMAAC and MGC seemed at a loss to pinpoint support. The schools were not concerned as long as the proposed changes did not affect them. Real estate people and downtown businesses generally were "passive," "disinterested," or "cautious." Two interviewees did mention that the Sacramento business group favored annexation. The attitudes of suburban businesses were not known. Several felt these enterprises were "receptive to change," or at least not opposed, while about the same number viewed that group as opposed to the MGC report because it might cost them money.

It was generally agreed that officials of all cities were opposed to the SMAAC-PAS plan. In the interviewees' minds, this attitude stemmed from their provincialism, "desire to protect financial interests of residents," and "fear of losing out by consolidation." Sacramento city officials were considered favorably disposed to less drastic MGC recommendations because they had finally realized that the city would have to take a prominent role in solving area-wide problems and in preventing the separate incorporation of the North Area. The interviewees varied widely in their evaluations of the county officials' attitudes to SMAAC and MGC—from "favorable—would be top dogs" to "negative" or "reserved—didn't even read the reports in some cases." Either there was a wide range of opinion among the county officials or the individuals interviewed interpreted the same actions in different ways. The League of Women Voters was seen by all questioned to be favorable to the SMAAC-PAS and the MGC proposal, although a few felt that most League members preferred city-county consolidation.

Although several respondents noted the lack of participation in the SMAAC and MGC movements of influential people in the Sacramento area, they could not identify who these people might be. They spoke of "important bankers and business people." These "influential people" were not involved because they could not see that a tangible problem existed. In addition, there were no established means for involving them—no distribution of information or effective use of communications media. One respondent recognized the underlying lethargy of government employees as holding down the development of effective community leaders and an active citizenry.

The interviewees found it easier to identify specific practical reasons for the failure of their movements to inspire effective action. The timing was poor in the case of SMAAC; its report was released just before a city council election in Sacramento. SMAAC should have established better public relations—perhaps employed a staff—and kept the community better informed. The sponsoring bodies should have been brought in on the final decision-making on recommended action. SMAAC was slanted toward outlying areas. The committee should have assigned subcommittees to study specific aspects of the report and not simply sat back and waited for a grand scheme to emerge in the final report. In fact, some felt that a consultant should not have been hired. PAS, they said, had little understanding of the community. It took an ivory tower approach and did not realize the attitude of the city and county; its recommendations threatened the existing power structure. As one individual said, "Keep the experts on tap, not on top." Another respondent felt, however, that "the climate for acceptance didn't exist to begin with."

Perhaps during SMAAC's work a climate congenial to governmental reorganization could have been created. One respondent felt that a daily paper in the northeast area that asked questions and took a straightforward approach would have helped a great deal to give citizens factual information on which to judge the issue. More participation by the public and wider publicity seemed to many persons a way of preventing the failure of SMAAC. One individual felt that the Metropolitan Information Panel was a good idea and should have had official backing.

A few persons involved in SMAAC felt personnel changes could have made a difference in SMAAC's fate. One called for a consultant board on city-county merger, using academicians and other resource persons. PAS, one interviewee said, did not demonstrate that the problems were real ones. The study should have been more "factual." One person thought that too many people on SMAAC made little or no contribution to the committee's work, inferring that the existing structure and membership of the organization might have been a stumbling block to effective action.

The same basic reasons—lack of support from the community or its generally acknowledged leaders, poor public relations and publicity—were cited by participants as causing the failure of MGC. Interestingly enough, two respondents thought the group should have been a citizens' committee, with experts available for advice, and complained that the city and county executives tended to dominate the committee and "dilute its effectiveness." "The officials are apathetic and they don't know too much about the problem." Some in SMAAC had felt that the exclusion of these officials from the committee's work had been a reason for its failure. Another from MGC felt that the study subcommittee approach used—and favored by some in SMAAC as an alternative to using experts—had not "made much of a contribution." Two thought that before another committee is formed, they ought to find out exactly what the community wants and what it is willing to do to obtain it. A few sharply criticized MGC membership for its apathy, for the poor attendance of representatives from small cities, and for its failure to be truly representative.

In general, the leaders of the movements felt that the SMAAC-PAS proposal was premature. The people did not see the need for change, did not know how they would be affected, and therefore did not see why established concepts of government ought to be disrupted. This general apathy, as the leaders called it, would have to be overcome before the public support, which was essential to the success of a reorganization movement, could be rallied. The work of a study committee in Sacramento such as SMAAC or MGC is, then, twofold. It must not only develop meaningful and practicable solutions to the problems it studies, but it must educate—and understand—the people whose problems it is considering.

CONCLUSIONS

Since the SMAAC and the MGC committees were the only two reorganization efforts that were sponsored and supported by existing governmental bodies—the others were founded and executed by private citizens—it seems most appropriate to conclude this chapter by exploring the implications of the SMAAC-MGC experience for questions of governmental sponsorship, and organizational structure and procedures. The kind of organization that is needed to bring about governmental change is revealed in part by locating the sources of motivation for change. If official governmental agencies are sincerely interested in governmental change, are willing to see the issue studied openly and objectively, and are willing to commit themselves to some kind of action, then those agencies ought to create, charter, and support a study group. If, on the other hand, the motivation for change arises outside the governmental structure, attempts may be made to arouse governmental agencies to appoint a study committee, or an independent group may be created that is committed to study and to bringing about change through political action. It is obviously easier, in almost any circumstances, to turn to government for creation and support of a study commission. If, as in Sacramento, agitation for change has not arisen from either government or a citizen group but from an institutional source such as a newspaper or a grand jury, the best and most feasible type of organization becomes a more troublesome problem. The constituted governmental authorities may feel that agitation is a direct attack on them rather than on the structure, and, while they may create a study commission, they are likely to be unsympathetic to the whole enterprise. This, in fact, seems to have been the case in Sacramento.

There are also important questions of employment and use of staff. SMAAC employed outside staff and the committee did not participate actively in the study. The members of MGC constituted themselves as study committees and wrote their own report. The weight of interviewee opinion favors the employment and full use of staff for fact-gathering, analysis, and reporting. Some MGC members showed marked dissatisfaction with its study procedure; the subcommittees had neither the time nor the competence to do their own staff work. A director of research, employed for a few months, was not adequate staff assistance. On the other hand, the even stronger criticism of the SMAAC-PAS relationship by participants indicates that when technical staff go beyond "fact-finding, analysis, and reporting" important problems of communication and authority can arise. Neither the committee nor the community was kept informed of PAS's progress or findings until the final report was revealed. Since the local government legislative bodies, which had financially sponsored the study, learned only after completion of a final report that it recommended their own liquidation, it is not surprising that councilmen and members of the county board of supervisors greeted the recommendation with less than enthusiasm.

It is equally apparent that the SMAAC-PAS relationship inhibited the development of community support for the plan by its policy of keeping everyone in the dark until the issuance of the final report. The conclusion implied is that an effective study group hoping to bring about governmental changes should employ an adequate trained staff but should also study and understand the problems itself and keep in touch with the work of the staff. Perhaps one of the most important activities of the study committee is to act as liaison between the technicians and the community they represent, informing each of the work and opinions of the other.

Significantly, although most members of both SMAAC and MGC credited the failure of their organizations primarily to lack of community support, they did not seem to be aware of this or other problems prior to or during the life of their group. Only through hindsight did they recognize the depth of citizen apathy, or organizational problems, or problems of technical assistance. The interviews, conducted after the fact, tapped this new insight.

Anything like the SMAAC-PAS plan, of course, would present almost insurmountable legal and constitutional problems to any community. When this community cannot obtain essential support from its city and county officials and the public, the plan becomes impossible to achieve. The MGC plan also required this support. As long as Sacramento County views annexations to the city of Sacramento as encroachments on its own power and prestige, and as long as the city is similarly fearful of functional consolidation as a device to transfer control of governmental functions to the county, neither the SMAAC-PAS plan for city-county merger, nor MGC's recommendations for mass annexation, seem likely to be put into effect.

3

The New City Incorporation Attempt

On June 17, 1958, a group known as the Citizens' Committee for Incorporation declared its intention to create a new 165-square-mile city, northeast of the city of Sacramento.[1] The area's estimated population of 150,000 was nearly equal to that of Sacramento, and it covered about five times as much territory as Sacramento. The committee's announcement occurred about five months after the last meeting of SMAAC and about four months after the first meeting of MGC.

Unknown to the public, discussion and preliminary planning for this move had begun late in 1957, paralleling the beginning of MGC. Its origins may have been nurtured much earlier, even before the Urban Government Committee of the Greater North Area Chamber of Commerce was formed. Members of the Carmichael Chamber of Commerce, many of whom also belonged to the Greater North Area chamber, historically have held an image of their community—a small unincorporated area about ten miles east of Sacramento—as a separate, rapidly growing, progressive community. They have sought to provide adequate governmental services for the area. In the early 1950's some citizens attempted to re-establish a community service utility district coterminous with the Carmichael fire district. This plan failed to gain the voters' approval, and some of the same citizens helped to form the Urban Government Committee of the Greater North Area Chamber of Commerce. Since the recommendations of this committee were similar to those subsequently presented in the SMAAC-PAS report, the Carmichael group—not all of whom were well known to each other—appeared willing to wait for action on the report during the summer of 1957. When city and county officials continued to delay action on the plan, this informal group turned to incorporation of the northeast area, of which Carmichael was only a small part, as a solution.

The informed citizens of Carmichael were aware of the serious nature of their local problems. The community is old. During the 1950's many of the four- and five-acre single-family lots and other larger acreages were subdivided. The numerous cesspools and open drainage ditches needed to be replaced by facilities that would provide more water and adequate sewage disposal. In addition, the county board of supervisors had allowed spot zoning for commercial purposes along the main thoroughfare, which bordered upper-middle-class homes on the south. Many considered the county maintenance of the roads and drainage system in the community entirely inadequate.

The final impetus for the 1958 incorporation attempt apparently grew out of informal conversations among members of the Carmichael Chamber of Commerce in late fall of 1957, after the SMAAC-PAS report had been all but shelved. Jack Moore, an insurance and investment broker and former president of the chamber, and Howard Craig, a retired general who had been a candidate for the governing board of the rejected utility district in Carmichael, were the first to take any action as individuals. In mid-January, 1958, they invited Carmichael resident Walter Isenberg, an employee of the State Department of Corrections, to lunch with them. Craig had recognized Isenberg's organizational abilities during his efforts in the utility district campaign. Jack Moore stated, "At this point we merely wanted to study our problems, but not again with the same people who had been working on them at the governmental level."

Some individuals in the movement reported a feeling expressed in the area that city-county merger could be attained through the "threat of incorporation," but Isenberg immediately took the view that if incorporation were attempted every effort should be made to insure success. At just what point the committee for incorporation as a whole accepted the view of incorporation as an end in itself is not entirely clear; apparently it became committee policy, although the members did not seem to agree unanimously on this point. This group first included only the Carmichael area in its plan for incorporation; then Arden was added, later Arcade; and finally, as new members were recruited to the study group, the area grew to 165 square miles. According to the minutes of the committee, the general boundaries had been agreed upon by the fourteenth of April, 1958 (see Map 3).

DEMOGRAPHIC SETTING

The population of the proposed new city in 1960 was 187,336.[2] The area was almost 99 per cent white, as compared with 87 per cent in the city of Sacramento. The population was comparatively young; most areas show a median age of about 25 years. By contrast, within the city of Sacramento the median

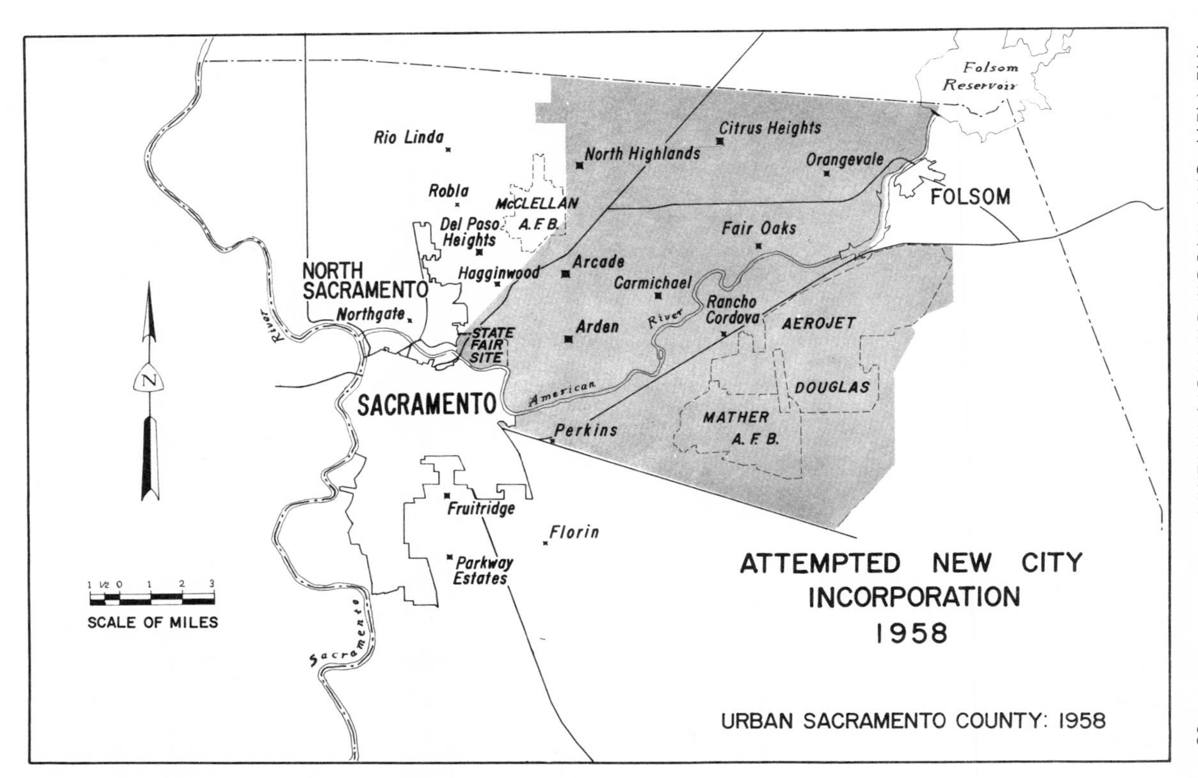
ATTEMPTED NEW CITY INCORPORATION 1958
URBAN SACRAMENTO COUNTY: 1958
Folsom Reservoir
FOLSOM
Citrus Heights
Orangevale
North Highlands
Rio Linda
Robla
McCLELLAN A. F. B.
Del Paso Heights
Fair Oaks
NORTH SACRAMENTO
Hagginwood
Arcade
Carmichael
Northgate
Rancho Cordova
AEROJET
STATE FAIR SITE
Arden
River
River
SACRAMENTO
American
DOUGLAS
MATHER A. F. B.
Perkins
Fruitridge
Florin
Parkway Estates
Sacramento
N
1 1/2 0 1 2 3
SCALE OF MILES

age exceeded 35 years in over one-half of the census tracts. Except for the Mather Air Force Base area, in which less than 3 per cent of the dwellings were occupied by the owner, owner-occupancy was registered at more than 62 per cent in most areas, exceeding the home ownership rate of 54 per cent in the city of Sacramento. In all but two of the United States census tracts within the proposed incorporation area, over 91 per cent of the houses were in sound condition. Only 85 per cent of the dwellings in the city were judged sound. It should be noted that the proposed new city contained large densely populated sections, but that it also contained much undeveloped land—particularly south of the American River.

LEADERSHIP

The incorporation movement was initiated and promoted by representatives of Carmichael. Only two of the 14 men who signed the letter of invitation to the first public meeting in June were outsiders. In late May, after much of the planning had been completed, David Yorton, chairman of the Arden-Arcade Council of Improvement Clubs, and Guy Fairchild, chairman of the Arden-El Camino park district board, joined the committee. They were to represent the Arden-Arcade area, the heavily populated territory of about 60 square miles between Carmichael and the city of Sacramento.

Moore, Craig, and Isenberg were joined primarily by friends and acquaintances who they knew were interested in governmental reform: Fred Boltres, an insurance securities trust fund representative; John Landry, manager of the Crocker-Anglo Bank of Carmichael; Owen Stewart, professor of political science at the American River Junior College (later the executive secretary of the group); and Martin Anderson, a specialist in revenue and finance in the State Controller's office. Anderson invited Eugene Campbell, a highway construction engineer. A building contractor, Win Johnson, and a dentist, Darol Rasmussen, were invited by Stewart. Horace Dunning, one of the first members, became attorney for the committee. All of these individuals had shown interest in or had participated in Carmichael community affairs. Isenberg also called upon Dr. Sam Wood for his expert opinion on the alternatives for governmental reorganization. Wood had been a governmental consultant and a professor of political science for several years and was at this time associated with Pacific Planning and Research, a consulting firm.

Nine of the 14 committee members were interviewed. (An additional 9 persons who favored incorporation but were not on the committee also were interviewed, as well as 2 persons who were opposed.) All claimed to have joined the committee because they were interested in planning, improved community organization, and better services and roads. Each ascribed the same motives for involvement to all 14 members, but some personal factors were also mentioned. The homes of two members had been affected by spot zoning; another member was about to subdivide a fair-sized acreage that he owned, and another member owning property along the main thoroughfare was having difficulty getting new offices built nearby. A proposed new freeway to run near a member's property helped spark his concern. One member allegedly had an interest in becoming planning officer for the new city and one was said to aspire to the position of mayor of the new city. On balance, however, it appears that personal considerations played a surprisingly minor motivational role. The scores of these individuals on the Allport-Vernon-Lindzey Study of Values underline their idealistic bent. The incorporation workers scored higher on the theoretical motivation scale and lower on the economic motivation scale than the leaders in any other movement studied.[3] Other data indicate that they were more highly educated than other leaders studied.

EARLY DISAGREEMENT AMONG LEADERS

Three main differences of opinion arose among the 14 leaders in the original citizens' committee. The first two were disagreements over goals. Dr. Sam Wood presented the greatest challenge to the Isenberg goal of incorporation. He was invited to join the group a month after the first ten members had organized officially, and at the first three meetings he attended, he strongly supported annexation rather than incorporation. He favored centralized planning for the total metropolitan area. If the city of Sacramento was willing to move "with sympathy" toward annexation, he felt this promised the best solution. The group agreed that statements of city leaders were sympathetic, but not likely to be acted upon. Despite this initial divergence of opinion with the chairman and with the group as a whole, Wood apparently was asked to prepare a feasibility study with his organization, Pacific Planning and Research. The April 14 minutes stated that Sam Wood agreed "to maintain records of the time devoted by his staff and depend upon the future for possible reimbursement."

On April 21, Wood presented a detailed report on the financing and feasibility of the new 165-square-mile city, boundaries for which were set at the April 14 meeting. He reported that the proposed city contained about 80 special service districts. Property tax rates varied from district to district, with the median about $6.81. A municipal tax rate of $4.235 would be needed to provide municipal services to the proposed new city. Sacramento's municipal tax rate was slightly less than $1.50. Wood's report concluded that incorporation of the proposed city was not economically feasible. Incorporation would require an extremely high property tax rate, with county and school district property taxes added to the municipal tax, or a low level of municipal services, or both.

Instead of ending the matter, this negative report from an expert caused the incorporation committee to "come to grips with the facts of life." It agreed that incorporation would provide improved planning, increased coordination of efforts, and more representative government, and even though the new city "would have to move slowly and increase its services as it was possible" the committee should continue to work toward incorporation. Several committee members reported that Wood's enthusiasm for the movement diminished after this time, but he remained a member of the committee and later supported incorporation in radio and television appearances.

Was incorporation a goal in itself or only a means to city-county merger? This second issue was often discussed but never resolved. Those who wanted merger felt that a separately incorporated city would have more leverage to push the city of Sacramento into merger. While the merger element on the committee never publicly gave up its ultimate goal, during the public campaign for incorporation that goal was sometimes dim indeed. Some of the original leaders interviewed still claimed that they were always working toward merger.

The third major difference of opinion was over the type of action program needed to put the new city idea across. Part of the group wanted to conduct a slow educational program. The committee would gradually be increased to about a thousand persons, representing the leadership of all of the organized activities within the proposed city—service clubs, churches, improvement clubs, chambers of commerce. A simultaneous public education program would emphasize the problems in the area and the practicability and desirability of forming a new city. After this educational effort a new citizens' committee was to be formed to set up the machinery to incorporate. The persons who opposed this view are not known by name, but their "crash" or "shock" approach finally won the day, perhaps by circumstance rather than by design.

No one dropped out of the movement because of these differences among the leaders. Some individuals lost their original enthusiasm but this cannot be attributed to Isenberg and Moore, who both worked extremely hard to bring about incorporation.

SMAAC AND MGC MEMBERS PARTICIPATE

Some SMAAC and MGC members played a part in the incorporation movement. A. F. Dredge, Jr., an influential member of SMAAC; James Cowan, chairman of SMAAC and an influential member of MGC; and Elmer Nelson, an Aerojet-General Corporation employee who was a member of SMAAC and for a time of MGC, were all aware of the activities of the incorporation committee well before the public meeting of June 17.

As early as January 30, 1958, when both MGC and, without publicity, the incorporation study group were getting under way, Dredge stated in his first speech as president of the Greater North Area Chamber of Commerce that "if substantial progress is not made in the next weeks we intend to take the lead as the one organization with universal representation in the north area to form a new committee to establish a new city north of the American River."[4] Three incorporation leaders stated that Dredge donated money to their group and was kept informed of its progress from the beginning. He was not officially a member of the group although he was invited to join the committee and later asked to provide names of people to be invited to the first public meeting.

James Cowan, approached for the same two purposes, was

never an active public supporter of incorporation, but by seeking governmental reorganization for the North Area, he gave it his blessing. He told the MGC in June that the incorporation effort must be taken seriously. He listed city-county merger (SMAAC-PAS) as his first choice, mass annexation to Sacramento as his second choice, and the new city as his third choice.

Elmer Nelson of Aerojet-General and John Goodman of Douglas Aircraft were officially approached by Martin Anderson and Howard Craig in the middle of May in order to explain the incorporation committee's plans and to obtain their "cooperative efforts." On May 26 it was reported in the minutes that Nelson and Goodman had expressed interest and appreciation upon being informed of the incorporation movement. City-county merger would be best, Nelson said, but probably could not be achieved. Both men saw the need for governmental change and stabilization of the tax structure and Nelson allegedly offered to help with the incorporation effort. (From later interviews it was learned that Douglas Aircraft had indeed helped, but that Aerojet-General became rather cool to the idea.)

ORGANIZATION

On March 3, when there were but ten members on the committee, Isenberg was elected temporary chairman, Stewart, temporary vice-chairman, and Mildred Welsch, temporary secretary. Mrs. Welsch was from the Arden-Carmichael school district headquarters; she dropped out of the committee in April. Many of the early meetings to study the problem were held at Isenberg's home and he could be credited with bringing the group together and forcing the pace of the original committee. The chairmanship fell to him rather naturally, and the officers were all Isenberg people committed to making the incorporation movement more than a mere foil for forcing the merger of city and county.

As part of the strategy for getting widespread support for the incorporation attempt, the original group of 14 decided that, after the first public meeting, most of them would play a minor public role. One of them was to sit on each of five subcommittees which were to be formed June 17. The remainder were to stay in the background so that it would not look as if Carmichael were running the whole show.

Some of the new members chosen to chair or sit on the new subcommittees were to be chosen prior to the public meeting. The records reveal very little about who was approached and who agreed to serve, but it is doubtful that much was done about developing new leadership before the public meeting. The transition in leadership was awkward and, as it turned out, unsuccessful. Of the 12 persons named to four subcommittees on June 23, four were from the original group and only two of the remaining eight—O. D. Kingsley, an Orangevale realtor, and Harold Bondeson, the manager of a large shopping center—were later considered to be activists. The executive committee, formed a week later, was composed of Isenberg (chairman), Landry, Dunning, Stewart, Moore, and Yorton of the original group, plus three new people, one of whom never became active, including Alvin W. Meyer of Citrus Heights, and Mrs. Howard Winslow, who was named canvassing chairman.

Kingsley was named finance chairman. Bondeson was expected to obtain the support of the big shopping areas. Alvin Meyer had made the motion to go ahead with the incorporation at the first public meeting. He was put on the executive committee primarily because of his position as president of the

Foothill Farms Improvement Association. He remained with the movement until the end. Mrs. Howard Winslow and her husband, who worked as a team on the gathering of signatures, had never had experience in politics or in canvassing. The leaders viewed both as hard-working, capable workers, but they were never recognized as policy makers. They dropped out after about 60 days of full-time work.

Two other women were also active in the movement, but did not hold official leadership positions. Josephine Brown, the public relations officer for the largest shopping center in the proposed incorporation, Country Club Centre, helped to arrange radio and television programs and to make contacts with the suburban newspapers. She also attempted, unsuccessfully, to get Mr. Blumenfeld, the developer of the Centre, to back the movement. Sign-up booths and loudspeakers were permitted in the Centre to promote signature gathering.

Mrs. Dorothy Orr was probably the most influential of all the newcomers. She was hired as secretary for the movement's office at Country Club Centre. Her immediate superior was Owen Stewart, who was hired as the executive secretary. When Stewart left the post less than a month after assuming it, Mrs. Orr exercised a great many of his responsibilities. She was the only person in the movement who could be reached regularly during business hours, and consequently became the focal point for communications among the leaders and executive committee from the middle of July through October.

In the middle of August several of the early incorporation proponents led a behind-the-scenes effort to oust Isenberg and place another member of the original incorporation group as chairman of the executive committee. The effort did not succeed. As a result, some members of the original group worked to control the incorporation movement and to pep up the lagging spirits of the executive committee and other workers. They were too late, however, and did not act with enough boldness to accomplish their goals.

FINANCES

In the first financial report to the original study committee on May 26, John Landry reported expenses of $186, which the incorporation committee was obligated to pay. Sam Wood's Pacific Planning and Research study probably accounted for $180 of this amount. A total of $5,000 was estimated as the budget for the incorporation effort. One thousand dollars was to be raised by asking each person invited to the first public meeting to pay $10. On June 2, each committee member present donated $5.00 to pay for mailing the public meeting invitations. Evidently nothing further was done to raise money until after the June 17 public meeting. Donations were mentioned at this meeting as the method of financing, but the people present were not solicited as originally planned.

On the twenty-third of June the chairman of the finance committee announced a goal of $5,000, $2,000 of which was to be a bank loan secured by 20 signatures. On July 3 the suburban papers carried a picture of Landry, manager of the Crocker-Anglo Bank of Carmichael, handing a check for $2,000 to the finance chairman.[5] All of the original group and several new committee members and supporters had signed for the loan.

On June 30, a week after the $5,000 budget was approved, a revised budget of $7,580 was adopted to be used for Stewart's salary ($75 per week), a part-time secretary, and rental of a suite in Country Club Centre. The larger budget was adopted in spite of a discouraging response during the first two weeks of public solicitation. One leader said that the original $2,000 gave out rather quickly and that Stewart was not getting paid. By mid-July Stewart stated that unless the incorporation group could get more money the movement would fail. Isenberg reported that they were not going to have the amount originally budgeted because contributions simply were not being made.

What money was raised came principally through the efforts of Jack Moore. The largest donation was $100, and there were several for $25. Few businessmen gave any money. Members of the Greater North Area Chamber of Commerce gave varying amounts as individuals, but not as much as expected. So it went with many of the people approached. One interviewee stated that only one merchant in the Town and Country Village and the Country Club Centre shopping centers contributed. At one point late in the campaign some of the firemen in the Arcade area were reported to have solicited funds door to door, but without much success. About $1,000 of advertising was sold to help finance a flyer for insertion in the *Suburban News-Shopper.*

Services and equipment were donated more readily than funds. Some printing was secured at cost, and an employee of Douglas Aircraft helped get an airplane and cameras for filming conditions in the North Area. The photographs were used at several meetings. One individual donated a mimeograph machine, and minor supplies were donated by a few other people. A local radio station donated a substantial amount of time, and at least one major television program was presented as a public service.

The leaders agreed that less than $3,000 had been spent for the entire effort and that about $1,600 of this had come out of the pockets of the people who had signed the bank note. Finally, everyone agreed that a great deal of money would have been needed to put the incorporation over; estimates ranged from $10,000 to $100,000.

THE DEBUT OF THE NEW CITY PROPOSAL

The incorporation committee, prior to the public meeting, was truly a study group. Some members were convinced from the beginning that incorporation was the only answer, but the group generally discussed issues in a rational, unemotional manner. Several experts were called in to explain procedures for action, and some members personally consulted others who could not be persuaded to speak to the group.

In mid-March, Horace Dunning visited the newly incorporated city of Fremont, California, where he talked with the assistant city manager. At a March meeting John Marshall, senior statistician of the State Board of Equalization, gave estimates of what a new city could expect from the gasoline tax, liquor license fees, and the sales tax. He suggested that they try to persuade the county board of supervisors to permit the city to retain all of the local sales tax collected within its boundaries for the first two years of its existence. About 70 per cent was being allotted to Sacramento County municipalities at that time. On April 7, Lewis Keller, a representative of the League of California Cities, spoke to the group. He explained that the League, which could offer only limited advisory services until after incorporation, favored annexation rather than separate incorporation.

Planning proceeded slowly and carefully until May 26, when the date for the public meeting was set. By contrast, haste and lack of thoughtful planning marked the activities of the last three weeks before the meeting. The committee largely ignored previous advice and policies on organization and procedure. Jack Moore had been designated by the committee to determine the timing of the movement, but if the group was really committed to a slow educational program, he was the wrong man for the job. According to several leaders, he convinced Isenberg that it would be best to spring the plan on the public at an open meeting in the near future.

He and his followers saw too much difficulty and valuable time lost in getting a great many people "educated." The group had the talent and the drive to get the job done. The time was right for swift action—the SMAAC-PAS plan was already dead and MGC seemed to be getting nowhere. Some advocates of moderation evidently were convinced, because a date, June 17, was set for a public meeting. Yet on the very eve of the public meeting the issue was not settled. The virtually unanimous support of the incorporation idea expressed at the meeting of the seventeenth permanently sidetracked the slower approach. The motion from the floor at the public meeting to file the incorporation papers forced the committee to work under the pressure of a deadline. Under California state law a governmental reorganization group has exclusive rights in an area for 90 days after filing a statement of intent to circulate petitions with the county board of supervisors, and the requisite number of signatures must be filed within that period or the exclusive jurisdiction ends.

About 100 persons attended the June 17 meeting. All had received a letter of invitation asking their help in evaluating "a concept for governmental change" developed by "a group of individuals, representing a cross section of the citizens of the North Area." The members of the committee were listed, and Walter Isenberg signed the letter. The plan was presented and explained to the audience and general discussion followed. Jack Moore argued against piecemeal annexation and contended that incorporation would give impetus to city-county merger. (MGC was, in fact, influenced by the incorporation movement to recommend mass annexation of the entire North Area.) "The basic imbalance," he said, "is between Sacramento city and the North Area. Sacramento has shown little interest in sharing their facilities with us. We have to build up our own strength, so that we will no longer be the poor relations they think we are now." A. F. Dredge, Jr., was quoted as saying that there would be "some aroused citizens around the county in the morning." Toward the end of the meeting Alvin Meyer moved to endorse the incorporation plan; it was approved without a dissenting voice. The motion instructed the committee to file with the county board of supervisors the next morning the required materials and documents relative to announcing the group's intention to begin official incorporation proceedings.[6]

The proposal itself was presented in a brief report entitled "Planning for Tomorrow," which was distributed at the June 17 meeting. The report contained a map of the proposed city and criticized the two primary types of government now serving the area. The county government was not organized to supply the many services needed in an urban area, and other services had to be provided by some 90 special districts. This multiplicity of districts was costly and prevented area-wide planning. Other possibilities for governmental change were discussed and declared unsatisfactory.

The new city would have to incorporate as a sixth-class general law municipality, but after a year could adopt its own charter. The traditional advantages of incorporating were listed: one government rendering area-wide services, more self-determination, a more prudent use of land and water, collection by the city of state subventions not now available, more stable financing, and the legal simplicity of incorporation as compared with other governmental changes. One item in the proposal provided that the new city would retain 90 per cent of the local sales tax instead of the 70 per cent usually allowed by the county supervisors. A budget of $4,411,885 was proposed for the first year—a little more than half of the amount estimated by Sam Wood as needed to provide adequate municipal services.

REACTIONS TO THE NEW CITY PROPOSAL

The incoporation plan provoked immediate discussion and interest from many sources within the Sacramento community. The day after the public meeting James Cowan stated that the inaction of city and county officials had "brought this on." "I think it is tragic," he said, "that the City of Sacramento has had so little foresight and has been so concerned with its own tax rate and cost pattern that it would allow this whole north area to grow without showing a willingness to assume some of the cost burden in the area outside the city."

City Manager Cavanaugh replied:

> In the first place the state law is very clear that a city such as Sacramento is prohibited from annexing any area until the people in that area make a formal request to the governing body of the incorporated municipality.
>
> No one in Sacramento's city government has ever stated he would not favor annexation of the area to the north.
>
> The city council never has refused to proceed with any annexation request where comprehensive areas are involved. The council always has welcomed any movement of people in outlying areas for consolidation with the City of Sacramento.[7]

County Executive Tarshes stated that the county government should not get involved, either pro or con. County Supervisor Garlick said that if the people of that area wanted to set up a municipality that was quite all right, but they still would have to pay county taxes. Another county supervisor, Kelley, felt that along with the loss of revenue (from the sales tax) as a result of incorporation the county would also be freed of a lot of problems. City planner Rathfon considered it regrettable that the metropolitan area of the capital city of California should become splintered into numerous city governments. Alvin Landis, chairman of MGC, said that the people of the area had the right to choose whatever governmental arrangements they wished, but that the creation of great numbers of cities, no matter how large, had never solved the problems of a metropolitan region.

On the evening of June 18, five members of the incorporation committee met at the home of Walter Isenberg. They discussed the previous evening's meeting and plans for the future. A news representative from a local television station also taped an interview with Isenberg for use on a news program later that evening. A reporter from the *Sacramento Bee* who was present quoted Isenberg as saying:

> I've never been so busy in my life answering telephones. People I know and people I've never heard of are offering support and asking how they can help.
>
> From every inkling that I've heard I believe we will get widespread support. If we can reach the people I'm certain they will support us.

Neighborhood chambers of commerce in the incorporation areas gave considerable attention and support to the incorporation movement. The Carmichael chamber endorsed the proposed incorporation on the assumption that the SMAAC-PAS proposal had no chance of being adopted. The North Highlands Chamber of Commerce voted 21 to 17 to support incorporation at a public meeting on the question.

The Greater North Area Chamber of Commerce also endorsed the new city movement but not until September 11.

The chamber's Urban Government Committee endorsed the idea, in a 20 to 3 vote, in late August, 1958. A number of the pro voters supported the idea reluctantly, preferring city-county merger. The final vote of approval by the board of directors of the Greater North Area Chamber of Commerce was 18 to 3. Their action had been preceded by a mail poll of chamber members in which 58 per cent of the number responding favored the new city. In any event, only one week remained between the chamber's endorsement and the expiration of the 90-day exclusive period.

Two days after the public meeting, Mayor Azevedo of Sacramento reacted to the proposal by inviting the entire North Area to seek annexation to the city. This, he thought, would be a better solution than separate incorporation, and he was optimistic about the chances of such annexation: "Since I have been on the council, for six years, we have done everything to encourage annexation of areas to Sacramento. We have not turned down a single request and the larger the area the easier it is to provide necessary services because there is a greater tax base." Other councilmen made similar statements, but councilman James McKinney doubted that the North Area favored such annexation and believed that any attempt by Sacramento to initiate it would result in accusations of "trying to swallow up the area." The same evening the Sacramento city council adopted the following resolution: "The council is receptive to annexation of any sizable area contiguous to the city. It is willing to extend the help of the administration and any department of the city."

Isenberg immediately rejected this plan. In a statement the following day, he said:

> It is assumed such interest [annexation of the North Area] is sincere and not the reaction to the earlier filing of a petition to incorporate that great area.

> The people of the committee were motivated solely in finding the best governmental system to meet the needs of the residents. We explored all avenues, including annexation. We presented our findings and recommendations to a large group who joined with us in favoring incorporation.
>
> Today, we have no intent to deviate from that path. We are neither qualified nor do we desire to participate in political maneuvering.
>
> We presented and substantiated our findings. It is our firm conviction that the great city we propose should be incorporated. It is our intent to organize immediately to accomplish this purpose.[8]

In later speeches the incorporation leaders charged Sacramento with making an about-face from its long held policy of self-containment, self-satisfaction, lack of interest, and complacency toward the metropolitan area to one of affection, cooperativeness, and willingness to change in the face of a threat to make a separate city. In the spring of 1960, when the leaders were interviewed, most retained their former opinion, that the "city fathers" had no interest in annexing the North Area. They were critical of the city manager, who, they said, lacked vision and feared the addition of a large area would change the political complexion of the city. If the city had a vigorous policy of annexing uninhabited territory, it could have gone out the Roseville Freeway and along both sides of the American River —in effect surrounding the most populous part of the North Area and preparing it for annexation.

In spite of this rejection of its offer, on June 26 the city council instructed the city manager to meet with North Area representatives to discuss questions of annexation, incorporation, and merger. Mayor Azevedo knew of many persons who preferred annexation to incorporation. He pressed the city manager to tell whether or not Sacramento had money available to provide the North Area with municipal services. The city manager replied that it depended upon the size of the area and exactly what services the residents wanted.

The *Sacramento Bee* opposed incorporation while most suburban newspapers strongly supported the movement. The *Bee* pointed out that two of the three reasons announced for separate incorporation no longer applied. The city council was now amenable to annexing the area; and since the proposal had been made public, the council had placed a city charter amendment on the November ballot so that, if annexation were approved, the new areas could have immediate representation on the city council. The amendment also did away with stringent residence requirements for employment with the city government. (The amendment was subsequently adopted.) As for the third reason—that the area would not vote to come in—this, said the *Bee*, was "an arrogant usurpation of the people's right to decide for themselves." The editorial claimed further that the whole incorporation proposal was developed in a series of secret meetings, and that the incorporation group had all of the legal papers ready for filing the very next morning after the proposal was made public—thus making it legally impossible for the residents to choose any other proposal.[9]

Isenberg replied that there was no legal way for the voters to choose between annexation and incorporation in the same election. He stressed that people in this country have the right to meet in private to discuss measures for their common welfare, and may make public statements if and when they please. Isenberg suggested that all keep their tempers and not resort to expressions such as "arrogant usurpations" and "secret meetings."

From the beginning the proposed new city idea received strong support from the *Suburban News-Shopper* and from several suburban newspapers such as the *San Juan Record*, the

Carmichael Westerner, and the *Citrus Heights Bulletin*. On June 26, these three newspapers carried the same editorial, under the title, "Area Incorporation Plan." The incorporation proposal was called a good plan, a workable plan, and one that challenged the imagination. The SMAAC-PAS proposal was best, but the "politicians" had shunted aside the plan of the "experts." MGC was accused of being unfriendly to city-county merger. The editorial writer expressed dislike of Sacramento's "smugness." Under the circumstances, the only feasible plan was separate incorporation. On the same date the *Suburban News-Shopper* carried an editorial under the heading, "Sheer Madness? Oh, Come Now," which defended the incorporation proposal and the 14 persons behind it. All preceding proposals to do anything about the area had been "left to lie there." The residents of the area should consider this proposal carefully and not be concerned particularly with the opinions of the city council of "a town located six miles away on the other side of a river." The *Carmichael Courier* also carried an editorial on June 26 which stated that the charge of "secret meetings" was unfounded. Incorporation had been "planned openly around a framework of the leading citizens of the area."

In a later article the *Courier* praised the incorporation group for its "yeoman service for the people of the Sacramento community" in moving the Sacramento city council to attempt to make the city charter more congenial to annexation. On the same date the *Carmichael Westerner* and the *San Juan Record* accused the *Sacramento Bee* and the Sacramento city council of trying frantically to create a new issue, but the real issue of whether the city and county officials and MGC would stop procrastinating and set up a consolidated government remained.

"Who Killed Cock Robin?" asked the *Citrus Heights Bulletin*. The SMAAC-PAS plan was dead. Why? The officials of Galt, Isleton, Folsom, and North Sacramento deserved part of the blame, but the chief blame belonged to officials of the city and county of Sacramento, who had created MGC to restudy what already had been studied thoroughly over a period of several months.

LAUNCHING THE NEW CITY CAMPAIGN

On the evening of June 23 a meeting was held at the home of Walter Isenberg. Only four members of the original committee were present—Jack Moore, Horace Dunning, Owen Stewart, and Walter Isenberg. Other "incorporation backers" attended, but newspaper accounts mention only three by name: O. D. Kingsley of Orangevale, John Goodman of Douglas Aircraft, and Mervyn A. Neumann, an oil company representative. According to newspaper accounts, the Interim Committee on Incorporation of Greater North-East Area was dissolved at this meeting, and the Citizens' Committee for Incorporation was created.

From early July on, the proponents of incorporation concentrated on selling the idea to the people of the area and getting the signatures of the necessary number of property owners. At a July 1, 1958, meeting at the incorporation group's headquarters at Country Club Centre, a publications committee was created, and it was decided that a series of articles, written by selected individuals but sponsored by the Citizens' Committee for Incorporation, would be printed in the local newspapers over a period of a few weeks. At the end of the series, all of the articles were to be combined into a brochure, to be paid for from the $1,500 budgeted for publicity. Walter Isenberg presided over this meeting, which was attended by five of the original committee and seven or eight others.

Some of the proposed articles were written and subsequently published in several newspapers, but, for lack of funds, the brochure was not printed. The Citizens' Committee for Incorporation sent speakers to community meetings and staged a number of public meetings of its own. Except for two or three of the earlier ones, the newspapers reported very small audiences at these meetings. For the most part, the few meetings called by opponents of incorporation had small attendance also. Generally the public did not seem to see any emergency that called for immediate and drastic action.

Several issues arose concerning signature gathering. The role of the executive secretary, Owen Stewart, in circulating petitions was in question. At least two of the incorporation leaders interviewed felt that Stewart was hired primarily to conduct the canvassing effort. He apparently did spend most of his first two or three weeks doing just that, but he also had to deal with the many other details—publicity, coordination, and organization—of getting the movement under way. All leaders agreed that he was assigned too many jobs to be really effective in any of them. He is reported to have been very discouraged about getting the number of signatures needed. After taking a vacation in the middle of July, he did not return to full-time work. The fact that his tenure as active executive ended within a month after he took over was not widely known.

A second issue involved the number of signatures actually needed to get the incorporation proposal on the ballot. The committee had originally assumed that around 10,000 to 12,000 would be sufficient. Later it was advised that the signatures of both husband and wife were legally required on community property, which meant that a considerably larger number were needed. This was discovered after the canvassing was well along, but still far short of the 10,000 goal.[10] The 24,000 figure was not revealed to the Winslows, who were now conducting the canvassing effort with little help from the executive committee.

About 80 per cent of the citizens approached did sign the petition, but Mrs. Winslow stated that it often took a half hour of explanation and discussion to get each signature. The workers could not get the necessary information from the executive committee to answer citizens' requests for facts or clarification of the leaders' public statements. None of the top people ever talked to the petitioners. Furthermore, the leaders themselves were not helpful in petitioning. One said that, "As an example of the poorness of the planning, one evening the executive board decided that the next day all members would come together and each bring five other people and they would try a saturation petitioning effort in a single area. Of the fifteen people who were each supposed to bring five, only the official petitioner from the area, Mrs. Winslow, and her followers, and Isenberg showed up."

Before long it became apparent that getting the necessary signatures was not going to be a simple task. By early August, community newspapers reported that 300 volunteers were circulating petitions, but that another 200 volunteers were needed. (Fewer than 200 persons ever worked on petitions according to reliable information.)

While the community chambers of commerce generally lined up in support of the proposed new incorporation, the Sacramento City-County Chamber of Commerce went on record in opposition. Its metropolitan development committee had recommended opposition in early July. In early August the executive committee stated that although approximately 40 per cent of its membership was living or working in the area proposed for incorporation, the chamber must place first importance on the welfare of the county as a whole. The Sacramento area could not expect to fare well in the competition between various sections of the state for industrial payroll and tax-producing industries if it did not present a unified approach. Intra-county disputes would harm all of the cities within the metropolitan area in the long run. The Sacramento Area Planning Association also took an official stand in opposition to the incorporation.

Newspaper reports of the public meetings during the summer of 1958 indicate other troublesome issues. Some sparsely populated fringe areas were being forced into the new city; Isenberg stated that after the necessary number of signers had been obtained the county board of supervisors could exclude such areas. Other small communities were concerned about being swallowed up by large city control. Many questions about finance also arose. The people really did not know how much service they would get at what cost, since the proposed budget was merely an estimate on the part of the incorporation backers.

As it became apparent that a special survey under MGC sponsorship might reveal that annexation would cost less than separate incorporation—just as Sam Wood's study had shown earlier—the backers of incorporation began to argue that other advantages justified a possible higher cost. The volunteer fire departments need not be destroyed under the proposed new fire department while annexation to the city of Sacramento would mean no volunteer workers. The job security of the employees of the various special districts would remain unimpaired. (Contrary to this view, the long-range goal of incorporation was the gradual abandonment of all special districts in the new city.) Separate incorporation would allow the residents to determine how their tax dollars were to be spent.

A. F. Dredge, Jr., a supporter of incorporation, was critical of other advocates for proposing an annual budget of only $2,000,000 for roads, when the county currently was spending $4,500,000 annually in the area for roads. Residents were being told that under incorporation they would not have to continue to pay the 25-cent road tax, without being informed that this "saving" would deprive them of $2,500,000 for the construction of streets and roads.

Other possible budget troubles in planning and police protection arose. The county planner stated that the proposed city budget provided only one-third as much for planning as the county already was spending for this purpose in the area. A professor of police science at Sacramento State College estimated law enforcement costs, based upon the expenditures for police departments in cities of about 150,000 population. A complete and fully organized police department for the new city would cost an estimated $1,500,000 annually, but the proposed incorporation budget provided only $800,000 for police. He suggested that it might be cheaper to contract with the county sheriff's office for police services. A month earlier the county board of supervisors had adopted a policy of providing services such as police protection, which the county was staffed to perform, to areas in the county on a contractual basis. (This policy has never been implemented.) No estimate of the contract cost for police services was made, however.

As the above issues and problems became public knowledge, the character of the incorporation meetings changed. At first there had been efforts to analyze what incorporation would mean for the area. Before long, however, a number of the incorporation proponents used the public meetings only to paint the city of Sacramento in unappealing colors. Sacramento maintained a condescending attitude toward the northeast area, annexed only for the tax income the new area would supply, and had a very high crime and tuberculosis rate. On the other hand, the northeast area had one of the highest-type populations in the United States and was willing to contribute time and money to the cause of good government.

EXCLUSIVE CONTROL PERIOD ENDS

The day before the incorporation group's exclusive control was to expire, the county counsel issued a legal opinion that renewal of exclusive control for another 90 days would be illegal. The incorporation group could continue to circulate its petitions, but it no longer would have exclusive control. In fact, another group could file another proposal for part or all of the area and receive 90 days' exclusive control for its proposal.[11] The next day, September 16, 1958, exclusive control was scheduled to expire at 5 P.M. The county counsel stated that if any legal maneuver to extend the period of exclusive control were attempted, the whole incorporation proceeding would be open to court challenge. He said that a petition for a 90-day exclusive period could be filed, but it would be considered a new period. Thus, the backers would have to start all over again to get signatures.

No signed petitions in support of the proposed new city were submitted. While the number of signatures obtained was not announced, failure to turn them in was taken generally to mean that the number was insufficient. In fact, fewer than 8,000 were ever gathered. However, at one minute after midnight of the expiration date, Jack Moore and Darol Rasmussen filed a new petition for incorporation with the clerk of the county board of supervisors at the clerk's home. Walter Isenberg stated that the petitions previously circulated would be withdrawn and held by the incorporation committee. For the time being, the incorporation group accepted the ruling that the exclusive control period was not extended; an entirely new movement was beginning in its place. The incorporation group might take legal action later to reinstate signatures previously obtained, according to Isenberg.

The legality of the second incorporation movement was questioned by the county boundary commission. The incorporation group argued that it did not need to file a map and a legal description of the same boundaries a second time. The county boundary commission refused to act, saying that legal petitions could not be circulated until it had reviewed the boundaries. Meanwhile, the county counsel argued, the second block of 90 days was being spent. Both sides agreed that a court contest might be necessary.[12]

Late in 1958 the city of Folsom posed another legal challenge to the incorporation group. Folsom wanted to annex 800 acres contained within the boundaries of the proposed new city. Isenberg contended that the incorporation group had exclusive rights in the area. The Folsom city attorney contended that the incorporation group had not obtained a second legal 90-day period of exclusive control because it had failed to file the necessary documents. Folsom completed the annexation in 1959, and no legal action was attempted against it.

OPPOSITION FORCES

Backers of the incorporation proposal confirmed in interviews that their most formidable opposition was the *Bee* and city officials of Sacramento. In the preceding narrative of events, the attitudes and positions of these opposition forces have been cited. Two citizen groups and some improvement clubs also presented opposition which, while not causing its failure, did weaken the incorporation movement.

One such group informally organized against incorporation was led by Edwin Morgan, sometime businessman and part-time teacher. On July 7 he made the statement to the press that some people south of the American River were unhappy about the proposed incorporation. He was referring to residents in the area of the proposed city which was designated for industrial development. About 40 per cent of the new city was on the south side of the American River. There was considerable undeveloped land here, and much of this was viewed as prime industrial area by the promoters of the new city. Many people living south of the American River felt they were to put up with increased dirt, noise, and smoke so that the new city could have an adequate economic base for municipal services for the more expensive residential areas on the other side of the river. The people south of the American River generally did not feel that their interests were akin to the interests of the North Area. Morgan himself owned land just to the south of the proposed new city. He stated that at first he only wanted to know why most of the incorporation leaders—the top five, as he put it—were from the Carmichael area, but his actions reflected a personal distrust of the Carmichael community.

Morgan brought together four merchants from the area and five other men. At the first meeting Morgan told them what he knew, and spoke of the "complete change in the area" that incorporation would bring. The group invited Jack Moore and Walter Isenberg to a later meeting and showed them figures they had worked out that varied radically from the estimates in the incorporation proposal. Morgan also approached 12 building contractors who wanted to build subdivisions and consequently were opposed to keeping the undeveloped area strictly industrial. They did not support Morgan's efforts, but later the Associated Home Builders came out against incorporation.

The Morgan group wanted to exclude its area from the proposed incorporation before the proposal went to the voters. The group finally filed protest petitions with 702 signatures of property owners—many fewer than needed to exclude the area. Morgan said that they had about three meetings per week from July 7 until the end of September, when the petitions were filed.

Mrs. Agnes Booe, the editor of a state legislative newsletter, evidently had talked with Morgan and attempted to organize an anti-incorporation campaign of her own. Mrs. Booe stated in an interview later that, as it turned out, she was a one-woman organization:

> I talked to people and they were completely indifferent one way or another. Most were against, but they wouldn't bother to come to a meeting. They asked me to do the work for them. I put out a few circulars on a multigraph, mostly in my own neighborhood. The boy who did the multigraphing also delivered them. The whole operation cost me only a few cents.

Mrs. Booe said that incorporation was not a formidable movement and that it really was not necessary to have anyone against it. If the movement had gotten serious, others would have joined in opposition.

Mrs. Booe's announcement that she was going to organize opposition was made on September 11. About two weeks earlier the Sierra Oaks Vista Improvement Club voted unanimously to seek exclusion of its area from the proposed new city. The Santa Anita Improvement Club, although it never was cited by the newspapers as an organized anti-incorporation group, presented perhaps the most powerful road block to effective action except for the *Bee* and Sacramento city.

Besides overt opposition, the growing interest in the creation of one single large city out of the entire Sacramento urban area presented the challenge of a counterproposal to the incorporation plan. The board of directors of the Sacramento City-County Chamber of Commerce in late September, 1958, voted unanimously in favor of one big city before MGC had presented its proposal to achieve the same end.[13] About ten days later the Associated Home Builders of Sacramento, in going on record in opposition to the proposed new city, came out in favor of one big city for the Sacramento area. The group stated that the incorporation movement was based upon emotion and upon incomplete facts.

THE DEMISE OF THE NEW CITY PROPOSAL

On November 18, 1958, Walter Isenberg announced that the incorporation effort was being abandoned. The task of getting the required number of signatures was too great. About 8,000 of the necessary 22,000 signatures had been obtained. Isenberg asserted that the incorporation group never had claimed that incorporation was the ideal answer, but in view of inaction and lack of interest on the part of county and city officials, it had seemed the best plan available. Isenberg acknowledged that implementing a sound plan for governmental reorganization was the function of MGC, but he charged that the committee had become sidetracked when it limited itself to studying merely annexation versus incorporation.

Earlier Isenberg had written to the four contenders in the November election for the two seats on the county board of supervisors—Fred Barbaria, Jack Mingo, Edgar Sayre, and Walter C. Kelley. Isenberg stated that if his group felt, after the election, that the county board would vigorously encourage and actively help to develop a city-county form of government, the group would endorse and enthusiastically support such a plan. He asked the candidates to answer some questions on the whole problem of government for metropolitan Sacramento. The letter and replies from Mingo and Barbaria were printed in the community newspapers of the northeast area.

The Mingo and Barbaria letters were considered the most satisfactory by incorporation proponents. Mingo favored city-county merger as the ultimate solution, but he would concentrate first on combining the city of Sacramento and the surrounding unincorporated urban territory—essentially the stand later taken by MGC. Barbaria avoided taking a stand on city-county merger, but he felt that the voters should have a chance to decide the issue in the near future. A week earlier he had been quoted in "unqualified opposition" to the proposed new city. The proponents of incorporation either were ignorant of city costs or were deliberately misleading the residents of the area, he said, and favored unification of Sacramento and the surrounding area as the first step toward city-county government. Mingo and Barbaria at this stage had similar views on metropolitan government; both men were subsequently elected. Sayre favored annexation of surrounding areas to the city of Sacramento and did not favor city-county merger. Isenberg said that Kelley "straddles the issue, and fails to come to grips with the fact that an immediate answer must be found."

Isenberg, in giving up the incorporation effort, proposed that the elected officials of the county and the cities publicly acknowledge the need to develop the best plan for governmental reorganization, which could then be submitted to the voters for their decision. The officials should appoint a small group (no more than three) to develop the general framework of the plan, and the group should have about two months to complete this task. A larger group would then review and develop the plan before taking the issue to the voters.

THE INCORPORATION MOVEMENT EVALUATED BY SELECTED INDIVIDUALS

Isenberg, Moore, Stewart, and Rasmussen were most often mentioned by leaders interviewed as the important decision-makers. Besides these persons, the remaining members of the original committee were mentioned, but as one interviewee said, "This was really a committee at work rather than a centralized organization." (Two of the active people who were added to the original group criticized this "committee system.")

Isenberg and Moore stood out about equally, but well ahead of the others in getting the group to accept their views. Anderson, Dunning, Rasmussen, and Wood evidently had much influence behind the scenes, but made few public statements after the movement got under way. Several of the leaders asserted that the group, as an action group, lacked the ability to proceed with clarity and decisiveness. As one of the petition managers stated, "The underlings did not actually know who the leaders were, or who made the decisions." Dorothy Orr, secretary in the Country Club Centre office from July to October, stated, "They had no machine, and no one even to tell me what to do." They needed an executive secretary to "run things on a professional basis."

In retrospect, the organization for action was poorly conceived, loosely constructed, and largely ineffective. New leadership from the public at large did not materialize and the responsibility for carrying the movement fell to the original planners, few of whom remained fully active. Some, like the Winslows, whose work was vitally important to the success of the movement, were unable to obtain help or decisions in spite of the fact that they were members of the executive committee.

The decision of the original group to play a small public role and support a larger, broadly representative citizens' committee did not work out. The new group could not be instilled with the same fervor and knowledge as the originators of the plan had had, nor could a sufficient number of persons be found to form an effective staff. The idea was abandoned almost immediately. When Owen Stewart was appointed executive secretary, many of the other original members evidently no longer felt a need to work as hard as before. One leader stated, "With a full-time person, many of us relaxed. . . . We should have continued to expend as much effort as we had up to that point." Stewart's resignation left a void that could not be filled with original members. The organization proposed on June 9 remained a paper one; the original group did not develop fully or effectively, and the coordinating and controlling function of the executive committee never worked well. Nevertheless, no leader interviewed was prepared to say that the incorporation movement would have succeeded if the organization had been clearer and more realistic. The primary reason for failure was public apathy, which could have been overcome only under extraordinary circumstances.

The incorporation movement really never got off the ground. The total number of people who were in any way active in the movement did not exceed 200, the majority of whom were individuals interested in governmental change and not members of potentially influential interest groups or organizations in the community.

The only organization mentioned that actively supported incorporation was the Greater North Area Junior Chamber of Commerce, which set up and manned the public address system in the Country Club Centre. A number of leaders of the incorporation attempt reported that some Arcade firemen conducted a house-to-house canvass of several areas for contributions to the movement, but the fire department organization—in contrast to departments involved in other reorganization movements—took no active role. This lack of organizational support was one of the most discouraging factors to the incorporation group. The Greater North Area Chamber did not furnish men and money for the movement, but only passed a resolution in support, after the movement was already doomed to failure. Service clubs such as Rotary, Lions, and Kiwanis were little more than interested audiences to speakers supporting incorporation, and none endorsed or worked for the movement. One leader said that service clubs were inactive because they dislike to endorse issues which may divide their members.

There was no doubt in the minds of the interviewees that the *Sacramento Bee* strongly opposed the incorporation movement. The *Bee*, they felt, wanted to protect "the integrity" of the city of Sacramento and its own political and social influence in the county. It favored annexation to Sacramento because "the *Bee* can get more national advertising if it can show a larger city for its location." The suburban papers, particularly the *Suburban News-Shopper*, published full coverage of all press releases, meetings, and other activities of the movement. The *Shopper* was the most dedicated editorially, while the community newspapers tended to support the movement primarily through attacks on the city and the *Bee*. The realtors located in the area to be incorporated were believed to be in favor of the movement, in contrast to the realtors based in Sacramento. None of the big realty companies gave support, and the Sacramento Home Builders' Association publicly opposed incorporation. According to the interviewees, the big realty companies were opposed because they were exercising major influence on city and county zoning and planning now and did not know what control they could exert on a new city. The downtown Sacramento businesses were said to be opposed because of their vested interests. "There are many empty rooms and stores down there. Taxes are being increased in Sacramento city because of the suburban trend in business." The suburban businesses were reported to have reacted in a manner similar to the realtors. The large businesses and the largest shopping centers would not support the movement, while many of the small independent businesses were said to favor the movement.

Officials of the city of Sacramento were believed by the interviewees to be opposed to the new city. The city, quite naturally, the leaders said, wanted the area left available for annexation at whatever time and in whatever amount Sacramento might desire. The county officials, especially the county executive, were opposed but gave fair treatment and information to annexation supporters. "One thing the county opposed was our idea of giving only 10 per cent of the sales tax back to the county." The county executive "was also disturbed by our accusations against the county government . . . we endangered his empire." Supervisor Kelley was the only supervisor to take a public stand and in the closing days "took a few jabs at us for no reason at all . . . and this might have cost him the supervisorial election." (Kelley was defeated in his bid for re-election.) The incorporation people did not try to force the supervisors into a position on the issue because "in the end they had to approve the petitions."

Several incorporation leaders voiced regret that the League of Women Voters had not endorsed the movement, although they were pleased that many League members worked at securing signatures. The League's official position supported the SMAAC-PAS proposal of city-county merger.

The owners of the big shopping centers and vacant land near Country Club Centre were mentioned most often as individuals who should have been but were not involved. Evidently most leaders originally had held high expectations of support from these people. Aerojet-General Corporation was the only other large business singled out as one which could have aided considerably but did not. Only two individuals were named as persons who should have given stronger support to the movement: Seldon Menefee, a North Area newspaper publisher, and James Cowan, a superintendent of schools and chairman of SMAAC and a member of MGC.

The greatest deficiency in the movement was that its appeal was to small property holders and individual citizens rather than to large businesses and influential interest groups and service clubs. The leaders were sure that some heads of businesses and social organizations would have helped if they had been brought into the planning committee during its early stages. It is significant that if the incorporation initiators had followed their original plans, just such people would have been involved.

The interviewees were questioned about why the incorporation movement failed. Five cited poor planning and lack of organization. Three others stated that it was "lack of money for promotion." Two said that the movement was poorly timed and premature, while four spoke of "apathy of the people" and "lack of enthusiasm for change." A sample of additional responses includes "taxes," "lack of personnel," "did not get enough people involved," and "failure to concentrate on the petitions."

Organizations and individuals did not support the group because they could not see any benefits in incorporation. "We could not provide them with better services, only higher taxes." Many of the residents of the incorporation area, in fact, preferred the capital city to a rival to it, if they must be in a city at all. One interviewee observed: "People who really are concerned on a theoretical basis, such as our group, cannot carry through a movement. There is no strong motive for an arduous campaign. You need some kind of economic incentive in order to get that kind of activity." Most supporters of incorporation concluded that it would take some kind of spectacular or catastrophic event to rally the residents to take any action at all.

CONCLUSIONS

The original idea of Moore and Craig to organize a citizens' group to study the alternatives for governmental improvement

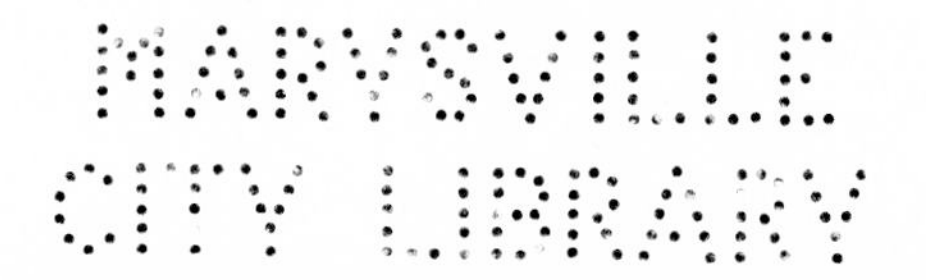

appears to have been timely. The SMAAC-PAS report was dead, and MGC, as it developed, had many of the same people on it and seemed at the time to lack a clear sense of direction. As it turned out, the incorporation movement probably helped give MGC a sense of direction, in forcing the committee to face up to and ultimately reject large-scale incorporation as one way of reorganizing metropolitan area government.

Isenberg's strong leadership had an important influence on the movement in all its stages. He worked vigorously on behalf of the committee during its formative stages, and because he took the initiative in calling meetings and in keeping the group active, he became the leader. He was respected and liked in this role, but his devotion to incorporation as an objective in itself seemingly prevented a full review of other alternatives.

Most of the committee believed that in order to achieve success an evolutionary plan of education and recruitment was necessary. Of the original members, at least seven were committed to this method. Examination of the minutes and the interview responses reveals little reason for the precipitous action finally taken by the committee. The decision to take the proposal to the public in a surprise move appears to have ended the careful planning and orderly action that first characterized the movement. The "action organization" was hastily conceived, poorly formed, and never completed. Functions and duties were not clearly delegated; communications and publicity lacked coordination; the organization during the action stage lacked the decisive leadership of the earlier planning stage.

The failure of the movement was virtually predictable. In a sense the opposition had little effect. The people in the North Area have in large part been undisturbed by predictions of future crises in relation to parks, sewers, water, streets, and police. The likelihood of success in incorporating such a large

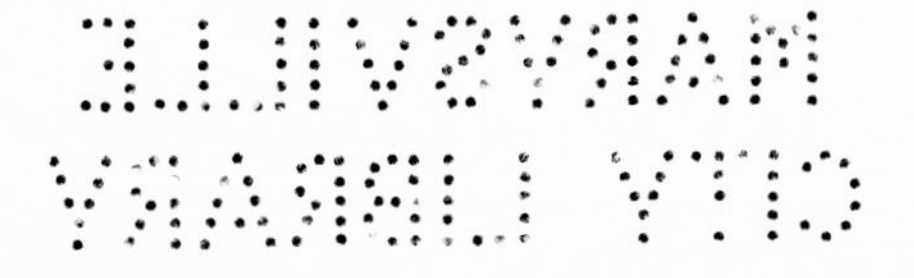

and diversified area would be remote, regardless of the quality of leadership and efficiency of organization.

Several indirect results may possibly be attributed to the incorporation movement. Sacramento County government, recently the subject of much criticism but by then infused with the leadership of a new, vigorous county executive, was revamped further by the unseating of several incumbent county supervisors. Expressions of dissatisfaction with county government that grew out of the incorporation effort added to an already sizable record of criticism and made it easy for the *Bee* and the public to abandon their traditional practice of endorsing the incumbent supervisors for re-election.

The city of Sacramento adopted an annexation policy, explaining that annexations to Sacramento "are encouraged and welcomed" and clarifying the effects on various services after annexation. The city also changed its charter so that residence in annexed areas would count as city residence in meeting requirements for election to city council and for city employment.

Finally, greater general interest in governmental affairs in the North Area seems to have been engendered by the movement. In later annexation elections, voters went to the polls with better awareness of the issues involved. The Hagginwood-Del Paso Heights attempt to annex to Sacramento was conceived when the area was excluded from the proposed incorporation. The Arden-Arcade attempt to annex to the city of Sacramento was begun primarily because a few persons in that area were opposed to incorporation. The new city incorporation attempt highlighted one method of modifying existing government, and incorporation is still discussed on occasion in some of the communities that were included in the proposed big new city.

4

The Hagginwood-Del Paso Heights Annexation Attempt

On July 3, 1958, about two weeks after the incorporation group began its public campaign, the Sacramento city council received a request for permission to circulate petitions for the annexation of certain communities in the county to Sacramento. The areas included were Hagginwood, Robla, Del Paso Heights, Northgate, Gardenland, and most of the McClellan Air Force Base. The area north of Sacramento but south of the American River, not then a part of the city of Sacramento, was also included. The annexation area, approximately 24 square miles, almost surrounded the city of North Sacramento. The new site for the California state fair, which touched on North Sacramento, was not part of the area. A week later permission was granted by the council, and the movement to annex the area commonly called Hagginwood-Del Paso Heights got under way.

The immediate impetus for the Hagginwood-Del Paso Heights annexation request was the attempt to incorporate the big new city discussed in the preceding chapter. Only part of the Hagginwood fire and sanitary districts was included in the incorporation area, but when some residents asked that all of Hagginwood and some neighboring communities be made part of the new city, their request was rejected, allegedly on the ground that boundary changes would cause too much delay. Several residents of the rejected areas stated bluntly that the proponents of the new city simply did not want them.[1] Rebuffed by the new city advocates, the residents of Hagginwood, Del Paso Heights, and neighboring communities looked to the city of Sacramento. The Sacramento city council had gone on record only a few days before as being willing to accept any of the surrounding areas wishing to be annexed.

The question of improving local governmental arrangements in the Hagginwood and Del Paso Heights communities had been discussed for several years. In the mid-1950's enough signatures had been obtained to authorize an election on a proposal to incorporate Del Paso Heights, but no further action was taken because of the work of SMAAC. A few months later spokesmen for the Del Paso Heights Chamber of Commerce threatened to initiate incorporation proceedings again if North Sacramento proceeded with the acquisition of the private water company serving the Del Paso Heights area. In early 1957 the Del Paso Heights-Robla Chamber of Commerce proposed that the Del Paso Heights, Robla, and Hagginwood communities combine to form a new city. Others argued for annexation to North Sacramento.[2]

The Hagginwood Improvement Association created a fact-finding committee to study the merits of the alternatives for governmental reform; their goal was securing municipal-type services for Hagginwood. This committee functioned throughout 1957 and 1958 until the request for annexation to the city of Sacramento was made. The results of a postcard survey, conducted by the committee in 1958, showed that only 23 per cent of the citizens responding favored the *status quo.*

DEMOGRAPHIC SETTING

The population of the Hagginwood-Del Paso Heights annexation area was approximately 35,000.[3] Of this, about 87 per cent was white. The city of Sacramento had the same percentage of white persons; the proposed big city area was 99 per cent white. About 60 per cent of the Negro population of the entire annexation area lived in one area east of North Sacramento.

The population of the Hagginwood-Del Paso Heights area is relatively young. The median age is under 29 years, as compared with a median of over 35 years in most of Sacramento. There is approximately as much rental housing as owner-occupied housing in the annexation area, and about one-fourth of the dwellings were in need of repair. The percentage of owner-occupancy and sound housing is higher in the city of Sacramento than in Hagginwood-Del Paso Heights. This area generally falls lower on the socio-economic scale than do the North Area communities included in the proposed big new city, and there are in fact extensive blighted pockets in it.

LEADERSHIP AND FINANCES

The Hagginwood Improvement Association was a major force for annexation. C. E. Cox, then president of the executive board of the association, and seven other board members signed the original request to circulate petitions. Just how and when the board decided to work for annexation is not clear.

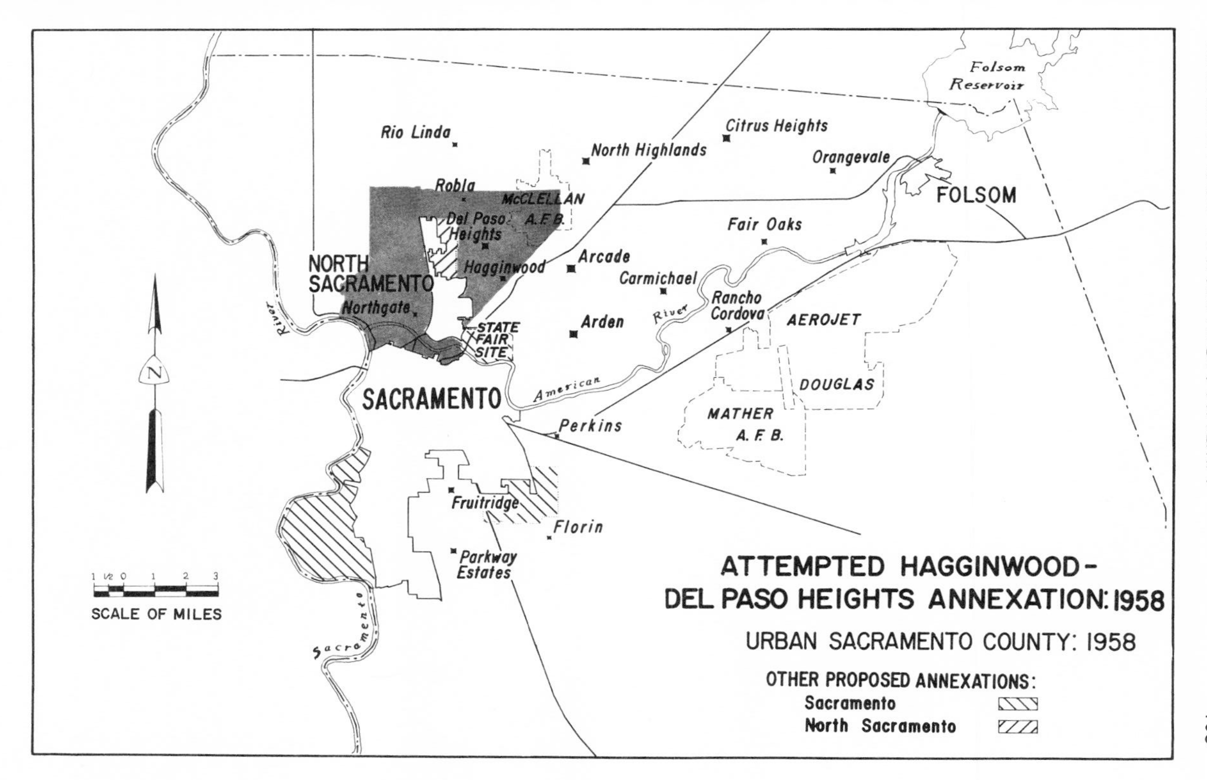
Folsom Reservoir
Rio Linda
North Highlands
Citrus Heights
Orangevale
FOLSOM
Robla
McCLELLAN A. F. B.
Del Paso Heights
Fair Oaks
NORTH SACRAMENTO
Hagginwood
Arcade
Carmichael
Northgate
Rancho Cordova
AEROJET
River
Arden
River
STATE FAIR SITE
N
SACRAMENTO
American
DOUGLAS
MATHER A. F. B.
Perkins
Fruitridge
Florin
Parkway Estates
1 1/2 0 1 2 3
SCALE OF MILES
Sacramento
ATTEMPTED HAGGINWOOD-DEL PASO HEIGHTS ANNEXATION: 1958
URBAN SACRAMENTO COUNTY: 1958
OTHER PROPOSED ANNEXATIONS:
Sacramento
North Sacramento

The chief of the Hagginwood Fire District, an active opponent of annexation, stated in an interview that at a meeting of the association a few days after the plan to incorporate the new city was made public, which he attended, the association had decided to ask that Hagginwood be included in the proposed new city. "To my surprise," he said, "one week later I learned that the Hagginwood group had requested the Sacramento city council to grant permission to circulate annexation petitions." It is not clear that any requests to be a part of the new city came from Hagginwood Improvement Association. In fact, two members of the executive board of the association stated that they had attended a meeting of the Del Paso Heights-Robla Chamber of Commerce and had "reversed their [Chamber of Commerce] position in regard to separate incorporation, and in 30 to 45 minutes the potential Hagginwood annexation became the Hagginwood-Del Paso Heights annexation movement."

Of the 10 or 12 persons who led the annexation proposal, all but three came from the Hagginwood Improvement Association's fact-finding committee, or from the association's board of directors, or both. The others included Francis Azevedo, who had attended one of the early public meetings on annexation and had "spoken vigorously" regarding the questionable quality of the schools in the area. He was later asked to become chairman of the publicity committee. Another, John Mogan, was appointed to the executive committee of the annexation committee because of his position as president of the Taxpayers Council of North Sacramento County. E. E. Hodgkinson, secretary of the Sacramento chapter of the Homeowners and Taxpayers Association, Inc., became identified closely with the annexation effort largely because of his frequent pro-annexation letters published by the newspapers.

C. E. Cox and John Mogan were picked out quite early as the leaders. These two divided the total canvass territory between them. Mogan reported that he "got about 24 long-time associates" to circulate petitions and that Cox "lined up" about 20. Newspaper accounts reported recruitment of approximately 100 petition circulators. About 60 of them made some effort, said Cox, but 30 of them "really did it." The active circulators were known as "community workers," and most of them were friends of annexation committee members.

Within this small leadership group there were internal differences. Mogan felt that the directors of the Hagginwood Improvement Association had "jumped the gun" and had made an "unfortunate choice of boundaries." He stated that his group had been waiting for the right moment and that after its premature decision to start the annexation drive the leadership provided by the Hagginwood group was not sufficiently aggressive. These differences of opinion increased significantly during the annexation campaign and the committee membership was revised, excluding anyone from the taxpayers' council. This was not publicized, however, and Mogan, the council's president, continued to support the annexation effort, particularly in letters to the newspapers. He was hard working and dedicated to the annexation idea, but personality conflicts led to the revising of the annexation committee membership.

The annexation proponents who were interviewed were sharply critical of their own leadership. They described it as amateurish and "less than dedicated," but most of them excepted C. E. Cox, whose leadership was considered to be high level but not backed up with adequate support.

An extremely limited budget severely hampered the annexation campaign. Apparently no one kept an exact record, but estimates ranged from $250 to $600. In spite of the city council's policy of active cooperation with annexation groups, it had not yet gotten past the stage of planning how information could be distributed. The work of research and publicity fell entirely on the annexation leaders. Only two or three annexation proponents devoted time to research and "digging up some facts." The publicity chairman assumed the task of "looking up tax bills." Late in the campaign a former newspaper reporter was employed for a period of four weeks at $50 a week to develop a tax breakdown and to prepare a brochure.

THE CAMPAIGN

Despite budget limitations and organizational deficiencies the annexation campaign moved along satisfactorily in the early stages. The Sacramento City Planning Commission recommended approval of the annexation request. The commission reported that the area, which contained about 35,000 inhabitants and had an assessed valuation of approximately $30,000,000, had five fire districts, four sanitary sewer districts, three water districts, three lighting districts, one park district, one flood control district, one high school district, parts of five elementary school districts, and a part of the American River Junior College district. The county boundary commission (with a few minor modifications) approved the proposed boundaries and on July 24 the city council announced approval. After 21 days signature solicitation was permissible.[4]

In mid-July a councilman proposed that the city council promise the people of the annexation area that if they voted favorably for annexation some representative of the area would be appointed to fill an anticipated council vacancy in January, 1959. The council so promised, subject to passage in November of a city charter amendment changing the definition of residency requirements for the city council.

Annexation proponents met on August 11 to plan for the circulation of annexation petitions. Within a month it was reported that 1,200 signatures had been secured. By mid-October about 3,300 signatures, well over the 2,906 signatures required, were registered. The campaign now proceeded to the property owners' protest hearing and, after that, to the election.[5]

The *Sacramento Bee* supported the annexation attempt from the start. The annexation request would test the city's willingness to adopt a "constructively affirmative" attitude toward annexation. The new city incorporation proponents charged that the city of Sacramento was opposed to large-scale annexation. Friendly action on the Hagginwood-Del Paso Heights request could remove all basis for that argument.

During the early weeks of the annexation campaign there was rather minor opposition. The *North Sacramento Journal* editorialized that the annexation area logically belonged with "friendly" North Sacramento and it should not be gobbled up by the "unneighborly big city," Sacramento. The *Journal* urged the affected residents to ask themselves: "Will annexation give me more for less than I can do for myself, or through my local geographical community?"[6] In the first public meetings a number of persons seemed undecided on whether or not to support annexation, but no vigorous opposition was expressed and most of the questions were about the schools in the area. A telephone poll conducted by the *Bee* in the annexation area revealed that about 65 per cent of those responding wanted change of some sort. Quite a few favored city-county merger, but an even greater number favored the proposed annexation.

OPPOSITION FROM THE SCHOOLS

Schools became the major issue in the annexation movement. Up to this time the policy of the city of Sacramento had been to make the schools of the area part of the Sacramento City Unified School District when an area was annexed. Since school merger was not required by law, the question of whether the schools were to be included in the Hagginwood annexation was brought up early in the campaign. Many assumed that the schools would be included, but a number of the annexation leaders stated in interviews that they had understood the reverse; some were quoted in newspaper stories to that effect.[7]

In September, 1958, the superintendent of the Grant Union High School district asked the county counsel for a legal opinion as to whether annexation for municipal governmental purposes must include annexation for school purposes. The counsel ruled that the city could require that the schools in the area annexed become a part of the unified school district, but it need not do so. The decision regarding the schools must be made before the annexation vote.[8] Editorially the *Sacramento Bee* advised the city council that to include the schools as a part of the annexation would invite the superintendents and trustees of the affected districts to spearhead a powerful force against the annexation, which probably would cause its defeat. Leaving out the schools would not prevent a redrawing of school boundaries later.

The city council held a number of meetings on the issue. Representatives of the Sacramento City Unified School District attended some of the meetings and argued for inclusion of the schools. The Sacramento city planner stated that if he had thought that the schools would be included he never would have recommended the annexation. As he put it, the annexation would cut "the guts out of too many school districts." The city councilmen were divided. Finally, on October 23, 1958, the council decided the school issue in a 7 to 2 vote in favor of including the schools in the annexation. The required number of signatures on the annexation petitions had already been secured, and those who signed the petitions had done so without knowing whether the annexation was to include the schools.

Once the decision to include the schools was made, immediate and vigorous opposition to the annexation arose. School superintendents and trustees from the affected school districts organized an anti-annexation group, commonly called the Hagginwood Anti-Annexation Committee or the Citizens' Committee Against Annexation. William L. Frazier, Jr., an employee of McClellan Air Force Base and former trustee of one of the affected school districts, was chairman of the group and public spokesman for it. The "power behind the throne," as seen by some leaders interviewed, was a building inspector in one of the school districts. Other influential members included the president of American River Junior College, three school superintendents, and a fire chief. As one interviewee said, "The organization in a hurry shook down to six or seven persons."

The group soon gave up a plan to block the annexation election through property owners' protests as hopeless. They decided to fight the issue in the annexation election campaign. Even so, the superintendent of one of the school districts secured many of the property owners' protests that were filed. These protests represented only about 5 per cent of the total assessed land valuation; over 50 per cent was required to prevent the election.[9]

The amount of money spent by the annexation opponents and its sources were never known. One interviewee who worked in the movement estimated that about $1,000 had been spent. The County Sheriffs' Association, he said, had contributed from $200 to $300, and he himself had contributed approximately an equal amount. Two or three "community-minded citizens" had been approached for money and had contributed. School suppliers had been asked to contribute and had done so. School teachers had not contributed directly, but "possibly in one or two instances school teachers' associations had earmarked funds in their budgets for this purpose." Another member of the anti-annexation organization, who estimated expenditures of approximately $1,000, claimed that about one-half of the amount was secured from numerous small donors—"teachers and other similarly interested persons"—who gave from $5 to $20 each. The remainder came from business contributions ranging from $50 to $150 each. He spoke of "a contribution from the county sheriff," probably referring to a contribution from the County Sheriffs' Association. A member of the school group reported that all of the $1,000 had come from school personnel, school architects, and school suppliers. He said that the annexation matter had been discussed with school administrative personnel—but not the teachers—and before "passing the hat" they were told that there was a serious likelihood of their losing their employment.

The proponents of annexation guessed that expenditures by the opposition ranged from $3,000 to $15,000. The higher estimates included the cost of a special report and the contributions in staff time by school districts, fire protection districts, and so on. Whatever the exact amount, the anti-annexation forces had more money than they needed. One of the interviewees acknowledged a surplus of funds, and he spoke of the difficulties involved in the effort to return the unused money to contributors on a pro rata basis.

The school issue caused one of the original annexation leaders to withdraw from the movement: "pressure was brought to bear upon me through the schools operating on my son by publicly embarrassing him." Numerous nuisance telephone calls led the father to obtain an unlisted telephone number. Another interviewee alleged that children were sent home with messages pinned to their clothing telling the parents that if the annexation succeeded the children's teacher would be out of a job and that "strange" teachers would be brought in. This was disputed by a teacher in the area who favored the annexation and said that she knew of only one teacher who had done this. Principals made direct collections of contributions for the opposition from teachers, going around to them with a clipboard containing the names of all the teachers and asking how much they wanted to give, another interviewee stated. He added that on the weekend before election "teachers were imported from as far away as Davis and Woodland to distribute anti-annexation literature."

Many of the school administrators of the annexation area were publicly involved in the opposition. It is impossible to tell how much pressure this put on the teachers. One teacher, who had been identified with the annexation movement, refused to be interviewed because of her fear of possible reprisal. However, another teacher, also publicly identified with the annexation movement, denied that there would be reprisals and stated that the reports of school teachers about forced contributions were grossly exaggerated. In her opinion, most of the teachers who contributed money or worked for the opposition did so out of self-interest. She did add that sound trucks owned by one of the school districts "were driven throughout the area on election day propagandizing the voters."

In an interview the school building inspector, acknowledged to have had a great deal of behind-the-scenes control, spoke freely of his part in the anti-annexation campaign. He had stressed the importance of "creating fighting issues" and not simply responding to the charges of the annexation group: "avoid personality involvement in every possible case." He had also opposed having teachers distribute propaganda: "They would need public support for salary increases and other matters at a later time." He summarized his role in the anti-annexation campaign as follows: "I set the stage, conditioned the outcome, and then deliberately kept out of sight."

The school annexation was, however, nearly always at the center of the controversy. Because of the schools the board of directors of the Greater North Area Chamber of Commerce voted unanimously in mid-November to oppose annexation.[10]

Only the public meetings that discussed what annexation would do to the schools drew sizable audiences. The *Sacramento Union* found the "spectacle" of North Area school officials fighting annexation "not enjoyable." As individuals they had the right to favor or to oppose the annexation, but they had no right to plunge the schools into a political fight.

On the insistence of annexation proponents, the city superintendent in mid-November issued a statement on the advantages of school annexation. "Without question" school taxes in the annexation area would be decreased, he said, and he promised reports from the research staff of the city district on how the schools of the annexation area would be administered.[11] The first of a series of three short reports on the school annexation matter was released about a week later. The report predicted a slight increase in the school tax rate within the Sacramento school district if annexation were approved, and a considerable decrease in the school tax rate in the areas to be annexed.[12] (The school tax in these areas was then from $0.61 to $1.45 higher than in the Sacramento school district.) In addition, the report presented policies designed to insure fair treatment for the pupils, the school teachers, other school employees, and the residents of the annexation areas generally.

The second report, released in early December, dealt with school personnel policies. Most of the incoming employees would receive at least comparable pay, and the report set out detailed methods for protecting the job security of both certified and noncertified teachers.[13] The third report, released in early January, 1959, further analyzed the school tax rates. The tax rate for bond interest and retirement, and for state loan repayment, was at least twice as high in the Del Paso Heights, North Sacramento, and Robla school districts as in the Sacramento school district. The North Area school districts would have to pay off their own bond obligations, but they would become full partners in the Sacramento school system without assuming any of the present bond obligations of the Sacramento district.[14]

THE KIMBER REPORT

The Sacramento City-County Chamber of Commerce sponsored a public round-table discussion on the effects of the proposed annexation upon the schools. Representatives of the Sacramento school district, the school districts of the annexation area, the office of county superintendent, and the state department of education were invited to participate. The superintendents of the various affected school districts spoke of the disruptive effects of annexation. Portions of several school districts would be added to the Sacramento school district while other portions would be left outside. Some of the portions not annexed would be left without any school buildings, while others would have too large facilities for the pupils remaining in the districts. All of the school districts other than Sacramento's would suffer a reduction in their property tax base.

The superintendent of the Sacramento school district presented figures to show that school taxes would be reduced by as much as $0.61 to $1.45 per $100 valuation. The president of the American River Junior College acknowledged that school taxes would be reduced, but he argued that the over-all tax burden in the annexation area would be increased. His figures came from the Kimber report.

In late November, 1958, the president of the American River Junior College reported to the college board of trustees that the pending annexation would be to the disadvantage of the junior college and that it would also reduce the effectiveness of the education system in the North Area generally. Upon the president's recommendation, Dr. George Kimber, a professional educator recently retired from the Sacramento Junior (now City) College, was employed to compile data concerning the effects of the proposed annexation upon the American River Junior College.

The Kimber report was released about a month later.[15] In his letter of transmittal, Dr. Kimber stated that his instructions had been to prepare a scientific collection and interpretation of facts without either supporting or opposing the annexation. Nevertheless, his report was unfavorable to annexation. He hoped that the voters, in spite of their understandable interest in city services, would not approve this annexation which would place the unannexed areas in constant uncertainty as to their school programs. Dr. Kimber acknowledged that this annexation would not affect adversely the educational program of the American River Junior College, but he felt it would be followed by other annexations that would have serious consequences for the junior college.

The Kimber report devoted about 50 pages to an analysis of school functions and population and age-group trends in the school districts which would be affected by the annexation. This was followed by a detailed analysis of taxation. School taxes in the annexed area would be reduced, but the total property tax burden would increase, since the cost of other municipal services for the area would exceed the savings in school taxes. The statements and figures presented in the Kimber report were used extensively by the annexation opponents during the final six weeks of the campaign. The report was the basis for the constantly repeated charge that "annexation will bring excessive taxation."

Later analysis of the economic claims made in the annexation campaign pointed out at least three deficiencies or errors in the Kimber analysis.[16] About 33 per cent had been added to the county property tax assessment figures to arrive at the assessed valuation that would be used for city tax purposes. The basis for this was the much higher average in the city of Sacramento property tax assessments than in the county. The assumption might be correct as to the average difference—city assessments ranged from 5 to 60 per cent higher—but the rate was dependent on the nature of the property, and most of the property of the annexation area was more nearly like that with 5 per cent than that with 60 per cent higher evaluation. Secondly, Dr. Kimber used the city tax rate applied to older city areas, which is higher than that applied to the more recently annexed areas. Finally, the report did not bring out the fact that, while most special districts have only the property tax as a revenue source, incorporated areas have access to sales tax and other sources of revenue. Increases in municipal-type services as a result of annexation would not be financed entirely from property taxes.

One of the school superintendents, who himself was actively opposed to the annexation, acknowledged that the author of the report "may unconsciously have lost his objectivity." However one may judge the report, it seems apparent that it had considerable influence on the outcome of this annexation attempt.

FIRE DEPARTMENT OPPOSITION

Four fire protection special districts were largely or wholly within the annexation area. Many members of these favored the *status quo* but did not take any action until the school group provided the leadership. The assistant chief of the American River fire department asserted that the department was neutral in the effort. He personally was involved in a sense, as secretary of the American River Property Owners' Association, which contributed funds to the anti-annexation forces. Practically all of the business and industrial property owners within the fire district were members of the association, but they were not opposed to annexation to the city of Sacramento. They favored annexation of the area immediately north of Sacramento but south of the American River, where property tax assessment evaluations were high. They were not enthusiastic about the Hagginwood-Del Paso Heights annexation as a whole because the proposal included areas in which slum conditions prevailed. They feared that areas of high assessment evaluations would be taxed inequitably to bring up the level of services in the poorer areas. (The Hagginwood-Del Paso Heights annexation area was excluded from the proposed new city for the same reason.)

The Natomas fire department actively fought the annexation through the Natomas Fire Association. Interviewees reported that the fire fighters prepared anti-annexation literature relating primarily to fire protection matters. Each association member was made responsible for a portion of the Natomas district; every household in the district was to be approached. The firemen distributed their own material and that of others from house to house. They claimed that they opposed annexation not as firemen but as public-spirited citizens.

The chief of the Del Paso-Robla fire department at the time of the annexation attempt said that he had worked actively against annexation. He stated that three of the four full-time members of the department and six or seven volunteers also had worked against it. Some of them worked under the direction of the Hagginwood Anti-annexation Committee as block captains, responsible for the distribution of anti-annexation material. The group tried to convert pro-annexation individuals through a follow-up visit or telephone call by a member of the anti-annexation committee.

The chief of the Hagginwood fire department said that he had taken only a small part in anti-annexation activities. He had appeared on a television program and had talked against annexation to persons in his neighborhood. No other member of his department took part at all, he said. One of the leaders of the Hagginwood Anti-annexation Committee attributed greater activity to this fire chief. The chief, he said, had attended the first anti-annexation meeting, had joined the anti-annexation committee, and had helped prepare anti-annexation literature. Hagginwood firemen had conducted a door-to-door anti-annexation campaign, according to this interviewee.

Although the effect of the opposition of fire department personnel cannot be measured, they certainly provided the mechanism for house-to-house coverage. The fire chiefs viewed the effectiveness of the school administrators as minimal and maximized their own activities as contributing to the defeat of annexation. One fire chief spoke of the strategy of the school administrators as ill-advised and in some instances improper. In his opinion, "All you have to do is talk with people, secure their trust, and enlighten them as to where their interests lie. If enough of this is done, you can win any campaign."

The school administrators in turn tended to downgrade the effectiveness of the fire departments in fighting annexation and to view their own efforts as crucial to the defeat of annexation. In substance, they felt that the fire departments' activities "were of some help, but unorganized. They did assist in the last-minute pamphlet distribution, and they were cooperative and they deferred to our group for leadership."

Two of the annexation proponents, who were long-time residents of the area, believed that the fire fighters' work had a great effect on the outcome. Between 60 to 80 volunteer fire fighters and members of their families had participated. They were interested in preserving the "neighborhood club" atmosphere of the fire stations. "There are pool tables, cards, friends at any hour of the day or night. If the service is professionalized, this home away from home would be destroyed."

OTHER OPPOSITION

At least three of the major annexation supporters spoke of opposition from members of the county sheriff's department, which, they said, believed that "annexation would affect adversely the number of employees required to staff the department." However, no direct evidence of such opposition was presented. Other annexation leaders insisted that county officials and employees worked indirectly against annexation, but presented no proof of such activity.

Pro-annexation interviewees identified activities of other groups to protect their job interests: employees of a sanitary district, bus drivers of a private transportation company, and the building trades people of North Sacramento. It was alleged that the latter hoped for annexation of the area by North Sacramento. No substantial evidence of overt activity against the annexation proposal from these groups was given.

McClellan Air Force Base, which was almost entirely within the annexation area, did not take an official position, but some annexation proponents felt the base had taken a position on the issue. Some annexation and some anti-annexation leaders were employed at the establishment. Base officials followed the annexation battle closely but impartially since Air Force property was potentially affected.

LAST-MINUTE CAMPAIGNING

On January 16, 1959, four days before the election, the *Bee* printed a prediction by annexation leader C. E. Cox that the opposition would level a "sneak punch" against annexation. The *Bee* editorial deplored this type of tactic and urged that informed voters firmly reject "the obvious assumption of the late release that the people can be made patsies, dupes and fall guys." The next day, Friday preceding the Tuesday election, some anti-annexation material was distributed door to door. Parents in the annexation area also received through the mail a mimeographed letter, under the school district letterhead, signed by the superintendent of one of the affected school districts. A postscript stated that the cost of preparing and mailing the letter had not been paid from school district funds. Parents were urged to oppose disruption of the school system. One supporter of annexation, although not a member of any official group, described this "avalanche of confusing and contradictory material" as evidence of "an ulterior motive" behind the anti-annexation campaign. The material was credited to the Citizens' Committee against Annexation.

One leaflet on pupil transportation compared in pictures and words the school buses of the North Area districts with the public transit buses used to transport city school children. The leaflet stressed the special safety precautions taken on school buses and claimed that school bus transportation was free while the public transit buses charged about $45 per pupil per school year. This leaflet also announced a half-hour local television program on annexation scheduled for Saturday afternoon preceding the Tuesday election. The people would learn that annexation meant two tax bills (city and county) instead of one. The other leaflet, "Who's Distorting the Facts?" showed that the city assessed valuation of a few major commercial properties was higher than the county assessed valuation of such property, but no mention was made of the tax rates applied by the city and county on this assessed valuation.

The 16-page pamphlet, "What Does the Hagginwood-Del Paso Heights Annexation Mean to You?," covered a wide range of material. At one point, a property owner in recently annexed Meadowview was quoted as saying that instead of the promised tax reduction his taxes were $211.86 more than in the year preceding the annexation. The pamphlet presented the Kimber report predictions of higher taxes and asserted that property owners could get street lights, curbs, and sidewalks only if they were willing to pay for them. The impact of annexation upon the North Area school districts and upon police and fire protection was also stressed.

Officials of the city of Sacramento immediately declared that they had not promised a tax reduction but only that the services needed in the area would be provided at less cost by the city of Sacramento. As to the higher tax bill in Meadowview, the officials pointed out that regardless of annexation the county tax had increased; that during the first year after annexation the bill contained an assessment for a sewer system installed before annexation; and that the county tax bill would have been even higher if it had included special districts' taxes, which were eliminated after annexation.

The Citizens' Committee against Annexation also ran an ad in the *North Sacramento Journal* headed "Don't Be Fooled." A shark was pictured gobbling up a fish labeled "Hagginwood, Del Paso Heights, Gardenland, Arden, Arcade, and Industrial Park." There were references to Kimber's predictions as to tax increases and to the effects upon the North Area schools. In addition to such printed propaganda, one school superintendent called a meeting of about 170 school principals and teachers and presented his views of the annexation to them.

Probably because of limited financial resources, the annexation advocates prepared and distributed only one piece of printed material, a document of four pages. It urged a "yes" vote to making the area a part of "the heart of California." Fourteen points in favor of annexation were listed, followed by a discussion of alleged benefits which annexation would bring to the educational system. Substantial reductions in the total property tax burden were promised for most of the annexation area. This pro-annexation material did not receive the widespread distribution which the anti-annexation material did.

THE VOTERS DECIDE

The day of election arrived—January 20, 1959. Sixty-nine per cent of the voters registered voted. The annexation was defeated by 159 votes. (This did not include 25 absentee ballots, but these could not affect the result.) The annexation proposal fared worst in the Gardenland area immediately west of North Sacramento and in the area south of the American River, where the issue lost by a margin of about 2½ to 1. In the Ben Ali area to the southeast of Hagginwood the issue lost by a 2 to 1 margin. In the northern portion of the Northgate area the issue lost by somewhat less than 2 to 1. In all other areas where annexation was rejected the margin of defeat was slim.

Annexation won in only five of the 16 precincts. Four of these precincts were contiguous and comprised the eastern three-fifths of Del Paso Heights, bordering McClellan Air Force Base on the southwest. The other precinct where annexation was favored was in the southern portion of the Northgate community.

In restrospect, William Frazier, an active anti-annexation leader, stated that it was the school issue that had defeated annexation. The people did not vote against the city of Sacramento but rather against this unplanned (as far as schools were concerned) annexation proposal. C. E. Cox, one of the pro-annexation leaders, said somewhat bitterly, "It was beaten by employees of the voters, at the voters' expense in some instances, with a direct appeal to ignorance at the last minute." His comment implied that the school officials spent taxpayers' money to fight the annexation proposal.

The school superintendents in the annexation area were spurred by the defeat to attempt to unify the school districts. Unification had been defeated by the voters on December 3, 1957, and at that time the school superintendents had not been enthusiastic about unification. One school superintendent stated that he had changed his mind about unification; it now seemed necessary if the schools were to be kept out of local politics.[17]

Sacramento's mayor, Clarence Azevedo, believed that a second attempt to annex the Hagginwood area would succeed. He noted a number of other instances (Elder Creek and Riverside, for example) in which a second attempt at annexation had succeeded. Annexation proponents would be ready a second time to counter the misinformation given right before the election.[18]

Some of the interviewees expressed surprise that annexation received strongest support in the poorer areas of Del Paso Heights. They had expected opposition from these residents because of their presumed fear that the city of Sacramento would compel them to spend money to improve their housing. Apparently they felt that taxes from the areas of higher assessed evaluation would be used to improve conditions and services in the poorer areas.

HAGGINWOOD-DEL PASO HEIGHTS ANNEXATION EVALUATED BY PARTICIPANTS

Fourteen persons were interviewed in connection with our study of this annexation attempt. Seven had been opposed to the annexation, six had been in favor, and one stated that he was completely neutral.

Annexation opponents agreed that the *Sacramento Bee* was an extremely potent force for annexation, while proponents considered the *Bee* to have been "responsible" but "passive" in the campaign. The *Bee*'s motives were variously appraised by anti-annexation leaders: the *Bee* became involved in order to prevent the emergence of a strong newspaper; the *Bee* was a crusader paper—"a bigger Sacramento is a better Sacramento."

The attitudes of school officials were summarized well by one annexation proponent:

> Unfavorable . . . school officials felt two things: first, that their school district boundaries were being violated in an unfortunate manner and that areas would be left out of

> the Sacramento Unified School District; and secondly, that their own status would be adversely affected if they became part of the city school system in the sense that they would be subordinate to a new layer of officials up above them.

An annexation opponent agreed generally, but added that "there was a genuine conviction that the area proposed for annexation would ruin the tax base necessary for effective school operation in those adjacent areas not annexed."

Few of the activists interviewed believed that the downtown businesses in Sacramento had been at all concerned, although businesses were hopeful for the broader tax base and expansion of the city that annexation would bring. Most suburban businessmen favored annexation, according to the annexation group, but were unwilling to state so publicly for fear of offending some customers. Annexation opponents asserted that suburban businesses had not taken stands on the matter but that some individuals had favored one side or the other.

With only one exception, leaders on both sides of the movement who were interviewed believed that city officials had strongly supported the annexation. Proponents felt that the city had several motives for favoring annexation. Sacramento did not want a number of small and inefficient neighboring cities to develop. One said, "Since the entire metropolitan area is called Sacramento, city officials quite properly are concerned about improving conditions in any blighted areas that might exist in the metropolitan area," and since the area in question contained at least one commonly acknowledged blighted region, the city was properly worried. One annexation leader said that city officials thought that a larger city would increase their political influence. Anti-annexation leaders gave only one reason for the support by city officials: city officials—and government officials everywhere—believe that "bigger is better."

Most leaders believed that county officials had opposed the annexation out of fear that county responsibilities and size of operation would be reduced. County officials resented the *Bee's* criticism of the county sheriff's office, and although these officials had not publicly opposed the annexation, they had supplied indirect help and encouragement to the anti-annexation forces.

All interviewees were asked, "If you were to attempt such a movement again, what changes would you make?" The annexation advocates had numerous suggestions. There was general agreement that the school issue should be kept out of a second annexation attempt. As one of them said, we must find a way to "tie the hands of the school people." Leaders emphasized the need for more money. A publicity man should have been hired early in the campaign. Some felt they should have sought out persons who had time to give to lead the movement, and involved many more organizations and groups. And finally, they came back to a way of dealing effectively with the opposition and preventing "the servants of the public from acting against the best interests of the people whom presumably they are serving."

The Hagginwood-Del Paso Heights annexation movement differed from others studied in that the public could see immediate, specific advantages from joining with the city. The area had a low level of municipal-type services, and it also had a comparatively low assessed valuation. By unification with the city of Sacramento, the city's greater tax resources could be used to help build up the level of governmental services in the annexation area. Perhaps for this reason the Hagginwood-Del Paso Heights annexation came closer to succeeding than any of the other reorganization efforts studied. (Nearly all of this area is a part of Sacramento city today.)

5

The Arden-Arcade Annexation Attempt

On November 18, 1958, the attempt to incorporate a big new city northeast of Sacramento was officially abandoned. Almost immediately thereafter a proposal was made for the city of Sacramento to annex the Arden-Arcade area. The area encompassed about 24.5 square miles, with an estimated population of 73,000 and an assessed property valuation of about $17,600,000. The Hagginwood-Del Paso Heights annexation was approximately the same size but included only one-half the number of residents. Both the Hagginwood-Del Paso Heights plan, which was nearing its election date of January 20, 1959, and the Arden-Arcade effort were on a much larger scale than any annexation previously attempted. The Oak Park and Eastern Sacramento annexation of 1911, which added 9.4 square miles and approximately 12,000 people to a city of 54,000, was the largest annexation to date. We should also keep in mind that during the period of the Hagginwood and Arden-Arcade annexation movements the Metropolitan Government Committee was seeking to develop a plan of government for the entire metropolitan area.

Although the Hagginwood-Del Paso Heights annexation effort presented an immediate precedent, and MGC was subsequently to propose large-scale annexation, the "new city" incorporation proposal actually triggered the Arden-Arcade annexation movement. From one perspective, the already existing annexation sentiment was organized into a counter-movement to incorporation. A group of citizens who had been considering annexation joined together informally, not publicly, to oppose incorporation. The same people later formed the annexation committee.

The earlier proposals to annex were known to many Arden-Arcade citizens, and the activities of SMAAC and MGC had kept the issue of metropolitan governmental reform alive. In addition, a core of leaders had formed, prior to 1958, primarily around efforts to improve services for local communities in the northeast area. Yet the fact that the catalyst for annexation was anti-incorporation, rather than pro-annexation sentiment had implications for the leadership structure and the history of the movement. Reformist leaders with similar long-range goals found themselves opposing each other rather than banding together to try to overcome the general apathy among the residents of the area. In addition, the timing and the strategy were based on reaction rather than on carefully developed planning. In summary, the Arden-Arcade annexation movement was fostered by negative reaction to the incorporation movement; the impractical proposals of SMAAC-PAS, and MGC's inaction; and a reservoir of experienced community leaders, who wished to avoid the multiplication of special districts and to improve government in Arden-Arcade.

DEMOGRAPHIC SETTING

The nine census tracts which compose the Arden-Arcade district, and which approximate the boundaries of the annexation area, had a population of 73,352 in 1960, making it one of the ten largest unincorporated districts in the United States.[1] It is, of course, not an official district at all but a group of subdivisions relying upon special districts for municipal services. However, in news releases from the United States Census Bureau, it sometimes is referred to as an unincorporated district. Ninety-nine per cent of the persons in Arden-Arcade were classified as white; in the city of Sacramento only 87 per cent are white. The population of Arden-Arcade is comparatively young. The median age does not exceed 35 years in any area and the median is under 30 years in most areas. In comparison, the age of citizens in the city of Sacramento varies much more widely. Slum housing is not a problem in Arden-Arcade. Well over half of the houses in the area are owner-occupied and over 90 per cent of all the dwellings are in sound condition.

In summary, the Arden-Arcade area is a middle-class, white, residential suburb; the population density is rather low, although few expanses of unimproved land still exist. Most of the homes are single-family dwellings occupied by the owners, with long-term mortgages. Since there is little industrial employment in the area and the commercial activity is primarily retail merchandising, most of the working residents commute to the city, to McClellan Air Force Base to the north, or to Aerojet-General Corporation and Mather Air Force Base to the southeast. Many state government workers, who have offices in the city, reside in Arden-Arcade.

Although the commuters put a strain on roadways, there is less traffic congestion in Arden-Arcade than in the city. The police force is limited to officers provided by the county sheriff's office and the California Highway Patrol, but the crime

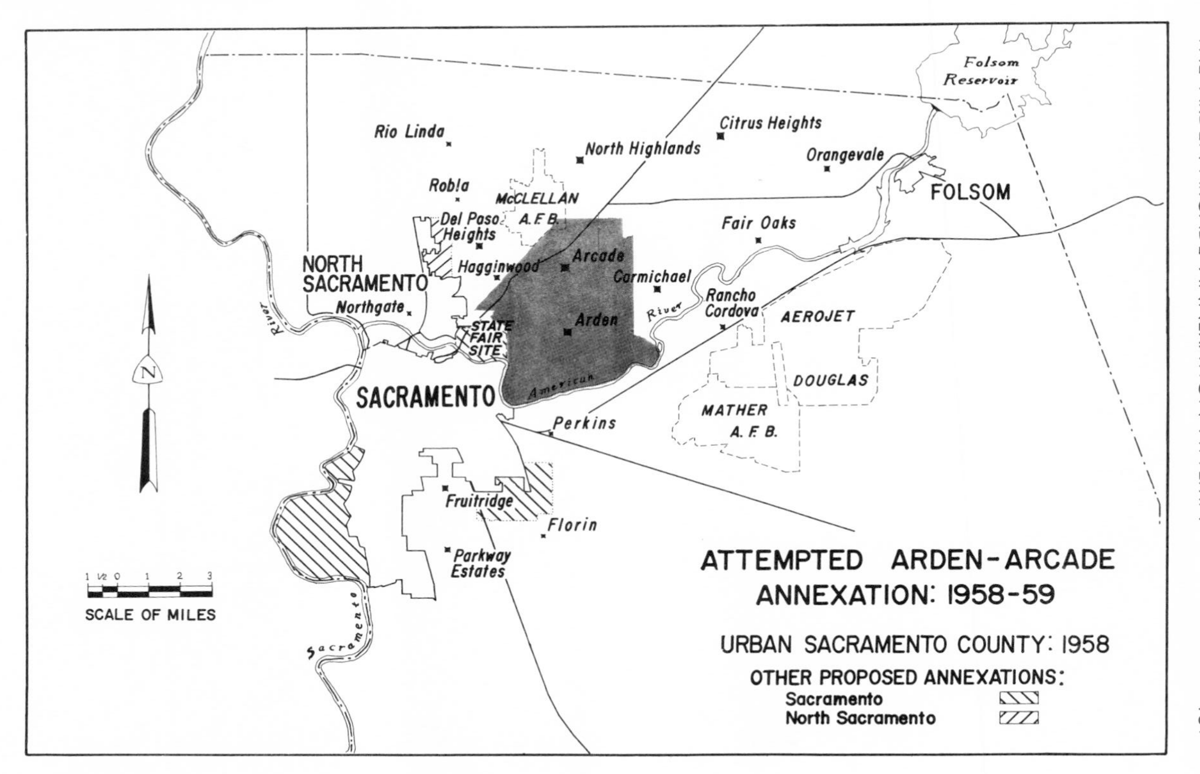
ATTEMPTED ARDEN-ARCADE
ANNEXATION: 1958-59
URBAN SACRAMENTO COUNTY: 1958
OTHER PROPOSED ANNEXATIONS:
Sacramento
North Sacramento
SCALE OF MILES
N
Folsom Reservoir
FOLSOM
Orangevale
Citrus Heights
Fair Oaks
AEROJET
DOUGLAS
MATHER A. F. B.
Rancho Cordova
Carmichael
North Highlands
Arcade
Arden
American
River
Perkins
Florin
McCLELLAN A. F. B.
Hagginwood
Del Paso Heights
Robla
Rio Linda
STATE FAIR SITE
SACRAMENTO
Fruitridge
Parkway Estates
NORTH SACRAMENTO
Northgate
River
Sacramento

rate is low. The inefficiency of the numerous special districts in supplying water, fire protection, and other services is criticized but has caused no serious difficulties. A suburban area with no slums, no serious traffic problems, and reasonably adequate services is likely to resist efforts to merge with the central city.

LEADERSHIP

The original impetus and the leadership for the annexation movement came from the Santa Anita Improvement Club, which represented about 1,500 residents. The club had promoted installation of street lighting and establishment of a park district for its area, and difficulties encountered in achieving these goals provoked discussion of the idea of annexation to the city of Sacramento. These citizens knew of the developments in SMAAC, and were aware of the PAS report and the work of MGC. Discussions between the club president, Frank Stipak, a civil engineer for the United States Bureau of Reclamation, and former president E. A. Pesonen, a conservationist for the same agency, led to the first overt efforts toward governmental change. Stipak and Pesonen began to get in touch with friends and other individuals active in civic clubs in the summer of 1958 in order to locate and assess the anti-incorporation sentiment. Stipak was later to become co-chairman, and Pesonen treasurer, of the annexation group. One person involved in annexation suggested that the notion of annexation had been born much before this. Several years earlier many citizens had worked to get a water district for the area. Some of the leaders in this movement later became active supporters of annexation. In addition, many residents who had moved to Arden-Arcade under the assumption that the area would soon become a part of the city were becoming impatient.

Besides Stipak and Pesonen, other leaders of the annexation group included Stanley Kronick, formerly an attorney for the United States Bureau of Reclamation and in private practice in the city at the time of the movement. He was an early member of the group and became co-chairman. His wife had been active in Democratic Party work in Sacramento County. Frank Bragg, another ex-president of the improvement club and then an employee of the California Department of Fish and Game; Wardon H. Moul, the Special Districts Supervisor of the County Department of Public Works; and Mrs. Doris Murray, who was active in women's affairs and the Democratic Party, complete the list of initial leaders for annexation. The core of leadership remained small throughout the course of the movement.

Why did the participants become active? Why did these citizens give so freely of their own time and in many cases, apparently, of their own money to seek to modify the metropolitan governmental structure? From interviews with 11 of the pro-annexation leaders (9 anti-annexation leaders also were interviewed) we get a partial answer to these questions. No campaign promises were made to those who became leaders. They were not offered positions or special benefits of any kind. Rather, the active participants were imbued with a genuine interest in civic improvement. They felt they ought to stop talking and do something about community problems. One respondent pointed out clear difficulties in zoning, parks and recreational needs, and sewer systems. He said that the Santa Anita Improvement Club had discussed the recommendations of SMAAC and MGC and had decided that annexation was a more practical and immediate answer to the needs of their area. Another interviewee felt that annexation was the natural solution to their problems: "there was a general realization that this rapidly urbanizing area needed a less haphazard form of government. The city already existed with an integrated structure for an urbanized area. Therefore, it follows that the city boundaries should be extended."

This respondent, turned theorist, suggested one reason for the failure of the movement: he thought that the leaders were "visionaries" who foresaw that the cost of government services under the existing multiple-district setup would continually increase. This type of person is in the minority, he said, and most people could dedicate themselves only to objectives more tangible than long-range good government. The leaders anticipated problems—taxation difficulties, road and sanitation inadequacies, recreational deficiencies. Immediate, urgent problems were not identified, nor were they manufactured to gain support for the campaign.

Pro-annexation leaders agreed that Stipak, Pesonen, and Kronick were the central leaders throughout the movement. One interviewee gave Stipak the label of "Mr. Annexation," and stated that he had done "more to arouse interest in something that would contribute to the community than anyone else." Kronick, a lawyer, was pushed before the public most often, one interviewee said, partly because the other annexation workers were not widely known and partly because most of them, as government employees, were afraid to be too much in the public view.

THE ANNEXATION MOVEMENT TAKES FORM

The Arden-Arcade District Council of Improvement Clubs, of which the Santa Anita club undoubtedly was among the most active, initiated the annexation effort. Seeking to stall the new city incorporation effort—or anticipating that the incorporation effort might fail—a few members of the council began to speak out in public for annexation of the Arden-Arcade area to Sacramento in July, 1958. The Santa Anita club adopted an official resolution requesting the city to make a study of suitable boundaries for an annexation proposal that would include Santa Anita. E. A. Pesonen went before the Sacramento city council on September 17 with the resolution. "We are now represented by a rash of districts," he said, and he asked the city's assistance in promoting annexation.[2] The council agreed to supply the information but emphasized that its intention was not to "harpoon the proposed new city." Approximately two weeks later the city manager and the city planning director were instructed to meet with the club members.

The evening before the incorporation effort was officially abandoned, the Arden-Arcade District Council of Improvement Clubs voted unanimously to call a public meeting to determine whether there was enough interest in annexing to Sacramento to justify further exploration. About 65 persons attended the meeting, which was held on December 18. The group was overwhelmingly in favor of getting annexation under way. Thirty-nine persons signed a petition to request annexation. The petition was rushed downtown, since the city council was meeting that same evening. The council accepted the petition for filing. The proposal was referred to the city planning commission for study and a report, and Sacramento's mayor promised that the city staffs would assist the annexation proponents in gathering data.

The annexation group was not formally organized until after the public meeting in December. In fact, according to Stipak, the organization did not crystallize until March, 1959. Stipak himself worked mainly with the mechanics of getting the movement started. The organization, as it finally emerged, was called the Arden-Arcade Annexation Committee and had four committees—research, publicity, precinct organization, and finance. Stanley Kronick and Frank Stipak were co-chairmen of the group, and E. A. Pesonen was treasurer. Kronick, William Carah (an employee of the California Water Commission and chairman of the publicity committee), Pesonen, David Yorton (a local attorney in private practice and chairman of the Arden-Arcade council), Stipak, and Mrs. Murray seem to have been the early planners of the annexation campaign. J. R. Liske, an architect and Santa Anita resident, later joined the leadership, representing the city-county chamber of commerce, which had offered its support for annexation. This group developed a two-phase approach—supplying the facts and disseminating these facts as live news. Members of the research committee were chosen as experts in taxation, governmental structure, and other technical fields, and not necessarily because of their support of annexation.

The group made an effort to reach the uncommitted, who would not normally attend public meetings but who might respond to speeches before their civic clubs and other organizations. Most of these talks were given by Stipak, Pesonen, and Kronick. Mrs. Murray tackled the second immediate problem, that of getting the necessary signed petitions in support of annexation. The leadership, structure, and purposes of the movement were easily established, probably because most of these people had worked together earlier and knew each other's interests and abilities. Some of the annexation committee members, however, did not participate at all or contributed very little.

While the annexation question was not a political party matter, charges of partisanship were made occasionally. Partisanship entered primarily because some of the annexation leaders happened also to be leaders in the Democratic Party organization in the area and found it convenient to use the party machinery and the registration lists. Measured by voter registration and by election results, the area is predominantly a Democratic Party area—as is generally true of metropolitan Sacramento as a whole.

On January 26, 1959 (the Hagginwood-Del Paso Heights annexation had been defeated on January 20), Sacramento's city planner announced that he would recommend approval by the city planning commission of the proposed Arden-Arcade annexation, with the schools excluded. The planning commission voted approval of the annexation, but referred the question of the schools to the city council since the commission members felt that they had insufficient information. The annexation proponents, when filing their petition, had suggested that the schools not be involved unless the forthcoming referendum to create the San Juan Unified School District was rejected. The city council, on February 6, agreed to this and waited until after the March 24 election to decide about the schools in the area. The school consolidation referendum was approved and the schools did not become involved in annexation.[3] The county boundary commission approved the boundaries on February 20, with the recommendation that a bordering commercial area, Arden Fair, be included. The suggestion was not accepted.

In early April, Sacramento's city council announced that the various legal requirements preceding the circulation of petitions had been met and that the effort to secure legal signatures could begin on April 18, 1959. It was definite by this time that the schools were not to be included.[4] In the meantime, Mrs. Murray had recruited a corps of petition circulators, primarily from those with whom she had worked as Democratic precinct worker. Between 200 and 300 persons signed up to solicit signatures, but many of these collected few if any signatures. Some individuals, when interviewed later about the movement, took it for granted that she used the Democratic

Party precinct worker cards, but did not involve the party as such.

During the four or five days immediately preceding the beginning of the drive to secure signatures, there was a house-to-house distribution by volunteer help of the newspaper tabloid, "Progress News: Arden-Arcade Annexation." The tabloid announced the opening of the campaign to secure signatures, and listed the qualifications required of the signers. It emphasized that the schools were not affected and that the Sacramento City-County Chamber of Commerce and MGC had endorsed the annexation proposal. Existing conditions within the annexation area were described as a "study in frustration." Included was a table of annual costs of operating a $15,000 home in the area compared with equivalent costs in the city of Sacramento. In one district the cost would be the same. In four districts the annual cost would be from $5 to $19 less, and in the remaining two, which had no sewers, the cost would be $7 more annually. Sacramento's new annexation policy was summarized. Finally, the tabloid included an appeal for funds.

At a public meeting, called for April 15 to explain how the signatures were to be obtained, Mrs. Murray reminded workers that "signing the petition does not obligate the resident in any way. It does not mean he has made up his mind how he'll vote. It merely makes it possible for us all to vote."[5] The canvassers received the registered voters list for the precincts, and only persons listed as registered voters were to be asked to sign. Invalid signatures were thus kept to a minimum. In addition to the precinct effort, booths were set up at some of the shopping centers in the area to collect signatures.

Petition circulation went surprisingly well; about 3,500 signatures were obtained during the first six days of the drive. In less than a month, signatures in excess of the total required had been obtained. On June 15 (the day of the last meeting of MGC), the county clerk certified that there were more than enough signatures on the annexation petitions.[6]

OPPOSITION EMERGES

On April 29, early in the drive to obtain signatures for the annexation petition, Benjamin Frantz announced the formation of the Arden-Arcade Preservation Committee—with himself as chairman. At that time he was also secretary of the Sierra Oaks Vista Improvement Club, which had gone on record as opposing the annexation. The preservation committee was committed to maintaining the local identity of Arden-Arcade, to protecting the suburban way of life, to collecting data relative to any and all governmental reorganization proposals affecting the area, and to assuring local representation in taxation matters. A week later the name of the opposition organization was changed to the Arden-Arcade Government Committee, with headquarters in the law offices of J. H. Tredinnick. It was the only formally organized opposition group throughout the annexation campaign.

The committee listed four major objections to annexation. Annexation would increase taxes. It would hinder achievement of city-county merger because the city of Sacramento opposes this solution to metropolitan problems. The large city would not take time from its own problems to consider those of Arden-Arcade. And finally, the group stated that the area would not be represented on the city council for five years; this last statement was almost immediately retracted.[7]

Benjamin Frantz announced that his group would seek to obtain signatures from over 50 per cent of the property owners in order legally to prevent the holding of an annexation election. He said that 20,000 protest forms had been printed and that approximately 100 persons had volunteered to conduct a door-to-door campaign for protest signatures.

Approximately 1,150 protest signatures had been filed by July 23, the date set by the city council for the property owners' protest hearing. A week later the council announced that the protests that had been filed represented only about 9 per cent of the total assessed land valuation of $17,566,600. The protest hearing was declared officially ended, but annexation opponents were authorized to file additional protests up to the close of the business day on August 10. By final tabulation only about 18 per cent of the total assessed land valuation was represented by the protest petition signatures.[8]

THE CAMPAIGN

In mid-July the annexation opponents selected two or three tax bills on property recently annexed to the city to "prove" what would happen to property taxes if annexation succeeded. Taxes on these properties increased by 30 per cent after annexation. It was claimed that these were typical tax bills sent to Frantz on the initiative of the taxpayers concerned. These property owners, in spite of paying higher taxes, had no greater police and fire protection than before. The two tax bills were made a part of a large paid political advertisement in the *Suburban News-Shopper*, which is distributed free throughout the annexation area.

About a month later, E. W. Chopson, an employee of the California State Personnel Board and a member of the anti-annexation group, told the Arden-Arcade Rotary Club that annexation would bring a 30 per cent tax increase. Fire protection would be no better, he said, and the area did not need better police protection. Sewers and street lights could be obtained at the same cost, with or without annexation.[9] Frank

Stipak responded to the *Bee*'s account of the speech in a letter to the editor. The taxation examples being used by anti-annexation spokesmen, he said, were the result of an error on the part of the county in making out the tax bills in that area. Part of the county tax was later refunded. Also, in the year before annexation to Sacramento city the fire district, anticipating annexation, levied no tax.

In the week following Chopson's appearance, Stanley Kronick spoke to the Arden-Arcade Rotary Club. He termed annexation the key to good government; the Sacramento area was one area, socially, culturally, and economically, and therefore should be one governmentally. He spoke of the downward trend of the property tax rate in Sacramento and the costs of the multiplicity of special districts outside of Sacramento. He stressed that only 50 per cent of Sacramento's expenditures was paid from the property tax.

At about the same time, J. R. Liske presented a policy statement on annexation to the Carmichael Rotary Club. After reviewing the process of urbanization and the need for governmental reform, and dismissing the tax issue, Liske presented some important arguments in support of the annexation. Sacramento, he said, is one of the most sound financially of the country's municipal organizations and it would be to the advantage of any contiguous area to become a part of it. This annexation would achieve 40 per cent of the task of unifying the entire urbanized area of Sacramento County and would provide the impetus for the remaining 60 per cent of the urbanized area to join Sacramento. This annexation would not interfere with MGC's recommendation for functional consolidation of services.

Benjamin Frantz wrote a long letter to the city council, on behalf of the Arden-Arcade Government Committee, which asked over 30 very specific questions about how annexation would be carried out—how many policemen will patrol the area, how many of the present firemen will be retained, and so forth. Sacramento's city manager replied by letter to Frantz's request for an immediate answer. The manager stated that some time would be required to get complete answers to the many detailed questions asked; in the meantime, he sent a copy of the city's official annexation policy, which would answer a number of the queries. The city manager also agreed to speak at a public meeting in the annexation area on September 22.[10] Members of the city council and some other city officials also attended this meeting. Since those strongly opposed to annexation dominated the meeting, this was a harrowing experience for most of the city officials. Many at the meeting seemed more devoted to trying to confound the city representatives than to seeking information.

The *Sacramento Bee* urged the city to move faster in this matter of presenting facts to the voters of the annexation area. In the September 1 issue an editorial stated that the voters were getting plenty of distorted, false, and self-serving misinformation, and this needed to be challenged by the true picture given by governmental officials. The presentation of facts in a precise manner should be a tireless and exhaustive effort, the *Bee* said. Sacramento was indeed supplying specific facts on annexation. During September pro-annexation leaders received reports on how city services are financed, and what annexed areas might expect as to street lighting, sewerage, water supply, storm drainage, and roads and streets.

During the month of September, the *Bee* devoted much space to the annexation issue. Each day at least one lengthy report about the advantages of annexation appeared in a prominent position in the paper. The following headlines typify the type of articles written: "City Offers Parks in North for Less," "Survey Refutes Arden Tax Fears," "Arden-Arcade Park Study Unit Is Urged," "City Promises Street Work If Annexing Wins,"

"City Offers Sewage, Drainage Program," "City Would Raise Fire, Police Protection in Arden-Arcade," "Suburb Will Get Trash Pickup," "Residents of Arden, Arcade Would Have Say in Planning."

About a week before the election the *Sacramento Union* ran a series of three articles on the annexation question, which concluded that there would be little difference in taxes, streets, sidewalks, gutters, sewers, and police and fire protection whether or not annexation succeeded. Only if this annexation were to be followed by other major ones, to unify the entire urban area, would the annexation be beneficial. Annexation opponents were enthusiastic about the *Union* articles, which seemed to confirm that the Arden-Arcade annexation would interfere with, rather than promote, unification of governmental agencies and services.

The several suburban weekly newspapers distributed in the area were openly opposed to annexation. The *Carmichael Westerner*, the *Carmichael Courier*, and the *Suburban News-Shopper*, for example, editorialized against annexation. One of the pro-annexation leaders complained during an interview that these suburban papers "wouldn't even publish the pro-annexation news releases although anti-annexation releases were published regularly."

Some of the Sacramento city councilmen and other city officials who were strongest for annexation stated off the record that the *Bee* presented the advantages of annexation too often and too conspicuously. They reasoned that voters resented being told so persistently what was good for them and began to suspect the motives of the *Bee* and city officials. Large numbers of suburbanites felt that the *Bee*, although liberal on national issues, took a provincial, pro-Sacramento position on local and regional issues. The *Bee*'s enthusiasm for annexation led some citizens to adopt the opposite position.

CAMPAIGN FINANCES

The records of the pro-annexation movement showed that contributions totaling slightly under $1,000 were received, and that expenditures were approximately $750. The surplus was transferred to the Sacramento City-County Chamber of Commerce, which had advanced an estimated $500 to $600 additional during the final days of the campaign, mainly for postage, mimeographing expense, and the retention of a public relations firm. The money was collected by a small group of the annexation leaders and came mainly from small individual contributions and from a few businesses in the annexation area and downtown Sacramento. For the most part, businessmen failed to contribute, and this represented a major difficulty in the annexation campaign. While some individual businessmen did give money, they wanted to remain anonymous and did not want to get their companies involved.

No dependable estimate could be obtained of the amount of money the anti-annexation forces collected, the sources from which it came, or the purposes for which it was expended. The treasurer of the anti-annexation forces was unwilling to divulge this information. Only one interviewee (pro-annexation) was willing to make a specific estimate of anti-annexation expenditures. His estimate was $3,000. Others made statements to the effect that one individual was paid $2,000 as a public relations expert and that the postage bill for the anti-annexation group must have exceeded $1,400.

ANTI-ANNEXATION STRATEGY

There were strange bedfellows in the anti-annexation movement. Some of the leaders favored merger of city and county government and concluded that piecemeal annexation would be a retrogressive step. Others were attracted to the movement because it provided the opportunity to forestall the growth of "big government." Some of these were primarily against a big Sacramento; some were, at least originally, against the whole concept of a metropolitan government on ideological grounds. For some, merger was the goal; for others, "home rule" was to be preserved. But whatever the original differences in goals, the leaders adopted a single-minded strategy for opposition.

Throughout the month of September the anti-annexation forces, with the help of a full-time paid public relations man, waged an active campaign. A few annexation opponents made numerous speeches and issued statements for the press almost daily. Newspapers were flooded with letters criticizing the "high and mighty" attitude of Sacramento city officials. Immediately before the election, the area was blanketed with anti-annexation literature which stressed home rule, grass-roots government, and taxes. A two-page "fact sheet" was entitled "If You Can't Afford Higher Taxes Vote No on Annexation." Defeat of annexation, the sheet read, would force city-county consolidation, which in turn would mean economy in government. Although the city's problems were never identified, the sheet stated that they were so numerous and complex that no time could be spent on improvements in Arden-Arcade. The opposition leaders believed that it was sufficient merely to remind the voter of life in the city, with its traffic congestion, slums, delinquency, crime, and racial strife. Apparently they were right.

Annexation opponents also distributed a four-page publication called "Urban Living"—with the subheading, "Dedicated to Progress Unlimited." This pamphlet repeated the charge that annexation means higher taxes. The readers' attention was called to the city's present preoccupation with a proposed freeway, and suggested that until this was resolved the city would have no time for Arden-Arcade's problems. The pamphlet concluded with a request for help in fighting the annexation and accused annexation proponents of making use of a public relations firm, hired to sell Arden-Arcade residents a bill of goods. (In fact, at the time this pamphlet was distributed, no public relations firm had been employed by the annexation organization, and the firm of Queale and Associates, which became involved later, was employed by the City-County Chamber of Commerce, not the annexation committee.) Printed in large block type was the statement, "We Need Annexation Like We Need A Hole in Our Head." Another brochure, mailed to "occupant," stressed the same arguments—home rule, taxation, and the evils of the big city. Anti-annexation posters were tacked on fences and posted on lawns along the major roads of the area during the last four or five days before the election.

PRO-ANNEXATION STRATEGY

In contrast to the great amount of literature distributed by the opponents, annexation advocates prepared only one newspaper tabloid, distributed during the final days before the election. The support of the League of Women Voters and the 8,000 citizens who had signed the petitions, and the endorsement by MGC and the Greater North Area Chamber of Commerce were emphasized. (It was not mentioned that the chamber endorsed the movement over the objections of its Urban Government Committee.) The pamphlet dealt with specific problems in the area such as water supply, sewers and storm drains, and taxation. To reinforce the tax reduction claims, the proponents summarized the results of a survey made recently by a Sacramento bank. The total property taxes paid on 40 homes in Sacramento was $10,355, as compared with $10,464 on 40 comparable homes in Arden-Arcade.

The literature also presented arguments for uniting the governments of two communities that were already one in most other aspects. The persons promoting annexation were introduced as men and women typical of the residents of the area as a whole. They owned modest homes, and were contributing their time plus dimes and dollars to achieve annexation.

The advantages of annexation boiled down to three major ones: Sacramento gets half of its revenue from sources not available to unincorporated areas; the industry and commercial developments within Sacramento would help pay for governmental services in Arden-Arcade; and a large, modern city government can render municipal-type services more efficiently and economically than can the special districts. The pro-annexation document concluded, "Tuesday We Vote; Let's Make Sacramento *Our* City." Unlike their opponents, the annexation leaders did little in the way of lawn signs or posters.

The pro-annexation leaders agreed that no significant differences in philosophy arose among themselves and only minor differences over strategy came up. "On one occasion," one leader stated, "we discussed whether to slug it out or to continue to be calm and objective." Since they were amateurs rather than professional organizers and publicists, however, some weak points in method were inevitable. There were the usual complaints of volunteer workers about their organization—too few did too much of the work with too little financial support. One participant felt that the organization needed the leadership of an experienced businessman and a large number of subordinates taking responsibility for the activities of the movement. Leaders also mentioned the lack of communication between precinct workers and the central organization. There was, in addition, the problem, basic to all reform groups, of effectively and dramatically presenting facts, while anti-reform groups need present only opinions and appeals to the citizens' emotions.

ORGANIZATIONAL SUPPORT FOR ANNEXATION

Annexation advocates made much of the fact that a number of organized groups had voted in support of the movement. The support given by the League of Women Voters and MGC has already been mentioned. The League prepared a leaflet in support of the movement, and League members distributed it. The League's support was based on a desire to reduce the bewildering number of governments, to obtain more services at a reasonable cost, to move toward merger, and to achieve economy in government through functional consolidation. In addition to directing the preparation and distribution of the pro-annexation leaflet, Mrs. Henry Long, president, participated in public forums as the League's spokesman. The Sacramento Area Planning Association also announced its support one week before the election, stating that annexation is "the best immediate approach to the solution of our metropolitan governmental problems."

The Urban Government Committee of the Greater North Area Chamber of Commerce had consistently opposed any extension of the city of Sacramento across the American River, a policy in which the chamber itself had concurred. However, about two weeks before the election, the Sacramento City-County Chamber of Commerce managed to obtain endorsement of annexation by its neighboring chamber. The city-county chamber pointed out that the Arden-Arcade annexation was the first step toward implementation of the recommendations embodied in MGC's final report, namely the annexation to Sacramento of the entire unincorporated, urbanized area of the county. The board of directors of the North Area chamber overrode the negative recommendation of its Urban Government Committee and voted on September 10 to support the annexation proposal. The directors included in their statement of support an endorsement of the MGC plan for mass annexation.[11]

Only the League of Women Voters and the City-County Chamber of Commerce contributed any active help other than public endorsement. The North Area chamber provided no working support and its endorsement was too late to be effective.

The only organizations reported to have given substantial support to anti-annexation workers were the County Sheriff's Association and various fire departments and firemen's associations in the annexation area. It is apparent that those associated with fire protection in the Arden-Arcade annexation area considerably influenced the negative vote, although the full extent of their contribution is not known. Whatever the motives of the individuals, the various fire fighting organizations had the potential time, talent, and other resources, plus personal financial and political incentive, to make them formidable factors in any political decision.

Besides job security and protection of retirement benefits, and the frankly expressed desire of fire chiefs to remain in their positions, reasons given by local fire department members for their opposition were various. They did not want to be controlled by the *Sacramento Bee* and the *Sacramento Union,* which some believed was controlled by the *Bee.* They also expressed fear that big government would lead to socialism or Communism. The fire fighters were convinced that the area was receiving better fire protection now than it would under annexation. The Arden Firemen's Association and several individual firemen and officials are reported to have contributed money to the anti-annexation forces, and many worked actively against annexation. One fireman stated that he and his wife had spent all of their free time in canvassing the area, telling the people about the evils of annexation. This fireman, along with others, had also attended many of the anti-annexation meetings and had worked closely with Frantz in soliciting signatures. A fire chief claimed to have visited every residence in the part of his district which was included in the annexation area. Another fireman reported devoting eight hours a day for three months to fighting annexation. Those persons in the fire departments who favored annexation tended to keep their opinions to themselves.

One interviewee associated a water district and a garbage collecting agency with the anti-annexation activities. He contended that the same addressograph plates used for water and garbage collection bills in the area were used to address anti-annexation literature.

ANNEXATION DEFEATED

The Arden-Arcade annexation was defeated by 11,410 to 6,170 in the September 29 election. The fact that there were about 1,400 more valid signatures on the annexation petitions than "yes" votes raises the interesting question—what motives did people have for signing the petitions? Did some agree to sign only to bring the issue to a vote or had some voters become converted to the opposition?

Thirty-nine per cent of the qualified voters did not vote. The annexation proposal received a majority of votes in only one of the 60 precincts.

Seemingly, the residents of Arden-Arcade did not feel that governmental reorganization was necessary. They already were receiving most of the customary municipal services. The fact that they had to rely upon several governmental units rather than upon one for these services, and that possibly they were paying more for these services than might be the case under annexation did not seem of particular importance to them. The emotional appeal of the opposition, based upon grass roots, local control, and home rule, received a more receptive hearing.

ARDEN-ARCADE ANNEXATION ATTEMPT EVALUATED BY INDIVIDUALS INVOLVED

Previous discussion of the backgrounds of the annexation leaders and their conduct of the annexation campaign indicated that they were, for the most part, concerned with a theoretical notion of better government, which made the abolishment of small, overlapping service districts necessary.

One primary motive of the anti-annexation leaders interviewed was a desire to keep the area "rural." Another interviewee preferred the *status quo* to annexation even if the latter would bring lower taxes. Another of the early leaders argued that annexation to Sacramento would mean a lessening of local representation as well as an increase in taxes. He saw the *Bee* and the "downtown power structure" as obstacles to effective government for Arden-Arcade. During the interview he made several references to a nation-wide "organization" which has repeatedly connected powerful metropolitan government with a Communist conspiracy to destroy democracy. One of the officers of the anti-annexation forces expressed the initial fear of higher taxes, but also expressed distrust of Sacramento merchants. He felt that annexation would result in Arden-Arcade's being taxed excessively so that downtown could be rehabilitated. "After all, redevelopment is a mistake; we've got to have slums. There must be cheap housing in a city. Why not leave it where it is?" Another leader said that even though the *Bee* favored bigness and centralization, and aid to downtown Sacramento, what was good for the city might not necessarily be good for the suburban area.

Leaders on both sides were asked to name community organizations that should have supported their efforts but did not. As might be expected, pro-annexation leaders listed several organizations from which they had expected more active support. Even those groups that had taken a part in the annexation effort, such as the League of Women Voters and the City-County Chamber of Commerce, were criticized for doing too little too late. Leaders felt that merchant groups, educators, PTA's, and improvement clubs ought to have joined in the movement. Lack of support from civic, professional, and service associations was also cited as weakening the cause. The indecision and division in the North Area chamber was most often mentioned as a decisive factor in the defeat of annexation.

> The North Area chamber should have been a spark plug to the annexation movement, but it failed to accept the Metropolitan Government Committee's report. This existing organization should have assumed leadership and there should not have been a superimposed annexation committee. It was the opposition of the North Area chamber that really defeated our movement. This organization is comprised mainly of small businessmen, but the attorneys therein apparently are the most interested in governmental reform and their recommendations have usually been followed. Most of these attorneys were opposed to annexation. Why the attorneys behaved this way, I am not sure, but I suppose they deem that it is not wise to be associated with a losing cause, and in thinking this way they, indeed, make governmental reform a losing cause. The small businessmen in and out of the chamber were the most timid of all I came across. We got very little money from small businessmen. Our $50 checks came from bigger businesses. The little businessman would not rock the boat. Maybe it was because he was just getting started in business. Also, much of the Northeast area business is handled by big chains, and the managers of the branches had no special community ties. Many of the businessmen feared change and this is a function of ignorance.

The answers reveal personal disappointment in the failure of organizations to rally in support of annexation, and the leaders' perceptiveness of attitudes and values in both the suburbs and the city. This understanding of the community, however, seems to have been gained through their experience in the annexation movement.

Surprisingly, the victors were not fully satisfied with the organized support they received. They too listed the PTA's, chambers of commerce, local improvement and service clubs, and professional associations as groups that should have lent support to their cause.

Pro-annexation leaders were agreed that there were influential citizens in the Sacramento area whose participation in the annexation movement might have been expected at the beginning and whose support would have improved the chances for the movement's success. But, for the most part, there is little consistency among the leaders in their selection of names of individuals or groups of persons that were generally regarded as influential. Downtown businessmen, the managers of two large shopping centers in the northeast area, bank officials, and industrial executives all were mentioned as potential sources of support that did not materialize. Influential citizens in these occupations were approached by annexation leaders but they expressed indifference and sometimes opposition to governmental reform. The conservatism of the business and industrial leaders was attributed to their desire to avoid any controversy that might alienate customers and to the stable employment situation in Sacramento, which gives these people a relatively secure income. Some businessmen who favored annexation did not participate in the movement because, in one person's words, "they did not realize the importance of annexation to their own self-interest." They were also overconfident that the movement would succeed. One annexation leader doubted that there were any "influential people" in Sacramento, an expression which gives further evidence to the fact that Sacramento lacks any widely recognized leadership.

The responses by anti-annexation leaders to the question concerning failure of influential people to participate in the countermovement were rather bland. One interviewee proclaimed, "Why, every influential person should have been involved," but, after a short pause, asked, "But who in the area are influential?" These people, whoever they may be, failed to support anti-annexation because of "business" reasons and general apathy. "Businessmen did not want to become embroiled. They felt their businesses would be better off if they stayed out." One leader estimated that no more than "10 to 12 per cent" of the area's inhabitants became involved in the annexation attempt. "The people are not interested. The issue must be close to them. They take government for granted. They gripe about taxes and weather but do nothing about either."

The leaders were asked about the reactions of specific groups and organizations in the area. All identified the *Bee* as favorable to annexation. Two annexation leaders credited the *Bee* with altruistic motives for its stand, but most said simply that the *Bee* had always maintained an annexation policy. The anti-annexation leaders explained the *Bee*'s stand by its traditional alignment with "downtown and the incumbent city government." Some also felt the *Bee* "didn't want another major newspaper" in the area. In spite of its position, one interviewee said that the *Bee* "actually did more to *get* votes for us than *lose* them." This leader did acknowledge the *Bee*'s "generally honest coverage," including news items "detrimental to its editorial policy." The rejection of annexation by the weeklies in the northeast area was generally interpreted as allegiance to suburban advertising. One reported that there was a desire to convert the weeklies into big city dailies. Another predicted that probably the *Suburban News-Shopper* would emerge as a daily paper to compete with the *Bee*. It was assumed that this could happen only if the North Area had its own identity as a separate big community rather than being a part of Sacramento city.

School personnel were regarded as "neutral" and "mixed" or "namby-pamby." One annexation leader reported that "teachers were afraid to sign petitions because of possible reaction of school administrators," but another felt that the officials were neutral while the teachers were opposed to annexation.

Most annexation leaders believed that among real estate people the decision about annexation was "primarily an individual matter." Probably the most perceptive comment was that "real estate brokers were favorable, developers unfavorable" for the reason that "although city and county building codes are similar, the city enforces its codes more effectively, and therefore it is understandable that the developers might not want annexation."

Only one annexation leader felt that the businessmen from Sacramento were unfavorably inclined, and that was because they "feared higher taxes in underwriting of the bedroom area." Another insisted that the businessmen did not make their position known. But all other annexation leaders interviewed felt that the downtown businessmen were pro-annexation; annexation would increase downtown prosperity. While acknowledging support, one leader complained that the businessmen's leadership "didn't show at all." The anti-annexation leaders interpreted the position of city businesses as a way of coping with suburban competition. Annexation would make it possible for the merchants to "flood the area with buses and to try to get the customers to go downtown." Another anti-annexation leader stated that downtown interests wanted to control outlying business through city government. Suburban businessmen failed to support either group in order to avoid taking a controversial stand, leaders on both sides of the issue said.

The pro-annexation leaders interviewed generally agreed that Sacramento city officials favored annexation. Reasons given for this support were both altruistic and protective. The officials felt a "sense of responsibility to the entire community"; they believed "the city, not the county or a new city in the northeast area, should be the base for the vast metropolitan area" and an annexation policy should be pursued even though it would cost the city money. From another point of view, "a bigger city has greater prestige." One interviewee stated, however, that in informal conversation city department heads had expressed "fear of tax increase within the existing city," and he also "guessed" that the city transit authority, aware of the costs of extending bus service to Arden-Arcade, had many doubts about the venture. These unofficial reservations about annexation were perceived by the opposition also: "Many city officials, especially those beneath the top ranks, didn't care, and many probably did not want an annexation." Most opposition leaders, however, felt that annexation had firm support from city officials, who, they said, hoped for "bigger empires."

County officials were unfavorably inclined toward the annexation movement. Annexation would mean a "waning prestige of county officials" and some county workers were severely reprimanded for becoming involved in the movement in any way. It was reported, though, that "the top appointive officials" thought annexation was a good step, but, as one anti-annexation leader put it, "they didn't dare express themselves openly." Nor were they willing to publicize their opposition to annexation.

The League of Women Voters supported annexation after a formal study of local government. Its previous promotion of the PAS report and city-county merger was not looked upon as inconsistent with support for annexation, and leading proponents interpreted its position as considered and sincere. Annexation opponents felt the League had taken a partisan position, however: League members are "radicals," "a bunch of Democrats." "The League was misguided although sincere. It turned partisan on this issue, although its constitution says it should not." One opponent admitted that "the League was unfavorable to us for academic and theoretical reasons." In spite of these references to "partisanship," it should be noted again that the annexation issue was not one of Democrats versus Republicans. By coincidence, some of the annexation leaders also were active in the Democratic Party while some of the annexation opponents were active in the Republican Party.

All of the annexation and the anti-annexation leaders interviewed were asked why the Arden-Arcade annexation proposal had been rejected by the voters. The annexation leaders, usually after a moment of embarrassed silence, gave the following range of explanations: "the shoe isn't pinching"; "there's a natural resistance to change"; "the citizens remained poorly informed." Annexation leaders also recognized that they had failed to counter the opposition on the tax issue in an effective way. "The anti-annexation forces knowingly confused issues with half-truths and mis-truths." Their group was hampered by too little time and money and poor organization. The suburban antagonism to the *Bee* was mentioned again as a factor leading to defeat.

Members of the defeated annexation group recognized that one basic cause of their failure was that the residents in the area were pretty well satisfied with their situation at the moment. "The people were just not unhappy enough," one person said. The annexation opponents readily admitted that this complacency was one decisive factor in their favor. This prevailing attitude, combined with an antagonism toward the city and a desire for self-identity, made up the ingredients for defeat of annexation. In addition, the opponents admitted to confusing the issues on purpose. Their slogan was, "If In Doubt, Vote NO." "Obviously," as one said, "we were quite successful."

CONCLUSIONS

There was a reasonably satisfactory level of municipal-type services in Arden-Arcade, although it was governed by a conglomeration of special districts. Those who anticipated an ultimate breakdown of services because of this system were unable to communicate their concern to the satisfied residents. In addition, the property taxation issue proved vital. The opposition pressed its tax claims—which were based on a few, carefully selected cases—much more vigorously than did the proponents. Although annexation leaders showed through their study that the effect of annexation would be more and better services for the same price, taxpayers tend to be fearful of tax increases and are more willing to believe those who forecast higher taxes. The effects of annexation upon governmental organization were also poorly explained. Annexation, as described by its opponents, meant adding another layer of government—and another tax collector—on top of the many already in existence. The proponents failed to emphasize that as special districts are eliminated the governmental structure would be simplified.

Citizens were suspicious of the annexation leaders' claims. They were told that they would get added services without increase in taxes, but surely this could not be. Annexation proponents failed to get across the fact that their area would have new sources of income after annexation—from the local sales tax, gasoline tax, liquor licenses, motor vehicle license fees, and so on.

Residents of this suburban area were very jealous of their local autonomy; a lot of small governments meant that they had grass-roots government. This, of course, is a basic principle of the American system, which some of the annexation opponents exploited by warning that a new form of government called "metro" was being advocated by Communist-tinged organizations. It did not matter that many of the voters apparently were unaware of the existence of all of the units of government to which they were subject, that they did not bother to vote for officials of these numerous governments, and that they did not know where the business of these governments was being transacted. It still was grass-roots government. One of the proponents, in diagnosing the defeat, suggested that, because FHA and other mortgages now provide for a monthly lump sum payment of both mortgage and local taxes, most citizens do not know how much the many districts collect in taxes.

Sacramento's image had been damaged by the many years when the city followed rigidly the letter of the law on annexation, which said that proposals must originate in the areas to be annexed. With the exception of the Oak Park annexation in 1911, there were no annexations to Sacramento until the middle 1940's. By the time Sacramento began to show active interest in annexation the population clusters beyond its borders had developed feelings of local identity. Sacramento, the city, can be distinguished from Sacramento, the community. When suburbanites travel, they undoubtedly speak of themselves as Sacramentans, but at home they look with suspicion upon the city of Sacramento. A number of persons conceded that annexation was more logical than whatever other arrangements they were advocating at the moment, but "Sacramento doesn't want us," "Sacramento isn't interested in us," or "Sacramento wants to grab only the highly assessed property." They also did not want to become involved in the city's problems such as the deteriorated west end and the rising percentages of minority groups in the city. Some pictured Sacramento as a "cold" city, which accepted police brutality as typical of the city policemen, and saw the city manager as an efficient but cold administrator.

The efforts of firemen in the northeast area were important. The firemen, operating from an already existing organization and fearful of the effect of annexation upon their jobs, provided the mechanism for getting house-to-house coverage with anti-annexation materials. They prepared and put up the "Don't Annex" posters and signs. They also contributed money to the cause. Although the proponents of annexation set up an elaborate organization on paper, their leadership base was narrow and only a very small number of persons really worked at it. Thus, success of the firemen's loosely knit organization reinforces the conclusion that few effective organizations exist in the northeast area.

The position taken by the Urban Government Committee of the Greater North Area Chamber of Commerce was influential. Neither the committee nor the chamber now exists, but the committee had an important effect on the thinking of the people northeast of the American River during its existence. The committee was wedded to the idea of city-county merger. It proposed something of the sort before PAS did and after that sat in judgment upon every proposal for governmental change in metropolitan Sacramento. It was sure to oppose any proposal that would enhance the position of the city of Sacramento. While the Greater North Area Chamber of Commerce finally endorsed the proposed Arden-Arcade annexation, its Urban Government Committee never did.

The leadership role of Dr. James R. Cowan was believed by some to have been crucial. Many of those who worked for annexation felt that if Cowan had supported this annexation wholeheartedly it would have received wider approval in the northeast area. Cowan, who had been a leader in the Urban Government Committee, SMAAC, and MGC, had supported first city-county merger and later large-scale annexation to Sacramento. As a prominent figure in the northeast area, Cowan could perhaps have swayed some citizens if he had expressed his approval more vigorously. It is difficult to judge the influence, manifest and potential, of one individual. While it is doubtful that active support from Cowan could have reversed the outcome of the election, the fact that some pro-annexation people believed that Cowan could have expressed such influence is another commentary on the lack of area-wide leadership.

The skill of modern public relations and propaganda utilized by annexation opponents was not matched by the proponents of annexation. Defenders of the *status quo* had the advantage —and retained it.

Perhaps more fundamental than these local issues is the changing character of American metropolises in general and the Sacramento metropolitan area in particular. Sacramento is a vivid example of extra-community influences which have become increasingly important in metropolises throughout the country. Many of the business leaders are hardly more than bureaucratic transients. Local bankers, long identified as stalwarts of the communities, are no longer local, but are career professionals in state-wide organizations. They may find themselves in strange, or at least new, communities from year to year. (California's larger bank organizations operate branch banks in many of the communities throughout the state.) More and more of the leading merchants are executives of chain stores, who are also likely to be transferred often. Conservatism, timidity, and calculated neutralism are thus built into the leadership in the business community. The reluctance of businessmen to take a stand on annexation was reported repeatedly by movement leaders, pro and con. This "neutrality," combined with Sacramento's dependence on state government rather than local enterprises, and on federal military and contract installations rather than local industries for the backbone of its economy, makes more apparent the reasons why neither effective leadership nor enthusiastic interest in governmental reform has been displayed.

The following letter, written to the *Sacramento Bee* about a week after the proposal was defeated, is perhaps the best summary of the situation:

> Now that the victorious opponents have gloated piously of a mandate for merger and the defeated proponents have swallowed their pride in brave rationalizations, let us consider the real reasons why annexation failed in the Arden-Arcade area. First is fear based on ignorance. Unable to resolve the tax question, people feared the worst and voted no. Second is suspicion. The motives of those who promote change are always suspect. People elect to live with the evils they have rather than take a chance on those they know not of. Third is self-interest. Jobs and prestige based on the status quo outweigh the public welfare. Finally there is the skill of "public relations" in exploiting these weaknesses—fear, ignorance, suspicion, and cupidity. So annexation was not defeated on its merits, and there was no mandate for merger or for anything else.[12]

In the Hagginwood-Del Paso Heights annexation area the potential advantages of annexation to the inhabitants were easily seen. In the Arden-Arcade annexation area the potential advantages did not stand out so clearly. This perhaps explains why the Hagginwood-Del Paso Heights proposal was defeated by a very narrow vote whereas the Arden-Arcade proposal was overwhelmingly put down. Practically all of the Hagginwood-Del Paso Heights annexation area is a part of Sacramento city today, while most of the Arden-Arcade area still is outside Sacramento's boundaries.

6

Sacramento Leaders and Metropolitan Affairs

This chapter describes and analyzes some of the characteristics, attitudes, and views of those persons who were considered to have been influential in the community but who did not participate directly in any of the governmental reorganization attempts. The same kind of information is then given about the leaders of the various study and reorganization attempts, and the movements are compared. In the third part, some important questions related to leadership in metropolitan Sacramento are considered.

GENERAL COMMUNITY LEADERS

Following Floyd Hunter's general approach, the authors located leaders by means of reputation—that is, the attribution of influence.[1] During personal interviews, 18 "judges" were asked two questions:

> Whom would you consider the ten most powerful and influential people in the whole Sacramento metropolitan area—that is, the real leaders?
>
> Thinking back over the 1950's, whom would you consider the most influential people in relation to the problem of changes in governmental organization in the Sacramento metropolitan area? Name five from the city of Sacramento. Name five from the remainder of the Sacramento metropolitan area.

The 18 judges, chosen by the authors, included a city official, a county official, a newspaperman, a legislator, a newspaper editor, two school superintendents, a judge, two officers of women's organizations, a Negro leader, a religious leader, a labor union official, a banker, a real estate man, two chamber of commerce officials, and a state official. They were assured that their nominations would be treated as confidential information.

The judges were individuals who the authors believed would be in the best position to provide an informed opinion about leadership in the area. One of the persons invited to be a judge declined. Another, who accepted, named only persons in official elective or appointive positions despite our instructions to avoid this. Since this approach to nomination of leaders did not agree with the intent of the question, the nominations by this judge in response to the first question were not considered. Several other judges expressed reservations about their ability to answer the first question about the "real leaders" although all eventually submitted names. A few commented that this would have been an easier question to answer 25 years ago. From the judges' responses the individuals to be interviewed as "general community leaders" and some of the leaders of particular movements were selected.

We did not find a clearly identified central power group of any sort—no small elite of businessmen such as Hunter identified in Regional City. While the results are not strictly comparable—different questions were asked and slightly different methods of obtaining nominations were employed—it seems clear that Hunter found much greater agreement among his judges than we did among ours.

In response to the first question about the "real leaders" only four of the judges named exactly ten individuals. Nine named more than ten; five named fewer than ten. The range was from zero to 37, with a median of 10.5. In all, 227 nominations were made by the 18 judges. The most frequent mention of one individual in response to the first question was eight times. One individual received seven nominations; two received six; five received five; eight received four; 98 received three or fewer nominations.

Of the 28 persons who were nominated by three or more of our judges, 14 were businessmen (including four bankers); three, department store executives; two, insurance executives; one, a newspaper publisher; one, an insurance agent; one, a realtor; one, a supermarket executive; and one, a defense industry executive. Eight nominees were government officials: two city officials, three county officials, one state official, one state assemblyman, and one congressman. The remaining six persons included two religious leaders, two attorneys, one school administrator, and one newspaper editor. Since state government is a major employer in the Sacramento area, it is significant that only two state officials were nominated, neither of whom was active in any of the governmental reorganization movements studied.

Of the 24 who received two or more nominations in response

to the question about governmental reorganization leaders, 13 were also nominated at least twice as general community leaders. Thus there is some overlap between the two groups, but the majority of those who are considered influential in the community have not been active in reorganization attempts, and conversely. The overlap between the responses to the two questions might have been caused by two aspects of the interviewing technique. The questions were asked during the course of a single interview, and the same names may have occurred to the respondent in answering both questions. Secondly, since the interviews were conducted after the reorganization attempts, some nominees, not acknowledged as leaders in the past, may have attained general community prominence as a result of their governmental reorganization activities.

All persons interviewed were asked to list names of individuals whose support would be especially important in securing voter approval of a county-wide referendum—"for example, a proposal to reorganize the governments of the Sacramento metropolitan area." Of the 21 individuals who were nominated by two or more interviewees, 13 were among the top 28 nominated independently by the 18 judges as general community leaders and seven were among the top 24 nominated by the judges as influential with respect to governmental reorganization. Here again we see partial agreement in identifying influential Sacramentans. Since the overlap is substantially greater for the "ten most powerful and influential leaders" than for reorganization leaders, this gives added support to the contention that the reorganization efforts were not able to attract the community's top leadership.

In using the "nomination" approach a great deal hinges upon the judges chosen. Their nominations reflect their own special points of view, and a different set of judges might very well result in a different set of leaders. The more homogeneous the judges the more likely their lists of leaders will agree. A rough measure of the degree of agreement among our judges was obtained by dividing the judges at random into two groups and by noting the degree of overlap obtained in identifying the top 15 Sacramento leaders. When this was done the two groups agreed on only six of the 15 leaders, a fact which indicates a low degree of agreement among the judges.

Of the 28 individuals receiving the greatest number of nominations as community-wide leaders, nine were known to have been active in connection with one or another of the reorganization movements. Since most of these nine persons active in the reorganization movements were to be interviewed in any case, we selected 14 of the remaining group to be interviewed as individuals influential in the community at large but not active in any of the movements studied.[2]

These 14 general community leaders included four bankers, two religious leaders, three department store executives, an insurance agent, a supermarket executive, an insurance company executive, a realtor, and a lawyer.

VIEWS OF METROPOLITAN PROBLEMS

All interviewees were asked, "What seem to you to be the most important problems in the Sacramento metropolitan area?" Responses of all 96 interviewees are summarized in Table 2. The five most important problems as seen by the 14 general community leaders were: traffic and parking (seven mentions), long-range area-wide planning (four mentions), and industrial development, schools, and sewage disposal and drainage (three mentions each). The general community leaders gave greater emphasis to sewage disposal and drainage, schools, core area development, industrial development, and traffic and parking, and less emphasis to multiplicity of special districts and to governmental reorganization than did the total group of interviewees. The most marked difference was traffic

TABLE 2

Sacramento Area Problems Viewed as Most Important by 96 Leaders

Problem	*Per Cent Mentioning*
Multiplicity of special districts	26
Long-range area-wide planning	26
Governmental reorganization	22
Consolidation of governmental services	17
Tax rates	14
Traffic and parking	13
Streets and roads	12
Problems generated by rapid growth	12
Tax base	12
Community spirit	10
Sewage disposal and drainage	10
Industrial development	8
Functions of city and county unclear	8
Continuing community leadership	8
Schools	6
Mass transportation	6
Water	6
Law enforcement	5
Parks and recreation	5
Freeways and arterials	3
Redevelopment	3

and parking, which 50 per cent of the general community leaders spontaneously mentioned as a problem, in contrast to only 12 per cent of the total group. The higher proportion of downtown business leaders among the general community leaders group probably explains this concern. The less frequent mention of problems connected with multiplicity of special districts and with governmental reorganization is consistent with that group's relatively inactive role in reorganization movements.

After the respondent indicated the problems that seemed most important to him, he was asked to select from a list of 23 problems each of the ones he considered an area-wide problem, and to assign the degree of importance of each problem selected. The general community leaders' scores were very similar to those for the group as a whole. There was only one item—zoning—for which the scores differed more than .5 of a point. This agreement was true even for the several items (sewage disposal and drainage, schools, multiplicity of special districts, and industrial development) on which they differed in the first question. This may have occurred because an open-ended question taps a somewhat different dimension than a question asking for degree of importance of each item on a list. The first elicits the salience of problems, the readiness with which they occur to the respondent; the second allows the person to assign a high rating to an item he had not thought of in replying to the open-ended question. Another complicating factor is the high scores—between 3.3 and 3.5—of most of the problems. The scoring method was as follows: "very important" equals 4, "moderately important" equals 3, "of little importance" equals 2, and "don't know" equals 1. Perhaps providing a larger number of alternative ratings would have produced more marked differences among problems.

The general community leaders were somewhat more satisfied with the county government than were the interviewees as a whole—71 per cent as compared with 60 per cent. More were dissatisfied with the county than with the city of Sacramento. Among the county's strong points eight specifically mentioned the county executive. "Excellent administrator," "realistic," "competent," "well trained," were typical comments. A number commented that the county government had been honest, free from graft and corruption. Among the weak points, six indicated that the "pay-as-you-go" approach failed to cope with the population boom in the area, and they were pleased that the county road-building bond issue had passed. A few commented unfavorably about supervisors who were poorly qualified or politically ambitious. Others felt that the caliber of supervisors had improved in recent years.

All of the general community leaders were satisfied with the government of the city of Sacramento, in contrast to 69 per cent of the interviewees as a whole. Eight of the 14 were particularly pleased with the city administration, and mentioned the department heads as well as the city manager. Five commented that the city provides good services, four that the city government is honest, three that the city has had good councilmen, and three that the city is financially sound, with reasonable taxes and low bonded indebtedness.

COMMUNITY LEADERS AND REORGANIZATION

Ten of the 14 general community leaders indicated that they had not been asked to take an active role in any of the reorganization movements. One commented that he lived in the city and so was not directly involved. Another businessman said he avoided becoming involved because of the bitter feelings and "we want to do business with all groups." He had considered running for the school board in another city but when his superior at the bank suggested that he might regret it later on, he decided not to run. He felt this was a wise decision since a school dispute later divided the community into opposing camps. One of the religious leaders, who had been asked to endorse some of the annexation proposals, had declined, feeling that this was not appropriate for him.

Many of the general community leaders felt that the movements helped citizens become aware of the metropolitan problems and showed that people were concerned. "We can't sit quietly and let the city become decadent. The movements are signs that something is going on; they are steps along the way." One respondent suggested that the movements had been great for the press, especially the *Sacramento Bee*. He described them as battles of the press. The public is not greatly concerned about the problem, but the *Bee* is, and he felt that many people were active in the movements in order to obtain the *Bee*'s support for their political ambitions.

Most of the general community leaders reported that they did not give any money or services to the reorganization movements or their opposition. Only two reported contributions to annexation proposals; another two reported contributions to the recent campaign for a county bond issue.

Nearly all of the general community leaders had discussed the reorganization movements with city and county officials, business associates, and friends but, for the most part, only in occasional, informal conversations. Only four of the respondents reported discussing the movements with newspaper publishers or editors.

When asked, "What has kept the governmental reorganization idea alive in spite of the failure of several major movements?" seven of the community leaders pointed out that the problems that the movements were designed to overcome continued to exist. The activity of concerned individuals who continue to call the community's attention to its unfinished business was mentioned by several leaders. Four respondents called attention to the actions of the news media, especially the *Sacramento Bee,* and three mentioned the chamber of commerce.

Some leaders, when asked about more effective methods of getting action on metropolitan problems, suggested greater use of existing organizations such as the chambers of commerce, the League of Women Voters, and various service clubs. Others stressed the need for a better cross section of supporters, a broader community base. A few felt that the reorganization movements had not drawn the top leadership in the community, and they contrasted them with the board of directors of the Sacramento Redevelopment Agency (in charge of downtown urban renewal activities), which was composed of citizens who, in their judgment, were widely known and respected.

Some stressed the need for more extensive public information programs through television, newspapers, and public forums, to clarify issues through more complete and less biased analysis. Successful governmental reorganization must be a long-range effort, perhaps requiring a well financed and well directed staff. Some felt that the schools could do a more effective job in imparting a sense of active civic responsibility instead of "viewing from the sidelines." Too often, one respondent said, those who plan programs are not those who must implement them, and more of the public officials should be involved in early planning discussions. Another respondent blamed the large number of absentee owners in Sacramento. The local managers cannot make decisions without checking with the home office, and are not very concerned about Sacramento's long-range interests anyway because they expect to be transferred to another city in a few years.

One respondent was very pessimisitic about achieving marked changes:

> When I was first elected to the city council, I had the naive idea that if a proposal was good it would be accepted. We hired lots of experts—engineers, traffic men, who would study a problem for months and make recommendations. However, someone would always oppose. They would show how the expert plan was all wrong, usually because it ran across the critic's property. I don't see how the press could do more than they are doing already. I'm afraid we can't do much more than we are now doing. I think we are progressing pretty well.

COMPARISONS AMONG SACRAMENTO LEADERS

Chapters Two through Five are based primarily upon personal interviews with movement leaders and upon newspaper accounts of the movements. The first section of this chapter reviews the results of the interviews with general community leaders. The present section compares the information obtained from the Leadership Questionnaire and the Allport-Vernon-Lindzey "Study of Values," which were left with the interviewee to be filled out at his convenience and mailed to the study director.

For purposes of this analysis the interviewees were assigned to one of the following nine groups:

SMAAC and MGC leaders
Pro-incorporation leaders
Anti-incorporation leaders
Pro-annexation leaders in Hagginwood-Del Paso Heights
Anti-annexation leaders in Hagginwood-Del Paso Heights
Pro-annexation leaders in Arden-Arcade
Anti-annexation leaders in Arden-Arcade
General community leaders
X group leaders

In addition to persons who were closely identified with particular movements and general community leaders, the authors interviewed a miscellaneous group of officials and reporters who were knowledgeable about several movements though not direct participants in any. This group has been called the X group. Five members of the X group were city or county officials, five were active in chamber of commerce efforts in the metropolitan government area, and two were newspapermen.

The reader should bear in mind that one-fourth of the interviewees did not complete the questionnaires. The losses were heaviest among the SMAAC-MGC and the X group leaders, and the results for these groups should be regarded with special caution. The returns from the other movements are reasonably complete. Since only two anti-incorporation leaders were interviewed, that movement has been omitted in the cross-movement comparisons included in this section, but data from these two individuals have been incorporated into the study of the characteristics of the leaders as a whole. The Hagginwood-Del Paso Heights group had a moderately high proportion of returns, but the small number of individuals who returned the questionnaires (4 for the pro-group and 5 for the anti-group) should be remembered when studying the results for these groups. (See Table 3.)

The interview and questionnaire material was obtained after rather than during the active period of the movements. Distortions of memory and hindsight presented as foresight are possible errors. On the other hand, the interviewees may have been more candid about certain past events than they would have been at the time of the event.

BIOGRAPHICAL CHARACTERISTICS OF LEADERS

Most of the leaders are between 40 and 60 years of age. The Arden-Arcade, new city incorporation, and X group members are younger than average. In the case of the first two, this difference probably reflects the relative youth of the residents generally in the areas. The leaders come primarily from the professional and managerial occupations, and more than half of the respondents have had professional or graduate training. SMAAC-MGC, the new city incorporation group, and the anti-annexation group in Arden-Arcade have very high proportions of leaders with professional or graduate training.

Almost all of the leaders are married and 58 per cent have one or more children under 18 years of age. A higher proportion of the pro-incorporation group and the anti-annexation leaders in Arden-Arcade have young children. About two-thirds of the leaders are Protestant. The general community

TABLE 3
Number of Leaders Interviewed and Number Returning Questionnaire Materials For Each Movement

Movement	*Number Interviewed*	*Number of Questionnaires Returned*	*Per cent of Questionnaires Returned*
SMAAC-MGC	16	9	56
Pro-incorporation	18	16	89
Anti-incorporation	2	2	100
Pro-annexation, Hagginwood-Del Paso Heights	6	4	67
Anti-annexation, Hagginwood-Del Paso Heights	7	5	71
Pro-annexation, Arden-Arcade	11	9	82
Anti-annexation, Arden-Arcade	9	9	100
General community leaders	14	12	86
X group	12	7	58
Total	95*	71	
Per cent of total			75

* Altogether 96 persons were interviewed. One of those interviewed concerning the Hagginwood-Del Paso Heights annexation movement was neither for nor against.

leaders have a higher proportion of Catholics than the group of leaders as a whole. With the exception of the general community leaders, the majority of the interviewees were not born in the city of Sacramento or even in California. Most of them, however, have resided in the Sacramento area for nine or more years. More of the pro-incorporation group and pro-annexation leaders in Arden-Arcade are relatively new to the Sacramento area.

For the most part, the annexation leaders were living in the area to be annexed at the time of the interviews. Most of the general community leaders lived in the city of Sacramento. Ninety per cent of the leaders doubted whether they would move out of the area in the next three to five years.

As one might assume, many of those who became active in governmental reorganization movements have an extensive background of participation in governmental affairs. Forty-four per cent of the leaders have held elective or appointive office. Public officeholding is less common among pro-annexation leaders in Arden-Arcade and pro-incorporation leaders and relatively more common among general community leaders, three of whom have held three or more such offices.

INTERESTS AND VALUES OF THE LEADERS

In addition to the Leadership Questionnaire, interviewees were asked to complete the Allport-Vernon-Lindzey "Study of Values," a 45-item inventory which is designed to measure the relative prominence of six basic interests or motives: the theoretical, economic, aesthetic, social, political, and religious. The classification is based directly upon Eduard Spranger's *Types of Men*. According to Spranger, a person can be better understood through knowledge of his interests and intentions rather than of his achievements.[3]

In comparison with the general norms provided by Allport, Vernon, and Lindzey, based on scores of 8,369 college students, our leaders have above average theoretical, political, and economic interests and below average social service, aesthetic, and religious interests. Since equal weight was given to the norms for the two sexes, and only seven of our interviewees were female, the norm for the college men group may be a more appropriate basis for comparison. Here the differences are much smaller. Our leaders have significantly lower social service scores and significantly higher theoretical scores. None of the other differences is statistically significant.

With these two exceptions, the mean AVL scores earned by our leaders as a whole are similar to those earned by the 5,894 college men in the AVL norm group. The differences in political, economic, aesthetic, and religious interests from the general norms appear to reflect primarily typical sex differences in interests. The higher theoretical interest found is consistent with the group's concern about long-range area-wide planning, consolidation of governmental services, and the problem of multiplicity of special districts. None of the 21 community problems most frequently volunteered by the leaders is of a social service nature. These leaders are not out to improve conditions at the county hospital, for instance, even though grand juries have found unsatisfactory conditions there.

Are there significant differences among the interests of the various movements? Analysis of variance of AVL scores, arranged by movement, reveals no statistically significant differences among movements. Though there are differences in mean scores between movements, there is also considerable variability within movements.

READING HABITS OF LEADERS

Most leaders reported reading some mass circulation news magazine (*Time*, *Newsweek*, or *U.S. News*). This was least true of the pro-annexation groups, however. A much smaller proportion of the leaders reported reading a liberal news magazine such as *The Reporter*, *The Nation*, or *The New Republic*. The groups that were less likely to read a mass circulation magazine were more likely to read a liberal news magazine, but the proportion reading a liberal magazine did not exceed

50 per cent in any of the movements. *Business Week, Wall Street Journal, Fortune,* and other business publications were read most frequently by the pro-annexation group in Arden-Arcade and the general community leaders, and least frequently by both Hagginwood groups. More people from the anti-annexation group in Arden-Arcade and the general community leader groups read intellectual magazines such as *Harpers* or *The Saturday Review,* and fewer from the anti-annexation group in Hagginwood and the X group. Individuals in all groups reported reading some out-of-city newspapers, but the largest proportions were among the general community leaders and the pro-incorporation group. Few in either Hagginwood group read out-of-town papers.

The leaders as a group seem well informed. Fifty-eight per cent read four or more periodicals regularly. Extensive reading is particularly characteristic of the general community leaders.

INTEREST IN COMMUNITY ACTIVITIES

To find out their interest in various community activities, we asked leaders to grade a list of various activities. Half of these were given mean ratings of "moderately interested" or higher, including political, redevelopment, industrial development, international relations, sports and recreational, banking and finance, and study and cultural. The respondents indicated little interest in fraternal activities and veterans' affairs.

The differences among the movements suggest that groups favoring reorganization of some kind have leaders who are more generally active and interested in a variety of community activities. The contrast between the two Hagginwood groups is pronounced. The general community leaders, who come primarily from the city of Sacramento, reiterate their strong interest in redevelopment while anti-annexation leaders in

Hagginwood and Arden-Arcade continue to show indifference to the problem.

VIEWS OF PROBLEMS IN SACRAMENTO

In the opening question in the personal interview, in which each leader was free to identify and describe whatever problems seemed most important to him, most of the leaders mentioned only a few problems. The interviewer took notes on these responses and the results were summarized and assigned, to the extent possible, to a list of categories previously prepared by the investigators. Table 2 indicates the percentage of interviewees selecting each of the 21 most frequently mentioned problems.

The relative lack of agreement on problems is evident in the interview results. The most frequently mentioned problem, multiplicity of special districts, is noted by only 28 per cent of the interviewees. In part, of course, this may be related to the small number of problems mentioned by the typical interviewee. Also, many of the problems have to do with governmental organization rather than with governmental function. For example, the problem perceived is that the Sacramento area has "too many special districts," not that certain functions are not performed at all or are performed inadequately or inefficiently by special districts. This tendency was most characteristic of the pro-annexation leaders in Arden-Arcade and least characteristic of the general community leaders.

After receiving the interviewee's free response answer to the question about problems facing Sacramento, the interviewer handed the respondent a previously prepared list of 23 problems. He was asked to rate each as Very Important, Moderately Important, Of Little or No Importance, or Don't Know. (See Table 4.) When the interviewees' responses to the prepared list of problems are analyzed, agreement among interviewees seems more apparent than suggested by the results in Table 2. The average rating on the first 18 problems in Table 4 is "moderately important" or higher.

TABLE 4
Leaders' Ratings of Degree of Importance of Area-wide Problems

Problem	*Mean Rating**
Long-range area-wide planning	3.8
Consolidation of governmental services	3.6
Freeways and arterials	3.6
Multiplicity of special districts	3.5
Sewage disposal and drainage	3.5
Streets and roads	3.5
Tax base	3.5
Zoning	3.5
Continuing community leadership	3.4
Industrial development	3.4
Tax rates	3.4
Mass transportation	3.3
Parks and recreation	3.3
Schools	3.3
Community spirit	3.1
Refuse disposal	3.1
Law enforcement	3.0
Public health	3.0
Core area development	2.9
Redevelopment	2.9
Cultural activity	2.8
Fire protection	2.8
Racial minority problems	2.5

* This is the mean score assigned the problem by 96 interviewees where "very important" equals 4, "moderately important" equals 3, "of little or no importance" equals 2, and "don't know" equals 1.

Long-range area-wide planning, consolidation of governmental services, and multiplicity of special districts are among the top four problems on both lists. Streets and roads, tax base, tax rates, sewage disposal and drainage, industrial development, continuing community leadership, community spirit, and schools are among the top 15 on both lists. There is only one marked shift in the rank ordering of the problems: freeways and arterials moved up from twentieth place in Table 2 to a second-place tie in Table 4.

The authors noted several differences among the classes of leaders in their responses to the prepared list of problems. The pro-annexation group in Hagginwood-Del Paso Heights and the X group saw a greater number of area-wide problems, but the general community leaders and the anti-annexation group in Hagginwood-Del Paso Heights saw fewer. In fact, the most clear-cut difference between a pro- and anti-annexation group is seen in the two groups involved in the Hagginwood-Del Paso Heights annexation attempt. Most problems the pro-annexation group favors as most important, the anti-annexation group sees as less important. The pro- and anti-annexation groups in Arden-Arcade show more agreement, differing mainly over multiplicity of special districts and mass transportation, which the pro-annexation group regards as more important. Both the general community leaders and the X group emphasize freeways and arterials and redevelopment.

There are some inconsistencies between the responses of the different classes of leaders to a request for personal description of community problems and to the prepared list of problems. For instance, the anti-annexation leaders in Hagginwood-Del Paso Heights indicated more concern with long-range area-wide planning in their personal evaluation, but less on the prepared list of problems. These inconsistencies weaken the characterizations of the particular movement involved, but since different methods were used in obtaining the data, the results do not have identical meaning even when the results are consistent. As mentioned earlier, a leader may not think of a particular problem when asked to describe what he considers to be area problems, but he may give the same problem a very high rating when it is brought to his attention. Also, because of the small number of problems mentioned by individual interviewees in response to the first question (Table 2), these results are probably more subject to sampling fluctuations than the results obtained from grading a list of problems. In other words, if leaders were asked to answer both questions a second time, the differences between the two answers to the first question would probably be greater than those between answers to the second question. The results from the second question (the list of problems) should probably be given more weight for that reason.

LEADERS' MEMBERSHIP AND REFERENCE GROUPS

Since community organizations often strongly affect a person's attitudes and behavior, the Sacramento leaders were asked about the relative influence of various groups and organizations in the county. When asked which groups would be effective in securing voter approval of some county-wide referendum, the leaders most consistently ranked the *Sacramento Bee* as the organization of greatest influence in the county. Elected and appointed officials of the county and the city, the chambers of commerce, and other news media are also seen as important. In all, 13 groups received mean ratings equivalent to "fairly important" or higher. Veteran, religious, political, and ethnic and racial groups are not so important, and senior military officers at local military installations received the lowest rating (see Table 5).

SMAAC-MGC leaders see political influence resting in the business community, news media, and city officials and discount the influence of political parties. The pro-incorporation leaders see political influence in much the same way as the group of lead-

TABLE 5
Leaders' Views of Influential Groups in Community

Organization	*Mean Rating**
Sacramento Bee	3.8
Elected and appointed officials of Sacramento County	3.6
Chambers of commerce	3.5
Sacramento Union	3.5
Elected and appointed officials of the city of Sacramento	3.5
Television and radio	3.5
Neighborhood newspapers, such as the *San Juan Record*, or the *Suburban News-Shopper*	3.3
Large manufacturing companies, such as Campbell Soup Company or Aerojet-General Corporation	3.3
Government employee associations (city, county, and state)	3.2
Nonpartisan groups, such as the League of Women Voters	3.2
Labor unions, such as AFL-CIO	3.2
Professional societies, such as the County Bar Association or the County Medical Society	3.1
Large commercial enterprises such as Breuner's or Sears	3.0
Service organizations such as Rotary or Kiwanis	2.9
Smaller manufacturing or commercial companies	2.7
Veterans organizations such as the American Legion or VFW	2.5
Catholic religious groups	2.4
Protestant religious groups	2.4
Local Republican Party	2.3
Local Democratic Party	2.3
Jewish religious groups	2.3
Ethnic and racial groups such as the NAACP or the Japanese American Citizens League	2.2
Senior officers at local military installations	2.0

* Mean of ratings assigned where 4 equals very important; 3, fairly important; 2, not so important; and 1, not at all important.

ers as a whole. The pro- and anti-annexation groups in Hagginwood exhibit some sharp contrasts once again, especially in their views of the influence of the League of Women Voters, ethnic and racial groups, and city officials, all of whom are seen as more influential by the pro-annexation group. They agree, however, that trade associations, neighborhood newspapers, and religious groups are less influential, although leaders generally found them more important. Pro-annexation groups in both Hagginwood and Arden-Arcade see the League of Women Voters as more influential than the average leader does. The anti-annexation leaders in Arden-Arcade tend to discount the influence of several organizations that the leaders as a whole regard as quite powerful. The general community leaders, some of whom are executives of large commercial enterprises, see such businesses as more influential than the average leader does. Political parties were judged by the X group to have very little influence on county issues.

In another question, the leaders were asked to say which organizations helped them make up their minds on issues. Again the *Bee* is seen as the most influential organization. Six other sources are given as more than moderately helpful: business associates, city officials, county officials, personal friends, the Sacramento City-County Chamber of Commerce, and the *Sacramento Union*. Service clubs seem to be of little help in this respect.

The leaders themselves are, of course, members of organizations (Table 6). Two-thirds or more of them are members of welfare organizations such as United Crusade or American Red Cross and of the YMCA and other youth groups, while only one-third or fewer are members of employee groups, business associations, and veterans' groups. In view of the relatively low social service interests of the general community leaders, we may assume that their participation in welfare organizations stems from religious or economic interests rather than social service interests. Leaders from the suburban areas are relatively less active in welfare organizations.

TABLE 6

Leaders' Participation in Community Organizations

Organizations (Arranged in order of over-all degree of participation)	SMAAC *and* MGC	*New city incorporation (Pro)*	*Hagginwood*		*Arden-Arcade*		*General community leaders*	*X Group*	*Over-all*
			Pro	*Anti*	*Pro*	*Anti*			
Welfare organizations	88%	69%	50%	67%	55%	44%	100%	87%	73%
Youth groups	55	81	50	33	77	55	88	87	69
Chambers of commerce	67	62	25	50	67	33	75	100	63
School organizations	55	87	50	67	44	44	58	50	63
Improvement clubs	44	56	100	50	100	77	50	25	62
Religious organizations	66	56	50	33	11	44	88	62	52
Professional organizations	66	44	0	33	66	55	33	87	49
Republican Party	44	69	100	33	22	66	58	62	46
Civic organizations	66	50	25	33	55	22	42	50	45
Service clubs	67	44	25	50	11	11	67	67	42
Labor unions and employee groups	55	25	50	0	55	44	12	25	33
Veterans' groups	22	31	0	33	22	22	16	50	26
Business associations	22	25	25	16	11	22	58	12	25
Democratic Party	33	31	0	16	11	0	0	37	18

Most of the X group, general community leaders, SMAAC-MGC, pro-annexation in Arden-Arcade, and pro-incorporation leaders belong to a chamber of commerce, and several are active in them. One-third or fewer of the pro-annexation group in Hagginwood and the anti-annexation leaders in Arden-Arcade belong to, or are active in, the chambers. A large proportion of the pro-incorporation leaders are active in school organizations while a relatively small proportion of the Arden-Arcade leaders on both sides have participated in such organizations.

A fairly striking difference among the movements occurs with respect to membership in improvement clubs. One hundred per cent of both pro-annexation groups were at least moderately active in improvement clubs. Less than half of the X group and SMAAC-MGC leaders belong to such organizations. A major objective of most improvement clubs is an increase in number of and an improvement in quality of the municipal services rendered. To the extent that the claim is accepted that annexation is the most effective means of achieving this, annexation becomes important to the club members. Membership in improvement clubs is more common in suburban neighborhoods than within the city, where many of the X group, SMAAC-MGC members, and general community leaders reside.

Forty-six per cent of the leaders report membership or activity in the Republican Party. The pro-annexation group in Hagginwood is made up entirely of Republicans; on the other hand, only 22 per cent of the Arden-Arcade pro-annexation group are Republicans, while 66 per cent of the Arden-Arcade anti-annexation group are Democrats. Over-all, only a small proportion (18 per cent) of the leaders studied is Democratic in a county that normally votes Democratic. Most of the leaders in the present study are employed in professional and managerial occupations—occupational groups which are more likely to be Republican than Democratic. However, there is no reason for believing that the leaders—whether Republican or Democrat—participated in the governmental reorganization struggles for partisan reasons.

Somewhat less than half of the leaders belonged to civic organizations. The variation from movement to movement in membership in civic organizations was smaller than for most other types of organizations listed in Table 6. Such membership is most characteristic of the SMAAC-MGC leaders.

MOTIVES FOR SUPPORTING REORGANIZATION

We wanted to learn what reasons leaders thought led to people joining one of the reorganization movements. As part of the personal interview, the leaders were asked to consider

TABLE 7

Motives for Joining Reorganization Movements As Seen by 96 Leaders

Motive	*Per cent**
Desire for more and better governmental services	63
A strong, continuing interest in achieving the most effective form of local government	59
Desire for lower taxes	45
Ownership of land or property which would be affected by reorganization	32
Occupation affected by reorganization	29
Desire for status and prestige	26
Desire for power and influence that could be exerted	23
Protection of a business which would be affected by reorganization	23
Attraction for opportunities for social interaction that the movement provided	22
A stepping stone to political office	18

* Percentage reported is the mean percentage given under the "very important" category.

those persons they knew who were active in one or more of the movements and, from a list of possible reasons, to give the proportion for whom they felt each was an important motivation. Table 7 shows that the first five reasons given by the 96 leaders for participation in the reorganization movements included desire for more and better governmental services, a continuing interest in achieving a more effective form of local government, desire for lower taxes, concern over property values, and concern over job security. Only a small proportion of activists, in the opinion of the interviewees, joined the movements as a means of gaining political office.

LEADERS' EXPECTATIONS FOR METROPOLITAN GOVERNMENT

All interviewees were asked to give their preference from a list of alternatives for metropolitan government reorganization (Table 8). They were then asked which alternative they felt

TABLE 8

The Future of Governmental Reorganization In Sacramento—The Leaders' Views

Reorganization Plan	*Preference*	*Expectation*
	Per Cent	
City-county merger	40	5
Large-scale annexation combined with consolidation of as many city-county functions as possible (MGC report)	18	20
City-county merger excluding rural area (PAS without the rural)	11	10
One large incorporation of the North Area and additional annexation to existing cities	10	2
Large-scale annexation to city of Sacramento	7	19
Other*	10	19
Status quo	4	25

*No suggestions were made to interviewees as to what "other" might include, but interviewees may have had in mind changes such as several small incorporations.

was likely to be put into effect in the next three to five years. The alternative preferred by most people (city-county merger) is felt to be among the ones least likely to be in effect in the near future and the least preferred alternative (the *status quo*) is felt to be the most likely to continue to exist. A substantial proportion of the leaders feel that large-scale annexation with integration of many city-county functions (MGC report) is both desirable and likely to be adopted. This opinion supports those who see large-scale annexation to the city of Sacramento as a probable solution.

EVALUATIONS AND CONCLUSIONS

There is little agreement among selected judges in identifying the key leaders of the area. Only 28 persons were nominated by three or more of 18 judges, only one by as many as eight. Certainly, in this light Sacramento does not have the kind of power elite that Floyd Hunter found in Regional City (Atlanta). Some interviewees felt that the concentration of power in Sacramento had decreased with the rapid population growth since World War II and the decline of local ownership of businesses. Leaders named by three or more judges come predominately from the business community in the city of Sacramento. They were not active in the reorganization movements. They are satisfied with city and county government and although many prefer merger or large-scale annexation they do not regard governmental reorganization as an urgent necessity. Traffic and parking problems seem more important. They are aware of the conflicting opinions over proposed changes and some wish to avoid taking a stand on controversial matters. Although the state government is a major employer in the area, state officials were not named as key leaders.

With the single exception of the publishers of the *Sacramento Bee*, the most influential Sacramento leaders did not initiate any of the reorganization movements. Only nine of Sacramento's 28 most influential leaders (nominated by three or more judges) were known to have been active in any of the movements. Several of these were elected or appointed public officials who participated by virtue of their official positions. The movement leaders were not well known throughout the metropolitan area. They were drawn from the upper middle class, and while professionally able and well educated, they had only moderate community-wide influence.

The three most important problems facing Sacramento chosen by the leaders—long-range area-wide planning, consolidation of governmental services, and multiplicity of special districts—reflect their own strong theoretical bent but are unlikely to find as ready acceptance among the voters at large. There was less, though still substantial, agreement among the leaders interviewed on the importance of streets and roads, tax base, tax rates, sewage disposal and drainage, industrial development, continuing community leadership, community spirit, and schools. Except for the streets and schools, these issues probably do not deeply concern the average voter.

In the areas concerned in the annexation and incorporation proposals, a high proportion of home owners paid their taxes along with monthly mortgage payments in one lump sum. As long as the monthly payments seem reasonable, many taxpayers do not worry about taxes. They may be aware that several homes in the area have septic tanks, but as long as their own plumbing works they are not greatly concerned.

Many citizens fail to see a relationship between industrial development and their personal welfare. They have little direct knowledge of industrial development efforts. Often they assume that such efforts will bring additional "industrial" sections to the community. Such views may be particularly common among Sacramento's high proportion of government workers.

There seems to be little support for the notion that Sacramento leaders, much less Sacramento voter-citizens, recognized a common set of community-wide problems requiring new governmental forms for adequate solution. To be sure, several leaders saw urgent problems within the metropolitan area, but they had difficulty winning community-wide support for doing something about them. The general community leaders were concerned with traffic and parking, but naturally this did not seem a pressing problem to the suburban leaders. The multiplicity of special districts that seemed such a critical problem to the pro-annexation leaders seemed much less important to the general community and the anti-annexation leaders. While there were differences among leaders in different movements, there were also large variations within movements. Our data do not support the notion of a highly homogeneous group of leaders joining together to form a metropolitan government movement.

The failure of several attempts to change the governmental structure in the Sacramento metropolitan area may be partially explained by the lack of agreement among the politically influential organizations. The *Bee* has vigorously supported SMAAC, MGC, and the annexation proposals, and vigorously opposed the proposal to incorporate a new city. County officials have tended to support the *status quo*. Until recently, city officials have not vigorously supported annexation. When they have, as in the Arden-Arcade attempt, there is reason to believe that they have been opposed by at least some county officials. The Sacramento City-County Chamber of Commerce has taken official positions on several proposals but has not forcefully supported them. The *Union* has been much less concerned than the *Bee* with metropolitan governmental problems.

While there was substantial agreement among all leaders in identifying the most influential organizations, there were some meaningful differences among movements. Pro-annexation leaders tended to see city officials as politically more influential and personally helpful than anti-annexation and pro-incorporation leaders. The pro-incorporation leaders found suburban newspapers, which tended to support the movement, more helpful, and downtown newspapers, which opposed the movement, less helpful than did the average group. The general community leaders, some of whom were executives of large commercial enterprises, saw such organizations as more influential than did the average leader. Leaders tended to see area-wide political influence residing in those organizations which they personally found helpful in deciding about political issues and in organizations that supported their particular movement. This attitude is another reflection of the lack of organizations as well as individuals that are widely acknowledged as influential forces in the community.

The bulk of our interview and questionnaire data suggests that a conscious desire to obtain more and better governmental services through a more effective form of local government was the most important surface inducement for action. For the most part, those who sought change were public spirited citizens dissatisfied with the level and efficiency of governmental services. Those who opposed change were somewhat more likely to be concerned with possible tax increases, but they were also likely to feel that the proposed change was a poor means to a desirable goal.

It seems clear that once a person has taken part in one such movement he is more likely to participate in a second one. There is clear continuity of concern on the part of the most active leaders. Originally they were asked to participate by a friend or an acquaintance. Frequently they became acquainted in the first place through participation in fund-raising drives, youth groups, or neighborhood improvement clubs.

Only a few activists were motivated by immediate economic vested interests. A small but active minority clearly was motivated by threats to jobs or positions. Noncity firemen, sheriffs' deputies, school administrators, and special district employees played important though usually supporting roles. There were probably cases where participation was motivated by anticipation of a new job or position as well as by a threat to one presently held, but this seemed to be much less common. This probably is related to the relative uncertainty of job reward and the relative certainty of job threat.

Also, there is evidence that proposed changes directly affecting property values lead to activity on the part of the property owner. For example, the Arden Fair property owners were concerned with the possible extension of the city transit lines to their shopping area in the event the Arden-Arcade annexation passed.

If we accept the modern view that leadership is a relationship involving the characteristics of the leader; the attitudes, needs, and other personal characteristics of the followers; and the social, economic, and political situation in which the movement takes place, we can begin to see why none of the reform movements succeeded.[4] The movement leaders lacked community-wide influence and did not have the resources to carry out a sustained campaign. Sacramento does not have a power elite that is able to dominate metropolitan governmental decisions. The followers (the voters) are not greatly concerned with metropolitan governmental problems. The most serious problems lie in the future and are not yet readily apparent. The issues are complex and require careful study, which the average voter is unwilling to give.

No group now has the legitimate status and the necessary resources to study problems of the entire metropolitan area.

The functions of the city and county governments overlap and the personal interests of the officials do not always encourage them to take a broad community-wide view. In addition, state laws and constitutional provisions make city-county merger difficult to achieve. As a consequence, the citizens resort to the two remaining options—annexation and the establishment of special districts—when a problem becomes acute. Such changes, taking place at regular intervals during the past several years, have resulted in a reasonably adequate level of metropolitan services administered by a complicated and illogical system of overlapping jurisdictions.

7

Government in the Sacramento Area: Retrospect and Prospect

In the Sacramento area, what metropolitan-wide organizations are available for reorganization action? What are the shape and content of the leadership structure in the area? What alternative approaches are available to metropolitan Sacramentans when considering governmental change? Is there any common awareness, or sense of community, that extends throughout the Sacramento metropolitan area? These and related questions are explored further in this chapter, based largely upon the data, observations, and conclusions of previous chapters.

THE SACRAMENTO COMMUNITY

On the surface, it appears that governmental change should have a greater chance of success in Sacramento than in many other metropolitan areas in the United States. Sacramento has relatively few separately incorporated municipalities—only four since the merger of Sacramento and North Sacramento—with their resident loyalties and vested interests in preserving the *status quo*. Metropolitan Sacramento also is young, and therefore has not had decade after decade of building up encrusted governmental traditions. Still another reason is that professionalism, the professional frame of mind so congenial to reform, is a dominant aspect of the milieu. Yet, the final impetus to governmental integration, a sense of community, seems to be lacking.

"Community," of course, is a word of many meanings. Sociologists have speculated at length on the status of "community" in the Western world and some have deplored what they call the loss of community. Out of the considerable writings on the subject, however, one can extract an appropriate definition of community, which will serve as a guide in determining presence or absence of community in the Sacramento area. The bond that brings a group of people into a community is a set of commonly held beliefs regarding activities that presumably contribute to the general welfare of all. These beliefs may be political, social, cultural, religious, or economic, but they must be held widely enough to create a feeling of group identity, purpose, and destiny—a sense of community.

It is clear from the preceding chapters that in postwar Sacramento County there is little agreement on the best form of government or even on the need for some different form of government for the area. Even under the very doubtful assumption that state legal obstacles to city-county merger can be surmounted, the incorporated communities, including the city of Sacramento, would be unlikely to give up their identities.[1] No plan like that of Dade County in Florida, which would create some super-metro agency while leaving existing cities intact, has been proposed.

THE BONDS OF COMMUNITY IN SACRAMENTO

Superficially at least, characteristics other than geography and traditional loyalties seem to provide a potential agent for building greater community cohesion. There are social, cultural, religious, and economic ties within the metropolitan area. The city of Sacramento remains the primary cultural and social center for the entire area. Suburbanites as well as city-dwellers take advantage of its art galleries; its theater groups; its symphony orchestra; its summer music circus; its civic auditorium; its stadium, which is connected with the city unified school district; and its zoo. Very few such facilities are available elsewhere in the county. Traveling road shows, operas, lectures, art shows, and musical events come to the city of Sacramento if they come to the urban area at all. In addition, some suburban citizens often commute considerable distances to the city's churches, parks, and movie houses. A great deal of communication among the many subcommunities in the county comes through the use of churches, colleges, and recreational facilities.

But these cultural and social ties to the city are not sufficient to create a feeling of community. Whether they can become so is questionable. At best the attachment is a loose one and represents that of only a limited segment of the county population. Even this segment is as likely to turn to the San Francisco Bay area, 90 miles away, for entertainment and cultural advantages. Further, an almost endless variety of recreational opportunities at lakes, mountains, and the seashore is readily available outside the metro area.

In some communities one religion predominates and unites,

but in the Sacramento area, as in most communities, no one of the many denominations and sects embraces a majority of the population. Catholics are the largest single group. A recent poll indicated that more than 50 per cent of the people in the area belonged to no church.

Some nascent stirrings of common interest appear in the economic sphere. Of a county working population of 195,000, as of January 1, 1961, there were 60,000 government employees, one for every two and one-quarter persons employed in non-governmental work. Aerojet-General Corporation, which employs an additional 15,000 (this total has been decreasing in 1965 because of a reduced volume of federal government contracts), is wholly dependent on federal government contracts. If, in addition to this, we consider all of the jobs in the county which are supported by governmental activity (construction, supplies, equipment, services, food service, transportation, and so on), it is clear that metropolitan Sacramento is predominantly a government-subsidized area. Families who are supported by government and government-related enterprises are scattered widely throughout the whole area, and they may have many common ideas regarding the kind of community that would best serve them. Whether they can be welded into an active force giving the community a conscious sense of direction is an open question.

At present this large group is not agreed on the political organization of the community. Likewise there seem to be few commonly held views regarding the economic destiny of the area. This disparity could be expected. There is little need for close communication and coordination among most governmental agencies in carrying out their assignments. Employees are often highly specialized technical and professional workers who may share certain values with others in their field but these are not necessarily values relating to local government, services, and ways of life. The division among them over the desirability of attracting new industry to the metro area is typical. Some government employees feel that a great deal of industry will change the character of the population and would interfere with the existing basically sound and secure government-based economy. On the other hand, some government employees and other citizens agree with the business view that industry would widen markets and provide a broader tax base. This division of opinion has strengthened the primary forces that have barred new industry from locating in the metropolitan area.

Whatever the potential, even among government employees a "community" spirit does not now exist in Sacramento. The PAS found that,

> Evident in every aspect of life in the area is the fact that the county of Sacramento is one community. It has a common history, common social and economic interests, similar political characteristics, and a similar set of long-range goals. The people of the area pride themselves on being "Sacramentans," not because they live in a county with this name, but because they are proud of an historic capital city that has carried forward in the twentieth century the spirit and quality that marked the latter part of the nineteenth.[2]

No evidence was given to support this bold assertion. Merely stating that "Sacramento is one community" does not make it so, and the fate of the PAS proposal is certainly evidence to the contrary. Our interviewees repeatedly called attention to the antagonism between the people in the North Area and those of the city of Sacramento. Smaller cities emphasized their desire to be left out of any large-scale governmental reorganization. Citizens in Sacramento feared higher taxes would follow large-scale annexation. Almost without exception our interviewees deplored the apathy and lack of interest toward metropolitan area problems, but this atmosphere was accepted as unchangeable.

Sacramento County then cannot confidently be called a community, even though, by census definition, it may be a metropolitan area. The latter, an arbitrary statistical definition, has nothing to do with the final determinants of community action—the attitudes, beliefs, and motivations of real people. Creation of a "community" rests on the ability to gain acceptance of certain common goals or aspirations as desirable. Ultimately the recognition of serious, area-wide problems might create bonds of community that would lead to such acceptance of goals.

A SEARCH FOR MEANINGFUL PROBLEMS

The population of metropolitan Sacramento has almost doubled in ten years. But no one problem or group of problems seems to be serious enough or urgent enough to provoke general discussion among the citizens. This is evident from the results of the interviews, in which not even one-third of a representative group of leaders and residents could agree on important problems common to the area. The most common problems spontaneously offered by leaders and residents were the multiplicity of special districts and the need for long-range area-wide planning, but these were selected by less than 30 per cent of those asked. The need for governmental reorganization and the need for consolidation of governmental services were mentioned by 20 per cent and 16 per cent of the interviewees respectively.

Are these truly common problems? Are they meaningful to the voter-citizen, who must ultimately give consent to new governmental forms designed to solve them? "Fractionated government," a poor term at best, is no more of a problem to the resident of the city of Sacramento than is west-end development of the city to the resident of the North Area. To be sure, both of these are the problems of someone in the metropolitan area, but so far, pleas that these be considered as community-wide problems have fallen on deaf ears.

Some of our interviewees blamed the rejection of SMAAC-MGC proposals, as well as others, on the committees' failure to pinpoint actual problems, to give them a sense of urgency, to use pocketbook appeal. Converting problems into tax losses and gains is insufficient, however. Special districts are established because certain areas need particular services and the average citizen, who sees problems in concrete terms, is not impressed by the argument that elimination of those districts is in the interest of the entire community. Taxes and costs of government are important, but citizens must first agree on what their problems are before favorable community-wide action toward solutions—which, as a side effect, might result in cost savings—is obtained.

Most of the leaders of reorganization movements insisted that problems were real enough but that the public was apathetic. While this may be a rationalization for their failure, the assertion is true in the sense that citizens are apathetic because they do not sense a need for change. As we have seen, some citizens—namely, those in the Arden-Arcade area—had opportunities to incorporate as a part of the large new city and to annex to Sacramento. A few of them have considered incorporation as a separate small community. To date, each of these attempts has ended in failure. The alternatives, which have been proposed and not yet brought to vote, afford little hope for success. City-county merger (SMAAC report) has been mentioned often by citizens in the northeast area as the solution most favored, but the objectives of such a movement would probably appear no more immediate and desirable than those of other movements.

Two other reasons have been given by the leaders of reorganization for the public apathy. The first, a common interpretation, is that most people already belong to so many social and recreational groups and have so committed their time that they have little left for local public affairs. The church organizations, PTA, boy and girl scouts, campfire girls, little leagues, bowling leagues, boat clubs, ski clubs, neighborhood improvement clubs, bridge clubs, and tennis and swimming clubs have more immediate importance to most heads of families than participation in a movement to do something about metropolitan government. The general citizenry do not look ahead but continue to vote for the *status quo,* confident that the future will take care of itself somehow. The citizen feels helpless to do anything about government anyway. It is too complex, too far removed from his immediate influence. One interviewee cited, as an example of this attitude, the lack of any complaint in the newspapers or to county officials when a county road rapidly deteriorated soon after its completion.

Upon analysis, however, these two explanations of citizen apathy lose much of their force. They do not explain why the busy people who belong to many organizations did not give money to the movements. In addition, if a situation is too complex for the ordinary citizen, he may support a group of leaders who appear to understand the complexities, but the organizations described here did not win this trust. In short, satisfaction with the *status quo,* not apathy, more accurately characterizes prevailing attitudes. Unplanned growth has not been disastrous in any material sense to the ordinary citizen; few obvious severe problems have developed during the postwar years.

Some students of metro affairs would insist that area-wide planning now would not only provide a much pleasanter way of life but could prevent many of the incipient problems from becoming serious. The jerry-built governmental service structures are sure to become severely strained as streets, sewers, and water systems wear out and new housing tracts become slums. This may be, but this is prevention of future problems, not identification of immediate and severe problems. This type of trouble-shooting is wise but does not stir up the citizenry to join together in a united movement. For the time being, most metropolitan residents seem to believe that they are receiving adequate municipal-type services. The special district device provides avenues for those who want more or better services, and it avoids the uncertainties involved in major governmental reorganization.

The active county government probably makes annexation and incorporation, or even a total reorganization of the metropolitan governments, seem much less urgent or appealing. And the existing, although partially disguised, rivalry between the county and the city government presents an obstacle to the solution of area-wide problems, even if they were to be widely recognized.

VEHICLES OF COMMUNITY ORGANIZATION

If critical problems were to become manifest, what type of community organization could be made available for area-wide action? Where might the leadership that seems to have been absent in reorganization efforts thus far be found?

Certainly the voluntary organizations established for the purposes of achieving mass annexations and large incorporation have been unsuccessful in mobilizing community opinion and action. None of the voluntary groups was well organized for accomplishing the ends sought. While the original committee for incorporating the big North Area city did a good job of studying the problems for that area, the action organization was not effective. The only successful efforts during the two major annexation attempts were ones to obtain the signatures necessary to put the proposals on the ballot. Arden-Arcade people placed great reliance on informal contacts made through Democratic Party workers. But in both cases organizations for obtaining community approval were ineffective and consisted of but a handful of people. All of these organizations were *ad hoc,* because no existing organization had interests coterminous with the geographic area involved.

The effective opposition which defeated the Hagginwood-Del Paso Heights annexation, the large incorporation attempt, and the SMAAC and MGC plans was centered in one or more of the large, powerful, existing institutions in the area—namely, the city and county governments, the City-County Chamber of Commerce, the *Bee,* and in one case the school districts. Support by one of these institutions for a specific governmental change does not guarantee its success, but opposition from a major or even a minor institution appears to bring certain defeat to a proposal. Major institutions are already mobilized for action and have the advantage of being well financed, well organized, and having experienced leadership. The leaders of the incorporation movement stated that they were unable to respond to the opposition tactics of the *Bee* and city officials. They did not have access to communications media, and they lacked positions of authority and influence. The opponents of incorporation, without really organizing, reached the people by crying "higher taxes," "protect our schools," "city domination," "double taxation," and so on.

The *ad hoc* citizens' groups that were formed to oppose each movement were led by people with roughly the same leadership prestige, experience, and capabilities as those possessed by the supporters. The opposition to the Arden-Arcade annexation probably had more money at its disposal than the supporters, but this was not true of those opposing large-scale incorporation. Unquestionably, those opposing the Hagginwood-Del

Paso Heights annexation had at their disposal more money and other resources than did the proponents. Emphasis on single, simple issues was the outstanding characteristic of the opposition—in both the major community institutions and the citizens groups. Their opposition was essentially negative; they did not need to make alternative proposals.

While there are no effective organizations for achieving specific governmental changes, are there any for studying and proposing action on metropolitan area problems? The Sacramento Area Planning Association, a volunteer group established in 1953, was composed of a small number of citizens who were challenged by a problem this study found most seriously regarded—the need for long-range area-wide planning. The association never became well known in the community, did not agree on concrete problems, and never attempted to take area-wide action. The Urban Government Committee may have been established under stimulation from SAPA, and the *Bee* studies and articles on governmental deficiencies in the unincorporated North Area may also have grown out of its work, but we found no evidence to support this interpretation. The association, which was disbanded in early 1960, was deterred from effective action primarily because its membership was not geographically or occupationally representative of the metropolitan area.

The *Sacramento Bee,* the area's largest and most powerful newspaper, was the first organization to study the unincorporated fringe around the city of Sacramento and to analyze the problems of the mushrooming suburban growth, and its work led directly to the creation of the Sacramento Metropolitan Area Advisory Committee. The paper specifically suggested that a commission be appointed to study the "crisis" that these articles had revealed. The Urban Government Committee of the Greater North Area Chamber of Commerce was also studying the problem—primarily from the viewpoint of the North Area—at the time of the *Bee* articles, but the public was not aware of its work, nor were local government officials influenced by it. Neither the Greater North Area Chamber of Commerce nor its Urban Government Committee is in existence today.

The SMAAC membership consisted entirely of persons selected by the governments in the county. SMAAC employed outside staff (Public Administration Service) to conduct the study, and SMAAC members did not participate in it. The subsequent report proposed a form of government virtually impossible to achieve because of constitutional provisions and its political unacceptability to sponsoring governments.

The Metropolitan Government Committee, which followed the demise of SMAAC, was appointed by essentially the same governmental bodies. By proposing mass annexation and recommending consolidation of certain functions, the report of MGC favored both the city and the county. Mass annexation has not taken place, however, and functional consolidation is still in the discussion stage. SMAAC and MGC recommendations have not found vigorous support from citizen groups or agencies backed up by widespread public support. Except for the *Bee*'s success in getting SMAAC started, all of these attempts to identify the area-wide problems and to mobilize public opinion and action have failed.

Sacramento, like many other communities, is overorganized. It has a plethora of professional and technical organizations with interests focused on specialized matters. Various health and welfare fund-raising organizations, California State Employees Association, United Crusade, service clubs, school boards, parent-teacher associations, and the legal, medical, and innumerable other professional associations all have active support. These competing specialized organizations impede creation and continuance of organizations with general metropolitan interests. There is also conflict and jealousy between the city of Sacramento and the giant, unorganized North Area and, paralleling this, rivalry between the city and the county. These opposing forces are probably the most serious impediment of all, because the proposal for one big city, which the powerful *Bee* and the City-County Chamber of Commerce have favored from the beginning, cannot even be studied objectively in a mutually competitive atmosphere. The representatives of these rival factions would come to an area-wide study and action group with biases and preconceived ideas of how their faction ought to fare.

Our interviewees' assertion that the *Bee* "pushed," "forced," or otherwise caused the city and county governments to act is significant. No comparable vehicle existed for initiating study of these problems, calling them to the attention of the people, and getting action under way. The newspaper was successful once, but the results of the action it inspired suggest that it did not truly represent community sentiment and generally recognized need.

No governmental agency has over-all responsibility to look after the problems of the area, and no private organization has assumed this responsibility. Who should do it? The City-County Chamber of Commerce, the League of Women Voters, political parties, a federation of service clubs, a permanent research organization, a new watchdog civic organization, or some combination of all of these might succeed, but plainly none of these is now effective. The civic graveyard is being filled with the remains of organizations that failed in the task.

THE MISSING LEADERS

Two important conclusions emerging from this study are that the traditional sources of leadership which has prestige and authority now produce few leaders in the Sacramento area, and that, for government reorganization, the metropolitan area is devoid of recognized, effective area-wide leadership.

Traditionally the primary leaders for most community activity have been the captains of business, finance, and industry, but these people have, with rare exceptions, refrained from taking a public position on metropolitan and local governmental reorganization issues in Sacramento. Study of the rosters of governmental reform ventures underlines this fact. The American community is so constituted that only business leaders have sustained power. When these individuals do not assume their traditional role and responsibility, a vacuum in the leadership of the community is created.

Why have business leaders failed to take up community leadership tasks? A partial answer lies in the changing business environment in Sacramento. Today the number of important, locally-owned firms is decreasing. The ones that do exist have little influence in the suburbs, where business is dominated by the corporate chain store. The growth of the chain store reveals another impediment to effective community leadership by businessmen. The local manager of a chain store, usually but one of many employees who hold similar positions in the company, competes within the company for a more important position. His life and welfare are wrapped up in the company. He is the modern organization man—more often than not, a junior-level executive. His participation in noncontroversial local fund-raising or cultural organizations is encouraged by the parent company to demonstrate his concern with local affairs. But he is definitely and firmly discouraged from making public statements or acting in controversial matters that might alienate a client or customer. At one time the local owner of a company or bank would run for public office or speak out on political issues. Today a comparable man is a branch manager, who is insecure in his position, eager for advancement, and who will not actively participate in political affairs or encourage his subordinates to do so. Aerojet-General Corporation in the Sacramento suburbs is an unusual exception to this corporate policy. Aerojet encourages its employees to be active in politics at local, state, and national levels and invites campaign speeches and party fund drives on company time and property. However, without some drastic reshifting of values, most of the local managers for large corporations, whose potential for local leadership is high, are lost to governmental affairs.[3]

In the traditional community a few political or labor leaders frequently supplement the leadership by business. However, both the county and the city of Sacramento have council-manager governments, which, while administratively desirable, do not provide strong political leadership. The managers, organization men themselves, are employed to carry out policy, not to initiate programs. Their bosses—councilmen or supervisors—usually act as a body. With few exceptions, county supervisors have not lent their support to any basic changes in the existing community structure. They pledge support to what the people want. This of course they must do, but as political leaders they can and ought to identify issues, point directions, and educate—as well as represent—the public. The city council and county board also lack individual political leaders. The group action that has been taken has naturally been in the interest of the established governmental structures.

Local state legislators, who more than any other political officials have positions of prestige and influence, have refrained from taking public stands on issues of community organization. Their motivation is undoubtedly the same as that of the private companies—they do not wish to alienate constituents. When the contending factions in the county can agree, the legislators are willing to support a program in the legislature, but they carefully avoid involvement in divisive issues of community organization. Their judgment is politically sound and is supported by the local political parties and political clubs, which do not encourage state legislative candidates to take stands on local issues.

The noninvolvement of political party organizations also conforms to the desires of most of the persons who were active in governmental reorganization movements. With few exceptions, the leaders felt that partisan politics had no place in local affairs. All local elections in California are nonpartisan and the *Bee* has long opposed partisan politics at the local level. Regardless of the reasons for inaction by experienced politicians and the political parties, groups seeking to make basic changes in the governmental organization are seriously hampered by lack of support from this part of the metropolitan community.

Labor leadership in community affairs in the Sacramento area is also practically nonexistent. Organized labor is not as strong as in most communities of this size because so many are employed by the government. Both the labor unions and their leaders have exercised little local political power.

One final traditional source of leadership is lacking in the Sacramento area. Although many persons are well-to-do, Sacramento does not have families with great fortunes, families that might devote a great deal of time to community affairs. The average personal income in the county is high, but the great majority of Sacramentans work for a living. There is no aristocracy of wealth which feels responsible for the welfare of the community.

If the traditional sources of personal community leadership have dried up or are nonexistent, what about the availability of alternative sources? Most of the other potentially strong leaders either do not have a stake in the community or, for policy reasons, are barred from overt activity which may be controversial. The heads of the three major military installations by tradition and decree must stay out of politics. Other federal employees can operate only under the limitations of the Hatch Act, which usually has been strictly construed, although employees are still free to engage in nonpartisan local activities.

Nearly all of the elected heads of the executive branch of state government reside in the Sacramento area, yet these and other nontenure state officials do not play a local role. Their interests center on state affairs and their usually short tenure makes them transients in the immediate community. Local citizens tend to be critical rather than appreciative of a state official's comments on a purely community matter, and discourage even those few who might become active locally.

Under the state civil service act, a career employee has considerable freedom to engage in political activity. But the kind of leadership needed in the community apparently is not nurtured in great governmental bureaucracies. Perhaps the upper-level career employees are so imbued with the tradition of civil service neutrality that they simply refrain from controversial public activity. Some are conscious of the negative attitudes of their superiors toward political involvement. Although a few have run for office, no state employee in Sacramento has served on a city council or the county board of supervisors in the postwar period, and none has been elected to the state legislature or to national office. State employees offer a good potential, but experience indicates that governmental officials or employees are not a fruitful reservoir of talent for local government matters.

PRESENT LEADERSHIP

Without exception, the persons who have led annexation and incorporation movements come from intermediate positions of prestige and influence. Most are professionally able persons with educational qualifications well above those of the traditional leader. They are idealistic. They have often had some experience in improvement clubs, political clubs, park districts, or local chambers of commerce. A few were state and federal employees. No person active in favor of one of the reorganization attempts was previously well known over the entire metropolitan area of Sacramento County, or even within the area for which change was sought. No one of them would have been mentioned as a county leader—in politics, business, or governmental affairs.

As actors in the reorganization attempts, each of these men was able to command the personal loyalty of only a handful of supporters. Their immediate followers rarely attracted more than a few helpers in petition and fund-raising attempts. The primary support for both the Hagginwood-Del Paso Heights and the Arden-Arcade annexation attempts came from a few persons associated with improvement clubs. Yet few members in these clubs were active in support of a movement. The incorporation leaders were not closely associated with any established organization, and the chambers of commerce, to which some of them belonged and in which they had held official positions, failed to respond to their call. None of the leaders was able to stimulate the business community to participate. Only the public relations officer from one large shopping center, the assistant manager of another, and a junior executive of a corporation ever joined one of the movements (incorporation); their company owners took no official position.

The conclusion that Sacramento lacks community-wide leadership is supported by the interviews with leading citizens. Most of the citizen-judges chosen had trouble naming ten community-wide leaders. Some, while readily admitting difficulty in identifying current community leaders, could recall names of leaders of 25 years ago—that is, the city of Sacramento's past generation of leaders. These were the leaders of an era when most of Sacramento County's population resided within the city of Sacramento and there was no suburban problem. Only one man was named as many as eight times in the replies of 18 judges. He is the recently retired Sacramento city manager, whose influence and power extended little beyond the city boundaries and at most to less than half the metropolitan population. As a professional manager he could not be said to wield the kind of political power with which we are here concerned and he did not presume to provide active leadership on controversial public issues. His power was properly that of an administrator, not a community innovator.

To obtain a list of ten community leaders we found it necessary to include people who had been mentioned only four times by the judges. Most of the list was composed of city dwellers, none of whom could be considered area-wide leaders; some of these frankly admitted in interviews that they did not consider themselves area-wide leaders. At the same time, most of them were unable readily to identify others who they believed were recognized community leaders. The interviewees expressed the same dismaying conclusion expressed above—that the potential leaders are not vitally interested in metropolitan area problems, are engrossed in their own affairs, and see no urgency in the so-called "crisis of growth."

POTENTIAL LEADERSHIP

Numerically the greatest potential source for leadership, able to tackle the tough and intricate problems of metropolitan growth, is the public employees. At the California capital a substantial percentage of the employees are well educated, capable, experienced in dealing with complex affairs, and able to establish and lead large-scale organizations. Public employees, although few in number, were more in evidence than representatives from any other groups on both sides of the mass annexation and incorporation movements. They were not, however, top administrators. But, regardless of their qualifications, government employees are politically impotent. However brilliant, they have rarely attained the community-wide status of the successful businessman or banker. Practically all urban Americans hold in highest esteem those who are most able to utilize for private purposes the economic resources of the community. The public servant's leadership status is low. The taxpayer-citizen resents his claim on the public pocketbook and looks upon him as a servant, not a leader. In this environment the career employee abstains from overt political action. He may be concerned with political affairs, but circumstances prevent him from seeking an important role.

The obstacles to providing civic leadership from high level state government employees are obviously great. A new image of public servants would have to be created—perhaps through radio, television, and other news media. The business community would have to help to build the prestige of this potential for community leadership that they, the traditional leaders, have been reluctant to assume. The governor, legislators, and department heads would have to permit their top echelon of employees to participate in local affairs, even controversial ones. Even if these external obstacles were surmounted, there are such additional, more subtle, problems as conflicts of interest and maintenance of neutrality of civil servants. Finally, the public employee himself would have to be willing to take on local problems. Merely listing some of the major obstacles to a change in status and role of public employees indicates that the chances of such a change are slim indeed.

School administrators and teachers, women's organizations, and minority groups are also unlikely sources of strong community leadership. Professional men and women—doctors, lawyers, clergy, college professors—have a prestige of sorts, but most are like the businessman, too concerned with the private world of "getting and spending," too afraid of damage to reputations, or too busy with their personal affairs and professional organizations to lend their energies to community-wide political activities.

AND WHAT OF THE FUTURE?

Metropolitan Sacramento has almost doubled in size in the last ten years and promises to redouble in the next 15. Future governmental changes are inevitable, and the rapid growth will surely hasten them. Will the changes tend to follow the basic pattern of the *status quo*—with minor incorporations here and there—or will there be a break with the present trend and development of substantial governmental consolidations, even of major governmental integration for the entire metropolitan area?

Some observers maintain that the present network of many special park, water, lighting, sewer, and other districts provides fairly adequate service and allows a greater number of citizens to participate directly as managers and board members. The voters may participate in and do exert real political influence on the establishment of the districts. They continue to affect policy through periodic elections of the boards. This personal involvement is impossible under total metropolitan government with its magnitude and complexities. Special districts and identifiable community entities broaden the base for democratic participation and provide a more intimate relationship between the citizen and his government.

Critics of the *status quo* emphasize the wastefulness in this arrangement and point to the wide variation in level of services. Under the special district system there has been no broad, long-range planning for the metropolitan area. They argue that "grass roots" democracy in district government is more illusion than reality. Certainly special districts, and to a limited extent separate cities, have been created without reference to any general plan for the total area. These developments are anathema to the well ordered mind and to those who see a need to plan and develop the resources of the total area for the common interest.

The benefits of truly metropolitan-wide planning have never been presented to the people, say some critics. The citizen knows what he has and seems fairly well satisfied in spite of the reorganizers' predictions of future problems in water, slums, crime, sewage, and streets. A more desirable physical and cultural way of urban life has never really been presented in concrete terms. The SMAAC-PAS report, emphasizing borough government and elimination of existing governmental structures, failed to go beyond the obvious problems of simplifying governmental structure or of providing some services more effectively. The average citizen is more inspired by an over-all plan for development of cultural centers, parks and open spaces, river-bank recreational facilities, and zoning regulations that insure more than the mere separation of residences from commerce and industry.

Another explanation for the failure of reorganization movements arises out of suggestions made during our interviews. While the officials and planners have studied desirable objectives for the community, the citizens have never participated in discussions of these objectives, nor have they been consulted on the results. The major studies of the area were done by government-sponsored agencies and by the newspaper staff, which prepared the articles on conditions in the unincorporated parts of the county. No study dealt with more than Sacramento

County, although parts of neighboring counties are clearly within the actual Sacramento metropolitan area.

These interviewees suggest establishing a "blue ribbon" committee made up of people of the highest prestige and respect, and broadly representative of community interests—business, labor, churches, parent-teacher associations, improvement clubs, service clubs. Thus, even if the members individually lacked community-wide prestige, all of the important interests of the community would be represented. The committee, which would be staffed to assure adequate research, communication, and publicity, would conduct a thorough inventory of the services and functions of the metropolitan area. It would function like a federal or state commission—that is, provide professional research supplemented by public hearings for fact-finding and determination of public attitudes. The results could then be compared with the best in other governmental systems—the best parks, the best water system, the best cultural facilities, the best health services, and so on. Superiorities as well as deficiencies in the Sacramento metropolitan area would become evident. The citizens' committee, in making positive recommendations, would have definite ideals in mind. The result of such a citizen effort might not conform to a planner's dream, but it would have the merit of providing standards against which the citizen could compare what he has now with what he wants and is willing to pay for.

Given the conditions found in Sacramento by this study, the prospect for any major change in the *status quo* is not bright. To date, at least, the people of the area do not think of themselves as a single community; no suitable community-wide organization for identifying issues and promoting action exists; there is rivalry between the city and the county, and between the city and the large North Area; area-wide leadership that could marshal cohesive sentiment around comprehensive metropolitan-wide objectives is lacking. In the meantime, since there are no obvious immediate problems in Sacramento, the general citizenry can hardly be blamed for seeming content with the *status quo.*

POSTSCRIPT

As of June 1, 1965, the Hagginwood-Del Paso Heights annexation (discussed in Chapter Four) has been virtually completed—not by one annexation as originally attempted but by a series of smaller annexations. These, plus numerous other annexations and the merger of North Sacramento with the major city, have increased the area of the city of Sacramento to 92.75 square miles. On January 1, 1961, the area was 50.25 square miles, as compared with 23.46 square miles on October 1, 1955, the approximate date of the *Bee*'s "crisis" articles. This growth can hardly be said to represent mass annexation as recommended by MGC, however.

Aside from the piecemeal but sizable growth of the city of Sacramento, reorganization of government in the metropolitan area has proceeded slowly. Agreement on consolidation of functions will result soon after 1965 when Sacramento County performs the property assessment and tax collection functions for all cities in the county and operates a new metropolitan airport as well as the former Sacramento Municipal Airport. The city of Sacramento seems destined to operate a consolidated city-county library. City and county prosecutor offices were consolidated in the county office. There have been inconclusive discussions regarding consolidating city and county animal pounds and the planning functions. The city-county consolidation proposal of SMAAC-PAS is rarely mentioned in the press. However, there is an obvious lessening of rivalry between city and county.

From time to time incorporation sentiment has sprung up in the heavily populated unincorporated suburbs. Residents of Carmichael, Rancho Cordova, Citrus Heights, Orangevale, and Fair Oaks have actively discussed incorporation. On the basis of these expressions of interest, the Metropolitan Development Committee of the Sacramento City-County Chamber of Commerce investigated the feasibility and economy of some proposed incorporations. The committee's report, issued in the summer of 1961, expressed serious reservations about the desirability of incorporation but urged the Sacramento city council and the county board of supervisors to reconsider MGC's proposals. The chamber report triggered the formation of a small voluntary committee of area businessmen associated with the chamber, called the Sacramento Citizens' Committee on Incorporation. This committee recommended that the city council annex a 115-square-mile area northeast of the city. The committee also suggested that the council select an "area-wide committee on better urban government." However, no specific request for annexation of this large area ever came before the Sacramento city council, and the area-wide committee has not materialized.

On June 2, 1964, the voters of the city of North Sacramento approved merger with the city of Sacramento by a narrow margin of 16 votes. This merger added 6.58 square miles and 13,219 people to the city of Sacramento, for a total population of 262,500. The merger was reported to be the first in California in a period of 28 years. A strong argument for the merger was that the city of Sacramento completely surrounded the smaller municipality.

One recent development may be of real significance for governmental reorganization—the establishment of the Sacramento Local Agency Formation Commission. This commission, and its authority, are new factors on the local governmental scene. A state statute, effective on September 20, 1963, provides that a local agency formation commission shall be established in every California county.[4] The law calls for a five-member commission consisting of two county officers appointed by the board of supervisors, two city officers and an alternate, and a public member. The city members and the alternate are appointed by a city selection committee of representatives of each city in the county, and the public member is appointed by the other regular members of the commission. The alternate serves whenever a proposal affecting the city of one of the regular city representatives is before the commission, the latter being disqualified. Sacramento County's Local Agency Formation Commission consists of two members of the county board of supervisors, the mayor of Sacramento, the mayor of Folsom, and a local lawyer. The alternate member is the mayor of Galt. The executive officer appointed by the commission is M. D. Tarshes, Sacramento County executive officer.

The commission's specific duty is to review and approve or disapprove proposals for the incorporation of cities, the creation of special districts, and the annexation of territory to existing cities and special districts. School districts are not included under this power. The more general charge to the commission is that it shall adopt standards and procedures for the evaluation of proposals for the creation of, and annexations to, cities and special districts. The statute stipulates certain factors that should be considered by the commission, including the number of people involved, assessed valuation of land, land area, topography, and drainage. The commission is also directed to study the need of the area for organized community services beyond those now being received, and the relationship of the proposal to mutual social and economic interests and to the local government structure of the county. The county board of supervisors is to furnish the commission with necessary supplies, equipment, and operating funds.

After the usual procedures of filing of documents of intentions and public hearings, the commission's decision on a reorganization proposal is final. This is the key to the commission's power. If the commission refuses to approve a proposal for governmental change, no action can be taken on it for at least one year. If the commission approves the proposal, the proceedings for annexation or incorporation follow in the usual manner.

It is too early to predict the influence of the commission on governmental organization. Its potential influence is limited by the fact that it cannot initiate proposals for change. The commission could, however, prevent the proliferation of special districts and small cities. Conceivably the commission's decisions could lead to further governmental integration in the metropolitan area. But, as with other organizations examined in this study, the commission can accomplish positive action only with the understanding and support of the citizens of the metropolitan area.

Appendix

RESEARCH METHODS

The study of the decision-making process in metropolitan affairs may be approached in two different ways. Both methods of approach were employed in this study. In one, the researchers seek to identify the individuals and organizations involved in particular decisions through case studies in depth. A well known and outstanding study of this type is *Politics, Planning, and the Public Interest,* in which Meyerson and Banfield examined a series of decisions on the location of public housing projects in Chicago.[1] Our surveys of the two efforts to study metropolitan problems in an organized manner and the three attempts to achieve specific governmental changes follow this general pattern.

In the other approach, the researcher attempts to identify the individuals with general influence in the community and to ascertain their attitudes and actions in relation to community issues. In the now classic study of this sort, *Community Power Structure,* Floyd Hunter sought to find the central decision makers for the entire community of Regional City (Atlanta).[2]

We were looking for the same group of persons in our investigation of the general community leadership in the Sacramento metropolitan area; our findings are presented in Chapter Six.

Three main sources of information were used for the depth studies (Chapters Two through Five): official records and documents, newspaper accounts, and personal interviews. The minutes of the meetings of SMAAC, MGC, and the Urban Government Committee of the Greater North Area Chamber of Commerce, the proponents of the big new city idea, and other groups were referred to extensively. Official reports were issued by Public Administration Service, SMAAC, MGC, and other groups and agencies, and these and other available official records and documents were collected and examined. This is true also of the propaganda literature issued by supporters and by opponents of specific proposals. Day-by-day clippings were made from the two daily newspapers and from the weekly suburban newspapers and shoppers' guides in Sacramento from the earliest stages of this research project on. We also made use of back issues of these newspapers.

The extensive personal interview program was the most important source of information for the depth studies. Three classes of persons were interviewed: those who had been closely identified with one or more of these movements; general community leaders—that is, persons who might be expected to be leaders in the community although they did not take part in the surveys and movements for change; and persons who had been involved in some manner with the study and action proposals without being closely identified with any particular one (referred to as the X group).

Having agreed upon the classes, we turned to the selection of the specific individuals to be interviewed. Eighteen "judges" were asked to name the ten "real leaders" in the metropolitan area and to list the most influential persons in the reorganization movements. (Chapter Six of this study gives details on the kinds of persons selected as "judges" and on their answers.) About one-third of the persons mentioned most frequently as the more powerful and influential people in the entire Sacramento area were known to have been active in a particular movement. The remaining two-thirds were placed in our general community leaders class, and 14 were interviewed.

We relied upon three sources for determining the activists in the movements: names given by the judges, names drawn from our own personal knowledge of the activities of various persons in the community, and names from a special card file of persons mentioned in newspaper accounts of governmental reform activities.

From 14 to 20 persons were interviewed from each of the attempted action programs and from SMAAC and MGC combined. In addition to our original 18 judges, 96 persons were interviewed. Table 3 in Chapter Six classifies the 96 interviewees as to the role in which they were interviewed. The response to requests for interviews was exceptionally good. Only three persons declined to cooperate. The interviews ran from two to four hours each.[3]

Several interview schedules were developed in order to provide a common structure for describing the movements and to assure a reasonable degree of uniformity. Interview Schedule A sought the views of all 96 interviewees on the local governmental picture in general, and their interest in and involvement with local government matters. All persons were also asked to judge the relative importance of several metropolitan area problems listed.

Persons closely connected with the specific study and action movements were asked questions from Interview Schedule B, which centered on the particular study or movement with which the interviewee had been involved. Who participated in the movement, what interests influenced the movement, how was it financed, what was right or wrong about it, and other similar questions were asked. The interviewees also were asked to grade a list of possible motives for becoming involved and a list of preferences for governmental arrangements in metropolitan Sacramento.

The general community leaders and the "X" group were asked about various efforts at change from a general and overall point of view (Interview Schedule C). These interviewees also were asked about the motives behind participation in these efforts and about their own desires for government in the Sacramento area.

Upon completion of each interview, a Leadership Questionnaire and a copy of the Allport-Vernon-Lindzey "Study of Values"[4] were left with the interviewee. He was asked to fill them in and return them to us with no personal identification in a prepared envelope. From a total of 96 interviewees, 75 per cent complied with this request. The Leadership Questionnaire sought detailed information about the interviewee's personal life—his occupation, family, religion, education, membership in organizations, and types of periodicals read.

As soon as possible after completing an interview the interviewer recorded (either by tape or by hand) the information and ideas obtained in the interview in considerable detail. The interviewer expanded the notes taken on the interview schedules during the course of the interview and these expanded notes were then transcribed on cards which served as a permanent record of the information, ideas, and allegations given in the interviews.

Initially we had planned that the sociologist in our team would provide a general survey of the demographic and sociological factors relevant to metropolitan Sacramento, along with specific items pertinent to each of the study and action movements under study by the political scientists. The economist was to prepare a similar analysis of economic characteristics. The psychologist was to conduct the study of community leadership. Later, when the minimum number of persons to be interviewed turned out to be so great, the economist and sociologist had to aid in conducting interviews and preparing a description and analysis of a particular action movement. Each interviewer conducted the study of a reorganization movement and wrote up his findings. Only the psychologist experienced the luxury of staying with his original assignment. Each member of the interdisciplinary research committee, however, carefully reviewed the completed study from the perspective of his own academic field.

Notes

Preface

1. For a more thorough discussion of the problems encountered in designing the research project, see Lyman A. Glenny, "The Problems of Project Design for the Study of Metro-Government Movements," *The Western Political Quarterly*, XII, No. 2 (June, 1959), 578.

Chapter 1

1. Allan Temko, "Sacramento's Second Gold Rush," *Architectural Forum* (October, 1960), 124.
2. *Ibid.*
3. The term "metropolitan area" in its common usage refers to a densely populated area of considerable size which socially and economically is one community, irrespective of governmental boundaries. Commonly, many separate local governments are contained within such an area. The U. S. Bureau of the Census defines a standard metropolitan area as a county or a group of contiguous counties containing at least one city of 50,000 inhabitants or more. Contiguous counties not having a 50,000 population city of their own are included only if they are essentially metropolitan in character and are socially and economically integrated with the large

city of the neighboring county. If any part of a county is considered part of a metropolitan area, then the entire county is so considered. U. S. Bureau of the Census, *Local Government In Standard Metropolitan Areas* (Washington, D. C.: U. S. Government Printing Office, 1957).

4. United States Census Reports.

5. "Legislators Cite Sacramento County in Report on Special Districts," *Sacramento Bee,* January 19, 1961.

6. Charter of the city of Sacramento, Art. III, Sec. 17.

7. State of California, Department of Employment, *Distribution of California's Civilian Employment by Major Industry and Area, 1946-1957,* Report 352A, No. 12 (November 25, 1958).

8. State of California, Department of Employment, *Community Labor Market Surveys—California, 1958* (unpaged).

9. Information on income supplied by Sacramento City-County Chamber of Commerce.

10. "North Area Sets 4 Way Fringe Study," *Sacramento Bee,* November 15, 1955.

11. *Problems of Local Government in the Greater North Area of Sacramento County,* an Interim Report of the Urban Government Study Committee of the Greater North Area Chamber of Commerce.

Chapter 2

1. The provision that SMAAC was to act in an advisory capacity to the governmental bodies later became the basis of argument as to whether the final report belonged to the survey agency employed by SMAAC, to SMAAC, to the city councils and the board of supervisors, or to the public.

2. The minutes for every meeting of SMAAC have been examined. Much of the information given in this chapter about the work and organization of SMAAC is based upon these minutes.

3. SMAAC minutes, March 6, 1956.

4. "Area Advisory Group, Caught in A Lull, Hunts A Job to Do," *Sacramento Union,* September 5, 1956.

5. "Metropolitan Area Study: First Interim Report," November 13, 1956.

6. "Metropolitan Area Study: Second Interim Report," February 19, 1957.

7. *The Government of Metropolitan Sacramento* (Public Administration Service, 1957), pp. 120-24.

8. SMAAC minutes, May 27, 1957.

9. "Amendment Is Called Key to Merger," *Sacramento Bee*, July 2, 1957.

10. Opinion of Edmund G. Brown, Attorney General; Paul M. Joseph, Deputy Attorney General. No. 57/140.

11. *Final Report by the Sacramento Metropolitan Area Advisory Committee on the Metropolitan Area Study*, September, 1957.

12. "Isleton Only County Chamber Opposed to Merger Proposal," October 18, 1957. The headline is misleading in that it conveys the idea that all others were in favor. Some had taken no stand.

13. "Solons Think Amendment to Permit City-County Merger is Long Way Off," *Sacramento Bee*, November 6, 1957.

14. Overheard by one of the authors of this study.

15. California Assembly Interim Committee on Municipal and County Government, *Transcript of Proceedings*, Sacramento, December 5, 1957.

16. "SMAAC Seeks Accord in Row with North City," *Sacramento Bee*, November 20, 1957.

17. "SMAAC Decides to Ask Sponsors for Extension," *Sacramento Bee*, December 3, 1957.

18. "City-County Favor New Merger Group," *Sacramento Union*, December 19, 1957; "Proposal for Merger Study Group Gets Okey," *Sacramento Bee*, December 19, 1957.

19. "Council Okehs Plan on New Area Group," *Sacramento Bee*, December 27, 1957.

20. "Cowan Says Small City Setup Will Weaken Metro Structure," *Sacramento Union*, January 10, 1958.

21. "SMAAC Has Last Meeting, End Nears," *Sacramento Bee*, January 14, 1958.

22. "Metropolitan Government Committee of 21 Organized," *Sacramento Bee*, February 20, 1958.

23. MGC minutes, March 12, 1958; "MGC Will Ask County-Cities' Planners Advice," *Sacramento Bee*, March 13, 1958.

24. "Interim Report Relating to Services and Costs of Municipal Government in the Northeast Area of Sacramento County, by Function and by Listed Services," typewritten and undated copy.

25. "Arden-Arcade Annex Backed by MGC Unit," *Sacramento*

Union, March 3, 1959. This annexation movement was begun December 18, 1958.

26. *Government Reorganization for Metropolitan Sacramento,* final report of the Metropolitan Government Committee, June, 1959.

27. "City Must Do Its Share to Justify Annexation," *Sacramento Bee,* August 20, 1959.

28. "Tarshes Weighs Possible Consolidation Methods for Area," *Sacramento Bee,* January 30, 1964.

Chapter 3

1. There is uncertainty as to the official name of the organization which sponsored the incorporation proposal. The file of minutes of all meetings from March 3 through June 20, 1958, is labeled "Citizens Committee for Incorporation." Until June 2, each set of minutes was labeled simply as Minutes of the Committee. The minutes for June 2, 9, and 18, and the "Planning for Tomorrow" pamphlet, used the title of "Interim Committee on Incorporation of Greater North-East Area."

2. Information based on 1960 United States census reports for 23 census tracts, which approximate the boundaries suggested for incorporation.

3. See Chapter Seven for analysis of this test.

4. "New City in North Area Is Prospect," *Sacramento Bee,* January 31, 1958.

5. *Carmichael Courier,* July 3, 1958.

6. "Northeast City of 150,000 Is Proposed," *Sacramento Bee,* June 18, 1958; "Move Launched to Form Big North Area City," *Sacramento Union,* June 18, 1958; "Incorporation of Large Area Seen As Only Answer to City Politicians," *Carmichael Westerner,* June 19, 1958.

7. "Move to Form City Is Traced to Inaction," *Sacramento Bee,* June 18, 1958.

8. "Northeast Group Spurns City Wooing," *Sacramento Bee,* June 20, 1958.

9. "Sacramento City Blasts Charges on Annexation," *Sacramento Bee,* June 23, 1958.

10. California state law requires the signatures of at least 25

per cent of the landowners owning at least 25 per cent of the total land value in the area proposed for incorporation.

11. "Exclusive Petition for New City Cannot Be Extended," *Sacramento Bee,* September 15, 1958.

12. "Second Petition for New City May Go to Court," *Sacramento Bee,* October 9, 1958.

13. "Chamber of Commerce Board Okehs Idea of Single Big City," *Sacramento Bee,* September 23, 1958.

Chapter 4

1. "Del Paso Heights Meeting Favors Annexation," *Sacramento Bee,* July 17, 1958.

2. "Suburban C of C Maps Fight to Form New City," *Sacramento Bee,* January 26, 1957.

3. Demographic material was computed from U. S. census data, 1960, for seven census tracts that closely approximate the annexation area.

4. "Annexation Papers Get City Okeh," *Sacramento Union,* July 25, 1958.

5. "Annex Signers Top 3000; Vote Seems Assured," *Sacramento Bee,* October 13, 1958.

6. "Annexation to Whom?" *North Sacramento Journal,* July 11, 1958.

7. "School Districts' Status in Annexation Plan is Sought," *Sacramento Bee,* September 18, 1958.

8. "Ruling Says City Can Annex or Bar Schools," *Sacramento Bee,* September 4, 1958.

9. "Annex Voting Will Be Set," *Sacramento Union,* November 16, 1958.

10. "North Area CC Directors Vote Against Annex," *Sacramento Bee,* November 21, 1958.

11. "Burkhard Says Annex Would Get Tax Benefits," *Sacramento Bee,* November 13, 1958.

12. Sacramento City Unified School District, November 24, 1958.

13. Sacramento City Unified School District, December 8, 1958; "Job Security Would Get Study in Annexation," *Sacramento Bee,* December 15, 1958.

14. Sacramento City Unified School District, January 8, 1959.

15. *A Study of the Proposed Annexation of the Hagginwood-Del Paso-Gardenland Area to the City of Sacramento and Its Effect on the Educational Program of the American River Junior College District,* December 15, 1958.

16. Robert Curry, "A Study of Tax Predictions Made During Annexation Election Campaigns in Sacramento County, 1958-1960," Master's thesis, Sacramento State College, Sacramento, California, December, 1960.

17. "North School Heads Call Annex Defeat Spur to Unifying," *Sacramento Bee,* January 21, 1959.

The 1959 session of the California State Legislature eased the pressure for unification by providing that political annexation to a city would not include school annexation without the consent of the school districts involved. This gave protection against carving up school districts, but it also removed the urgency for school districts to unify. The school districts in the annexation area have not unified.

18. "Azevedo Thinks Second Annex Vote Would Pass," *Sacramento Union,* January 22, 1959.

Since the failure of this large-scale annexation attempt, several smaller efforts in the same area have been successful. The entire area, except for McClellan Field, of this attempted large annexation now is a part of Sacramento city.

Chapter 5

1. All 1960 census material presented in this section was obtained from U. S. Census Bureau, "Advance Final Data from the 1960 Census" for Sacramento County.

2. "North Area Group Wants Area Study," *Sacramento Bee,* September 18, 1958.

3. "City Approves Arden-Arcade Annex Start," *Sacramento Bee,* February 6, 1959.

4. "Council Okehs Arden-Arcade Annex Sign-up," *Sacramento Bee,* April 3, 1959.

5. "Arden-Arcade Annex Signup Will Be Explained," *Sacramento Bee,* April 15, 1959.

6. "Arden-Arcade Annex Move Is Validated," *Sacramento Bee*, June 15, 1959.

7. "Annexation Foes List 4 Main Objections," *Sacramento Bee*, June 12, 1959.

8. "Arden-Arcade Annex Vote Is Assured," *Sacramento Bee*, August 14, 1959.

9. "Foe Says Taxes Would Rise with Annexation," *Sacramento Bee*, August 19, 1959.

10. "City's Annexation Policies Are Listed," *Suburban News-Shopper*, September 10, 1959.

11. "Annex, Says Northern CC," *Sacramento Union*, September 11, 1959.

12. From a carbon copy of the letter, dated October 5, 1959.

Chapter 6

1. See research methods in Appendix. See also, Floyd Hunter, *Community Power Structure* (Chapel Hill: University of North Carolina Press, 1953).

2. In addition to the questions asked of all interviewees, a special set of questions was prepared for this group. One of the nominees was very busy during the period in which the interviews were being made and it proved impossible to arrange a satisfactory time for the interview. A substitution was made by selecting a person in a comparable position.

3. Eduard Spranger, *Types of Men*, trans. by Paul J. W. Pigors (Halle: Max Niemeyer Verlag); Gordon W. Allport, Philip E. Vernon, and Gardner Lindzey, *Study of Values* (New York: Houghton-Mifflin Co., 1960).

Following is a brief characterization of Spranger's types: the dominant interest of the theoretical man is the discovery of truth; the economic man is characteristically interested in what is useful; the aesthetic man sees his highest value in form and harmony; the highest value for the social man is love of people; the political man is interested primarily in power, although his activities are not necessarily within the narrow field of politics—whatever his vocation, he reveals his concern with power; the highest value of the religious man may be called unity—he is mystical, and seeks to comprehend the cosmos as a whole, to relate himself to its embracing totality. Spranger does not imply that a given man belongs exclusively to one or another of these types of values. His depictions are entirely in terms of "ideal types." Most individuals have mixed interests.

4. See, for example: Cecil A. Gibb, "Leadership," in Gardner Lindzey, *Handbook of Social Psychology*, v. II (Reading, Mass.: Addison-Wesley Publishing Co., 1954) and Douglas McGregor, *The Human Side of Enterprise* (New York: McGraw-Hill Book Co., 1960), pp. 182-85.

Chapter 7

1. North Sacramento, however, has done so. See Postcript to this study.

2. *The Government of Metropolitan Sacramento* (Chicago: Public Administration Service, 1957), p. 36.

3. The U. S. Chamber of Commerce and the American Medical Association have recently urged greater political activity by their members. How successful these urgings will be remains to be assessed.

4. California, *Statutes of 1963*, c. 1808.

Appendix

1. Martin Meyerson and Edward C. Banfield, *Politics, Planning, and the Public Interest* (Glencoe, Ill.: The Free Press, 1955).

2. Floyd Hunter, *Community Power Structure* (Chapel Hill: University of North Carolina Press, 1953).

3. In our interviews we were not able, of course, to record every single word uttered by the interviewees. Thus the quoted comments in the study are not always literally precise, though the essence and nuances of comments are retained.

4. Gordon W. Allport, Philip E. Vernon, and Gardner Lindzey, *Study of Values* (New York: Houghton-Mifflin Co., 1960). See Chapter Six for results of this test.